Golden Oldies in General Relativity

Hidden Gems

Andrzej Krasiński • George F. R. Ellis •
Malcolm A. H. MacCallum
Editors

Golden Oldies in General Relativity

Hidden Gems

Springer

Editors
Andrzej Krasiński
Polish Academy of Sciences
N. Copernicus Astronomical Center
Warszawa, Poland

George F. R. Ellis
Mathematics Department
University of Cape Town
Cape Town, South Africa

Malcolm A. H. MacCallum
Department of Mathematics
University of Bristol
Bristol, United Kingdom

ISBN 978-3-642-34504-3 ISBN 978-3-642-34505-0 (eBook)
DOI 10.1007/978-3-642-34505-0
Springer Heidelberg New York Dordrecht London

Library of Congress Control Number: 2012955858

Printed on acid-free paper

Springer is part of Springer Science+Business Media (www.springer.com)

Preface

The Golden Oldies series of reprints of important old papers in general relativity was launched in the *General Relativity and Gravitation* journal in 1997, following a proposal formulated by Alan Held in 1993 [1]. The inspiration came from the publication of an English translation of the paper by Jürgen Ehlers [2], which is included in Chapter 9 of this volume. Our guiding principle was that these should be papers containing results or ideas of extraordinary value, which, for one or another unfortunate reason, are today not easily accessible or not properly appreciated. Difficult accessibility may be caused by the original paper having been published in a language different from English (12 of the 18 papers included in this volume are examples), in a journal that was never intended for wide circulation (the Novikov paper in Chapter 13 is an example), or in a journal that ceased publication long ago (the Bronstein paper in Chapter 10 is an example). In a few cases, the original text has never been properly published, but only circulated in the form of a preprint or even handwritten seminar notes, like the new approach to the Bianchi classification, reprinted in Chapter 5 here.

We also adopted the principle that Golden Oldies should be, at the time of their re-publication, at least 30 years old. This principle was meant, on the one hand, to ensure that we have a mature historical perspective for evaluating the ideas in the paper and their (actual or potential) influence on later research, and, on the other hand, to prevent attempts to treat our series as a means of bestowing a distinction on the author.

Each of the Golden Oldies was accompanied by an editorial note, written by a knowledgeable modern author, that explained the significance of the work and where it has led to subsequently, and also by a brief biography of each of the original authors.

At the moment of writing this introduction, Golden Oldie # 57 has entered the production pipeline,[1] and 6 more are at various stages of preparation. A complete list is given at

[1] The Ehlers paper counts as # 0 because it was printed before the series began to be published systematically. The English translations of the papers by Thirring and Lense, Chapter 6 of this volume, were published long before the Golden Oldies idea was announced. However, we recognised them as our precursors, consistent with all our ideas, and, *ex post*, included them in our series as # (-1).

http://www.mth.uct.ac.za/~cwh/goldies.html
and in the appendix in this volume. We have a feeling that the series is slowly approaching its natural end – the supply of brilliant papers that are not easy to access and have not been as well known as they should be is drying up. We have thus realised that it would serve a useful purpose if the GOs and editorial notes were printed in handy volumes, serving as reference works for researchers in the field. They are, however, too many to reprint in a single volume,[2] so this volume contains only a selection from those that have been published. The general principle for this selection was that their original versions are the most difficult to get access to. Depending on how well this volume is received, we may continue with further volumes reprinting more of the GOs.

We thank the Springer company for agreeing to this idea: we believe it will be valuable to the general relativity community.

The Editors
Andrzej Krasiński *akr@camk.edu.pl*
George Ellis *George.Ellis@uct.ac.za*
Malcolm A. H. MacCallum *m.a.h.maccallum@qmul.ac.uk*

References

1. Held, A. (1993). Editorial. *Gen. Relativ. Gravit.* **25**, 1211 (1993).
2. Ehlers, J. (1961). Beiträge zur relativistischen Mechanik kontinuierlicher Medien [Contributions to the relativistic mechanics of continuous media], *Abhandlungen der Mathematisch-Naturwissenschaftlichen Klasse der Akademie der Wissenschaften und Literatur Mainz*, No 11; English translation: *Gen. Relativ. Gravit.* **25**, 1225 (1993).

Acknowledgements

We thank all the authors and publishers for their kind permissions to re-use in this book the texts that were previously published in the *General Relativity and Gravitation* journal. Most of all, we thank those authors of the editorial notes and biographies who took the trouble to update the earlier texts, or even write brand new versions for this book. Our special thanks go to the following persons:

- Alexei Starobinsky and Sergei Sushkov for their help in tracing the current whereabouts of the copyright holders to the papers by A. Z. Petrov and I. D. Novikov.
- Friedrich W. Hehl for his help in obtaining the permission to reprint the Thirring – Lense papers.
- Bahram Mashhoon for his help in contacting the authors of the Thirring and Lense biographies.

[2] A complete set of the GOs and editorial notes would occupy about 2500 pages.

Contents

Part I
Basic results in differential geometry and general relativity

Chapter 1
Władysław Ślebodziński 1931, the definition of the Lie derivative

Original paper: Władysław Ślebodziński, "Sur les équations canoniques de Hamilton", *Bulletins de la Classe des Sciences, Acad. Royale de Belg.* (5) **17**, 864 – 870 (1931).

Reprinted with the kind permissions of the Académie Royale de Belgique, Andrzej Trautman and Witold Roter.

Paper translated by Marie-Noëlle Célérier, Laboratoire Univers et Théories (LUTH), Observatoire de Paris, CNRS, Université Paris-Diderot, 5 place Jules Janssen, 92190 Meudon, France, email: marie-noelle.celerier@obspm.fr

Previously published in *Gen. Relativ. Gravit.* **42** (2010), pages 2525 – 2528 (the editorial note + biography), 2529 – 2535 (the paper).

Gen Relativ Gravit (2010) 42:2525–2528
DOI 10.1007/s10714-010-1056-7

Editorial note to: Władysław Ślebodziński, On Hamilton's canonical equations

Andrzej Trautman

Published online: 12 August 2010

Keywords Tensor calculus · Lie derivative of a tensor field · Integral invariants · Golden Oldie

The article by W. Ślebodziński, included here in English translation, has become a classic in tensor calculus mainly because it contains, for the first time in print, the general formula for—what is now called—the Lie derivative[1] $\mathcal{L}(X)T$ of a tensor field T with respect to a vector field X. In special cases, the formula was known and used before. In particular, D. Hilbert had in [1] an expression equivalent to the Lie derivative of a metric tensor g. Around 1920, É. Cartan defined the Lie derivative of exterior differential forms; see equation (5) on p. 84 in [2]. In connection with this definition, Ślebodziński gives a reference to a paper by Th. Lepage published in 1929. D. van Dantzig recognized the priority of Ślebodziński in defining the Lie derivative in the general case, introduced the name *Liesche Ableitung* and complemented the work of Ślebodziński by pointing out that the Lie derivative with respect to a vector field X of a geometric object T can be defined as the difference between the value of T at a point and the value of that object at the same point obtained by an infinitesimal 'dragging along' the vector field X [3]. In contemporary notation this is expressed by

$$\mathcal{L}(X)T = \frac{d}{dt}\varphi_t^* T_{|t=0},$$

[1] In this note, I transcribe equations from the form given by the Author to the notation in current usage.

The republication of the original paper can be found in this issue following the editorial note and online via doi:10.1007/s10714-010-1057-6.

A. Trautman (✉)
Instytut Fizyki Teoretycznej, Uniwersytet Warszawski, Hoża 69, Warsaw, Poland
e-mail: andrzej.trautman@fuw.edu.pl

where $\varphi_t^* T$ is the pull-back of T by the flow $(\varphi_t, t \in \mathbb{R})$ generated by X. In view of the equation

$$\frac{d}{dt}\varphi_t^* T = \varphi_t^* \mathcal{L}(X)T,$$

the vanishing of $\mathcal{L}(X)T$ is equivalent to the invariance of T with respect to the flow generated by X: local invariance is reduced to the infinitesimal one; see, e.g., §24 in [4] or §4.3 in [5]. In his influential book J. A. Schouten, in footnote 1 on p. 102 of [6], lists the 1931 paper by Ślebodziński as the first reference for the notion of Lie differentiation.

In the first section of the article, the most important is equation (4) defining the Lie derivative with respect to X of a mixed tensor. The author notes that the Lie derivative satisfies the Leibniz rule for tensor products and that it commutes with contractions over pairs of contragredient tensor indices.

Section 2 introduces the main theme of the article: Hamilton's canonical equations and the integral invariants in the sense of Poincaré [7]. The Hamiltonian is considered there as a function H on the (phase) space $\mathbb{R}^{2n}$ with coordinates $(x^i)_{i=1,\dots,2n}$ and the (symplectic) 2-form (10), written here as

$$\omega = \frac{1}{2} A_{ij} \mathrm{d}x^i \wedge \mathrm{d}x^j, \tag{*}$$

where $A_{ij} + A_{ji} = 0$, $A_{i\,n+i} = -A_{n+i\,i} = 1$ for $i = 1, \dots, n$ and, if $|i - j| \neq n$, then $A_{ij} = 0$. Physicists usually write $(q^1, \dots, q^n) = (x^1, \dots, x^n)$ and $(p_1, \dots, p_n) = (x^{n+1}, \dots, x^{2n})$ so that $\omega = \mathrm{d}q^\mu \wedge \mathrm{d}p_\mu$ (sum over $\mu = 1, \dots, n$). The Hamiltonian vector field X_H, given by the Author in eq. (9'), is nowadays defined by its contraction with ω

$$X_H \lrcorner\, \omega = \mathrm{d}H, \quad \text{thus, in physicists' notation,} \quad X_H = \frac{\partial H}{\partial p_\mu}\frac{\partial}{\partial q^\mu} - \frac{\partial H}{\partial q^\mu}\frac{\partial}{\partial p_\mu}.$$

The Lie derivatives with respect to X_H of ω, of the tensor A defined in (*) and of its inverse B all vanish. The flow of X_H solves the canonical equations of motion.

Section 3, entitled *Generalizations*, is devoted to a few simple results in the theory of integral invariants. The invariants are given by tensors with vanishing Lie derivatives with respect to X_H. Using the tensors A and B to raise, lower and contract indices of such tensors, and by forming their products, one obtains new tensors with the same property of invariance. This procedure generalizes the results described by de Donder [8].

Further information on the history of the notion of Lie differentiation and its generalizations, and on their applications in mathematical physics, can be found in [9–12] and in the bibliographies given there.

Władysław Ślebodziński—a brief biography

By Witold Roter

Władysław Ślebodziński was born on 6th February 1884 in Pysznica, district Nisko. He studied mathematics at the Jagiellonian University in Kraków in the years 1903–1908. After graduation he was a teacher of mathematics for the next 5 years. Owing to Kretkowski's foundation, he spent the academic year 1913/14 in Göttingen. The outbreak of the first world war forced him to return home. At the beginning of 1919 he moved from Kraków to Poznań, where he began to lecture in the State Higher School of Machine Construction and Electric Technology. In Poznań he prepared the doctor and habilitation dissertations, defended at Warsaw University in 1929 and 1934, respectively.

During the German occupation he was expelled from Poznań and began clandestine teaching. Denounced, he was arrested in 1942. In January 1943 he was transported to the concentration camp Auschwitz, from where he was sent to Gross-Rosen and afterwards to the then-newly established concentration camp in Nordhausen (Thüringen). There he lived to see freedom. In July 1945 he returned to Poland. Settling in Wrocław, he took up the post of a professor at the, at that time, joint Wrocław's schools of higher education: the University and the University of Technology. Here he was one of the organizers of the new Wrocław mathematical center.

Professor W. Ślebodziński was an excellent mathematician. His scientific interests included the theory of Riemannian spaces, the theory of surfaces in affine spaces, exterior forms, problems of equivalence, integral invariants, infinite Lie groups and manifolds of almost complex structure. He wrote about 50 scientific papers, two monographs on exterior forms and their applications (a 2-volume monograph in French, and its revised and expanded version in English), as well as many review articles. His results always dealt with the main trends in differential geometry.

One of his greatest achievements was the introduction (in the paper [13], reprinted in this issue) of a new differential operator which can be applied to scalars, tensors and affine connections. Later (1932) the operator was called the Lie derivative by van Dantzig [3].

The outstanding Japanese geometer Kentaro Yano devoted to the notion a monograph [9], where one can find the definition of the Lie derivative (of general geometric objects), its basic properties and many applications. Apparently Yano's book contains all the results published on the subject up to 1955. He studied there groups of motions, groups of affine motions, groups of projective and conformal motions. Moreover, he considered also the Lie derivatives in general affine spaces of geodesics, in compact orientable Riemannian manifolds, and almost complex manifolds.

Thus, the Lie derivative contributed to quick development of many areas of differential geometry. It is an important and powerful instrument in the study of groups of automorphisms and has extensive and important application in geometry. It belongs to very useful and important notions in differential geometry.

It seems that D. van Dantzig was the first who applied the notion of the Lie derivative to physics. More precisely, he applied the Lie derivative to thermo-hydrodynamics of ideal fluids [14]. The Lie derivatives play now an important role in mathematical physics in connection with conservation laws and the study of symmetries of solutions of field equations such as Einstein's equations of general relativity.

For his contributions Prof. Ślebodziński was many times rewarded. He received honorary doctorates from the Wrocław University and the Technical Universities of Wrocław and Poznań.

He died on the 3rd January 1972 in Wrocław.

Professor Władysław Ślebodziński's contributions have recently received a new recognition. In February 2009 the city of Wrocław honoured Ślebodziński by naming one of the new streets after him.

Acknowledgements I would like to thank Professors Andrzej Trautman and Andrzej Krasiński for their helpful comments made while I was preparing this text for publication.

Witold Roter, Professor Emeritus
Institute of Mathematics and Computer Science,
Wrocław University of Technology,
Wybrzeże Wyspiańskiego 27, 50-370 Wrocław,
Poland

References

1. Hilbert, D.: Die Grundlagen der Physik (Erste Mitteilung). Nachr. Göttingen (1915), pp. 395–407
2. Cartan, É.: Leçons sur les invariants intégraux, based on lectures given in 1920–21 in Paris. Hermann, Paris (1922). Reprinted in 1958
3. van Dantzig, D.: Zur allgemeinen projektiven Differentialgeometrie. Proc. Roy. Acad. Amsterdam **35** Part I, 524–534; Part II, 535–542 (1932)
4. Lichnerowicz, A.: Géométrie des groupes de transformations. Dunod, Paris (1958)
5. Marsden, J.E., Ratiu, T.S.: Introduction to Mechanics and Symmetry. 2nd edn. Springer-Verlag, New York (1999)
6. Schouten, J.A.: Ricci-Calculus. 2nd edn. Springer-Verlag, Berlin (1954)
7. Poincaré, H.: Les méthodes nouvelles de la Mécanique céleste, t. III. Gauthier-Villars, Paris (1899)
8. de Donder, Th.: Théorie des invariants intégraux. Gauthier-Villars, Paris (1927)
9. Yano, K.: The Theory of Lie Derivatives and its Applications. North-Holland, Amsterdam (1955)
10. Olver, P.J.: Applications of Lie Groups to Differential Equations. 2nd edn. Springer-Verlag, New York (1993)
11. Kolář, I., Michor, P.W., Slovák, J.: Natural Operations in Differential Geometry. Springer-Verlag, Berlin (1993)
12. Trautman, A. : Remarks on the history of the notion of Lie differentiation. In: Krupková, O., Saunders, D.J. (eds.) Variations, Geometry and Physics, Nova Science Publishers, New York (2008)
13. Ślebodziński, W.: Sur les equations de Hamilton. Bull. Acad. Roy. de Belg. **17**, 864–870 (1931)
14. van Dantzig, D.: One the thermo-hydrodynamics of perfectly perfect fluids. Nederl. Akad. Wetensch. Proc. **43**, 157–171 (1940)

Gen Relativ Gravit (2010) 42:2529–2535
DOI 10.1007/s10714-010-1057-6

GOLDEN OLDIE

Republication of: On Hamilton's canonical equations

Władysław Ślebodziński

Published online: 12 August 2010

An editorial note to this paper and a biography can be found in this issue preceding this Golden Oldie and online via doi:10.1007/s10714-010-1056-7.

Original paper: Władysław Ślebodziński, "Sur les équations canoniques de Hamilton", *Bulletins de la Classe des Sciences, Acad. Royale de Belg.* (5) **17**, 864–870 (1931). Reprinted with the kind permission of the Académie Royale de Belgique. Translated by Marie-Noëlle Célérier, Laboratoire Univers et Théories (LUTH), Observatoire de Paris, CNRS, Université Paris-Diderot, 5 place Jules Janssen, 92190 Meudon, France, e-mail: marie-noelle.celerier@obspm.fr.

Editorial responsibility: A. Krasiński, e-mail: akr@camk.edu.pl.

W. Ślebodziński (Deceased January 3, 1972)
Poznań, Poland
e-mail for correspondence: akr@camk.edu.pl

MATHEMATICAL ANALYSIS.

On Hamilton's canonical equations,

by W. ŚLEBODZIŃSKI (*)

In the present article, I apply the method of tensor calculus to the theory of integral invariants of Hamilton's canonical equations. I obtain thus the generalisation of theorems due to H. Poincaré, Th. De Donder, E. Cartan and E. Schuntner.

1. *The X and $\bar{X}$ operators.* — Let an infinitesimal transformation be given

$$\boxed{X(f) \equiv X^r \partial_r f} \tag{1}$$

where

$$\partial_r f \equiv \frac{\partial f}{\partial x_r}.$$

The coefficients X^r are functions of the variables x_i $(i = 1 \ldots m)$. *Extending* (2) this transformation (1) to the *mixed* components

$$A^{l_1 \ldots l_t}_{k_1 \ldots k_s}$$

of the tensor $\bar{A}$, we shall have

$$X(A^{l_1 \ldots l_t}_{k_1 \ldots k_s}) \equiv -\sum_{i=1}^{s} A^{l_1 \ldots l_t}_{k_1 \ldots k_{i-1} r k_{i+1} \ldots k_s} \partial_{k_i} X^r + \sum_{j=1}^{t} A^{l_1 \ldots l_{j-1} r l_{j+1} \ldots l_t}_{k_1 \ldots k_s} \partial_r X^{l_j}. \tag{2}$$

$$k_1 \ldots k_s, l_1 \ldots l_t = 1 \ldots m.$$

We are thus led to introduce the *operator* (3)

$$\left.\begin{array}{c} \bar{X}(A^{l_1 \ldots l_t}_{k_1 \ldots k_s}) \equiv X(A^{l_1 \ldots l_t}_{k_1 \ldots k_s}) \\ + \sum_{i=1}^{s} A^{l_1 \ldots l_t}_{k_1 \ldots k_{i-1} r k_{i+1} \ldots k_s} \partial_{k_i} X^r - \sum_{j=1}^{t} A^{l_1 \ldots l_{j-1} r l_{j+1} \ldots l_t}_{k_1 \ldots k_s} \partial_r X^{l_j}. \end{array}\right\} \tag{3}$$

(1) Communicated by Th. De Donder.

(2) See, e.g., chapter XI of the *Théorie des Invariants intégraux,* by TH. DE DONDER (Paris, Gauthier-Villars, 1927). In the following, we will refer to this book as *T. I. I.*

(3) In the case of integral forms, i.e. when the tensor $\bar{A}$ is skew symmetric, this operator has already been introduced and studied by TH. LEPAGE (*Mém. in-8º de l'Acad. roy. de Belg.*, Cl. des Sc., 1929.)

We see, from (2), that the $A^{l_1...l_t}_{k_1...k_s}$ will be covariant with respect to the indices k_1 ... k_s and contravariant with respect to the indices l_1 ... l_t if we have identically

$$\bar{X}(A^{l_1...l_t}_{k_1...k_s}) \equiv 0; \tag{4}$$

conversely, if we have (2) identically, we will have of course (4) identically.

The operator $\bar{X}$ possesses remarkable properties from the point of view of variance. Let us also remark that the formal law of differentiation of a product still applies to this operator; thus, we will have, e. g.,

$$\bar{X}(B_{\alpha\beta}C^{\gamma}) = \bar{X}(B_{\alpha\beta})C^{\gamma} + B_{\alpha\beta}\bar{X}(C^{\gamma}). \tag{5}$$

$$\alpha, \beta, \gamma = 1 \ ... \ m.$$

Let us also remark that if we have (4) and likewise

$$\bar{X}(B^{k_1...k_s}_{l_1...l_t}) \equiv 0, \tag{6}$$

we shall have

$$\bar{X}(A^{l_1...l_t}_{k_1...k_s} B^{k_1...k_s}_{l_1...l_t}) \equiv 0, \tag{7}$$

where we have adopted the summation convention upon repeated indices, as is usual in vector calculus.

Assuming (7), we set

$$I \equiv A^{l_1...l_t}_{k_1...k_s} B^{k_1...k_s}_{l_1...l_t}. \tag{8}$$

Instead of (7), we can equally write

$$\bar{X}(I) \equiv 0. \tag{8'}$$

2. *Hamilton's canonical equations.* — Let us consider *the canonical system*

$$\frac{dx_i}{\partial_{n+i}H} = \frac{dx_{n+k}}{-\partial_k H} \tag{9}$$

$$i, k = 1 \ ... \ n$$

where H is a function of the variables x_α ($\alpha = 1, 2 \ ... \ 2n$).(†)

Let us denote by $H(f)$ the infinitesimal transformation

$$H(f) \equiv \sum_{i=1}^{n} \partial_{n+i}H \,.\, \partial_i f - \sum_{i=1}^{n} \partial_i H \,.\, \partial_{n+i} f. \tag{9'}$$

(†) The original text said $\alpha = 1, 2 \ ... \ n$; corrected in translation (editor).

We know that the system (9) possesses the Poincaré *2-form integral invariant*

$$\boxed{\delta x_1 \delta x_{n+1} + \delta x_2 \delta x_{n+2} + \dots \delta x_n \delta x_{2n}} \tag{10}$$

which we will write

$$\sum_{i=1}^{n} A_{i,n+i} \delta x_i \delta x_{n+i}. \tag{10'}$$

$$i = 1 \dots n.$$

We will have therefore

$$\left.\begin{array}{c} A_{i,n+i} \equiv -A_{n+i,i} \equiv 1 \\ A_{\alpha\beta} \equiv -A_{\beta\alpha} \equiv 0 \quad \text{if} \quad |\alpha - \beta| \neq n. \end{array}\right\} \tag{10''}$$

Denoting by $\bar{H}$ the operator $\bar{X}$ defined by (3), when we consider the infinitesimal transformation (9'):
We shall have, from (10) to (10″),

$$\bar{H}(A_{\alpha\beta}) \equiv 0. \tag{10'''}$$

Besides the *covariant* components (10″) of the tensor $\bar{A}$, let us consider the *contravariant* components

$$\left.\begin{array}{c} B^{i,n+i} \equiv -B^{n+i,i} \equiv 1 \\ B^{\alpha\beta} \equiv -B^{\beta\alpha} \equiv 0, \quad \text{if} \quad |\alpha - \beta| \neq n \end{array}\right\} \tag{11}$$

of the tensor $\bar{B}$. *We shall still have*

$$\bar{H}(B^{\alpha\beta}) \equiv 0. \tag{11'}$$

The proof results immediately from (3) applied to (11) and (9). One can also derive this property of the contravariance of $B^{\alpha\beta}$ with respect to (9), by noticing that these $B^{\alpha\beta}$ are the *minors* (with signs) of $A_{\alpha\beta}$ considered in the determinant

$$A \equiv \|A_{\alpha\beta}\|$$

$$\alpha, \beta = 1, 2, \dots 2n$$

each minor being *divided* by the determinant A. Thanks to this remark, we will be able to say that the $A_{\alpha\beta}$ and $B^{\alpha\beta}$ are the covariant components of *the same* tensor $\bar{A}$ or $\bar{B}$.

The covariant components $A_{\alpha\beta}$ of the tensor $\bar{A}$ and the contravariant components $B^{\alpha\beta}$ of the tensor $\bar{B}$ will play a role analogous to that of the components $g_{\alpha\beta}$ and $g^{\alpha\beta}$ of the tensor g in special relativity.

Thus, *by applying the rules of tensor algebra* and, in particular, *the rule of index saturation, one will be able to derive from a given tensor* a new tensor of *lower rank.*

3. *Generalisations.* — Let us generalise some theorems of the theory of the integral invariants.

a) Let us consider a vector $\bar{U}$ with covariant components U_α and assume that we have

$$\bar{H}(U_\alpha) \equiv 0, \tag{12}$$

where $\bar{H}$ is the *operator* $\bar{X}$ defined by (3), for the case of the infinitesimal transformation (9'). Thanks to the tensor B defined by (11), we will derive from the U_α the contravariant(†) components

$$U^\alpha \equiv \sum_{b=1}^{2n} U_\beta B^{\alpha\beta}. \tag{12'}$$

We have

$$\bar{H}(U^\alpha) = 0. \tag{13}$$

This theorem generalizes the theorem [1] which applies to the case where U_α is given by

$$U_i \equiv \frac{\partial U}{\partial x_i}, \qquad U_{n+i} = \frac{\partial U}{\partial x_{i+n}} \tag{14}$$

the symbol U representing an *invariant* of (9').

More generally, we can formulate the following theorem:

If we have a relation of the form

$$\bar{H}\left(U_{k_1\ldots k_r}^{\cdot\;\cdot\;\cdot\lambda_1\ldots\lambda_s}\right) \equiv 0, \tag{15}$$

$U_{k_1\ldots k_r}^{\cdot\;\cdot\;\cdot\lambda_1\ldots\lambda_s}$ *being the mixed components of a tensor of rank* $r+s$, *this relation* (15) *does not stop being verified if one raises or lowers one or more indices, using the above given rules.*

(†) Translator's note: The original text said "covariant", which was an obvious typo.
(1) See *T. I. I.*, p. 89.

b) If U_α and V_α are the covariant components of the tensors $\bar{U}$ and $\bar{V}$, so that we have

$$\bar{H}(U_\alpha) \equiv 0, \qquad \bar{H}(V_\alpha) \equiv 0, \tag{16}$$

then the scalar product

$$\bar{U} \cdot \bar{V} \equiv A^{\alpha\beta} U_\alpha V_\beta \tag{17}$$

is an invariant (or an integral) of the canonical equations (9).

This theorem possesses as a particular case *the Poisson theorem.*

Our rule can be extended, without difficulties, *to tensors of higher rank.*

If, e.g., $U_{\alpha\beta}$ and $V_{\alpha\beta}$ are the coefficients of two absolute *integral* invariants of the set of equations (9), we derive from them three first integrals

$$\sum_{h,i=1}^{n} (U_{h\ i} V_{n+h\ n+i} - U_{h\ n+i} V_{n+h\ i} - U_{n+h\ i} V_{h\ n+i} + U_{n+h\ n+i} V_{h\ i}),$$

$$\sum_{h,i=1}^{n} (U_{h\ i} U_{n+h\ n+i} - U_{h\ n+i} U_{n+h\ i}), \qquad \sum_{h,i=1}^{n} (V_{h\ i} V_{n+h\ n+i} - V_{h\ n+i} V_{n+h\ i}).$$

We have obtained them by calculating the scalar product and the squares of the tensors with components $U_{\alpha\beta}$ and $V_{\alpha\beta}$.

Example. — If $n = 2$, the three integrals can be written

$$2U_{13}V_{13} + 2U_{24}V_{24} + U_{12}V_{34} + U_{34}V_{12} + U_{14}V_{23} + U_{23}V_{14},$$
$$U_{13}^2 + U_{24}^2 + U_{12}U_{34} + U_{14}U_{23}, \qquad V_{13}^2 + V_{24}^2 + V_{12}V_{34} + V_{14}V_{23}.$$

c) Let us consider an absolute integral invariant I_{2m} of even rank $2m$ ($m < n$) of the set (9) and denote its coefficients by $U_{\alpha_1\alpha_2\ldots\alpha_{2m}}$. Let us apply to this tensor m times the index saturation operation; we obtain thus a sequence of skew symmetric tensors with covariant components

$$\left.\begin{array}{c} \sum_{h_1=1}^{n} U_{\alpha_1\ldots\alpha_{2m-2}h_1 n+h_1}, \qquad \sum_{h_1,h_2=1}^{n} U_{\alpha_1\ldots\alpha_{2m-4}h_2 n+h_2 h_1 n+h_1}, \ldots, \\ \sum_{h_1,\ldots,h_m=1}^{n} U_{h_m n+h_m \ldots h_2 n+h_2, h_1 n+h_1}. \end{array}\right\} \tag{18}$$

Every tensor of the sequence (18) is obviously invariant under the transformation $\bar{H}(f)$. The last of them, being a scalar function, is a first integral of the system (9); we obtain thus the theorem of E. Schuntner (1) which is the generalization of a theorem by Th. De Donder. A sequence of *absolute integral invariants* of the canonical system (9)

$$I_{2m-2}, \qquad I_{2m-4}, \ldots, I_2$$

corresponds to the tensors of the sequence (18), save the last one.

The same process can be applied to an absolute integral invariant I_{2m+1} of odd rank $2m+1$. We thus see that the *operation of index saturation allows one to derive from any absolute integral invariant of even rank* $2m$ *a sequence of absolute integral invariants of ranks* $2m-2$, $2m-4$, ..., 2, *and a first integral, and from any absolute integral invariant of odd rank* $2m+1$, *a sequence of absolute integral invariants of ranks* $2m-1$, $2m-3$, ..., 1.

It is obvious that the operations employed in this issue can be applied in various ways.

---

(1) See *T. I. I.* [eqs. (61), p.29, and (313), p. 91].

Chapter 2
Karl Stellmacher 1938, the Cauchy problem for the Einstein equations

Original paper: Karl Stellmacher, "Zum Anfangswertproblem der Gravitationsgleichungen", *Mathematische Annalen* **115**, 136 – 152 (1938).

Paper translated by Henk van Elst, Karlshochschule International University, Karlstraße 36–38, 76133 Karlsruhe, Germany, email: hvanelst@karlshochschule.de

Previously published in *Gen. Relativ. Gravit.* **42** (2010), pages 1765 - 1767 (the editorial note + biography), 1769 – 1789 (the paper).

Gen Relativ Gravit (2010) 42:1765–1767
DOI 10.1007/s10714-010-0959-7

GOLDEN OLDIE EDITORIAL

Editorial note to: Karl Stellmacher, On the initial value problem of the equations of gravitation

Helmut Friedrich

Published online: 21 April 2010

Keywords Solutions of Einstein's equations · Initial value problem for Einstein's equations in vacuum and with a dust source · Golden Oldie

The general relativistic notion of 'domain of dependence' is important in the (global) analysis of gravitational fields. It is technically convenient because it is a geometrical concept defined, irrespective of any field equations, purely in terms of the metric. Its fundamental importance, however, comes from the fact that it gives precise meaning to the statement that the evolution process of the Einstein equations respects the notion of causality defined by the solution metrics of these equations.

This is taken for granted nowadays but it was not obvious in the early days of General Relativity. It is the merit of Stellmacher's article that it clarifies this issue for the first time. Previous work by A. Einstein, D. Hilbert, G. Darmois, C. Lanczos and others, which revealed important properties of the field equations, was based on the study of linearizations, formal expansions and Cauchy–Kowalevskaja type arguments. These techniques did not allow one to draw any conclusion of the type given in Stellmacher's article. The new ingredient he uses is an argument based on 'energy integrals' which was proposed by K. O. Friedrichs and H. Lewy (cf. [1] for earlier work in this direction). It is discussed in section 2 (we note that the inequality following equations (3) is slightly misleading and should be replaced by the statement that the mean value theorem is used to obtain these equations).

The republication of the original paper can be found in this issue following the editorial note and online via doi:10.1007/s10714-010-0960-1.

H. Friedrich (✉)
Max-Planck-Institut für Gravitationsphysik, Am Mühlenberg 1, 14476 Golm, Germany
e-mail: hef@aei.mpg.de

It appears to be the first time that this powerful argument has been used in the context of general relativity. It has largely been generalized by now and 'energy estimates' represent an indispensable tool in the existence theory.

Stellmacher analyses the Einstein–Maxwell equations in harmonic and Lorentz gauge respectively, arguing that these gauge conditions themselves respect the causality relations defined by the solution metric. In this case the proof that the domains of (geometric) uniqueness for the field equations coincide with the domains of dependence defined by the metric is short, clear, and clean. He also analyses the case of pressure free matter (dust) which is more complicated and his results are in fact not so clear and complete because they only refer to regions where neither caustics nor matter-vacuum interfaces occur. This should not be held against him, the situation is complicated and poses problems even nowadays (cf. the discussion in [2]). He finally also remarks that he did not succeed in the case of more general matter fields. Again, this is not surprising.

While Stellmacher's results and methods are standard now they were certainly new at the time and marked in a sense a kind of 'phase transition' in the analysis of the field equations. Had the times been better (K. O. Friedrichs who had suggested the work had left the country for political reasons half a year before the article was submitted) it might have led to a much earlier clarification of the local Cauchy problem.

Karl Ludwig Stellmacher—a brief biography

By Hubert Goenner, based on Ref. [3].

Karl Ludwig Stellmacher (1910–2001) studied mathematics, physics and chemistry at the University of Göttingen from 1927–1933 in order to become a teacher at a Gymnasium (high school/first two college years). His final exam did not recommend him for a PhD. Nevertheless, Richard Courant seems to have given him a problem in hyperbolic differential equations. After having lost Courant as an advisor, in 1935 the mathematician Gustav Herglotz (1881–1953) took him on and proposed propagation of gravity as a subject. Stellmacher finished his dissertation in 1937; it is the paper reproduced here. Surprisingly, he credited Kurt Friedrichs as responsible for the theme who, in the same year left his position in Braunschweig/ Germany as a consequence of his political views and his jewish fiancée (whom he later married).

After his dissertation, Stellmacher worked as an assistant to Max Schuler (1882–1972), director and successor of the famous pioneer of aerodynamics, Ludwig Prandtl (1875–1853), in the Kaiser-Wilhelm Institute for Fluid Dynamics, on problems concerning gyroscopes. During the 2nd world war, he had to be a soldier from 1939 to 1944. After the war, his results on the gyroscope found no interest so that he went back to pure mathematics.

In 1948 he became both an assistant and a Privatdozent (lecturer) in the Mathematics Institute of the University of Göttingen. After having received the title of professor (but not the pay) in 1955, he left Göttingen immediately to become a research associate in the Institute for Fluid Dynamics and Applied Mathematics of the University of

Maryland in College Park. Within a year, he became full professor and stayed there until retirement in 1977.

References

1. Courant, R., Hilbert, D.: Methods of Mathematical Physics II. Interscience, New York (1962)
2. Choquet-Bruhat, Y., Friedrich, H.: Class. Quantum Grav. **23**, 5941 (2006)
3. Rammer, G.: Die Nazifizierung und Entnazifizierung der Physik an der Universität Göttingen, p. 512. Dissertation, Göttingen, 2004

Gen Relativ Gravit (2010) 42:1769–1789
DOI 10.1007/s10714-010-0960-1

GOLDEN OLDIE

Republication of: On the initial value problem of the equations of gravitation

Karl Stellmacher

Published online: 26 May 2010

An editorial note to this paper and a biography can be found in this issue preceding this Golden Oldie and online via doi:10.1007/s10714-010-0959-7.

Original paper: Karl Stellmacher, "Zum Anfangswertproblem der Gravitationsgleichungen", *Mathematische Annalen* **115**, 136–152 (1938). Reprinted with the kind permission of Springer-Verlag, the current owner of the publication rights. Translated by Henk van Elst, Karlshochschule International University, Karlstraße 36–38, 76133 Karlsruhe, Germany, e-mail: hvanelst@karlshochschule.de.

Editorial responsibility: A. Krasiński, e-mail: akr@camk.edu.pl.

K. Stellmacher (Deceased March 22, 2001)
Göttingen, Germany
e-mail for correspondence: akr@camk.edu.pl

On the initial value problem of the equations of gravitation

By

Karl Stellmacher at Göttingen*):

For the solutions of the field equations of general relativity it is shown in the following that they possess the property of depending in a world point P only on that part of the remaining world, which lies inside the past of the temporal divide defined by the point P.

If, hence, the state variables are changed outside of the temporal divide, these variables will keep their values in P, respectively they can be returned to their original values by a transformation.

1.

In general relativity one defines: a world point P is "earlier"[1]) than a world point Q, if P and Q can be connected by an everywhere timelike line and if Q is associated with a larger value of the x^0-coordinate than P.

"Simultaneous" are called two points, which can be connected by an everywhere spacelike line.

By these definitions a certain causal connectedness is imprinted on the world, as one will have to demand that for simultaneity no cause–effect relationship can exist.

The value of a state variable in a world point P can therefore depend only on those world points, which can be connected to P by everywhere timelike lines; i.e., only on those points, which lie in the interior of the temporal divide constructed in P, namely in that part which is directed into the past. Hence, all field actions are allowed to propagate at most at the speed of light.

However, the cause–effect relationship is determined independently of these definitions by certain partial differential equations for the physical state variables.

Therefore, the question arises whether the cause–effect relationship which is fixed by the field equations is in agreement with the cause–effect

*) This work is part of a dissertation accepted by the Faculty of Mathematics and the Natural Sciences of the University of Göttingen. It was suggested to me by Herr Professor Friedrichs. For this favour, as well as for his supporting interest, I feel obliged to thank him dearly.

[1]) See Hilbert, Math. Annalen **92**, p. 11ff. — Ges. Abh. Bd. III, p. 268ff.

relationship, which is imposed on the world by the introduction of an indefinite metric.

In the case of special relativity it is known of the Maxwell–Lorentzian equations that they yield the correct cause–effect relationship for the electromagnetic field strength.

That is, if the Cauchyian initial value problem is posed, i.e., if the field strength is specified at a certain time $x^0 = 0$, then its value in a later world point P depends only on that part of the initial values, which is cut out of the initial surface by the characteristic cone of this equation system to be constructed in P; the characteristic cone coincides with the temporal divide. Here, the demanded cause–effect relationship arises as a very special property of a solution to the Maxwell–Lorentzian equations.

Also in the case of general relativity we need to investigate if the field equations, and the gravitational equations in particular, possess the respective property to depend only on a part of the initial values.

This issue has already been raised a number of times. Initially it was Einstein, who settled the problem for weakly gravitating fields in first approximation[2]).

Next, it was shown by Vessiot[3]) that the characteristic cone of the equations of gravitation coincides with the temporal divide, i.e., that all discontinuities of derivatives of second or higher order of the field components propagate at the speed of light, wherein these singularities cannot be removed by a transformation, and wherein the respective derivatives of lower order are continuous in the point of concern.

A further step towards settling this problem was taken by de Donder[4]), who carried over the coordinate system used by Einstein for weak fields to the case of arbitrary gravitational fields; in such a coordinate system the equations of gravitation assume a form in which the propagation of gravitation at the speed of light and the correct causal structure of the world already become quite plausible. (This coordinate system will be the foundation of one of our two proofs. For further details, see Chapter 3.)

(In not so immediate connection to the present question are the famous

---

[2]) Ber. d. Berl. Akad. Wiss. 1916, p. 688, and 1918, p. 154; see thereto the comments and concerns of Eddington, Relativitätstheorie in math. Behandlung (Berlin 1925), § 57.

[3]) Compt. rend. **166** (1918); see on this also Levi–Civita, Atti d. Linc. **11** (1930), p. 1.

[4]) La gravifique Einsteinienne (Paris 1921), p. 40–41.

Hilbertian investigations on the causality problem[5]). Based on the theorem by Cauchy–Kowalewsky it is shown there, that the values of the field components and their first derivatives, to be analytically given on a space-like initial manifold, *uniquely* induce, up to transformations, an analytical solution in a neighbourhood of the initial surface.

The question which is of interest to us is barely touched upon hereby, as no statement is made on the domain of dependence of the solution, and, by means of the method of Cauchy–Kowalewsky, such a statement is impossible in principle. Also, uniqueness is not guaranteed inasmuch as the case is conceivable that a second non-analytical system of solutions may exist.)

In contrast to this we will conduct the proof that field actions propagate at most at the speed of light, thereby considering two cases; 1. arbitrary electrical and gravitating fields in empty space, 2. pure gravitational fields, which are superimposed on continuously distributed matter.

2.

The pursuit of our proof for the validity of the postulate of causality succeeds on the basis of applying a method devised by Friedrichs and Lewy[6]).

First we prove as a lemma that the uniqueness theorem by Friedrichs and Lewy can be carried over without difficulty to the case of an arbitrary, no longer necessarily linear hyperbolic differential equation. This allows us to repeat, simultaneously, the course of thoughts of the Friedrichs–Lewyian proof. It shall be shown that the value of the solution of a given differential equation of totally hyperbolic character depends only on that part of the initial values, which is cut out from the initial surface by the characteristic conoid to be constructed in P.

Let an equation be given of the type[7]):

$$L[u] \equiv a^{ik} u_{ik} + f = 0 \qquad \left(u_{ik} = \frac{\partial^2 u}{\partial x^i \partial x^k} \right). \tag{1}$$

[5]) See the work cited in footnote 1.

[6]) Math. Annalen **98**, p. 192ff.

[7]) One overlooks easily, that the present simple generalisation can also be carried over to the case that an arbitrary differential equation of the most general form $F(u) = 0$ is given. For then the considerations can be applied, under respective assumptions of differentiability, to the differentiated equation $\frac{\partial F}{\partial x^0} = \frac{\partial F}{\partial u_{ik}} u_{ik0} + W = 0$. See on this the work by Schauder, Fundam. Mathem. **34** (1935), p. 213.

In a domain G, which contains a part of the initial surface A, let a^{ik} and f be continuously differentiable functions of the four independent variables x^i, as well as of u and its first derivatives, the u_i. Equation (1) shall be totally hyperbolic for a certain given solution u, i.e. the quadratic form associated with a^{ik} shall possess the index $(---+)$; furthermore, the a^{ik} shall transform contravariantly, and the coordinate system shall be chosen such that the conditions of spacelikeness are satisfied:

$$(2) \qquad a^{00} > 0, \qquad \begin{vmatrix} a^{11} & a^{12} & a^{13} \\ a^{21} & a^{22} & a^{23} \\ a^{31} & a^{32} & a^{33} \end{vmatrix} < 0\,.$$

Then the matrix (a^{ik}) $(i, k = 1, 2, 3)$ is evidently non-degenerate negative definite; since the discriminant of the form is negative by assumption, and so only the indices of inertia $(++-)$ or $(---)$ are possible. However, since by assumption only *one* positive sign can occur, the claim thus follows.

Within the domain G there shall be given a well determined second solution $\breve{u}$[8]) of the differential equation (1), which coincides on a spacelike[9]) initial manifold A with respect to its value as well as the values of the first derivatives with the solution u. Let $\breve{u}$ in G, besides its derivatives of first and second order, be bounded from above; likewise, of course, u:

$$|u|,\ |u_i|,\ |u_{ik}|,\quad |\breve{u}|,\ |\breve{u}_i|,\ |\breve{u}_{ik}| < M \qquad i, k = 0, 1, 2, 3.$$

M is a fixed well determined number. Thus, it also holds:

$$L\,[\breve{u}] = \breve{a}^{ik}\,\breve{u}_{ik} + \breve{f} = 0,$$

where $\breve{a}^{ik}$ and $\breve{f}$ mean the functions a^{ik}, f, upon substitution therein of u and its derivatives by $\breve{u}$ and its derivatives. Under these premises there holds the

[8]) That we consider *one particular* second solution, therein lies the new aspect compared to Friedrichs and Lewy.

[9]) Let A be the surface $f = \text{const.}$ Then f is called spacelike, if

$$a^{ik}\frac{\partial f}{\partial x^i}\frac{\partial f}{\partial x^k} > 0.$$

T h e o r e m. *Within a still to be determined neighbourhood of A both solutions are identical.*

To begin with the proof we form $L(u) - L(\breve{u})$:

$$a^{ik}(u_{ik} - \breve{u}_{ik}) + \breve{u}_{ik}(a^{ik} - \breve{a}^{ik}) + (f - \breve{f}) = 0\,.$$

If I now set:

$$\begin{aligned} u - \breve{u} &= v, \\ u_i - \breve{u}_i &= v_i, \\ u_{ik} - \breve{u}_{ik} &= v_{ik}, \end{aligned}$$

then I obtain an in v linear and homogeneous differential equation with vanishing initial values on A:[10]

$$(3) \qquad a^{ik}\, v_{ik} + b^\rho\, v_\rho + c\, v = 0 \qquad \left\{ \begin{aligned} b^\rho &= \frac{\partial a^{ik}(\overline{u})}{\partial u_\rho}\,\breve{u}_{ik} + \frac{\partial f(\overline{u})}{\partial u_\rho} \\ c &= \frac{\partial a^{ik}(\overline{u})}{\partial u}\,\breve{u}_{ik} + \frac{\partial f(\overline{u})}{\partial u} \end{aligned} \right\}$$

$$\breve{u} \leq \overline{u} \leq u\,.$$

From now on we can transfer the considerations by Friedrichs and Lewy (in particular l.c. Chapter 4) without further ado.

One constructs in a point P within G the characteristic cone B, of which we make the simplifying assumption that, together with the initial surface A, it defines a simply connected domain G'; this domain shall be completely contained within G. B in turn shall cut out from A a domain A' which is also simply connected. Furthermore, without loss of generality we assume that A be the surface $x^0 = 0$ (see Fig. 1).

We then imagine the interior of the cone to be filled by a foliation of surfaces $x^0 =$ const, which we assume to be spacelike in the interior of G'. By integration of equation (3), multiplied by v_0, over a domain G'', which is bounded by A', $x^0 = c > 0$ and B', one obtains the equation

$$(4) \qquad \iiint\limits_{G''} \left(a^{ik}\, v_{ik} + b^\rho\, v_\rho + c\, v\right) v_0\, d\tau = 0.$$

[10] In the last $\partial f/\partial u$ in the equation below the argument of f was corrected from u to $\overline{u}$ by the translator.

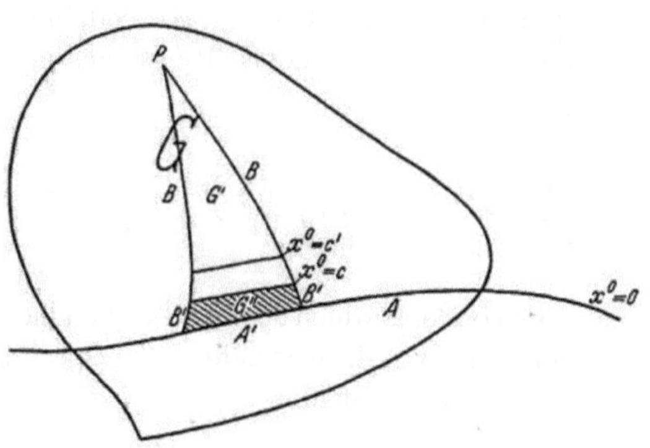

Fig. 1.

The integrand may be represented by a divergence and an additional quadratic form of the first derivatives of v and of v itself:

$$\iiint\limits_{G''} \left[\left(a^{ik}\, v_i\, v_0\right)_k - \tfrac{1}{2}\left(v_i\, v_k\, a^{ik}\right)_0\right] d\tau = \iiint\limits_{G''} Q(v_i, v)\, d\tau.$$

The left-hand side is converted into a surface integral by the Gaußian integral theorem:

$$\iint\limits_{O} a^{ik}\left(v_i\, v_0\, \xi_k - \tfrac{1}{2}\, v_i\, v_k\, \xi_0\right) d\omega = \iiint\limits_{G''} Q(v_i, v)\, d\tau.$$

The ξ_i denote the direction cosines of the inward-pointing normals. The part of the surface integral extending over A' vanishes; the remaining part subdivides into a surface integral over a strip M' of the mantle of the conoid M and a surface integral over the domain C of the surface $x^0 = c$. We recast the integrand of the surface integral and obtain:

$$-\frac{1}{2}\iint\limits_{C+M'} \frac{1}{\xi_0}\left[a^{ik}\left(v_i\, \xi_0 - v_0\, \xi_i\right)\left(v_k\, \xi_0 - v_0\, \xi_k\right) - a^{ik}\, \xi_i\, \xi_k\, v_0^2\right] d\omega$$

$$= \iiint\limits_{G''} Q(v_i, v)\, d\tau.$$

Under the surface integral on the left-hand side we have evidently, as on C $\xi_i = \delta_i^0$ and therefore $a^{ik}\, \xi_i\, \xi_k = a^{00}$ holds, as integrand of the part over

C a non-degenerate negative definite form of all first derivatives of v. The integrand of the part extending over M' certainly does not become negative, as on M $a^{ik}\xi_i\xi_k = 0$ holds. By suppressing this latter part, we can evidently estimate:

$$\iint\limits_{C} \sum v_i^2\, d\omega \leq D \iiint\limits_{G''} \left(\sum v_i^2 + v^2\right) d\tau \leq E \iiint\limits_{G''} \sum v_i^2\, d\tau.$$

E and D are constants which do not depend on the choice of the parameter c[11]). By integration of the inequality with respect to x^0 from $x^0 = 0$ up to $x^0 = c$ one obtains eventually:

$$\iiint\limits_{G''} \sum v_i^2\, d\tau \leq cE \iiint\limits_{G''} \sum v_i^2\, d\tau.$$

Now by choosing $c < 1/E$ it follows evidently that $v = 0$ within G''.

If, with the method of proof just given, I now proceed piecewise from the surface $x^0 = c$ up to the surface $x^0 = c'$, then from c' up to c'' and so on, I can eventually reach the tip P of the characteristic cone, as one can give an a priori estimate for E which holds evenly for every domain strip between $x^0 = c^{(n)}$ and $x^0 = c^{(n+1)}$, irrespective of how these two positive constants $c^{(n)}$ and $c^{(n+1)}$ may be chosen (as long as $c < c(P)$).

Accordingly, we have proven the following theorem: every within and on the boundary of G'' twice continuously differentiable function $\breve{u}$, which there, besides its derivatives of first and second order, is bounded, and which in the subdomain A' of the initial surface A (including its boundary), besides its first derivatives, agrees with u, is identical to u within G'.

Therefore, it follows, too: $u(P)$ does not change when arbitrary changes of the initial values on A are introduced outside of A'.

3.

In order to be able to apply this method to the equations of gravitation, we use a coordinate system in which the equations of gravitation appear considerably simplified with respect to those parts which contain the second derivatives of the g_{ik} with respect to the coordinates.

[11]) For further details see the original work.

In this regard we are considering a generalisation of the coordinate system used by Einstein when integrating the equations for non-stationary weak metric fields.

(The possibility of introducing such a coordinate system was first discovered by de Donder[12]); a little later Lanczos[13]) discovered it independently and gave the equations to be satisfied a particularly elegant and concise form (see on this also the remark by Darmois[14])).

According to Lanczos, in a coordinate systems in which the four relations

(1)
$$\frac{1}{\sqrt{-g}}\frac{\partial\sqrt{-g}\,g^{ri}}{\partial x^r}=0 \qquad (i=0,1,2,3)$$

are satisfied, the components of the contracted Riemannian tensor simplify to the following form $\left(\text{we write } \Box \text{ for } g^{\lambda\mu}\frac{\partial^2}{\partial x^\lambda\,\partial x^\mu}\right)$:

(2)
$$R_{ik}=\tfrac{1}{2}\Box\, g_{ik}+\Gamma_{rp,i}\Gamma_{qs,k}g^{rs}g^{pq}-\frac{\partial g_{pi}}{\partial x^r}\frac{\partial g_{qk}}{\partial x^s}g^{pq}g^{rs}.$$

We show that in the neighbourhood of an arbitrary spacelike hypersurface ($x^0=0$) we can always find a new proper[15]) coordinate system such that the coordinate surfaces $x^i=$ const are connected with the metric field in an invariant way, and that there the equations (1) are satisfied.

We thus look for a non-singular transformation

(3)
$$\overline{x}^i=\varphi^{(i)}(x^0,x^1,x^2,x^3) \qquad (i=0,1,2,3)$$

such that in the new coordinate system it is true that:[16]

(1a)
$$\frac{1}{\sqrt{-\overline{g}}}\frac{\partial\sqrt{-\overline{g}}\,\overline{g}^{ri}}{\partial\overline{x}^r}=0 \qquad (i=0,1,2,3).$$

For this purpose we consider the "generalised potential equation":

(4)
$$\frac{1}{\sqrt{-g}}\frac{\partial\sqrt{-g}\,g^{rs}\dfrac{\partial\varphi}{\partial x^s}}{\partial x^r}=0.$$

[12]) La gravifique Einsteinienne (Paris 1921), p. 40/41.
[13]) Phys. Zeitschr. **23** (1921), p. 537ff.
[14]) Mém. des Sc. Math. **25** (1926), p. 14–19.
[15]) See Hilbert, l. c.
[16]A typo in the formula below was corrected by the translator: the bar over x^r was missing.

It is invariant, if φ constitutes a scalar function. Then we determine four scalar functions $\varphi^{(i)}$ such that each one of them satisfies equation (4), and that in addition the initial conditions

$$\left\{\begin{aligned} &(\varphi^{(i)})_{x^0=0} = x^i, \\ &\left(\frac{\partial\varphi^{(i)}}{\partial x^0}\right)_{x^0=0} = \delta^i_0, \end{aligned}\right. \tag{5}$$

hold, from which follows immediately:

$$\frac{\partial\varphi^{(i)}}{\partial x^k} = \delta^i_k.$$

The solution of this initial value problem can be obtained following Hadamard[17]).

On the four equations obtained in this way,

$$\frac{1}{\sqrt{-g}}\frac{\partial\left(\sqrt{-g}\,g^{rs}\,\dfrac{\partial\varphi^{(i)}}{\partial x^r}\right)}{\partial x^s} = 0 \qquad (i = 0, 1, 2, 3),$$

we apply the transformation (3); then there evidently results:[18]

$$\frac{1}{\sqrt{-\overline{g}}}\frac{\partial(\sqrt{-\overline{g}}\,\overline{g}^{rs}\,\delta^i_r)}{\partial\overline{x}^s} = 0 \qquad (i = 0, 1, 2, 3),$$

i.e., the equations (1a).

When the old coordinate system was a proper one, in particular on the initial surface, then, by reasons of continuity, this also holds in a certain neighbourhood of the initial surface for the new one.

A de Donderian coordinate system is determined uniquely by the property that, along a certain spacelike hypersurface, it coincides with an arbitrarily given second one up to the first derivatives of the transformation functions. Then the components of an arbitrary tensor agree on this initial

[17]) Le problème de Cauchy, App. 1. – There the existence proof is given only for equations with analytical coefficients. However, as Hadamard remarks himself, there are no essential difficulties involved in overcoming this premise.

[18]A typo in the formula below was corrected by the translator: the bar over g^{rs} was missing.

surface in both coordinate systems. The new coordinate surfaces are connected in an invariant way with the metric, since the equation (4) possesses an invariant form.

4.

Now the means have been made available to settle the problem. We imagine a world piece G, free of matter, in which, hence, the equations hold:

(1)
$$R_{ik} = kS_{ik} \qquad (i, k = 0, 1, 2, 3).$$

R_{ik} denotes the contracted Riemannian tensor, S_{ik} the electromagnetic energy tensor:

$$-S_{ik} = F_{i\alpha}F^{\alpha\cdot}_{\cdot\,k} - \tfrac{1}{4}g_{ik}F_{\alpha\beta}F^{\alpha\beta} \qquad (i, k = 0, 1, 2, 3)$$

with $S = 0$. The skew symmetrical tensor F_{ik} of the electric field strength can be represented by the vector potential Φ_i.

(2)
$$F_{ik} = \frac{\partial}{\partial x^i}\Phi_k - \frac{\partial}{\partial x^k}\Phi_i \qquad (i, k = 0, 1, 2, 3),$$

and there hold the Maxwell–Lorentzian field equations:

(3)
$$\frac{\partial\sqrt{-g}\,F^{i\alpha}}{\partial x^\alpha} = 0 \qquad (i = 0, 1, 2, 3).$$

It is well known that by (2) and (3) the vector Φ_i is determined only up to an arbitrary additive gradient, so that in general one imposes the additional constraint

(4)
$$\frac{\partial\sqrt{-g}\,\Phi^\alpha}{\partial x^\alpha} = 0.$$

From (2), (3) and (4) one obtains by simple algebra[19]):

(5)
$$\frac{1}{\sqrt{-g}}\frac{\partial}{\partial x^l}\left(g^{kl}\sqrt{-g}\,\frac{\partial\Phi_i}{\partial x^k}\right) + \frac{1}{\sqrt{-g}}\frac{\partial}{\partial x^k}\left[\Phi_l\frac{\partial}{\partial x^i}\sqrt{-g}\,g^{kl}\right]$$
$$+g_{ih}F^{\cdot k}_{j\cdot}\frac{\partial g^{hj}}{\partial x^k} = 0 \qquad (i = 0, 1, 2, 3).$$

[19]) v. Laue, Phys. Zeitschr. **21** (1920), p. 659ff.; see also Rel. Theor. Bd. II, 1st ed., p. 148.

Within G we now construct in analogy with Chapter 2 (Fig. 1) the characteristic conoid B (that is the temporal divide), which, together with the initial surface $x^0 = 0$, bounds the simply connected domain G'. Now suppose that in G' there exist two systems of solutions of equations (1), (5)

$$g_{ik},\ \Phi_i \quad \text{and} \quad \breve{g}_{ik},\ \breve{\Phi}_i \qquad (i, k = 0, 1, 2, 3),$$

which we assume both to be twice continuously and bounded differentiable within G' and on its boundary surfaces. Both systems of solutions may agree on the domain A' cut out by the temporal divide B from the surface $x^0 = 0$, including their first derivatives.

Under these premises we prove the T h e o r e m: *Within G' both systems of solutions are transferable into each other by a coordinate transformation*, hence, are physically equivalent.

For the proof we introduce in the world with the metric g_{ik} a de Donderian coordinate system x_{I}. In this world there thus holds:

$$\frac{\partial\sqrt{-g}\,g^{ir}}{\partial x^r_{\mathrm{I}}} = 0 \qquad (i = 0, 1, 2, 3).$$

Respecting the initial conditions (5) of Chapter 3, this coordinate system shall, besides the x^0_{I}-axes emerging from the initial surface, agree with that proper one on the initial surface $x^0 = 0$ in which the initial values are given.

Analogously we choose in the second world with the metric $\breve{g}_{ik}$ a de Donderian coordinate system x_{II}, which satisfies the conditions:

$$\frac{\partial\sqrt{-\breve{g}}\,\breve{g}^{ir}}{\partial x^r_{\mathrm{II}}} = 0 \qquad (i = 0, 1, 2, 3).$$

Again, on $x^0 = 0$ the coordinates x_{II} shall agree with the coordinates in which the initial values are given, likewise thereupon the x^0-axes, too.

If we then map the world II by the equations

$$x^i_{\mathrm{I}} = x^i_{\mathrm{II}} \qquad (i = 0, 1, 2, 3)$$

onto the world I, then in this mapping all invariant equations which existed in the world II remain valid. On the initial surface there then holds:[20]

$$(6) \qquad (g_{ik})_{x^0=0} = (\breve{g}_{ik})_{x^0=0}$$

[20] Two typos in the formula below were corrected by the translator: (i) The sign ˘ over g on the right-hand side was missing in the original; (ii) Equation-number (6) did not exist anywhere; most probably it should be here and therefore it was added.

(see p. 9).[21] But also the first derivatives of the g_{ik} still agree on the initial surface, as the second derivatives of the transformation functions $\varphi_{\mathrm{I}}^{(i)}$ and $\varphi_{\mathrm{II}}^{(i)}$ agree on the initial surface, which can be easily confirmed with the help of (4), Chapter 3.

In this way we then have gained in the world I, according to Ch. 3, equation (2), the two equations:

$$(7)\qquad \begin{aligned} &\Box\, g_{ik} + A_{ik} = kS_{ik} \qquad &&\left(\Box = g^{\lambda\mu}\frac{\partial^2}{\partial x^\lambda\,\partial x^\mu}\right),\\ &\breve{\Box}\, \breve{g}_{ik} + \breve{A}_{ik} = k\breve{S}_{ik} \qquad &&\left(\breve{\Box} = \breve{g}^{\lambda\mu}\frac{\partial^2}{\partial x^\lambda\,\partial x^\mu}\right). \end{aligned}$$

The A_{ik} are expressions that only contain the g_{ik} and their first derivatives. This also holds for the $\breve{A}_{ik}$[22]).

In addition we have the equations (5) for the electromagnetic potentials. As in these the second derivatives of the g_{ik} occur only in the combination $\frac{1}{\sqrt{-g}}\frac{\partial\sqrt{-g}\,g^{ie}}{\partial x^e}$, respectively in derivatives of these quantities, we can write for these equations:

$$(8)\qquad \begin{aligned} \Box\, \Phi_i + B_i = 0,\\ \breve{\Box}\, \breve{\Phi}_i + \breve{B}_i = 0, \end{aligned}$$

where the B_i only contain first derivatives of the g_{ik} and the Φ_i. Also, the S_{ik} in the equations (7) contain at most first derivatives of the Φ_i. Similarly for the quantities flagged with a hook.

When we now set

$$g_{ik} - \breve{g}_{ik} = l_{ik} \quad \text{and} \quad \Phi_i - \breve{\Phi}_i = f_i,$$

then we obtain from forming differences of (7), (8):

$$\begin{aligned} L_{ik} &\equiv \Box\, l_{ik} - D_{ik} = 0,\\ L_i &\equiv \Box\, f_i - D_i = 0. \end{aligned}$$

[21]The page numbers referred to are at the bottom of each page [editor].

[22]) Although in Ch. 2 we did not need for the second solution $\breve{u}$ the assumption that for it the coefficients a^{ik} satisfy the conditions of total hyperbolicity, we here need to also assume of the $\breve{g}_{ik}$ that its matrix possesses the index of inertia $(---+)$ (in the proof of the next section this is not necessary), as otherwise in the world II the transformation of de Donder need not be feasible.

In analogy to equation (3), p. 5, the D are linear functions of the l_{ik} and their first derivatives, and of the f_i and their first derivatives. As coefficients of these linear forms D serve certain polynomial forms of the g_{ik}, $\breve{g}_{ik}$, Φ_i, $\breve{\Phi}_i$ and their first and second derivatives. The particular appearence of these forms is of no importance for the proof. On the equation system thus obtained, we can apply the Friedrichs–Lewyian considerations. Evidently, the g^{ik} satisfy all preconditions that we set in Chapter 2 for the a^{ik} with regard to the index of inertia and, in particular, also with regard to conditions (2), Section 2, which are equivalent to the Hilbertian demand for a proper coordinate system[23]).

According to the conclusion of Chapter 2, the demanded proof is thus completed. Instead of the integral (4), p. 5, one only has to form:

$$\iiint\limits_{G''} \left(\frac{\partial l_{ik}}{\partial x^0} L_{ik}\right) d\tau = 0 \quad \text{resp.} \quad \iiint\limits_{G''} \left(\frac{\partial f_i}{\partial x^0} L_i\right) d\tau = 0$$

and then proceed just as in Chapter 2. We thus obtain the domain of dependence of the solution in a point P by envisaging the temporal divide in this point. The domain of dependence is then cut out from the initial manifold by the temporal divide.

In the language of relativity theory we can also express this theorem by:

In parts of space which are free of matter, "simultaneously" positioned world points cannot influence one another.

5.

The proof just given cannot be extended to the case where, besides the state variables g_{ik} and Φ_i, there is also continuously distributed matter present[24]).

By a special trick, however, one can treat the case of incoherent matter in the absence of any electrical fields.

We then have the equation system:

$$\mathfrak{R}_{ik} = k(\mathfrak{T}_{ik} - \tfrac{1}{2} g_{ik} \mathfrak{T}) \tag{1}$$

with

$$\mathfrak{T}_{ik} = \mathfrak{m} u_i u_k.$$

[23]) l. c.

[24]) Attempts in this direction fail due to the form of the divergence equation of the matter, the treatment of which, in our sense, proves altogether unpleasant.

As usual, $\mathfrak{m}$ denotes the mass *density* of the matter; u^i the tangent vector of length 1 to the world lines of the matter. For determining these quantities we have the equations:

$$\frac{du^i}{ds} + \Gamma^i_{\alpha\beta} u^\alpha u^\beta = 0, \qquad g_{ik} u^i u^k = 1 \tag{2}$$

and

$$\frac{\partial(u^i \mathfrak{m})}{\partial x^i} = 0. \tag{3}$$

In G'[25]) let there be given two systems of solutions (twice continuously and bounded differentiable)

$$g_{ik},\ u^i,\ \mathfrak{m} \quad \text{and} \quad \breve{g}_{ik},\ \breve{u}^i,\ \breve{\mathfrak{m}}$$

of these equations, which agree in the domain A' of the initial surface $x^0 = 0$; in particular, the g_{ik} shall agree up to their first derivatives, while the functions u^i and $\mathfrak{m}$ must agree only themselves with $\breve{u}^i$ and $\breve{\mathfrak{m}}$ on $x^0 = 0$.

Again, we then have the T h e o r e m: under the stated assumptions both systems of solutions are transferable into one another by transformation.

The settling of the problem, i.e. the proof of the theorem, succeeds due to the introduction of a "rest coordinate system" (see Hilbert, l. c.); that is, a coordinate system in which the contravariant tangent vector u^i possesses the components $(1\ 0\ 0\ 0)$. In contrast to the de Donder coordinate system the rest coordinate system is fixed by differential equations of *first* order for the coordinates. This is as in the new coordinate system there shall hold:

$$\overline{u}^i = \frac{\partial \overline{x}^i}{\partial x^\alpha} u^\alpha = \delta^i_0 \qquad (i = 0, 1, 2, 3). \tag{2a}$$

Those world positions in which there is no matter present shall be viewed as filled in a continuously differentiable way by a unit vector field that also satisfies the differential equations of the geodesic lines.

In our rest coordinate system the continuity equation of the matter assumes the form:

$$\frac{\partial \mathfrak{m}}{\partial x^0} = 0. \tag{3a}$$

[25]) Cf. Ch. 2, Fig. 1.

Moreover, it follows from the equations of motion (2) by respecting the value for the vector u^i:

$$\Gamma^i_{00} = 0 \qquad (i = 0, 1, 2, 3),$$

therefore, too:

$$(4) \qquad \Gamma_{00,i} = 0 \qquad (i = 0, 1, 2, 3).$$

Moreover, because of $u^i\, u_i = 1$ (unit vector!) it holds:

$$g_{00} = 1.$$

Therefore, we obtain from the equations (4)

$$(5) \qquad \frac{\partial g_{0i}}{\partial x^0} = 0 \qquad (i = 0, 1, 2, 3).$$

In each of the two envisaged worlds we introduce a rest coordinate system x_{I}, respectively x_{II} (according to the procedure given on p. 11), which agree on $x^0 = 0$ with the original coordinate system. Due to the agreement of the initial values of the u^i, on the initial surface there then also agree, because of (2a), the derivatives

$$\frac{\partial x^i_{\mathrm{I}}}{\partial x^k} \quad \text{resp.} \quad \frac{\partial x^i_{\mathrm{II}}}{\partial x^k}$$

with the unit tensor δ^i_k. For the same reason also the second derivatives of the x_{I}, respectively x_{II} agree on the initial surface.

If we map again the world II by the equations $x_{\mathrm{I}} = x_{\mathrm{II}}$ onto the world I, i.e. if we let the world lines of the matter of the two worlds sit on top of one another, then we obtain agreeing initial values of the g_{ik} and $\mathfrak{m}$ with the $\breve{g}_{ik}$, respectively $\breve{\mathfrak{m}}$; moreover, there follows in the interior of G', due to (3a),

$$(6) \qquad \breve{\mathfrak{m}} = \mathfrak{m}$$

and from (5)

$$(7) \qquad g_{0i} = \breve{g}_{0i} \qquad (i = 0, 1, 2, 3).$$

Respecting the simplifications that our coordinate system yields, we obtain for the two solutions, instead of (1), the systems of equations:

$$\begin{aligned} R_{ik} &= k\,\frac{\mathfrak{m}}{\sqrt{-g}}\,(g_{0i}\,g_{0k} - \tfrac{1}{2}\,g_{ik}) \\ \breve{R}_{ik} &= k\,\frac{\breve{\mathfrak{m}}}{\sqrt{-\breve{g}}}\,(\breve{g}_{0i}\,\breve{g}_{0k} - \tfrac{1}{2}\,\breve{g}_{ik}). \end{aligned}$$

Employing (6) and (7), we form the differences:

(1a) $$R_{ik} - \breve{R}_{ik} =$$

$$k \cdot \mathfrak{m} \left[\left(\frac{1}{\sqrt{-g}} - \frac{1}{\sqrt{-\breve{g}}} \right) (g_{0i}\, g_{0k} - \tfrac{1}{2}\, g_{ik}) + \frac{1}{\sqrt{-\breve{g}}} (-\tfrac{1}{2})(g_{ik} - \breve{g}_{ik}) \right].$$

We set again $g_{ik} - \breve{g}_{ik} = l_{ik}$ and can express the right-hand side of (1a) as a homogeneous linear function of the *six* quantities l_{ik} $(i, k = 1, 2, 3)$. We will also express the left-hand side as a linear form of the l_{ik} and their first and second derivatives. In line with Section 2, we will be most interested in that part which contains the second derivatives of the l_{ik}.

Moreover, for the following it will be important that in the equation (1a) with index 0 0, the left-hand side allows for such a representation of the first derivatives of the l_{ik} in which only derivatives with respect to x^0 occur. Because of the equations (4), it evidently holds:

$$\begin{aligned} R_{00} &= -\frac{\partial \Gamma^r_{0r}}{\partial x^0} - \Gamma^r_{0s}\Gamma^s_{0r}, \\ \breve{R}_{00} &= -\frac{\partial \breve{\Gamma}^r_{0r}}{\partial x^0} - \breve{\Gamma}^r_{0s}\breve{\Gamma}^s_{0r}, \end{aligned}$$

and from this by subtraction[26]

$$R_{00} - \breve{R}_{00} = -\tfrac{1}{2}\, g^{rs}\, \frac{\partial^2 (g_{rs} - \breve{g}_{rs})}{(\partial x^0)^2} + \Gamma^r_{0s}\, (\Gamma^s_{0r} - \breve{\Gamma}^s_{0r})$$

$$+ \breve{\Gamma}^s_{0r}\, (\Gamma^r_{0s} - \breve{\Gamma}^r_{0s}) + l_{\rho\mu}\, A^{\rho\mu}.$$

The $A^{\rho\mu}$ are polynomials of the g_{ik}, g^{ik} and the $\Gamma_{ik,l}$, as well as of the $\breve{g}_{ik}$, $\breve{g}^{ik}$ and the $\breve{\Gamma}_{ik,l}$, and of their first derivatives.

By equation (7) $l_{0i} = 0$ holds; it then follows:

$$R_{00} - \breve{R}_{00} = -\frac{1}{2}\, g^{rs}\, \frac{\partial^2 l_{rs}}{(\partial x^0)^2} + \frac{1}{2}\frac{\partial l_{rt}}{\partial x^0}\, (\Gamma^t_{0s} g^{sr} + \breve{\Gamma}^t_{0s} \breve{g}^{sr}) + l_{\rho\mu}\, \overline{A}^{\rho\mu}.$$

By integration with respect to x^0 from the initial surface up to x^0, and a suitable subsequent recasting by means of integration by parts, we obtain

[26] A typo in the equation below was corrected by the translator: the indices "r" and "s" in the last Γ of the first line were interchanged.

eventually:

$$(8)\quad \int_0^{x^0} (R_{00} - \breve{R}_{00})\, dx^0 = -\frac{1}{2}\frac{\partial l_{rs}}{\partial x^0} \cdot g^{rs} + l_{\rho\mu}\, B^{\rho\mu} + \int_0^{x^0} l_{\rho\mu}\, C^{\rho\mu}\, dx^0,$$

wherein the coefficients $C^{\rho\mu}$ are again formed by polynomial algebraic operations, the $B^{\rho\mu}$ in addition also by integration with respect to x^0, from the g_{ik}, $\breve{g}_{ik}$ and their first and second derivatives.

Before we write down the remaining components, we remark that the contracted Riemannian tensor can be written with respect to its second derivatives as follows:[27]

$$2\, R_{ik} = -\, g^{rs}\frac{\partial^2}{\partial x^r\, \partial x^s} g_{ik} + \frac{\partial}{\partial x^i}\beta_k + \frac{\partial}{\partial x^k}\beta_i + L_{ik}\ ;$$

herein we set

$$(9)\qquad \beta_i = g^{rs}\left(\frac{\partial g_{ri}}{\partial x^s} - \frac{1}{2}\frac{\partial g_{rs}}{\partial x^i}\right),$$

while the L_{ik} only contain first derivatives of the g_{ik}. By subtraction it follows from this:

$$(9a)\qquad 2(R_{ik} - \breve{R}_{ik}) = -\, g^{rs}\frac{\partial^2 l_{ik}}{\partial x^r\, \partial x^s} + \frac{\partial \gamma_k}{\partial x^i} + \frac{\partial \gamma_i}{\partial x^k} + M_{ik}$$

with

$$(9b)\qquad \gamma_i = g^{rs}\left(\frac{\partial l_{ir}}{\partial x^s} - \frac{1}{2}\frac{\partial l_{rs}}{\partial x^i}\right).$$

Then, respecting equations (8), (9a), (9b), equations (1a) can be written:[28]

$$(10)\qquad \begin{aligned} &\text{a)}\ \ -2\,\gamma_0 = l_{\rho\mu}\, B^{\rho\mu} + \textstyle\int_0^{x^0} l_{\rho\mu}\, \overline{C}^{\rho\mu}\, dx^0, \\ &\text{b)}\ \ -\frac{\partial \gamma_0}{\partial x^i} - \frac{\partial \gamma_i}{\partial x^0} = \Lambda_i \qquad (i = 1, 2, 3), \\ &\text{c)}\ \ \Box\, l_{ik} - \frac{\partial \gamma_i}{\partial x^k} - \frac{\partial \gamma_k}{\partial x^i} = \Lambda_{ik} \qquad (i, k = 1, 2, 3). \end{aligned}$$

[27] A typo in the equation below was corrected by the translator: a missing index "k" in the third term on the right-hand side was added.

[28] Several identical typos were corrected by the translator: in the original text, the upper limit in the integral was x_0 in the first equation below, in the first equation after (11), in the text above (15), in eqs. (13) – (16) and in the next one after (16).

The Λ are again linear forms of the l_{ik} and their first derivatives.

By eliminating γ_0 between equations (10a) and (10b), we obtain the following three equations:

$$-\frac{\partial \gamma_i}{\partial x^0} = \Lambda_i - \int_0^{x^0} \frac{\partial}{\partial x^i} l_{rs} \overline{C}^{rs} \, dx^0 - \frac{\partial}{\partial x^i} l_{rs} B^{rs},$$

or

$$\frac{\partial \gamma_i}{\partial x^0} = \Lambda_i - \int_0^{x^0} V_i \, dx^0, \tag{11}$$

where Λ_i and V_i are again linear forms of the l_{ik} and their first derivatives.

Now we multiply, respectively, the equations (10c) by $\frac{\partial l_{ik}}{\partial x^0}$ and (11) by $2\frac{\partial l_{ik}}{\partial x^k}$, add these up, and obtain:

$$\Box(l_{ik}) \cdot \frac{\partial l_{ik}}{\partial x^0} - \frac{\partial l_{ik}}{\partial x^0}\left(\frac{\partial \gamma_i}{\partial x^k} + \frac{\partial \gamma_k}{\partial x^i}\right) + 2\frac{\partial l_{ik}}{\partial x^k}\frac{\partial \gamma_i}{\partial x^0} = L + \int_0^{x^0} N \, dx^0.$$

The left-hand side is a divergence up to quadratic forms of the $\frac{\partial l_{ik}}{\partial x^r}$; this is because the following holds (see Friedrichs and Lewy)

$$2\frac{\partial^2 l_{ik}}{\partial x^r \partial x^s}\frac{\partial l_{ik}}{\partial x^0} = \left(\frac{\partial l_{ik}}{\partial x^r}\frac{\partial l_{ik}}{\partial x^0}\right)_s - \left(\frac{\partial l_{ik}}{\partial x^r}\frac{\partial l_{ik}}{\partial x^s}\right)_0 + \left(\frac{\partial l_{ik}}{\partial x^s}\frac{\partial l_{ik}}{\partial x^0}\right)_r.$$

We thus obtain:

$$\left(g^{rs}\frac{\partial l_{ik}}{\partial x^r}\frac{\partial l_{ik}}{\partial x^0}\right)_s - \frac{1}{2}\left(g^{rs}\frac{\partial l_{ik}}{\partial x^r}\frac{\partial l_{ik}}{\partial x^s}\right)_0 - \frac{\partial\left(\frac{\partial l_{ik}}{\partial x^0}\gamma_i\right)}{\partial x^k} - \frac{\partial\left(\frac{\partial l_{ik}}{\partial x^0}\gamma_k\right)}{\partial x^i} + 2\frac{\partial\left(\frac{\partial l_{ik}}{\partial x^k}\cdot\gamma_i\right)}{\partial x^0} = \overline{L} + \int_0^{x^0} \overline{N} \, dx^0. \tag{12}$$

Furthermore, we also add to (12) the equations (11) correspondingly multiplied by $2P\gamma_i$:

$$+P\frac{\partial(\gamma_i)^2}{\partial x^0} = 2P\Lambda_i\gamma_i + 2\gamma_i P\int_0^{x^0} N \, dx^0. \tag{13}$$

(P is a constant, the choice of which we still keep open.) We integrate the sum of the left-hand sides of (12) and (13) over the domain G'' (see p. 6, Fig. 1). Then there results upon application of the Gaußian integral theorem under the surface integral a quadratic form of the $4 \cdot 6$ derivatives of first order of the l_{ik} and the three γ_i, namely

$$g^{rs}\left(\frac{\partial l_{ik}}{\partial x^r}\frac{\partial l_{ik}}{\partial x^0}\xi_s - \frac{1}{2}\frac{\partial l_{ik}}{\partial x^r}\frac{\partial l_{ik}}{\partial x^s}\xi_0\right) - \frac{\partial l_{ik}}{\partial x^0}\gamma_i\,\xi_k - \frac{\partial l_{ik}}{\partial x^0}\gamma_k\,\xi_i$$
$$+2\frac{\partial l_{ik}}{\partial x^k}\gamma_i\,\xi_0 + 2\,P\,(\gamma_i)^2\,\xi_0.$$

The part deriving form the □-expressions is positive definite on C (cf. Ch. 2). Hence, this quadratic form can be made positive definite on C by a sufficiently large choice of the constant P. By substituting for the γ_i again their values (9b),[29] we gain an equation:

$$\text{(14)}\qquad \iint\limits_{M'+C} Q\left(\frac{\partial l_{ik}}{\partial x^r}\right)d\omega$$
$$= \iiint\limits_{G''}\left[\overline{Q}\left(\frac{\partial l_{ik}}{\partial x^r}\,l_{st}\right) + E^{irmn}\frac{\partial l_{mn}}{\partial x^r}\int_0^{x^0} V_i\,dx^0\right]d\tau.$$

Q and $\overline{Q}$ are quadratic forms of the 24 first derivatives of the l_{ik}, $\overline{Q}$ also of the l_{ik}, Q is on C non-degenerate positive definite, on M' possibly degenerate positive definite. Besides the quadratic form $\overline{Q}$, we also have under the volume integral on the right-hand side an expression which may be viewed as a quadratic form of the l_{ik} and their first derivatives, as well as of the three quantities $\int_0^{x^0} V^i\,dx^0$. Accordingly, we gain from (14) the inequality:

(15)
$$\iint\limits_{C}\left(\frac{\partial l_{ik}}{\partial x^0}\right)^2 d\omega \le D\iiint\limits_{G''}\left[\left(\frac{\partial l_{ik}}{\partial x^r}\right)^2 + (l_{ik})^2 + \left(\int_0^{x^0} V_i\,dx^0\right)^2\right]d\tau.$$

With the aid of the Schwarzian inequality one easily gains:

$$\text{(16)}\qquad \left(\int_0^{x^0} V_i\,dx^0\right)^2 \le E\int_0^{x^0}(V_i)^2\,dx^0.$$

[29] This reference to the equation-number was corrected by the translator, in the original text it was (9a).

Furthermore, it evidently holds:

$$\int_0^c dx^0 \int_0^{x^0} (V_i)^2\, dx^0 \leq W \int_0^c (V_i)^2\, dx^0.$$

D, E and W are constants, which may be chosen independently of the parameter c. Then we obtain from (15):[30]

$$\iint\limits_{C} \left(\frac{\partial l_{ik}}{\partial x^r}\right)^2 d\omega \leq F \iiint\limits_{G''} \left[\left(\frac{\partial l_{ik}}{\partial x^r}\right)^2 + (l_{ik})^2\right] d\tau, \tag{17}$$

again F can be chosen independently of c (cf. p. 7). From equation (17) then follows, according to what was said in Chapter 2, the uniqueness within G'. Thus, the theorem is proven.

Therefore, also for the case of a world with arbitrarily distributed incoherent matter the correct causal structure is guaranteed.

6.

So far, I have not succeeded in the settling of the general case, in which the matter is in addition charged. However, on the basis of results from Chapter 4, some statements can be made on causality also in this case, given the charged matter is distributed discontinuously on several very thin world tubes. With the aid of the final result of Chapter 4 one can then easily derive the theorem:

All *actions* that a matter particle *generates* in a world point P lie within the future-directed part of the temporal divide to be constructed in P.

To show this, one only has to use in the course of the proof of Chapter 4 as the boundary of the domain G' instead of the temporal divide B' a hypersurface which is obtained as follows: one erects in P the future-directed temporal divide B and forms the intersection of it with an arbitrary space-like surface R. Then the envelope of all those temporal divides, which are to be envisaged in points of the intersection manifold B, R, yields the desired new boundary surface of the domain G'.

(Received on 7.6.1937)

[30] A typo in (17) was corrected by the translator: the volume element on the right-hand side was $d\omega$ in the original text.

Chapter 3

Richard Arnowitt, Stanley Deser and Charles W. Misner 1962, the dynamics of general relativity

Original paper: Richard Arnowitt, Stanley Deser and Charles W. Misner, "The dynamics of general relativity". In *Gravitation: An Introduction to Current Research*, Ed. Louis Witten, (1962).

Previously published in *Gen. Relativ. Gravit.* **40** (2008), pages 1989 – 1995 (the editorial note + biographies), 1997 – 2027 (the paper).

Gen Relativ Gravit (2008) 40:1989–1995
DOI 10.1007/s10714-008-0649-x

GOLDEN OLDIE EDITORIAL

Editorial note to R. Arnowitt, S. Deser, C. W. Misner The dynamics of general relativity

Jorge Pullin

Published online: 23 July 2008

Editor's note

Although the general theory of relativity was introduced in 1915, many of the concepts and ideas associated with it that we today take for granted were only introduced many years later. In some cases a significant amount of work and controversy had to take place to end up with ideas that, when presented today, appear so elegant as to be mistaken as obvious. An example of this would be the notion of a black hole. Another example is the unravelling of the true Hamiltonian dynamics of general relativity. These two milestones occurred in the 1960s.

Unravelling the dynamics of general relativity was a daunting task. The equations of the theory are long, nonlinear and complicated. The theory is invariant under diffeomorphisms, a concept that took a while to be fully appreciated. Early work on the initial value problem was done by Lichnerowicz, culminating with the celebrated developments of Choquet-Bruhat and York. In spite of this, experience with other theories showed that perhaps the best route to understanding the true dynamics of a theory is to cast it in canonical form. Progress in the canonical formulation of field theories with complex symmetries was only achieved in the late 1950s by Dirac [1] with a somewhat different (4-covariant) route followed by Bergmann's group [2] at Syracuse as well. However, a concrete and clear application of canonical methods to general relativity remained somewhat elusive for a few years, given that one can make several different choices in how to set up the canonical theory. In a short period around

The republication of the original paper in corrected version can be found via doi:10.1007/s10714-008-0661-1.

J. Pullin (✉)
Department of Physics and Astronomy, Louisiana State University,
Baton Rouge, LA 70803-4001, USA
e-mail: pullin@lsu.edu

1960 Arnowitt, Deser and Misner proceeded in a series of papers [3–15] to lay out what we now recognize as the modern canonical formulation of general relativity, in particular identifying the physical energy functional and dynamical variables of the theory in an asymptotically flat context. A summary of these papers appeared in the volume edited by Louis Witten [16]. This volume presented what time has proved to be a remarkably influential snapshot of the status of research in gravitation in the early 1960s. Arnowitt, Deser and Misner were awarded the Dannie Heinemann Prize of the American Physical Society in 1994 in large part in recognition of the influence of their joint work.

The "ADM paper" appearing in the book is a delight to read, given how concisely it describes things. It presents the canonical formulation of general relativity as we would recognize it today, in terms of the three-metric and the extrinsic curvature with the lapse and shift as Lagrange multipliers. The paper identifies the time evolution as a canonical transformation and the role of the coordinate time as a parameter. It bases the formulation on the Palatini action principle, which makes things considerably simpler than in previous treatments. It uses the formulation to define energy and momentum for a gravitational field in a rigorous way, including gravitational radiation and to establish the Lorentz invariance of the energy four momentum vector. It also includes an influential argument about the finiteness of the energy of a point particle in classical general relativity that suggests that non-perturbatively quantum gravity should be finite. The argument is still invoked as a motivation in modern treatments of quantum gravity (see for instance [17]). In the immediate aftermath, the ADM paper was influential in the first attempts by DeWitt [18] to construct a non-perturbative canonical quantization of the gravitational field. DeWitt's work however, uncovered many of the significant difficulties that a canonical non-perturbative quantization of gravity faces even today and work in this area stalled for almost two decades. Given that some of these difficulties simplify in minisuperspaces, applications in quantum cosmology flourished in the 1970s and 1980s. One of the most important followups of the ADM paper was the proof by Schoen and Yau [19] that mass is positive in general relativity, further elaborated for supergravity by Deser and Teitelboim [20] and later by Witten [21]. Another beautiful development was the Hamiltonian formulation that is Poincaré covariant at infinity by Regge and Teitelboim [22]. Taking a longer view, it can also be argued that the ADM papers are influential in the renewed interest in the canonical approach that took place following the introduction of the Ashtekar variables in the mid 1980s (see [23] for a review including a historical perspective). The ADM paper was also influential in the pioneering numerical efforts of DeWitt, Smarr and others [24], particularly with the aim of simulating binary black hole collisions. In this area, an additional ingredient was needed to achieve significant progress, namely having well posed hyperbolic formulations. This had to wait till the 1990s [25]. Although the original ADM formulation with fixed lapse and shift is not manifestly hypberbolic, it is commonly used as a building block in modern hyperbolic formulations of the Einstein equations. The notions of ADM mass and momentum are also routinely used in numerical simulations. The formulation is also of great use in the related area of approximate solutions to the problem of motion in relativity (for a recent example see [26]). Numerical simulations and approximate solutions to the problem of motion have become the central current areas of research in gravitational physics given the distinct possibility of detection of gravitational waves by interferometric type detectors, and keep the ADM techniques as a common currency among new generations of researchers in gravitational physics.

Summarizing, it is unusual to find a paper that is still playing such a direct and influential role in such a broad range of modern research topics close to fifty years after its publication, as this one is.

Note by the series editor The text of the ADM paper was photocopied from the version published in the gr-qc archive in 2004 (gr-qc/0405109). Some typographical corrections were made in it by the authors; otherwise this is the same text as in the book of 1962.

Acknowledgments This work was supported in part by grant NSF-PHY-0554793, CCT-LSU, and the Horace Hearne Jr. Institute for Theoretical Physics at LSU. I wish to thank Richard Arnowitt, Abhay Ashtekar, Stanley Deser, Charles Misner, Ted Newman, Bob Wald and Richard Woodard for discussions and comments on the manuscript.

Richard Arnowitt: a brief biography

By Richard Arnowitt

Name: Richard Arnowitt
Date of Birth: May 3, 1928

Education

B. S. Physics, Rensselaer Polytechnic Institute 1948
M. S. Physics, Rensselaer Polytechnic Institute 1948
Ph.D. Physics, Harvard University, 1953

Positions Held

1952–1954 Research Associate, Radiation Laboratory (LBL) University of California, Berkeley
1954–1956 Member Institute For Advanced Study
1956–1959 Assistant Professor, Syracuse University
1959–1962 Associate Professor, Syracuse University
1962–1986 Professor, Northeastern University
1986–1988 Professor, Texas A&M University
1986–2004 Distinguished Professor of Physics, Texas A&M University
2004– Distinguished Professor Emeritus, Texas A&M University
1986–1995 Director, Center For Theoretical Physics, Texas A&M University
1987–1993 Head, Department of Physics, Texas A&M University

Honors

Fellow, American Physical Society
Guggenheim Fellow, 1975–1976
Dannie Heineman Prize, American Physical Society, 1994
Burgess Chair of High Energy Physics, 1997–2004

Dr. Arnowitt has worked widely in many branches of theoretical physics including general relativity, quantum field theory, high energy particle phenomenology, supersymmetry and supergravity, and liquid Helium theory. His work with Stanley Deser and Charles Misner on general relativity was awarded the Dannie Heineman Prize by the American Physical Society. His work in particle theory with Marvin Friedman and Pran Nath led to the formulation of effective Lagrangian theory (to characterize the Lagrangian of a system holding below a given symmetry breaking scale) and has had applications over the full energy range of particle theory from the electroweak scale to string theory at the Planck scale. His work on supergravity with Ali Chamseddine and Pran Nath developed the theory of supergravity grand unification with gravity mediated breaking and the mSUGRA model, which has become a benchmark theory used to search for new supersymmetry physics at high energy accelerators. His work on the many body theory of liquid Helium with Marvin Girardeau has stimulated many applications in that field.

Stanley Deser: a brief biography

Edited by Andrzej Krasiński, from data sent by Stanley Deser

Born 19 March 1931 in Rovno, Poland

Education

1949: BS Summa Cum Laude, Brooklyn College
1950: MA, Harvard University
1953: PhD Harvard University (Adviser: J Schwinger)

Professional Record (only longer-term positions)

1953–55, 1988, 1993: Member, Institute for Advanced Study
1954: Research Associate, University of California, Radiation Laboratory
1955–57: Member, Niels Bohr Institute for Theoretical Physics, Copenhagen
1957–58: Lecturer, Harvard University
1958–65: Associate Professor, Brandeis University
1961–62, 1976, 1980–81, 1994: Visiting Scientist, CERN, Geneva
1964–70: Co-Director and Co-Editor, Brandeis Summer Institutes in Theoretical Physics
1965–80: Professor of Physics, Brandeis University
1980– Ancell Professor of Physics, Brandeis University
1966–67: Visiting Professor, Institut Henri Poincare of the Sorbonne

1969–71, 1976–77: Chairman, Physics Department, Brandeis University
1977: All Souls Oxford Fellow
1971–72: Visiting Professor, Faculty of Science, University of Paris – Orsay
1984–85: Co-Director, Field Theory Program, Institute for Theoretical Physics, Santa Barbara
1987–88: Visiting Professor, Boston University
2005–Visiting Professor, California Institute of Technology

Honors and Awards

National Academy of Sciences, USA
American Academy of Arts & Science
Foreign member, Torino Academy of Science, Italy.
Dannie Heineman Prize with R. Arnowitt and C. W. Misner
Guggenheim & Fulbright Fellow
Honorary Doctorate Degrees from Stockholm Univ and Chalmers Institute of Technology
Festschrift: "Deserfest," World Pub (2006). J T Liu, M J Duff, K S Stelle and R P Woodard, eds.

Among better-known work, besides ADM: SUGRA in 1976 (with Zumino), topologically massive 3D theories with Jackiw 1982, 3D Einstein with Jackiw & 't Hooft, more recently a series on partially massive gauge theories in (A) dS with Waldron.

Charles W. Misner: a brief biography

By Charles W. Misner

Personal

Born Jackson, Michigan, 13 June 1932; married 13 June 1959; four children; Roman Catholic.

Education

B.S. Physics, Notre Dame 1952 (mathematics mentor Arnold Ross);
Ph.D., Princeton 1957 (mentor John Archibald Wheeler).

University Positions

Princeton 1956–63,
University of Maryland 1963–2000, now Emeritus.
Visitor at Max-Planck Einstein Institute (Potsdam) 2000, 2002, 2005;
UC Santa Barbara ITP 1980–81;
All Souls College, Oxford 1973;
Caltech 1972; Princeton 1969; DAMTP Cambridge UK 1966–67.

Honors and Awards

Fellow of the American Academy of Arts & Sciences 2000;
Dannie Heineman Prize for Mathematical Physics (APS, shared with Arnowitt and Deser) 1994;
Guggenheim Fellowship 1972–73;
Festschrift (Papers in honor of Charles Misner, Cambridge Univ. Press 1993, Hu et al. editors).

Selected Publications

Arnowitt, Deser and Misner (ADM) "The dynamics of general relativity" (This paper; originally pp 227–265 in L. Witten, ed., Gravitation: An Introduction to Current Research, Wiley, New York, 1962), gr-qc/0405109;

Misner, Thorne and Wheeler (MTW) Gravitation (Freeman 1973) [Now a classic, never revised, still selling over 500 copies per year.];

Misner and Finkelstein "Some New Conservation Laws" (Annals of Physics 6, 230–243, 1959);

Misner, "Wormhole Initial Conditions" (Physical Review 118, 1110–1111, 1960)

Misner (with D L Beckedorff), "Infinite Red-Shifts in General Relativity" (pp 75–89 in T Gold, ed., The Nature of Time—Cornell Conference 1963, Cornell University Press 1967); http://hdl.handle.net/1903/4280;

Misner, "Quantum Cosmology" (Physical Review 186, 1319–1327, 1969);

Misner, "Minisuperspace" (pp 441–473 in J. Klauder, ed., Magic Without Magic—J. A. Wheeler 60th Anniversary Volume, Freeman 1972);

Misner, "The Mixmaster Cosmological Metrics" (pp 317–328 in Deterministic Chaos in General Relativity, Hobill et al., eds., Plenum Press 1994);

Misner, van Meter and Fiske "Excising das All: Evolving Maxwell waves beyond scri" (Phys.Rev. D74, 064003, 2006), gr-qc/0603034

Students

Twenty-two Ph.D. students, of whom half became full professors in physics or mathematics, while several others had equally important positions.

References

1. Dirac, P.A.M.: The theory of gravitation in Hamiltonian form. Proc. R. Soc. Lond. A **246**, 333 (1958)
2. A good historical account is given in P. Bergmann "The canonical formulation of general relativistic theories: the early years 1930–1959". In: Howard, D., Stachel, J. (eds.) Einstein and the History of General Relativity. Birkhäuser, Boston (1989)
3. Arnowitt, R., Deser, S.: Phys. Rev. **113**, 745 (1959)
4. Arnowitt, R., Deser, S., Misner, C.: Phys. Rev. **116**, 1322 (1959)
5. Arnowitt, R., Deser, S., Misner, C.: Phys. Rev. **117**, 1595 (1959)
6. Arnowitt, R., Deser, S., Misner, C.: Nuovo Cimento **15**, 487 (1960)
7. Arnowitt, R., Deser, S., Misner, C.: Phys. Rev. Lett. **4**, 375 (1960)

8. Arnowitt, R., Deser, S., Misner, C.: Phys. Rev. **118**, 1100 (1960)
9. Arnowitt, R., Deser, S., Misner, C.: J. Math. Phys. **1**, 434 (1960)
10. Arnowitt, R., Deser, S., Misner, C.: Phys. Rev. **120**, 313 (1960)
11. Arnowitt, R., Deser, S., Misner, C.: Phys. Rev. **120**, 321 (1960)
12. Arnowitt, R., Deser, S., Misner, C.: Ann. Phys. **11**, 116 (1960)
13. Arnowitt, R., Deser, S., Misner, C.: Nuovo Cimento **19**, 668 (1960)
14. Arnowitt, R., Deser, S., Misner, C.: Phys. Rev. **121**, 1556 (1960)
15. Arnowitt, R., Deser, S., Misner, C.: Phys. Rev. **122**, 997 (1961)
16. Arnowitt, R., Deser, S., Misner, C.: The dynamics of general relativity. In: Witten, L. (ed.) Gravitation: An Introduction to Current Research, Wiley, New York (1962)
17. Ashtekar, A.: Lectures on Non-perturbative Canonical Gravity. World Scientific, Singapore (1991)
18. DeWitt, B.S.: Phys. Rev. **171**, 1834 (1968)
19. Schoen, R., Yau, S.-T.: Phys. Rev. Lett. **42**, 547–548 (1979)
20. Deser, S., Teitelboim, C.: Phys. Rev. Lett. **39**, 249 (1977)
21. Witten, E.: Comm. Math. Phys. **80**, 381 (1980)
22. Regge, T., Teitelboim, C.: Ann. Phys. (NY) **88**, 286 (1974)
23. Rovelli, C.: Quantum Gravity. Cambridge University Press, Cambridge (2005)
24. The early work in numerical relativity is well captured in the articles in the book. In: Smarr, L. (ed.) Sources of Gravitational Radiation. Cambridge University Press, Cambridge (1979)
25. For a review of hyperbolic formulations of the Einstein equations see O. Reula. Liv. Rev. Relat. lrr-1998-3 (1998)
26. Damour, T., Jaranowski, P., Schafer, G.: arXiv:0711.1048 [gr-qc]

Gen Relativ Gravit (2008) 40:1997–2027
DOI 10.1007/s10714-008-0661-1

GOLDEN OLDIE

Republication of: The dynamics of general relativity

**Richard Arnowitt · Stanley Deser ·
Charles W. Misner**

Published online: 8 August 2008

Abstract This article—summarizing the authors' then novel formulation of General Relativity—appeared as Chap. 7, pp. 227–264, in *Gravitation: an introduction to current research*, L. Witten, ed. (Wiley, New York, 1962), now long out of print. Intentionally unretouched, this republication as Golden Oldie is intended to provide contemporary accessibility to the flavor of the original ideas. Some typographical corrections have been made: footnote and page numbering have changed–but not section nor equation numbering, etc. Current institutional affiliations are encoded in: arnowitt@physics.tamu.edu, deser@brandeis.edu, misner@umd.edu.

An editorial note to this paper and biographies of the authors can be found via doi:10.1007/s10714-008-0649-x.

Original paper: R. Arnowitt, S. Deser and C. W. Misner, in Gravitation: an introduction to current research (Chap. 7). Edited by Louis Witten. John Wiley & Sons Inc., New York, London, 1962, pp. 227–265. Reprinted with the kind permissions of the editor, Louis Witten, and of all three authors.

R. Arnowitt (✉)
Department of Physics, Texas A&M University, 4242 TAMU,
College Station, TX 77843-4242, USA
e-mail: arnowitt@physics.tamu.edu

S. Deser (✉)
Mailstop 057, Brandeis University, 415 South Street,
Waltham, MA 02454, USA
e-mail: deser@brandeis.edu

Lauritsen Laboratory, California Institute of Technology,
Pasadena, CA 91125, USA

C. W. Misner (✉)
Gravitation Theory Group, Department of Physics, Room 4205D,
University of Maryland, College Park, MD 20742-4111, USA
e-mail: misner@umd.edu

R. Arnowitt,[i] *S. Deser,*[ii] *and C. W. Misner*[iii]

The Dynamics of General Relativity

1. Introduction[iv]

The general coordinate invariance underlying the theory of relativity creates basic problems in the analysis of the dynamics of the gravitational field. Usually, specification of the field amplitudes and their first time derivatives initially is appropriate to determine the time development of a field viewed as a dynamical entity. For general relativity, however, the metric field $g_{\mu\nu}$ may be modified at any later time simply by carrying out a general coordinate transformation. Such an operation does not involve any observable changes in the physics, since it merely corresponds to a relabeling under which the theory is invariant. Thus it is necessary that the metric field be separated into the parts carrying the true dynamical information and those parts characterizing the coordinate system. In this respect, the general theory is analogous to electromagnetic theory. In particular, the coordinate invariance plays a role similar to the gauge invariance of the Maxwell field. In the latter case, this gauge invariance also produces difficulties in separating out the independent dynamical modes, although the linearity here does simplify the analysis. In both cases, the effect

[i]Physics Department, Syracuse University, Syracuse, NY. (Research supported in part by the National Science Foundation and the United States Air Force—Aeronautical Research Laboratory, WADD, and Office of Scientific Research.)

[ii]Physics Department, Brandeis University, Waltham, MA. (Research supported in part by the National Science Foundation and the United States Air Force Office of Scientific Research.)

[iii]Palmer Physical Laboratory, Princeton University, Princeton, NJ. (Alfred P. Sloan Research Fellow.)

[iv]The work discussed in this chapter is based on recent research by the authors. In the text, the original papers will be denoted by Roman numerals, as given in the references.

of invariance properties (both Lorentz and "gauge" invariance) is to introduce redundant variables in the original formulation of the theory to insure that the correct transformation properties are maintained. It is this clash with the smaller number of variables needed to describe the dynamics (*i.e.*, the number of *independent* Cauchy data) that creates the difficulties in the analysis. In Lorentz covariant field theories, general techniques (Schwinger, 1951, 1953) (valid both in the quantum and classical domains) have been developed to enable one to disentangle the dynamical from the gauge variables. We will see here that, while general relativity possesses certain unique aspects not found in other theories, these same methods may be applied. Two important advantages are obtained by proceeding in this fashion. First, the physics of Lorentz covariant field theory is well understood. Consequently, techniques which relate back to this area of knowledge will help one to comprehend better the physics of general relativity. Second, insofar as quantization of the theory is concerned, a formulation closely associated with general quantization techniques which include consistency criteria (*i.e.*, the Schwinger action principle) will be more appropriate for this highly non-linear theory. Direct correspondence principle quantization (which is suitable for linear theories without constraints) may well prove inadequate here.

A precise determination of the independent dynamical modes of the gravitational field is arrived at when the theory has been cast into canonical form and consequently involves the minimal number of variables specifying the state of the system. At this level, one will have all the relevant information about the field's behavior in familiar form. The canonical formalism, involving only the minimal set of variables (which will turn out to be four), is also essential to the quantization program, since it yields directly simple Poisson bracket (P.B.) relations among these conjugate, unconstrained, variables. Two essential aspects of canonical form are: (1) that the field equations are of first order in the time derivatives; and (2) that time has been singled out so that the theory has been recast into 3+1 dimensional form. These two features are characteristic of Hamilton (or P.B.) equations of motion, in contrast to the Lagrange equations. The first requirement may be achieved in general relativity, since its Lagrangian may be written in a form linear in the time derivatives (which is called the Palatini form). The type of variable fulfilling the second requirement is dictated by the desire for canonical form, and will be seen also to possess a natural geometrical interpretation.

The use of the Palatini Lagrangian and of 3+1 dimensional notation does not, of course, impair the general covariance of the theory under arbitrary coordinate transformations. In possessing this covariance, general relativity is precisely analogous to the parameterized form of mechanics in which the Hamiltonian and the time are introduced as a conjugate pair of variables of a new degree of freedom. When in parameterized form, a theory is invariant under an arbitrary re-parameterization, just as general relativity is invariant under an arbitrary change of coordinates. The action of general relativity will thus be seen to be in "already parameterized" form. The well-known relations between the usual canonical form and the parameter description will thus provide the key for deriving the desired canonical form for the gravitational field. We will therefore begin, in Section 2, with a brief review of parameterized particle mechanics. In Section 3, the Lagrangian of general relativity will be cast into Palatini and 3+1 dimensional form, and the geometrical significance of the variables will be discussed. We will see then that relativity has a form identical to parameterized mechanics. Section 4 completes the analysis, to obtain the canonical variables and their relations as well as the P.B. equations of motion.

Once canonical form is reached, the physical interpretation of quantities involved follows directly as in other branches of physics. Thus, the canonical variables themselves represent the independent excitations of the field (and hence provide the basis for defining gravitational radiation in a coordinate-independent way). Further, the numerical value of the Hamiltonian for a particular state of the system provides the primary definition of total energy (a definition which amounts to comparing the asymptotic form of the spatial part of the metric with that of the exterior Schwarzschild solution). Similarly, the total momentum is defined from the generator of *spatial* translations. The energy and momentum are invariant under coordinate transformations not involving Lorentz rotations at spatial infinity, and behave as a four-vector under the latter. It is also possible to set up the analysis of gravitational radiation in a fashion closely analogous to electrodynamics by introducing a suitable definition of the wave zone. In this region, gravitational waves propagate as free radiation, independent of the strong field interior sources. The waves obey ordinary (flat-space) wave equations and consequently satisfy superposition. The Poynting vector may also be defined invariantly in the wave zone. In contrast, the Newtonian-like parts of the metric cannot be determined within the wave zone; they depend strongly on the interior non-linearities. These points are discussed in Section 5.

When the analysis is extended to include coupling of other systems to the gravitational field (Section 6), the above definition of energy may be used to discuss self-energy questions. In this way, the static gravitation and electromagnetic self-masses of point particle systems will be treated rigorously in Section 7. Here the canonical formalism is essential in order that one may recognize a pure particle (no wave) state. The vanishing of the canonical variables guarantees that there are no independent field excitations contributing to the energy. The total clothed mass of a classical electron turns out to be finite, independent of its bare mass and completely determined by its charge. Further, a "neutral" particle (one coupled only to the gravitational field) has a zero clothed mass, showing that the mass of a particle arises entirely from its interactions with other fields. The physical origin of these finite results is discussed at the beginning of Section 7 in terms of equivalence principle considerations. The self-stress $\mathfrak{T}^{ij}$ of the electron vanishes, showing that the particle is stable, its repulsive electrostatic self-forces being precisely cancelled by gravitational attraction without any *ad hoc* compensation being required. Thus, a completely consistent classical point charge exists when gravitation is included. These rigorous results are in contrast to the higher and higher infinities that would arise in a perturbation analysis of the same problem.

Whether gravitational effects will maintain the finiteness of self-energies in quantum theory (and if so, whether the effective cutoff will be appropriate to produce reasonable values) is at present an open question. In the final section (8), some speculative remarks are made on this problem. Since a complete set of P.B.'s has been obtained classically in Section 4, it is formally possible to quantize by the usual prescription of relating them to quantum commutators. However, the non-linear nature of the theory may necessitate a more subtle transition to the quantum domain. Section 8 also discusses some of these questions.

2. Classical Dynamics Background

2.1. *Action principle for Hamilton's equations.* As was mentioned in Section 1, general relativity is a theory in "already parameterized form." We begin, therefore, with a brief analysis of the relevant properties of the parameter formalism (see also Lanczos, 1949). For simplicity, we deal with a system of a finite number M of degrees of freedom. Its action may be written as

$$I = \int_{t_2}^{t_1} dt\, L = \int_{t_2}^{t_1} dt \left(\sum_{i=1}^{M} p_i \dot{q}_i - H(p,q) \right) \tag{2.1}$$

where $\dot{q} \equiv dq/dt$ and the Lagrangian has been expressed in a form linear in the time derivatives.

(This will be referred to as the first-order form since independent variation of p_i and q_i gives rise to the first-order equations of motion.) The maximal information obtainable from the action arises when not only p_i and q_i are varied independently, but t is also varied and endpoint variations are allowed. Postulating that the total δI is a function only of the endpoints $[\delta I = G(t_1) - G(t_2)]$ leads to: (1) the usual Hamilton equations of motion for p_i and q_i; (2) conservation of energy $(dH/dt = 0)$; and (3) the generating function

$$G(t) = \sum_i p_i \, \delta q_i - H \, \delta t \tag{2.2}$$

Here $\delta q_i = \delta_0 q_i + \dot{q}_i \, \delta t$ where $\delta_0 q_i$ denotes the independent ("intrinsic") variation of q_i. The generating function can easily be seen to be the conventional generator of canonical transformations. Thus $G_q = \sum_i p_i \, \delta q_i$ generates changes $q_i \to q_i + \delta q_i$, $p_i \to p_i$ while $G_t = -H \, \delta t$ generates the translation in time. That is, for G_q one has $[q_j, G_q] = \sum_i [q_j, p_i] \delta q_i = \delta q_j$, where $[A, B]$ means the Poisson bracket (P.B.), and for G_t one has $[q_i, G_t] = -[q_i, H]\delta t = -\dot{q}_i \, \delta t$ by the P.B. form of the equations of motion. The above elementary discussion may be inverted to show that, for the action of (2.1), if every variable occurring in H is also found in the $p\dot{q}$ term, then the theory is in canonical form and p_i and q_i obey the conventional P.B. relations. This is the classical equivalent of the Schwinger action principle (Schwinger, 1951, 1953).

2.2. *The action in parameterized form.* The motion of the system (2.1) is described in terms of one independent variable t (the "coordinate"). The action may be cast, as is well known, into parameterized form, in which the time is regarded as a function q_{M+1} of an arbitrary parameter τ:

$$I = \int_{\tau_2}^{\tau_1} d\tau \, L_\tau \equiv \int_{\tau_2}^{\tau_1} d\tau \left[\sum_{i=1}^{M+1} p_i q_i' \right] .$$

Here, $q' \equiv dq/d\tau$, and the constraint equation $p_{M+1} + H(p, q) = 0$ holds. One may equally well replace this constraint by an additional term in the action:

$$I = \int_{\tau_2}^{\tau_1} d\tau \left[\sum_{i=1}^{M+1} p_i q_i' - NR \right] \tag{2.3}$$

where $N(\tau)$ is a Lagrange multiplier. Its variation yields the constraint equation $R(p_{M+1}, p, q) = 0$, which may be any equation with the solution (occurring as a simple root) $p_{M+1} = -H$. The theory as cast into form (2.3) is now generally covariant with respect to arbitrary coordinate transformations $\bar{\tau} = \bar{\tau}(\tau)$, bearing in mind that N transforms as $dq/d\tau$. The price of achieving this general covariance has been not only the introduction of the $(M+1)$st degree of freedom, but, more important, the loss of canonical form, due to the appearance of the Lagrange multiplier N in the "Hamiltonian," $H' \equiv NR$. (N occurs in H' but not in $\sum_{i=1}^{M+1} p_i q_i'$.) A further striking feature which is due to the general covariance of this formulation is that the "Hamiltonian" H' vanishes by virtue of the constraint equation. This is not surprising, since the motion of any particular variable $F(p, q)$ with respect to τ is arbitrary, *i.e.*, F' may be given any value by suitable recalibration $\tau \to \bar{\tau}$.

2.3. *Reduction of parameterized action to Hamiltonian form—intrinsic coordinates.* As we shall see, the Lagrangian of general relativity may be written in precisely the form of (2.3). We will, therefore, be faced with the problem of reducing an action of the type (2.3) to canonical form

(2.1). The general procedure consists essentially in reversing the steps that led to (2.3). If one simply inserts the solution, $p_{M+1} = -H$, of the constraint equation into (2.3), one obtains

$$I = \int d\tau \left[\sum_{i=1}^{M} p_i q_i' - H(p,q) q_{M+1}' \right] . \tag{2.4}$$

All reference to the arbitrary parameter τ disappears when I is rewritten as

$$I = \int dq_{M+1} \left[\sum_{i=1}^{M} p_i (dq_i / dq_{M+1}) - H \right] \tag{2.5}$$

which is identical to (2.1)with the notational change $q_{M+1} \to t$. Equation (2.5) exhibits the role of the variable q_{M+1} as an "intrinsic coordinate." By this is meant the following. The equation of motion for q_{M+1} is $q_{M+1}' = N(\partial R/\partial q_{M+1})$ from (2.3). Also, none of the dynamical equations determine N as a function of τ. Thus N and hence q_{M+1}, are left arbitrary by the dynamics (though, of course, a choice of q_{M+1} as a function of τ fixes N). One is therefore free to choose $q_{M+1}(\tau)$ to be any desired function and use this function as the new independent variable (parameter): $q_i = q_i(q_{M+1})$, $p_i = p_i(q_{M+1})$, $i = I \ldots M$. The action of (2.5), and hence the relations between q_i, p_i, and q_{M+1} are now independent of τ. They are manifestly invariant under the general "coordinate transformation" $\bar{\tau} = \bar{\tau}(\tau)$ (for the simple reason that τ itself no longer appears). The choice of q_{M+1} as the independent variable thus yields a manifestly τ-invariant formulation and gives an "intrinsic" specification of the dynamics. This is in contrast to the original one in which the trajectories of $q_1 \ldots q_{M+1}$ are given in terms of some arbitrary variable τ (which is extraneous to the system).

In practice, we shall arrive at the intrinsic form (2.5) from (2.4) in an alternate way. Since the relation between q_{M+1} and τ is undetermined, we are free to specify it explicitly, *i.e.*, impose a "coordinate condition." If, in particular, this relation is chosen to be $q_{M+1} = \tau$ (a condition which also determines N), the action (2.4) then reduces (2.5) with the notational change $q_{M+1} \to \tau$; the non-vanishing Hamiltonian only arises as a result of this process. [Of course, other coordinate conditions might have been chosen. These would correspond to using a variable other than q_{M+1} as the intrinsic coordinate in the previous discussion.]

This simple analysis has shown that the way to reduce a parameterized action to canonical form is to insert the solution of the constraint equations and to impose coordinate conditions. Further, the imposition of coordinate conditions is equivalent to the introduction of intrinsic coordinates.

In field theory it will prove more informative to carry out this analysis in the generator. We exhibit here the procedure in the particle case: The generator associated with the action of (2.3) is

$$G = \sum_{i=1}^{M+1} p_i \, \delta q_i - N R \, \delta\tau \tag{2.6}$$

Upon inserting constraints, the generator reduces to

$$G = \sum_{i=1}^{M} p_i \, \delta q_i - H \, \delta q_{M+1} \, . \tag{2.7}$$

Imposing the coordinate condition $q_{M+1} = t$ then yields (2.2) From *this* form, one can immediately recognize the M pairs of canonical variables and the non-vanishing Hamiltonian of the theory.

One can, of course, perform the above analysis for a parameterized field theory as well. Here the coordinates appear as four new field variables $q^{M+\mu} = x^\mu(\tau^\alpha)$, and there are four extra momenta $p_{M+\mu}(\tau^\alpha)$ conjugate to them. Four constraint equations are required to relate these momenta to the Hamiltonian density and the field momentum density, and correspondingly, there are four Lagrange multipliers $N_\mu(\tau^\alpha)$ for a field. An example in which the scalar meson field is parameterized may be found in III.

3. First-Order Form of the Gravitational Field

3.1. *The Einstein action in first-order (Palatini) form.* The usual action integral for general relativity[1]

$$I = \int d^4x\, \mathcal{L} = \int d^4x \sqrt{-g}\, R \tag{3.1}$$

yields the Einstein field equations when one considers variations in the metric (*e.g.*, $g_{\mu\nu}$ or the density $\mathfrak{g}^{\mu\nu} = \sqrt{-g}\, g^{\mu\nu}$). These Lagrange equations of motion are then second-order differential equations. It is our aim to obtain a canonical form for these equations, that is, to put them in the form $\dot{q} = \partial H/\partial p$, $\dot{p} = -\partial H/\partial q$. As a preliminary step, we will restate the Lagrangian so that the equations of motion have two of the properties of canonical equations: (1) they are first-order equations; and (2) they are solved explicitly for the time derivatives. The second property will be obtained by a 3 + 1 dimensional breakup of the original four-dimensional quantities, as will be discussed below. The first property is insured by using a Lagrangian linear in first derivatives. In relativity, this is called the Palatini Lagrangian, and consists in regarding the Christoffel symbols $\Gamma_\mu{}^\alpha{}_\nu$ as independent quantities in the variational principle (see, for example, Schrödinger, 1950). Thus, one may rewrite (3.1) as

$$I = \int d^4x\, \mathfrak{g}^{\mu\nu} R_{\mu\nu}(\Gamma) \tag{3.2}$$

where

$$R_{\mu\nu}(\Gamma) \equiv \Gamma_\mu{}^\alpha{}_{\nu,\alpha} - \Gamma_\mu{}^\alpha{}_{\alpha,\nu} + \Gamma_\mu{}^\alpha{}_\nu \Gamma_\alpha{}^\beta{}_\beta - \Gamma_\mu{}^\alpha{}_\beta \Gamma_\nu{}^\beta{}_\alpha\,. \tag{3.3}$$

Note that these covariant components $R_{\mu\nu}$ of the Ricci tensor do not involve the metric but only the affinity $\Gamma_\mu{}^\alpha{}_\nu$. Thus, by varying $g^{\mu\nu}$, one obtains directly the Einstein field equations

$$G_{\mu\nu} \equiv R_{\mu\nu} - \tfrac{1}{2}\, g_{\mu\nu}\, R = 0\,. \tag{3.4a}$$

These equations no longer express the full content of the theory, since the relation between the now independent quantities $\Gamma_\mu{}^\alpha{}_\nu$ and $g_{\mu\nu}$ is still required. This is obtained as a field equation by varying $\Gamma_\mu{}^\alpha{}_\nu$. One then finds

$$\mathfrak{g}^{\mu\nu}{}_{;\alpha} \equiv \mathfrak{g}^{\mu\nu}{}_{,\alpha} + \mathfrak{g}^{\mu\beta}\Gamma_\beta{}^\nu{}_\alpha + \mathfrak{g}^{\nu\beta}\Gamma_\beta{}^\mu{}_\alpha - \mathfrak{g}^{\mu\nu}\Gamma_\beta{}^\beta{}_\alpha = 0 \tag{3.4b}$$

which, as is well known, can be solved for $\Gamma_\mu{}^\alpha{}_\nu$ to give the usual relation $\Gamma_\mu{}^\alpha{}_\nu = \{{}_{\mu}{}^{\alpha}{}_{\nu}\} \equiv \frac{1}{2} g^{\alpha\beta}(g_{\mu\beta,\nu} + g_{\nu\beta,\mu} - g_{\mu\nu,\beta})$.

The Palatini formulation of general relativity has a direct analog in Maxwell theory, the affinity corresponding to the field strength $F_{\mu\nu}$ and the metric to the vector potential A_μ. Here the Lagrangian is

$$\mathcal{L} = A_{\mu,\nu} F^{\mu\nu} + \tfrac{1}{4}\, F_{\mu\nu} F^{\mu\nu} \tag{3.5}$$

[1] We use units such that $16\pi\gamma c^{-4} = 1 = c$, where γ is the Newtonian gravitational constant; electric charge is in rationalized units. Latin indices run from 1 to 3, Greek from 0 to 3, and $x^0 = t$. Derivatives are denoted by a comma or the symbol ∂_μ.

with A_μ and $F_{\mu\nu}$ to be independently varied. The field equations then become

$$F^{\mu\nu}{}_{,\nu} = 0 \tag{3.6a}$$

and

$$A_{\nu,\mu} - A_{\mu,\nu} = F_{\mu\nu} \tag{3.6b}$$

which correspond to (3.4*a*) and (3.4*b*).

The next step in achieving canonical form is to single out time derivatives by introducing three-dimensional notation. For the Maxwell field, we thus define

$$E^i \equiv F^{0i} . \tag{3.7a}$$

Since the canonical form requires equations of first order in time derivatives (but not in space derivatives), we may use the equation $F_{ij} = A_{j,i} - A_{i,j}$ to eliminate F_{ij}. In terms of the abbreviation

$$B^i \equiv \tfrac{1}{2}\,\epsilon^{ijk}(A_{k,j} - A_{j,k}) \tag{3.7b}$$

the Lagrangian reads

$$\mathcal{L} = -E^i\,\partial_0 A_i - \tfrac{1}{2}\,(B^i B^i + E^i E^i) - A_0 E^i{}_{,i} . \tag{3.8}$$

At this stage, the Maxwell equations are obtained by varying $\mathcal{L}$ with respect to the independent quantities E^i, A_i, and A_0.

3-2. *Three-plus-one dimensional decomposition of the Einstein field.* The three-dimensional quantities appropriate for the Einstein field are (as will be discussed in detail later)

$$g_{ij} \equiv {}^4g_{ij} , \qquad N \equiv (-{}^4g^{00})^{-1/2} , \qquad N_i \equiv {}^4g_{0i} \tag{3.9a}$$

$$\pi^{ij} \equiv \sqrt{-{}^4g}\,({}^4\Gamma_p{}^0{}_q - g_{pq}\,{}^4\Gamma_r{}^0{}_s g^{rs})g^{ip}g^{jq} . \tag{3.9b}$$

Here and subsequently we mark every four-dimensional quantity with the prefix ^{4}, so that all unmarked quantities are understood as three-dimensional. In particular, g^{ij} in (3.9b) is the reciprocal matrix to g_{ij} . The full metric ${}^4g_{\mu\nu}$ and ${}^4g^{\mu\nu}$ may, with (3.9a), be written

$${}^4g_{00} = -(N^2 - N_i N^i) \tag{3.10}$$

where $N^i = g^{ij}N_j$, and

$${}^4g^{0i} = N^i/N^2 , \qquad {}^4g^{00} = -1/N^2 \tag{3.11a}$$

$${}^4g^{ij} = g^{ij} - (N^i N^j/N^2) . \tag{3.11b}$$

One further useful relation is

$$\sqrt{-{}^4g} = N\sqrt{g} . \tag{3.12}$$

In terms of the basic quantities of (3.9), the Lagrangian of general relativity becomes

$$\begin{aligned}\mathcal{L} = \sqrt{-{}^4g}\,{}^4R = -g_{ij}\,\partial_t\pi^{ij} - NR^0 - N_i R^i \\ -2(\pi^{ij}N_j - \tfrac{1}{2}\pi N^i + N^{|i}\sqrt{g})_{,i}\end{aligned} \tag{3.13}$$

where

$$R^0 \equiv -\sqrt{g}\,[{}^3R + g^{-1}(\tfrac{1}{2}\,\pi^2 - \pi^{ij}\pi_{ij})] \tag{3.14a}$$

$$R^i \equiv -2\pi^{ij}{}_{|j} \,. \tag{3.14b}$$

The quantity ${}^3R$ is the curvature scalar formed from the spatial metric g_{ij}, $|$ indicates the covariant derivative using this metric, and spatial indices are raised and lowered using g^{ij} and g_{ij}. (Similarly, $\pi \equiv \pi^i{}_i$.) As in the electromagnetic example, we have allowed second-order space derivatives to appear by eliminating such quantities as $\Gamma_i{}^k{}_j$ in terms of $g_{ij,k}$.

One may verify directly that the first-order Lagrangian (3.13) correctly gives rise to the Einstein equations. One obtains

$$\partial_t g_{ij} = 2Ng^{-1/2}(\pi_{ij} - \tfrac{1}{2}\, g_{ij}\pi) + N_{i|j} + N_{j|i} \tag{3.15a}$$

$$\begin{aligned}\partial_t \pi^{ij} = &- N\sqrt{g}\,({}^3R_{ij} - \tfrac{1}{2}\, g^{ij}\, {}^3R) + \tfrac{1}{2}\, Ng^{-1/2} g^{ij} (\pi^{mn}\pi_{mn} - \tfrac{1}{2}\pi^2) \\ &- 2Ng^{-1/2}(\pi^{im}\pi_m{}^j - \tfrac{1}{2}\pi\pi^{ij}) + \sqrt{g}\,(N^{|ij} - g^{ij}\, N^{|m}{}_{|m}) \\ &+ (\pi^{ij}N^m)_{|m} - N^i_{|m}\pi^{mj} - N^j{}_{|m}\pi^{mi}\end{aligned} \tag{3.15b}$$

$$R^\mu(g_{ij}, \pi^{ij}) = 0 \,. \tag{3.15c}$$

Equation (3.15a), which results from varying π^{ij}, would be viewed as the defining equation for π^{ij} in a second-order formalism. Variation of N and N_i yields equations (3.15c), which are the ${}^4G^0{}_\mu \equiv {}^4R^0{}_\mu - \tfrac{1}{2}\,\delta^0_\mu\, {}^4R = 0$ equations, while equations (3.15b) are linear combinations of these equations and the remaining six Einstein equations (${}^4G_{ij} = 0$).

3.3. *Geometrical interpretation of dynamical variables.* Before proceeding with the reduction to canonical form, it is enlightening to examine, from a geometrical point of view, our specific choices (3.9) of three-dimensional variables. Geometrically, their form is governed by the requirement that the basic variables be three-covariant under all coordinate transformations which leave the t=const surfaces unchanged. Any quantities which have this property can be defined entirely within the surface (this is clearly appropriate for the 3+1 dimensional breakup). One fundamental four-dimensional object which is clearly also three-dimensional is a curve $x^\mu(\lambda)$ which lies entirely within the 3-surface, *i.e.*, $x^0(\lambda) =$ const. The vector $v^\mu \equiv dx^\mu/d\lambda$ tangent to this curve is therefore also three-dimensional. The restriction that the curve lie in the surface t=const is then $v^0 = 0$, and conversely *any* vector V^μ, with $V^0 = 0$ is tangent to some curve in the surface. Three such independent vectors are $V^\mu_{(i)} = \delta^\mu_i$. Given any covariant tensor $A_{\mu\ldots\nu}$, its projection onto the surface is then $V^\mu_{(i)} \ldots V^\nu_{(j)} A_{\mu\ldots\nu} = A_{i\ldots j}$. Thus, the *covariant* spatial components of any four-tensor form a three-tensor which depends only on the surface[2] (in contrast to the contravariant spatial components which are scalar products with gradients rather than tangents, and hence depend also on the choice of spatial coordinates in the immediate neighborhood of the surface). This accounts for the choice of g_{ij}, rather than ${}^4g^{ij}$. In contrast, N and N_i do not have the desired invariance and, in fact, by choosing coordinates such that the $x^i =$ const lines are normal to the surface, one obtains $N_i = 0$. (If x^0 is arranged to measure proper time along these lines, one has also $N = 1$.) By the same argument, one can see that A_i and F_{ij} are appropriate three-dimensional quantities in a general relativistic discussion of the Maxwell field.

The quantity which plays the role of a momentum is more difficult to define within the surface, since it refers to motion in time leading out of the original $t =$ const surface. Such a quantity

[2] Of course, any three-dimensional operations (such as contraction of indices with the three-metric) on a three-tensor yield quantities defined on the three-surface.

is, however, provided by the second fundamental form K_{ij} (see, for example, Eisenhart, 1949), which gives the radii of curvature of the $t =$ const surface as measured in the surrounding four-space. These "extrinsic curvatures" describe how the normals to the surface converge or diverge, and hence determine the geometry of a parallel surface at an infinitesimally later time. Since K_{ij} describes a geometrical property of the $t =$ const surface, as imbedded in four-space, it again does not depend on the choice of coordinates away from the surface. This may also be seen from a standard definition, $K_{ij} = -n_{(i;j)}$, which expresses K_{ij} as the covariant spatial part of the tensor $n_{(\mu;\nu)}$ (the four-dimensional covariant derivative of the unit normal, $n_\mu = -N\delta^0_\mu$ to the surface).[3] For convenience in ultimately reaching canonical form, we have chosen, instead of K_{ij}, the closely related variable $\pi^{ij} = -\sqrt{g}\,(K^{ij} - g^{ij}K)$. Thus, the geometrical analysis defines g_{ij} and g_{ij} as suitable quantities, unaffected by the choice of coordinates later in time, while N and N_i describe how the coordinate system will be continued off the $t =$ const surface.

3.4. *Initial value problem and dynamical structure of field equations.* Returning to the field equations (3.15), we may now analyze them from the point of view of the initial value problem. If one specifies g_{ij}, π^{ij} and N, N_i initially, it is clear that the equations uniquely determine g_{ij} and π^{ij} at a later time, while N and N_i remain undetermined then. Since the latter merely express the continuation of the coordinates, the intrinsic (coordinate-independent) geometry of space-time is determined uniquely by an initial choice of g_{ij} and π^{ij}. This choice is restricted, however, by the four constraint equations (3.15c) which relate these twelve variables at the initial time. Subject to these compatibility conditions, then, the (g_{ij}, π^{ij}) constitute a complete set of Cauchy data for the theory.

The maintenance in time of these constraints is guaranteed by the Bianchi identities (${}^4G_\mu{}^\nu{}_{;\nu} \equiv 0$). Hence

$$ {}^4G_\mu{}^0{}_{,0} = -\,{}^4G_\mu{}^i{}_{;i} - {}^4G_\mu{}^\nu\,{}^4\Gamma_\nu{}^0{}_0 + {}^4G_\nu{}^0\,{}^4\Gamma_\mu{}^\nu{}_0\,. \tag{3.16} $$

Thus, by virtue of the dynamic equations ${}^4G_{ij} = 0$ (and consequently of their spatial derivatives) at $t = 0$, the constraints ${}^4G_\mu{}^0 = 0$ hold at all times if they hold initially.

In electrodynamics, the constraint equation corresponding to (3.15c) is obtained by varying A_0 in (3.8), and is $F^{0i}{}_{,i} \equiv \boldsymbol{\nabla} \cdot \boldsymbol{E} = 0$. The identity ensuring its maintenance in time is just $F^{\mu\nu}{}_{,\mu\nu} \equiv 0$, which may be rewritten as $(F^{0i}{}_{,i})_{,0} = -(F^{i\nu}{}_{,\nu})_{,i}$. The right-hand side then vanishes by the dynamic equations $F^{i\nu}{}_{,\nu} = 0$.

While the twelve variables (g_{ij}, π^{ij}) constitute a complete set of Cauchy data, they do not provide a minimal set (which the canonical formalism will eventually give to be two pairs, corresponding to two degrees of freedom). We may now count the number of minimal variables. Of the twelve g_{ij}, π^{ij}, we may eliminate four by using the constraint equations (3.15c). There will correspondingly be four "Bianchi" identities among the twelve equations of motion (3.15a) and (3.15b). As we have seen, N and N_i determine the continuation of the coordinate system without affecting the intrinsic geometry (*i.e.*, the physics of the field). For every choice of N and N_i as functions of the remaining eight Cauchy data (which represents a choice of coordinate frame), there will result four equations stating that the time derivatives of four of the remaining eight (g, π) variables vanish. [More precisely, a choice of coordinate frame is made by specifying four functions q^μ of (g_{ij}, π^{ij}) as the coordinates x^μ. The equations for $\partial_t q^\mu$ $(= \delta^\mu_0)$ then determine N and N_i.] Thus, after these coordinate conditions are imposed, we are left with four dynamic equations of the

[3] Thus one has $K_{ij} = -n_{(i;j)} = -n_{(i,j)} + n_\mu \Gamma_i{}^\mu{}_j - N\Gamma_i{}^0{}_j$.

form $\partial_t u_a = f_a(u)$ $(a = 1, 2, 3, 4)$. These equations govern the motion of a system of two degrees of freedom. This is to be expected, since the linearized gravitational field is a massless spin two field, and the self-interaction of the full theory should not alter such kinematical features as the number of degrees of freedom.

3.5. *The gravitational field as an already parameterized system.* To conclude this section, we point out the characteristic properties of the Einstein Lagrangian of (3.13). We reproduce it here with a divergence[4] and a total time derivative[5] discarded:

$$\mathcal{L} = \pi^{ij}\,\partial_t g_{ij} - NR^0 - N_i R^i\,, \qquad R^\mu = R^\mu(g_{ij},\, \pi^{ij})\,. \tag{3.17}$$

Equation (3.17) is thus precisely in the form of a parameterized theory's Lagrangian as in (2.3). This form just expresses the invariance of the theory with respect to transformations of the four coordinates x^μ and hence the x^μ are parameters in exactly the same sense that τ was in the particle case. That the N and N_i are truly Lagrange multipliers follows from the fact that they do not appear in the pq' (*i.e.*, $\pi^{ij}\,\partial_t g_{ij}$) part of $\mathcal{L}$. Their variation yields the four constraint equations R^μ, $= 0$. The "Hamiltonian" $\mathcal{H}' \equiv NR^0 + N_i R^i$ vanishes due to the constraints. The true non-vanishing Hamiltonian of the theory will arise only after the constraint variables have been eliminated and coordinate conditions chosen. The analysis leading to the canonical form is carried out in the next section.

4. Canonical Form for General Relativity

4.1. *Analysis of generating functions.* We are now in a position to cast the general theory into canonical form. The geometrical considerations of Section 3 were useful in obtaining the Lagrangian in the form (3.17), which, in the light of Section 2, we recognized as the Lagrangian of a parameterized field theory corresponding to (2.3).

The reduction of (3.17) to the canonical form analogous to (2.1) requires an identification of the four extra momenta to be eliminated by the constraint equations (3.15c). To this end we consider the generator arising from (3.17):

$$G = \int d^3x\, [\pi^{ij}\,\delta g_{ij} + T^0{}_\mu{}'\,\delta x^\mu]\,. \tag{4.1}$$

The $T^0{}_\mu{}'\,\delta x^\mu$ term comes from the independent coordinate variations. However, $T^0{}_\mu{}'$ vanishes as a consequence of the constraint equations.[6] For example, $T^0{}_0{}' = -NR^0 - N_i R^i = 0$. When the constraints are inserted, G reduces to

$$G = \int d^3x\, \pi^{ij}\,\delta g_{ij} \tag{4.2}$$

[corresponding to (2.7)] where four of the twelve $(g_{ij},\, \pi^{ij})$ are understood to have been expressed in terms of the rest by solving $R^\mu = 0$ for them. This elimination exhausts the content of the

[4]A divergence in a theory with constraints cannot necessarily be discarded, since such a term may cease to be a divergence upon elimination of constraints. In IIIA, it is shown that the divergences neglected in this work are indeed to be discarded.

[5]The addition of a total time derivative in a Lagrangian does not, of course, alter the equations of motion and corresponds to a canonical transformation in the generator. The properties of canonical transformations in general relativity are discussed in IVA.

[6]More precisely $T^0{}_i{}'$ reduces to an irrelevant divergence.

constraint equations. Finally, as in the particle case, coordinate conditions (now four in number) must be chosen and this information inserted into (4.2), leaving one now with only four dynamical variables (π^A, ϕ_A). If, in fact, the generator at this stage has the form

$$G = \int d^3x \left[\sum_{A=1}^{2} \pi^A \, \delta\phi_A + \mathfrak{T}^0{}_\mu(\pi^A, \phi_A)\delta x^\mu \right] \tag{4.3}$$

then the theory is clearly in *canonical* form with π^A and ϕ_A as canonical variables and $\int \mathfrak{T}^0{}_\mu \, \delta x^\mu$ the generator of translations δx^μ. In (4.3), $\mathfrak{T}^0{}_\mu[\pi^A, \phi_A]$ has arisen in the elimination of the extra momenta $p_{M+\mu}$, by solving the constraint equations, and the x^μ, now represent the four variables chosen as coordinates $q_{M+\mu}$.

For the Maxwell field, the generator corresponding to (4.2) is, by (3.8),

$$G = \int d^3x[-E^i \, \delta A_i - \tfrac{1}{2}(E^i E^i + B^i B^i)\delta t + (\mathbf{E}\times\mathbf{B}) \cdot \delta\mathbf{r}]$$

with the constraint $E^i{}_{,i} = 0$ understood to have been eliminated. The solution of the constraint equation is, of course, well known in terms of the orthogonal decomposition $E^i = E^{iT} + E^{iL}$ ($\nabla\cdot\mathbf{E}^T \equiv 0 \equiv \nabla\times\mathbf{E}^L$). The longitudinal component E^{iL} of the electric field vanishes, and therefore the canonical form is reached when one inserts this information into G:

$$G = \int d^3x[-E^{iT} \, \delta A_i{}^T - \tfrac{1}{2}(E^{iT} E^{iT} + B^i B^i)\delta_t + (\mathbf{E}^T\times\mathbf{B}) \cdot \delta\mathbf{r}] \, .$$

Note that $A_i{}^L$ has automatically disappeared from the kinetic term (by orthogonality), and never existed in $\mathbf{B} \equiv \nabla\times\mathbf{A} = \nabla\times\mathbf{A}^T$. Thus, the two canonically conjugate pairs of Maxwell field variables are $-E^{iT}$ and $A_i{}^T$.

4.2. *Analysis of constraint equations in linearized theory-orthogonal decomposition of the metric.* In order to achieve the form (4.3) for relativity, it is useful to be guided by the linearized theory. Here one must treat the constraint equations to second order since our general formalism shows that the Hamiltonian arises from them. Through quadratic terms, equations (3.15c) may be written in the form

$$g_{ij,ij} - g_{ii,jj} = \mathcal{P}_2{}^0[g_{ij}, \pi^{ij}] \tag{4.4a}$$

$$-2\pi^{ij}{}_{,j} = \mathcal{P}_2{}^i[g_{ij}, \pi^{ij}] \tag{4.4b}$$

where $\mathcal{P}_2{}^0$ and $\mathcal{P}_2{}^i$ are purely quadratic functions of g_{ij} and π^{ij}. These equations determine one component of g_{ij} and three components of π^{ij} in terms of the rest. The content of equations (4.4) can be seen more easily if one makes the following linear orthogonal decomposition on g_{ij} and π^{ij}. For any symmetric array $f_{ij} = f_{ji}$ one has

$$f_{ij} = f_{ij}{}^{TT} + f_{ij}{}^T + (f_{i,j} + f_{j,i}) \tag{4.5}$$

where each of the quantities on the right-hand side can be expressed uniquely as a linear functional of f_{ij}. The quantities $f_{ij}{}^{TT}$ are the two transverse traceless components of $f_{ij}(f_{ij}{}^{TT}{}_{,j} \equiv 0,\ f_{ii}{}^{TT} \equiv 0)$. The trace of the transverse part of f_{ij}, *i.e.*, f^T, uniquely defines $f_{ij}{}^T$ according to

$$f_{ij}{}^T \equiv \tfrac{1}{2}[\delta_{ij} f^T - (1/\nabla^2) f^T{}_{,ij}] \tag{4.6}$$

(and clearly, ${f_{ij}}^T{}_{,j} = 0$, ${f_{ii}}^T = f^T$. The operator $1/\nabla^2$ is the inverse of the flat space Laplacian, with appropriate boundary conditions. The longitudinal parts of f_{ij} reside in the remaining part, $f_{i,j} + f_{j,i}$. Decomposing f_i into its transverse and longitudinal (curl-less) parts, one has $f_i = f_i^T + \frac{1}{2} f^L{}_{,i}$ $(f_i^T{}_{,i} \equiv 0)$. The remainder then becomes $f_i^T{}_{,i} + f_j^T{}_{,i} + f^L{}_{,ij}$. One may express f_i, f^T and ${f_{ij}}^{TT}$ in terms of f_{ij} by

$$f_i = (1/\nabla^2)[f_{ij,j} - \tfrac{1}{2}(1/\nabla^2) f_{kj,kji}] \tag{4.7a}$$

$$f^T = f_{ii} - (1/\nabla^2) f_{ij,ij} \tag{4.7b}$$

$${f_{ij}}^{TT} = f_{ij} - {f_{ij}}^T[f_{mn}] - \{f_{i,j}[f_{mn}] + f_{j,i}[f_{mn}]\}. \tag{4.7c}$$

The foregoing orthogonal decomposition for a symmetric tensor is just the extension of the usual decomposition of a vector into longitudinal and transverse parts employed in electromagnetic theory.

Returning to (4.4), one has

$$-\nabla^2 g^T = {\mathcal{P}_2}^0 \tag{4.8a}$$

$$-2\nabla^2(\pi^{iT} + \pi^L{}_{,i}) = {\mathcal{P}_2}^i\,. \tag{4.8b}$$

To first order, one sees that g^T and π^i vanish.[7] These structures begin, therefore, at second order where $g^T = -(1/\nabla^2)\mathcal{H}_{\rm lin}$ and $-2(\pi^{iT} + \pi^L{}_{,i}) = (1/\nabla^2)\mathfrak{T}^{0i}_{\rm lin}$. Here $\mathcal{H}_{\rm lin}$, and $\mathfrak{T}^{0i}_{\rm lin}$ are obtained from $\mathcal{P}^0_2$ and $\mathcal{P}^i_2$ by setting g^T and π^i equal to zero there. They are just the linearized theory's Hamiltonian and field momentum densities.[8] The analysis has thus shown that the constraint equations can be solved for g^T and π^i (in terms of the remaining variables) as the four extra momenta. To see explicitly that $\mathcal{H}_{\rm lin}$, and $\mathfrak{T}^{0i}_{\rm lin}$ generate the appropriate time and space translations, one must return to the generator. Inserting the orthogonal decomposition (4.5) for both g_{ij} and π^{ij} into (4.2), one obtains

$$G = \int d^3x\, [\pi^{ijTT}\, \delta {g_{ij}}^{TT} + \pi^{ijT}\, \delta {g_{ij}}^T + 2(\pi^i{}_{,j} + \pi^j{}_{,i})\delta g_{i,j}]\,. \tag{4.9}$$

The cross-terms in (4.9) have vanished due to the orthogonality of the decomposition (*e.g.*, $\int d^3x\, \pi^{ijTT}\delta g_{i,j} = -\int d^3x\, \pi^{ijTT}{}_{,j}\, \delta g_i = 0$). We have also used here the fact that taking the variation of a quantity does not alter its transverse or longitudinal character in such a linear breakup and that the derivatives commute with the variation. Equation (4.9) may be brought to the desired form by further integration by parts and addition of a total variation:[9]

$$G = \int d^3x\, \{\pi^{ijTT}\, \delta {g_i}^{TT} - (-\nabla^2 g^T)\delta[-(1/2\nabla^2)\pi^T] + [-2\nabla^2(\pi^{iT} + \pi^L{}_{,i})]\}\delta g_i \tag{4.10}$$

4.3. *Imposition of coordinate conditions.* Equation (4.10) is now in the form of (2.7). The final step in reduction to canonical form is to impose coordinate conditions. The structure of (4.10)suggests that one choose as coordinate conditions

$$t = -(1/2\nabla^2)\pi^T \tag{4.11a}$$

[7]We use boundary conditions such that g^T and π^i vanish asymptotically. Note also that π^{iT} and π^L are independently determined by (4.8b).

[8]More precisely, $\mathcal{H}_{\rm lin}$ differs from the linearized Hamiltonian density by a divergence which vanishes when coordinate conditions (4.11) are imposed.

[9]Addition of a total variation to a generator corresponds to the addition of a total time derivative in the Lagrangian, as stated earlier.

$$x^i = g_i \,. \tag{4.11b}$$

Alternately, these coordinate conditions can be written in more conventional form[10] by eliminating π^T and g_i via (4.7):

$$\pi^{ii}{}_{,jj} - \pi^{ij}{}_{,ij} = 0 \tag{4.11c}$$

$$g_{ij,j} = 0 \,. \tag{4.11d}$$

One can see that these coordinate conditions are acceptable by looking at those of the field equations that involve $\partial_t g_i$ and $\partial_t \pi^T$ The linear part of equation (3.15a) gives, as the equations for the longitudinal part of g_{ij},

$$\partial_t(g_{i,j} + g_{j,i}) = N_{i,j} + N_{j,i} \,. \tag{4.12}$$

The Lagrange multipliers $N_i \equiv g_{0i}$ are functions determined only when coordinate conditions are imposed and must vanish at infinity where space is flat. Inserting (4.11b) into (4.12) gives, consistent with the boundary conditions, $N_i = 0$ everywhere. Similarly, from (3.15b), one has

$$\partial_t[-(1/2\nabla^2)\pi^T] = N \,. \tag{4.13}$$

Condition (4.11a) implies $N \equiv (-g_{00})^{-1/2} = 1$, again consistent with the required asymptotic limit.

Alternately, one can see that equations (4.11) are physically appropriate coordinate conditions by a direct comparison with the known results of linearized theory as discussed in I. Thus, as mentioned above, $\mathcal{H}_{\text{lin}} = -\nabla^2 g^T$ and $\mathfrak{T}^{0i}_{\text{lin}} = -2\nabla^2(\pi^{iT} + \pi^L{}_{,i})$ are the linearized theory's Hamiltonian and momentum densities and so their coefficients in the generator (4.10) must be δt and δx^i respectively, in order that the form (4.3), also obtained in I, be reproduced.

4.4. *Hamiltonian form and independent variables for linearized theory.* Since the generator is now

$$G = \int d^3x \, [\pi^{ijTT} \, \delta g_{ij}{}^{TT} - \mathcal{H}_{\text{lin}}(\pi^{ijTT}, \, g_{ij}{}^{TT})\delta t + \mathfrak{T}^{0i}_{\text{lin}}(\pi^{ijTT}, g_{ij}{}^{TT}) \, \delta x^i] \tag{4.14}$$

the linearized theory has been put into canonical form, with $g_{ij}{}^{TT}$ and π^{ijTT} as the two canonically conjugate pairs of variables.

We will now see the usefulness of the linearized theory in suggesting the choice of canonical variables for the full theory. Since the identification is made from the bilinear part of the Lagrangian $\pi^{ij} \, \partial_t g_{ij}$, which is the same as for the linearized theory, the greater complexity of the full theory, *i.e.*, its self-interaction, is to be found only in the non-linearity of the constraint equations. Even in the constraint equations, the linearized theory will guide us in choosing g^T and π^i as the four extra momenta to be solved for.

4.5. *Hamiltonian form and independent variables for the full theory.* The full theory can now easily be put into canonical form. The generator of (4.10) is, of course, also correct for the full theory since it comes from the bilinear part of the Lagrangian. The constraint equations (3.15c) now read (in a coordinate system to be specified shortly)

$$-\nabla^2 g^T = \mathcal{P}^0[g_{ij}{}^{TT}, \, \pi^{ijTT}; \, g^T, \, \pi^i; \, g_i, \, \pi^T] \tag{4.15a}$$

$$-2\nabla^2(\pi^{iT} + \pi^L{}_{,i}) = \mathcal{P}^i[g_{ij}{}^{TT}, \, \pi^{ijTT}; \, g^T, \, \pi^i; \, g_i, \, \pi^T] \tag{4.15b}$$

[10]The differential form (4.11c,d) of the coordinate conditions may be integrated to yield (4.11a,b) either by imposing appropriate boundary conditions or by a procedure given in Appendix B of III.

where $\mathcal{P}^\mu$ are non-linear functions of g_{ij} and π^{ij}. One can again solve these (coupled) equations (at least by a perturbation-iteration expansion) for g^T and π^i. Thus, one can again choose $-\nabla^2 g^T$ and $-2\nabla^2(\pi^{iT} + \pi^L{}_{,i})$ as the four extra momenta to be eliminated. We denote the solutions of equations

$$-\nabla^2 g^T = -\mathfrak{T}^0{}_0[g_{ij}{}^{TT},\, \pi^{ijTT},\, g_i,\, \pi^T] \tag{4.16a}$$

$$-2\nabla^2(\pi^{iT} + \pi^L{}_{,i}) = -\mathfrak{T}^0{}_i[g_{ij}{}^{TT},\, \pi^{ijTT},\, g_i,\, \pi^T]\,. \tag{4.16b}$$

These equations are the counterpart of $p_{M+1} = -H$ in the particle case.

As we have seen, the four constraint equations are maintained in time as a consequence of the other field equations. Hence, after inserting equations (4.16) into (3.15a,b), one finds that four of these twelve (those for $\partial_t g^T$ and $\partial_t \pi^i$) are "Bianchi" identities, leaving eight independent equations in the twelve variables $g_{ij}{}^{TT}$, π^{ijTT}, g_i, π^T and N, and N_i. These equations are linear in the time derivatives of the first eight variables.

We now impose the coordinate conditions (4.11) which determine π^T and g_i. The $\partial_t g_i$ and $\partial_t \pi^T$ equations become determining equations for N and N_i [the full theory's analog of (4.12) and (4.13)]. N and N_i are no longer 1 and 0 respectively, but now become specific functionals of $g_{ij}{}^{TT}$ and π^{ijTT}, which could (in principle) be calculated explicitly.[11] In the last four equations, then, N and N_i may, in principle, be eliminated, leaving a system of four equations involving only $g_{ij}{}^{TT}$ and π^{ijTT}, and linear in their time derivatives. We will now see that this reduced system is in Hamiltonian form.

The generator (4.10) reduces to canonical form [with coordinate, conditions (4.11) imposed and constraints (4.16) inserted]:

$$G = \int d^3x\, [\pi^{ijTT}\, \delta g_{ij}{}^{TT} + \mathfrak{T}^0{}_0\, \delta t + \mathfrak{T}^0{}_i\, \delta x^i] \tag{4.17a}$$

while the Lagrangian now becomes

$$\mathcal{L} = \pi^{ijTT}\, \partial_t g_{ij}{}^{TT} + \mathfrak{T}^0{}_0\,. \tag{4.17b}$$

It can be shown that the solutions $\mathfrak{T}^0{}_\mu$ of the constraint equations do not depend explicitly on the coordinates x^μ of (4.11) (see III). This is not unexpected, since the variables g_{ij} and π^{ij} appearing on the right-hand side of (4.15) do not depend explicitly on the coordinates in this frame. (Thus, only $g_{i,j} = x^i{}_{,j} = \delta^i{}_j$ and $\pi^T = -2\nabla^2 t = 0$ appear in g_{ij} and π^{ij}.)

4.6. *Fundamental Poisson brackets and P.B. equations of motion for general relativity.* With the generator now in canonical form, we can immediately write down the fundamental equal time P.B. relations for $g_{ij}{}^{TT}$ and π^{ijTT}. These are[12]

$$[g_{ij}{}^{TT}(\mathbf{x}),\, \pi^{mnTT}(\mathbf{x}')] = \delta^{mn}{}_{ij}(\mathbf{x} - \mathbf{x}') \tag{4.18a}$$

$$[g_{ij}{}^{TT}(\mathbf{x}),\, g_{mn}{}^{TT}(\mathbf{x}')] = 0 = [\pi^{ijTT}(\mathbf{x}),\, \pi^{mnTT}(\mathbf{x}')]\,. \tag{4.18b}$$

The $\delta^{mn}{}_{ij}(\mathbf{x})$ in (4.18a) is a conventional Dirac δ-function modified in such a way that the transverse-traceless nature of the variables on the left-hand side is not violated. Note that the definition of this modified δ-function does not depend on the metric; it is symmetric, transverse, and traceless on each pair of indices:

[11] An example of such an explicit determination of N and N_i for coordinate conditions (4.22) is given in (7.7) and (7.9).

[12] The variables being used as the coordinates x^μ (4.11a,b) of course have vanishing P.B. with *all* variables.

$$\delta^{mn}{}_{ij} = \delta^{nm}{}_{ij} = \delta^{mn}{}_{ji} = \delta^{ij}{}_{mn} \tag{4.19a}$$

$$\delta^{mm}{}_{ij} = 0 = \delta^{mn}{}_{ii} \tag{4.19b}$$

$$\delta^{mn}{}_{ij,j} = 0\,. \tag{4.19c}$$

From the form of equations (4.17), one also has the P.B. equations of motion:

$$\partial_t g_{ij}{}^{TT} = [g_{ij}{}^{TT},\, H] = \delta H/\delta\pi^{ijTT} \tag{4.20a}$$

$$\partial_t \pi^{ijTT} = [\pi^{ijTT},\, H] = -\delta H/\delta g_{ij}{}^{TT} \tag{4.20b}$$

where $H \equiv -\int d^3x\, \mathfrak{T}^0{}_0$ is the Hamiltonian. The last equalities in equations (4.20) follow from equations (4.18). That (4.18) and (4.20) are consistent with the Lagrangian equations obtained by varying (4.17b) is now immediate.[13] Corresponding to the time translation equations (4.20), one also has for spatial displacements

$$\partial_k g_{ij}{}^{TT} = [P_k, g_{ij}{}^{TT}] = -\delta P_k/\delta\pi^{ijTT} \tag{4.21a}$$

$$\partial_k \pi^{ijTT} = [P_k, \pi^{ijTT}] = \delta P_k/\delta g_{ij}{}^{TT} \tag{4.21b}$$

where $P_k \equiv +\int \mathfrak{T}^0{}_k\, d^3x$ is the total momentum operator. With the canonical momentum $P_k{}^c \equiv \int d^3x\, \mathfrak{T}^0{}_k{}^c \equiv -\int d^3x\, \pi^{mnTT} g_{mn}{}^{TT}{}_{,k}$, equations (4.21) are obvious. The quantities P_k and $P_k{}^c$ actually coincide, since their respective integrands differ by a divergence of canonical variables, as shown in IIIA.

4.7. *Alternate canonical form arising from different coordinate conditions.* The canonical form of equations (4.17) is, of course, not unique. First one can make the usual type of canonical transformations among the canonical variables $g_{ij}{}^{TT}$ and π^{ijTT}, leaving the coordinate conditions fixed. A more general class is available, however: the generator (4.2) can also be reduced to canonical form with coordinate conditions other than (4.11). That the reduction can always be carried out is shown in Appendix A of III. An example which will be of use later is given by the coordinate conditions

$$x^i = g_i - (1/4\nabla^2) g^T{}_{,i} \tag{4.22a}$$

$$t = -(1/2\nabla^2)(\pi^T + \nabla^2\pi^L) \tag{4.22b}$$

or, in differential form,

$$g_{ij,jkk} - \tfrac{1}{4} g_{kj,kji} - \tfrac{1}{4} g_{jj,kki} = 0 \tag{4.22c}$$

$$\pi^{ii} = 0\,. \tag{4.22d}$$

In this frame, the canonical variables are $g_{ij}{}^{TT}$ and π^{ijTT} (where the ^{TT} symbol now refers to the coordinate system (4.22) and hence these canonical variables are *different* from the set previously considered). The $\mathfrak{T}^0{}_\mu$ in this frame are the solutions of the constraint equations (4.16) for $\nabla^2 g^T$ and $-2\nabla^2(\pi^{iT} + \pi^L{}_{,i})$, which will be *different* functions of the new canonical variables, since again ^{T} and ^{L} are referred to (4.22) rather than (4.11). This frame is of interest since, for $g_{ij}{}^{TT} = 0$, the metric g_{ij} reduces to the isotropic form, *i.e.*, $g_{ij} = (1 + \frac{1}{2} g^T)\delta_{ij}$. We will discuss, in Section 5, a preferred class of physically equivalent frames which includes the two given here.

[13] The primary consistency criterion on the canonical reduction is that equations (4.20) be identical to the Einstein equations in the coordinate frame (4.11); this is established in IIIA.

It is perhaps worth noting that, with the coordinates of (4.22), the orthogonal decomposition of g_{ij} takes on the simple form

$$g_{ij} = g_{ij}{}^{TT} + \delta_{ij}(1 + \tfrac{1}{2}\, g^T\, [g_{mn}{}^{TT},\, \pi^{mnTT}]) \,. \tag{4.23}$$

Thus, the dynamical aspects of the theory are, as expected, to be found in the deviation of the metric from its flat space value. By tracing (4.23), one obtains a local relation giving the canonical variable $g_{ij}{}^{TT}$ at a point in this coordinate system in terms of g_{ij} at that point: $g_{ij}{}^{TT} = g_{ij} - \frac{1}{3}\,\delta_{ij} g_{kk}$. The g^T term, as we have seen, is directly given by the Hamiltonian density, and, as will be shown in the next section, carries the asymptotic (Newtonian) m/r term in the metric.

7-5. Energy and Radiation

5.1. *Expressions for the energy and momentum P_μ of the gravitational field.* The canonical formalism developed in the previous section has brought out the formal features of general relativity which have their counterpart in usual Lorentz covariant field theories. As a consequence, the physical interpretation of the gravitational field may be carried out in terms of the same quantities that characterize other fields, *e.g.*, energy, momentum, radiation flux (Poynting vector).

The energy E of the gravitational field is just the numerical value of the Hamiltonian for a particular solution of the field equations. In obtaining this numerical value, the form of the Hamiltonian as a function of the canonical variables is irrelevant, and one may make use of the equation, $\mathfrak{T}^0{}_0 = \nabla^2 g^T$ to express E as a surface integral. One has then[14]

$$\begin{aligned} P^0 \equiv E = -\int d^3\mathbf{r}\, \nabla^2 g^T = &= -\oint dS_i\, g^T{}_{,i} \\ &= \oint dS_i\, (g_{ij,,j} - g_{jj,i}) \end{aligned} \tag{5.1}$$

where dS_i is the two-dimensional surface element at spatial infinity. Similarly, the total field momentum P_i may be written as

$$P^i = -2 \oint (\pi^i{}_{,j} + \pi^j{}_{,i}) dS_j = -2 \oint \pi^{ij}\, dS_j \,. \tag{5.2}$$

In (5.1) and (5.2) we have assumed, as throughout, that the coordinate system is asymptotically rectangular. The basic requirement for an energy to be at all defined is, of course, that space-time become flat at spatial infinity (so that the integrals in (5.1,2) are well defined). The rectangularity requirement may then be conveniently imposed to avoid unnecessarily complicated expressions.

5.2. *Coordinate invariance of energy-momentum expression; Lorentz four-vector nature of P_μ.* The usefulness of formulas (5.1) and (5.2) is enhanced by the fact that, in evaluating the surface integrals at infinity, it is not necessary to employ there the original canonical coordinate frame. At spatial infinity, where the metric approaches the Lorentz value, coordinate transformations, which preserve this boundary condition, must approach the identity transformation (excluding, for the moment, Lorentz transformations at infinity). For the transformation $\bar{x}^\mu = x^\mu - \xi^\mu$ then, one has

[14] It should be emphasized that, while the *energy* and *momentum* are indeed divergences, the integrands in the *Hamiltonian* and in the space translation generators $\mathfrak{T}^0{}_\mu[g^{TT},\, \pi^{TT}]$ are *not* divergences when expressed as functions of the canonical variables.

$\xi^\mu{}_{,\nu} \to 0$ (since only $\partial\bar{x}^\mu/\partial x^\nu$ occurs in the transformation laws of $g_{\mu\nu}$ and $\Gamma_\rho{}^\alpha{}_\sigma$). Let us first keep only the linear terms in the transformation:

$$\bar{g}_{ij} = g_{ij} + \xi^i{}_{,j} + \xi^j{}_{,i} \tag{5.3a}$$

$$\bar{\pi}^{ij} = \pi^{ij} - (\delta_{ij}\xi^0{}_{,ll} - \xi^0{}_{,ij}) \,. \tag{5.3b}$$

For the entire formalism, it is necessary that $g_{\mu\nu} - \eta_{\mu\nu}$ and $\Gamma_\rho{}^\alpha{}_\sigma$ fall off at least as $\sim 1/r$ at spatial infinity. Hence, $\xi^\mu{}_{,\nu}$, must have a similar behavior. As can be seen from (4.5), these transformations affect only g_i and π^T leaving $g^T{}_{ij}$, $,\pi^i{}_{,j}$, and the canonical variables invariant to $O(1/r)$. However, in (5.1) and (5.2), one needs to know $g^T{}_{,i}$ and $\pi^i{}_{,j}$, to $O(1/r^2)$, so that one must also investigate the quadratic terms in the transformation law. These present a problem only if a derivative acting on $\xi^\mu{}_{,\nu}$ does not alter its order, *i.e.*, if one has a "coordinate wave" $\xi^\mu{}_{,\nu} \sim e^{ipx}/r$ whose derivatives are again $\sim 1/r$. A detailed investigation (IVC) shows that expressions (5.1) and (5.2) for P^μ *are* invariant under *all* transformations preserving $\bar{g}_{\mu\nu} - \eta_{\mu\nu} \sim 1/r$, provided one averages the surface integrals over the oscillatory terms. The prescription of averaging is justified, since the energy and momentum can be defined directly from the leading $\sim 1/r$ terms of g^T and π^i, in the coordinate frame (4.11). [For instance, from (4.16a), one sees that $g^T = (1/\nabla^2)\mathfrak{T}^0{}_0 \sim E/4\pi r$.] However, as shown in IVC, these $1/r$ terms are in fact coordinate invariant and the averaged P^μ integrals agree with them. Thus, by either prescription, our formula for P_μ is unaffected by changes of coordinates not involving Lorentz transformations at spatial infinity.

The definition of energy and momentum, as given by the canonical formalism, may be shown to coincide with that arising from the stress energy pseudotensor of Landau and Lifshitz. In this definition, the energy and momentum can also be given in surface integral form. (Here, too, one must average over any oscillatory terms to obtain a coordinate-independent result.) The linear terms in this pseudotensor are then just the surface integrals in (5.1) and (5.2), while the higher powers are negligible at infinity (using the averaging). This agreement also enables us to note directly that P^μ behaves as a Lorentz vector under rigid Lorentz transformations of the coordinate system, since this property is manifest for the pseudotensor expression. That the energy of a system should be defined from the asymptotic metric alone follows from the equivalence principle, the total gravitational mass being measurable by the acceleration of a test particle at infinity.

Since P^μ is constant, due to the fact that $\mathfrak{T}^0{}_\mu$ does not depend on x^μ, explicitly, it can be evaluated at any given time. For this purpose, one should need only those initial Cauchy data required to specify the state of the system uniquely. Prior to the imposition of coordinate conditions, these are g_{ij} and π^{ij} (and not, for example, $g_{0\mu}$) as we have seen in Section 3. The last members of (5.1) and (5.2) are so expressed. When any (asymptotically rectangular) coordinate conditions are imposed, one needs only the two pairs of canonical variables of that frame to specify the state of the system and, of course, to calculate P^μ (using the formula $P_\mu = \int d^3r\, \mathfrak{T}^0{}_\mu$).

5.3. *Conditions for equality of Hamiltonian and energy—"Heisenberg representation."* As was discussed in the previous section, there are an infinite number of coordinate conditions that may be invoked to put the theory in canonical form. The question therefore arises as to whether the above physical results are affected by the coordinate conditions chosen. While, for example, we have seen that the P_μ of (5.1) and (5.2) is invariant under change of frame, this does not prove that different canonical forms, arising from different frames, will yield the same integral for P_μ. Indeed, not every canonical form of the theory has the same value of the Hamiltonian, as one is free to perform Hamilton–Jacobi (H.-J.), or partial H.-J., transformations. One must therefore, take as primary, frames and canonical variables which are in the "Heisenberg representation." More precisely, we require that in the expression giving the full metric $g_{\mu\nu}$ at any time in terms of the canonical variables at that time, there be no explicit coordinate dependence. The preferred role is given to the metric in this requirement since basic measurements (rods and clocks) refer to $g_{\mu\nu}$ directly. Similarly, in standard particle mechanics, the directly measured quantities (which appear in the original Lagrangian) are the basic "Heisenberg" variables, and one requires that any other "Heisenberg" set be related to them in a time-independent fashion. More generally, even if the original Lagrangian (which always defines the Heisenberg representation) is not in canonical form, the same requirement is to be imposed. For relativity, the basic variables in the Einstein Lagrangian are $g_{\mu\nu}$. (Note that the question of whether a given canonical form is Heisenberg thus does not require a comparison with some particular canonical form, and also that each of the two forms in Section 4 is indeed Heisenberg.) With this criterion, it is clear that the Hamiltonian densities will also have no explicit coordinate dependence, since the canonical equations can be obtained by substituting $g_{\mu\nu}[g^{\rm can}, \pi^{\rm can}]$ into the original field equations. This step produces no explicit coordinate dependence in the Heisenberg representation. It can be shown (see IVA) that all Heisenberg forms are linked by coordinate transformations of the type $\bar{x}^\mu = x^\mu + f^\mu[g^c(x), \pi^c(x)]$ where the f^μ do not depend on x^μ explicitly. Thus the relation between Heisenberg frames is determined by the intrinsic geometry and does not contain functions of the coordinates, which are independent of the physics of the state. As a result of this form of the transformations, one finds that P_μ is numerically the same in any Heisenberg frame for a given physical situation. Thus (5.1) gives correctly the numerical value of the Hamiltonian of any Heisenberg frame and could have been derived in any Heisenberg canonical form.

5.4. *"Waves" as excitations of canonical variables.* Excitations of the canonical modes of a field provide a definition of what one calls waves. This is, of course, the same definition as that given in electrodynamics. In electrodynamics, the gauge invariance partly obscures the fact that only the transverse parts $\mathbf{A}^T$ and $\mathbf{E}^T$ of the vector potential and electric field need be examined to recognize waves. These variables, as we saw in Section 4, are just the canonical ones for the Maxwell field. Correspondingly, in general relativity we may utilize the non-vanishing of the canonical variables as the criterion of the existence of waves. As in electrodynamics, when sources (among which are included the non-linear self-interactions) are present, the radiation and induction effects can be meaningfully separated only in the "wave zone"; but also as in electrodynamics, the wave concept can be employed consistently nearer the sources where it is of heuristic value, although the definition is somewhat arbitrary.

An example of an application of the wave criteria is furnished by the time-symmetric case ($\pi^{ij} = 0$ initially). In the frame of (4.22), it was seen that $\pi_{ij}{}^{TT} = 0$ if, and only if, the metric is isotropic, $g_{ij} = (1 + \frac{1}{2} g^T)\delta_{ij}$. One need not explicitly transform to this frame to verify whether the canonical variables vanish in it; it is sufficient to examine the three-tensor (see, for example, Eisenhart, 1949, Section 28):

$$R_{ijk} \equiv R_{ij|k} - R_{ik|j} + \tfrac{1}{4}\left(g_{ik}R_{,j} - g_{ij}R_{,k}\right) \tag{5.4}$$

which vanishes if, and only if, $g_{ij}{}^{TT}$ vanishes in the frame[15] of (4.22). A class of time-symmetric

[15] In terms of the canonical formalism, the time-symmetric situation $\pi^{ij} = 0$ can be viewed as the requirement that excitations in the canonical momentum variable π^{ijTT} vanish. This follows from the fact that if $\pi^{ijTT} = 0$, then $\pi^i = 0$ by the constraint equations (3.15c); consequently, $\pi^T = 0$ in the frame of (4.22). Thus, in this frame, $\pi^{ijTT} = 0$ implies $\pi^{ij} = 0$.

waves is furnished by the initial conditions recently discussed by Brill (1959).

5.5. *Radiation and definition of wave-zone.* The concept of radiation embodies the idea of field excitations escaping to infinity and propagating independently of their sources. In linear field theories, the region in which this occurs (the wave zone) is simply one which is many wavelengths away from the source. In general relativity, for strong fields, nonlinear effects are important, and act as effective sources, so that merely going beyond the matter sources is not, in general, adequate to define free radiation. To have a wave zone situation, one must set up further conditions such that the non-linear self-interaction terms do not influence the propagation of the waves. Such further conditions are also necessary in quantum electrodynamics due to vacuum polarization. The latter brings in effective non-linear contributions to the Maxwell action, so that in the presence of an arbitrary external electromagnetic field, the self-coupling can produce distortion of the "waves" (Delbrück scattering); again, scattering of two photons can occur, even in the absence of external fields, thus affecting superposition, that is, free propagation. The usual definitions of radiation would thus fail in such situations and, in fact, the "wave zone" is then defined only when the self-coupling is negligible. One must therefore require that in the wave zone, the various field amplitudes be small when the theory is non-linear. On the other hand, the process of going to the asymptotic wave zone region is not an *a priori* equivalent to taking the linearized approximation. In general relativity, we have previously seen that $(g^T, \pi^i) \sim P^\mu/r$ asymptotically, where P^μ is clearly determined by the interior region. Thus, components of the metric other than the wave modes survive asymptotically, and in fact may be comparable in size to the wave modes.

To define the wave zone, we consider a general situation in which the gravitational canonical modes behave as fe^{ikx}/r asymptotically in some region (with wave numbers k up to some maximum $k_{\max}$); beyond some point (the wave front) they are assumed to vanish very rapidly. (This last assumption is needed to make the energy of the waves finite.) The first wave zone requirement is the familiar one that $kr \gg 1$; this implies that gradients and time derivatives acting on the canonical modes are also $\sim 1/r$. Second, we demand that for all components of the metric, $|g_{\mu\nu} - \eta_{\mu\nu}| \sim a/r \ll 1$. This criterion can always be met (for radiation escaping to infinity) by taking r large enough, and waiting for the wave to reach this distance. This condition implies that, not only is the wave disturbance weak ($f/r \ll 1$), but also the Newtonian-like parts of the field (*e.g.*, g^T) are small ($P^\mu/r \ll 1$). It insures that quadratic (and higher) terms in $g_{\mu\nu} - \eta_{\mu\nu}$, are negligible compared to the linear ones. Finally, we impose the condition that $|\partial g/\partial(kx)|^2 \ll |g - \eta|$. In terms of $k_{\min}$, the minimum frequency (or wave number) to be considered (with, of course, $k_{\min} r \gg 1$), one has $k_{\min} r \gg k_{\max}(a/r)^{1/2}$. Again, for large enough r, this inequality can always be achieved. [For fixed r, it sets a lower bound on frequencies which may be treated in the wave zone.] The purpose of this requirement is to make the non-linear structures containing gradients also small compared to the leading linear terms. The last two conditions together are arranged to guarantee the absence of all self-interactions in the wave zone.

5.6. *General structure of gravitational radiation; superposition, coordinate-independence.* With the above definition of the wave zone, one finds (see IVB) that the field equations (3.15a,b) reduce to

$$\partial_t g_{ij} = 2(\pi^{ij} - \tfrac{1}{2}\delta_{ij}\pi^{ll}) + N_{i,j} + N_{j,i} + O(1/r^2) \tag{5.5a}$$

$$\begin{aligned}\partial_t \pi^{ij} = {}& \tfrac{1}{2}(g_{ij,kk} + g_{kk,ij} - g_{ik,kj} - g_{jk,ki}) \\ & - \tfrac{1}{2}\delta_{ij}(g_{kk,ll} - g_{kl,kl}) - (\delta_{ij}N_{,kk} - N_{,ij}) \\ & + O(1/r^2)\end{aligned} \tag{5.5b}$$

where $O(1/r^2)$ is much less than the leading terms. Using a result of IVB that the orthogonal parts of $\sim 1/r$ and $\sim 1/r^2$ structures are also at most $\sim 1/r$ and $\sim 1/r^2$ respectively, one finds that

$$\partial_t g_{ij}{}^{TT} = 2\pi^{ijTT} \tag{5.6a}$$

$$\partial_t \pi^{ijTT} = \tfrac{1}{2}\nabla^2 g_{ij}{}^{TT} \tag{5.6b}$$

to the order $1/r$. Thus the *rigorous* dynamical modes obey the linearized theory's equations in the wave zone. Equations (5.6) are, of course, source-free, so that the radiation propagates independently of its origin, and without any self-interaction or dependence on the Newtonian-like parts g^T and π^i of the field. The flat-space wave equation with constant coefficients, $\Box^2{}_{\text{flat}} g_{ij}{}^{TT} = 0$, represented by (5.6), indicates the absence of curvature effects on the radiation. Further, there are no coordinate effects left, since the coordinate-dependent components g_i, π^T, and $g_{0\mu}$, have disappeared.

We have seen previously that $g_{ij}{}^{TT}$ and π^{ijTT} are invariant to the order $1/r$ under coordinate transformations leaving the metric asymptotically flat and not involving Lorentz transformations. Thus, in a fixed Lorentz frame, g^{TT} and π^{TT} represent a coordinate-independent description of the radiation. The detailed analysis of IVB shows how any coordinate waves which may be present can be identified unambiguously.

The above derivation may be carried through equally well in the presence of coupling to matter (*e.g.*, the Maxwell field), provided of course, the matter system is bounded within the interior region (and has finite energy content). Under these circumstances, any gravitational waves in the wave zone are still independent of the matter sources. If, however, electromagnetic radiation is also propagating in the wave zone with small enough amplitude (and $\sim 1/r$), the gravitational radiation is still free, being unaffected to $\sim 1/r$ by the electromagnetic waves, since the latter couple quadratically ($\sim 1/r^2$) in the field equations. Hence, in the common wave zone, the systems are correctly independent.

5.7. *Measurables in the wave zone: the Poynting vector, radiation amplitudes. Relation to curvature tensor.* Since the dynamics of the rigorous canonical modes is identical, in the wave zone, to that given by linearized theory, any radiation arising from strong interior fields can be precisely simulated by a corresponding solution of the linearized theory with sources. Such an analog metric can be made everywhere weak, so that linearized theory is everywhere valid, and one may arrange the sources so that in the wave zone, the correct g^T and π^i result. An absorber of gravitational radiation in the wave zone clearly cannot distinguish between the true and analog situations. One is therefore free to use the Poynting vector of linearized theory to measure the energy flux. From the symmetric stress-tensor of I, one finds straightforwardly that

$$T^{i0}_{\text{lin}} = \pi^{lmTT}(2g^{TT}_{li,m} - g^{TT}_{lm,i}) \tag{5.7}$$

in the coordinate frame of (4.11). However, it is easy to see that the general form of T^{i0}_{lin}, is coordinate-invariant through terms of $0(1/r^2)$, as is the right-hand side of (5.7). Hence (5.7) provides an invariant formula for the wave-zone Poynting vector. We remark that, as in electrodynamics, the physical part of the Poynting vector is obtained by averaging over oscillatory terms. This averaging is of use in proving that T^{i0}_{lin} is coordinate-independent, since it also eliminates "coordinate wave" effects. It may also be noted that, in (5.7), no terms involving g^T, π^i have survived to $O(1/r^2)$, showing that the Newtonian-like parts do not affect the energy flux of radiation. Similarly, in electrodynamics $\mathbf{E}^L$ does not contribute to $\mathbf{E}\times\mathbf{B}$ since $\mathbf{E}^L \sim 1/r^2$.

The canonical variables ${g_{ij}}^{TT}$ and π^{ijTT} are, in general, non-local functions of the metric and its first derivatives. However, it is possible to obtain their $1/r$ terms by measurements entirely within the wave zone. From the condition $kr \gg 1$, it is possible to isolate sufficiently accurately the k Fourier component of g_{ij} and $\partial_t g_{ij}$, and to obtain ${g_{ij}}^{TT}(k)$, $\pi^{ijTT}(k)$ algebraically from these. Equivalently, one may measure the k-component of the curvature tensor ${R_\alpha}^\mu{}_{\nu\beta}$ to obtain these physical quantities since it may be shown that in the wave zone

$$ {g_{ij}}^{TT} = (2/\mathbf{k}^2) {R_i}^m{}_{mj} \tag{5.8a} $$

$$ \pi_{ijTT} = -(ik_\ell \mathbf{k}^2) {R_i}^0{}_{lj} \,. \tag{5.8b} $$

Such measurements find a parallel in Maxwell theory, where $\mathbf{E}^T$, $\mathbf{A}^T$ are found from $\mathbf{E}$ and $\mathbf{B}$ by

$$ \mathbf{E}^T = \mathbf{E} - \mathbf{k}[(\mathbf{k}\cdot\mathbf{E})/\mathbf{k}^2] \tag{5.9a} $$

$$ \mathbf{A}^T = i(\mathbf{k}\times\mathbf{B})/\mathbf{k}^2 \,. \tag{5.9b} $$

In fact, electromagnetic wave measurements are commonly of this type.

It should be noted that just as the definition of electromagnetic radiation is not restricted to outgoing radiation (but may include mixtures of outgoing and incoming waves), this also is true of the above definition of gravitational radiation. For the purely outgoing case, the $1/r$ part of the curvature tensor is of type II—Null in Petrov classification (Petrov, 1954, and Pirani, 1957). We have seen in this section that the leading ($\sim 1/r$) physical terms in the field past the wave front are g^T and π^i, and that these yield invariantly the energy-momentum vector of the system. In the wave zone itself, the dynamical modes specifying the radiation, as well as the Poynting vector, could be obtained invariantly. Actually, a good deal of detailed information about the strong fields in interior regions is obtainable by purely asymptotic ("S-matrix") measurements,[16] so that a large portion, at least, of the description of a system is manifestly frame-independent, and can be gotten without the need for introducing apparatus into strong field regions and hence having to analyze its behavior there.

6. Extension to Coupling with Other Systems

6.1. *Lagrangian for coupled Einstein-Maxwell-charged particle system.* Up to now we have considered the free gravitational field, sketching in the process the main physical features which have emerged there. The extension to coupling enables us to verify that these characteristics remain unchanged and provides a basis for the self-energy calculations of the next section. The analysis is completely analogous to the discussion of the free field and so we limit ourselves to a brief outline for the case of coupling to electrodynamics with point charges (more details may be found in V and VA). The addition to the Einstein Lagrangian is

$$ \begin{aligned} \mathcal{L}_M &= \tfrac{1}{2}(A_{\mu,\nu} - A_{\nu,\mu})\mathfrak{F}^{\mu\nu} + \tfrac{1}{4}(-{}^4g)^{1/2}\, \mathfrak{F}^{\mu\nu}\mathfrak{F}^{\alpha\beta}\, {}^4\mathfrak{g}_{\mu\alpha}\, {}^4\mathfrak{g}_{\nu\beta} \\ &+ \int ds\, e(dx^\mu(s)/ds)A_\mu(x)\,\delta^4(x - x(s)) \\ &+ \int ds\, \{\pi_\mu(dx^\mu/ds) - \tfrac{1}{2}\lambda'(s)(\pi_\mu\pi_\nu\, {}^4g^{\mu\nu} + {m_0}^2)\}\delta^4(x - x(s)) \,. \end{aligned} \tag{6.1} $$

[16] J. Plebanski (private communication) has shown how properties of orbits of test particles may be examined by purely asymptotic measurements (using geodesic projections) even if the orbits remain entirely within the strong field interior region.

We are here employing a first-order formalism for the particle as well as for the Maxwell field. The Maxwell part of $\mathcal{L}_M$ is the straightforward covariant generalization of (3.5), with the field strength $\mathfrak{F}^{\mu\nu}$ now a tensor density. Thus, A_μ and $\mathfrak{F}^{\mu\nu}$ are varied separately, as are $x^\mu(s)$, the mechanical momentum $\pi_\mu(s)$, and the Lagrange multiplier $\lambda'(s)$. The particle Lagrangian is in parameterized form (with arbitrary parameter[17] s), as is required for manifest covariance. The δ^4-function is a scalar density, invariantly defined in a metric-independent way by $\int \delta^4(x)d^4x = 1$; the δ^3-function is similarly defined in three-space, *i.e.*, $\int \delta^3(\mathbf{r})d^3r = 1$. Introducing the notation $\mathcal{E}^i \equiv \mathfrak{F}^{0i}$ and $\mathcal{B}^i \equiv \frac{1}{2}\epsilon^{ijk}(A_{k,j} - A_{j,k})$ (where $\epsilon^{ijk} = \epsilon_{ijk} = 0, \pm 1$ is totally antisymmetric) and using the gravitational variables of Section 2, one obtains

$$\begin{aligned}\mathcal{L}_M = {} & A_i\, \delta_0 \mathcal{E}^i + \delta^3(\mathbf{r} - \mathbf{r}(t))[(\pi_i(t) + eA_i)(dx^i/dt) + \pi^0] \\ & + A_0[e\, \delta^3(\mathbf{r} - \mathbf{r}(t)) - \mathcal{E}^i{}_{,i}] \\ & -\tfrac{1}{2}\lambda[g^{ij}\pi_i\pi_j - N^{-2}(\pi_0 - N^i\pi_i)^2 + {m_0}^2]\delta^3(\mathbf{r} - \mathbf{r}(t)) \\ & -\tfrac{1}{2} N g^{-1/2} g_{ij}[\mathcal{E}^i\mathcal{E}^j + \mathcal{B}^i\mathcal{B}^j] + N^i[\epsilon_{ijk}\mathcal{E}^j\mathcal{B}^k]\,. \end{aligned} \tag{6.2}$$

We have also performed the s integration which led to the replacement of λ' by $\lambda \equiv [\lambda'(s)(ds/dx^0(s)]_{x^0(s=t)}$. It is convenient next to the eliminate the non-gravitational constraints obtained from varying A_0 and λ. These are:

$$\mathcal{E}^i{}_{,i} = e\, \delta^3(\mathbf{r} - \mathbf{r}(t)) \tag{6.3a}$$

$$\pi^\mu \pi_\mu + {m_0}^2 = 0 \tag{6.4a}$$

whose solutions are

$$\mathcal{E}^{iL} = -\nabla(e/4\pi|\mathbf{r} - \mathbf{r}(t)|) \tag{6.3b}$$

$$\pi_0 = N^i\pi_i - N(g^{ij}\pi_i\pi_j + {m_0}^2)^{1/2}\,. \tag{6.4b}$$

Here, we have again used the orthogonal breakup of $\mathcal{E}^i$ in the same flat-space sense as in Section 4. Therefore, $\mathcal{E}^i$ is to be regarded subsequently as an abbreviation for $\mathcal{E}^{iT} + \mathcal{E}^{iL}$, with $\mathcal{E}^{iL}$ expressed from (6.3b).

6.2. *Canonical reduction of coupled system.* At this stage,the matter Langrangian is in the reduced form

$$\begin{aligned}\mathcal{L}_M = {} & (-\mathcal{E}^{iT})\, \partial_0 A_i^T + [p_i\, dx^i/dt]\delta^3(\mathbf{r} - \mathbf{r}(t)) \\ & - \tfrac{1}{2} N g^{-1/2} g_{ij}[\mathcal{E}^i\mathcal{E}^j + \mathcal{B}^i\mathcal{B}^j] \\ & - N[g^{ij}(p_i - eA_i^T)(p_j - eA_j^T) + {m_0}^2]^{1/2}\, \delta^3(\mathbf{r} - \mathbf{r}(t)) \\ & + N^i[\epsilon_{ijk}\mathcal{E}^j\mathcal{B}^k + (p_i - eA_i^T)\delta^3(\mathbf{r} - \mathbf{r}(t))] \end{aligned} \tag{6.5}$$

where $p_i \equiv \pi_i + eA_i^T$. The first two terms of $\mathcal{L}_M$ are in the standard $\Sigma p\dot{q}$ form and thus $(A_i^T, -\mathcal{E}^{iT})$ are the canonical variables for the electro-magnetic field, while $p_i(t)$, $x^i(t)$ are those of the particle. Note that the variables A_i, $\mathcal{E}^i$, p_i or x^i, which arose naturally in obtaining canonical form, are also the appropriate variables from the geometrical (Cauchy problem) considerations of Section 3. Thus, A_i, p_i, π_i are covariant spatial components of the corresponding four-vectors, while $x^i(t)$ locates the particle within the three-surface. The three-vector density character of $\mathcal{E}^i$ is established

[17]Equation (6.1) is covariant against any reparameterization $\bar{s} = \bar{s}(s)$ with λ' transforming as a "vector": $\bar{\lambda}' = \lambda'(ds/d\bar{s})$ (just as N does in Section 2).

from its relation to the covariant spatial components of the four-dimensional dual ${}^*\!F_{\mu\nu}$ of the field strength:

$$\mathcal{E}^i = \tfrac{1}{2}\,\epsilon^{ijk}\,{}^*\!F_{jk} \equiv \tfrac{1}{2}\,\epsilon^{ijk}(\tfrac{1}{2}\,\epsilon_{jk\mu\nu}\mathfrak{F}^{\mu\nu})$$

The total Lagrangian, which is the sum of $\mathcal{L}_M$ and the gravitational part (3.13), now has precisely the parameterized form of (2.3), as a consequence of the general covariance of each part. The coefficients of N and N^i now yield the extended constraint equations

$$gR+\tfrac{1}{2}\pi^2-\pi^{ij}\pi_{ij}=\sqrt{g}\,\delta^3(\mathbf{r}-\mathbf{r}(t))[g^{ij}(p_i-eA_i{}^T)(p_j-eA_j{}^T)+m_0{}^2)]^{1/2}+\tfrac{1}{2}g_{ij}[\mathcal{E}^i\mathcal{E}^j+\mathcal{B}^i\mathcal{B}^j] \quad (6.6a)$$

$$-2\pi_{ij}{}^{|j}=\delta^3(\mathbf{r}-\mathbf{r}(t))(p_i-eA_i{}^T)+\epsilon_{ijk}\mathcal{E}^j\mathcal{B}^k\,. \quad (6.6b)$$

Equations (6.6) can again be solved for

$$\nabla^2 g^T = \mathfrak{T}^0{}_0$$

and

$$-2\nabla^2(\pi^{iT}+\pi^i{}_{,i}) = \mathfrak{T}^0{}_i\,.$$

When the coordinate conditions (4.11) are again imposed, the theory is in canonical form with the *same* gravitational canonical variables. One sees most clearly now how the gravitational canonical variables characterize independent gravitational field excitations. All the other components of the metric, *e.g.*, g^T and N, N^i depend on the non-gravitational variables as well. In contrast, only $g_{ij}{}^{TT}$ and π^{ijTT} can be specified initially, irrespective of the excitations of the matter system.

The quantity $-\mathfrak{T}^0{}_0$ now represents the Hamiltonian density of the total system[18] and depends only on the canonical variables of all three parts. Further, the expressions (5.1) and (5.2) now give the energy-momentum of the total system. The fact that the total energy of the interacting system is obtainable from purely metric quantities is analogous to the expression in electrodynamics of the total charge in terms of the integral of the longitudinal electric field. It is worth noting that the choice of coordinate conditions is unaffected by coupling. Finally, we record the equations of motion of the gravitational field including coupling. The extension of the constraint equations (3.15c) has been given by (6.6) above; the $\partial_t g_{ij}$ equations (3.15a) are unchanged, since they are of the type $m\dot{x}=p$ (namely, they are defining equations for Christoffel symbols). However, one must now add a term $\frac{1}{2}N\mathfrak{T}_M{}^{ij}$ to the right-hand side of (3.15b) for $\partial_t\pi^{ij}$, where $\mathfrak{T}_M{}^{ij}$ is the symmetric spatial stress-energy tensor density of the matter:

$$\begin{aligned}\mathfrak{T}_M{}^{ij} &\equiv g^{ij}g^{jk}\sqrt{g}\,{}^4\mathfrak{T}_{Mlk}\\ &= \delta^3(\mathbf{r}-\mathbf{r}(t))(p^i-eA^{iT})(p^j-eA^{jT})[(\mathbf{p}-e\mathbf{A}^T)^2+m_0{}^2]^{-1/2}\\ &\quad+g^{-1/2}[\tfrac{1}{2}\,g^{ij}(\mathcal{E}^m\mathcal{E}_m+\mathcal{B}^m\mathcal{B}_m)-\mathcal{E}^i\mathcal{E}^j-\mathcal{B}^i\mathcal{B}^j]\,. \end{aligned} \quad (6.7)$$

7-7. Static Self-Energies

7.1. *Physical basis for finiteness of self-energy when gravitation is included.* The canonical formalism developed for coupled systems in the previous section allows one to define pure particle states by the vanishing of the canonical variables referring to field excitations. For particles at rest, therefore, the energies of such states are just the total rest and interaction energies, which include, of course, their self-energies.

[18] In the flat space limit, this Hamiltonian density is correctly the usual one for electrodynamics.

The aim of classical point electron theory, since its inception, has been to obtain a finite, model-independent electromagnetic self-energy, and if possible, to dispense with mechanical mass altogether. Thus the total mass of the particle would arise from its coupling to the field. Such a program, however, was not feasible, since the self-energy diverged linearly, with no realistic compensation possible. In terms of renormalization theory, this implied an infinite "bare" mechanical mass. Since gravitational interaction energy is negative on the Newtonian level, it may be expected to provide compensation. Indeed a simple argument yields a limit on the self-energy due to just such a compensation. Consider a *bare* mass m_0 distributed in a sphere of radius ϵ. In the Newtonian limit, the total energy (*i.e.*, the clothed mass) m is given by $m = m_0 - \frac{1}{2}\gamma {m_0}^2/\epsilon$. For sufficiently small ϵ, m could become zero and then negative. In general relativity, the principle of equivalence states that it is the total energy that interacts gravitationally and not just the bare mass. Thus, as the interaction energy grows more negative, were a point reached where the total energy vanished, there could be no further interaction energy. Consequently, there can be no negative total energy, in contrast to the negative infinite self-energy of Newtonian theory. General relativity effectively replaces m_0 by m in the interaction term: $m = m_0 - \frac{1}{2}\gamma m^2/\epsilon$. Solving for m yields $m = \gamma^{-1}[-\epsilon + \epsilon^2 + 2\gamma m_0 \epsilon)^{1/2}]$, which shows that $m \to 0$ as $\epsilon \to 0$.

More interesting is the fact that the gravitational interactions produce a natural cutoff for the Coulomb self-energy of a point charge. Here the self-mass resides in the Coulomb field $\frac{1}{2}\int (e/4\pi r^2)^2\, d^3\mathbf{r}$. By the general argument above on gravitational compensation, one expects that the Coulomb energy near the origin (which is, in fact, "denser" than the neutral particle's δ-function distribution) will have a very strong gravitational self-interaction, resulting in a vanishing *total* contribution to the self-mass from the region near the origin. Thus, the integral effectively extends down only to some radius a, yielding $m_{\mathrm{EM}} = \frac{1}{2}\int_a^\infty (e/4\pi r^2)^2 d^3r = (e^2/4\pi)/2a$. We can determine this effective flat-space cutoff a by the same equivalence principle argument. Without the gravitational contribution, the mass is $m_0 + \frac{1}{2}(e^2/4\pi\epsilon)$, so the total clothed mass is determined by the equation $m = m_0 + \frac{1}{2}\, e^2/4\pi\epsilon - \frac{1}{2}\gamma m^2/\epsilon$. This yields

$$m = \gamma^{-1}\{-\epsilon + [\epsilon^2 + 2m_0\epsilon\gamma + (e^2/4\pi)\gamma]^{1/2}\}\,. \tag{7.1}$$

We will see below that this formula is a rigorous consequence of the field equations. In the limit $\epsilon \to 0$, we have $m = (e^2/4\pi)^{1/2}\gamma^{-1/2}$. The bare mechanical mass m_0 again does not contribute to the clothed mass. Our result is then equivalent to a cutoff $a = \frac{1}{2}\,(e^2/4\pi)^{1/2}\gamma^{1/2}$ on the flat-space Coulomb integral.

7.2. *Calculation of charged and neutral particle self-energies from initial data.* We begin the self-energy calculation, then, by obtaining a solution of the field equations for a pure one-particle state, *i.e.*, a state containing no independent excitations of the gravitational or electromagnetic fields in the rest frame. This requires that ${g_{ij}}^{TT} = \pi^{ijTT} = A^{iT} = \mathcal{E}^{iT} = p_i = 0$ on the $t =$ const surface where the energy is being computed. With the coordinate conditions of (4.22), we are therefore dealing with the time-symmetric situation $\pi^{ij} = 0$. According to (4.23), then, the metric is isotropic; it is convenient to write it as $g_{ij} = \chi^4(r)\,\delta_{ij}$. From Section 7-5, the energy is the coefficient of $1/(32\pi r)$ in the asymptotic form of $\chi(r)$. The field equation determining g^T and hence χ is (6.6a). One has[19]

$$\sqrt{g}\;{}^3R = -8\chi\nabla^2\chi = m_0\,\delta^3(\mathbf{r}) + \tfrac{1}{2}\,\chi^{-2}\mathcal{E}^{iL}\mathcal{E}^{iL} \tag{7.2a}$$

[19]Professor L.N. Cooper has pointed out to us that (7.2a) with $e = 0$ $[-8\chi\nabla^2\chi = \rho_0(r)$, where ρ_0 is any bare mass distribution] can be obtained by simple equivalence principle arguments starting from Newtonian theory. The Poisson equation $\nabla^2\phi = 4\pi\gamma\rho_0 = \frac{1}{4}\,\rho_0$ for the gravitational potential ϕ, must be corrected to include the particle's

and the electric field $\mathbf{E}^L$ by (6.3b) is

$$\mathcal{E}^{iL} = (-e/4\pi r)_{,i} \,. \tag{7.3}$$

A formal solution of (7.2a) may be found by setting $\chi^2 = \psi^2 - \phi^2$ (Misner and Wheeler, 1957). One finds

$$\begin{aligned} -8\chi\nabla^2\chi &= 8(\phi\nabla^2\phi - \psi\nabla^2\psi) + 8(\psi^2 - \phi^2)^{-1}(\phi\nabla\psi - \psi\nabla\phi)^2 \\ &= m_0\,\delta^3(\mathbf{r}) + \tfrac{1}{2}(\psi^2 - \phi^2)^{-1}(\mathbf{E}^L)^2 \,. \end{aligned} \tag{7.2b}$$

If one then makes the assumption

$$\mathbf{E}^L = 4(\phi\nabla\psi - \psi\nabla\phi) \tag{7.4}$$

with $\phi = e/16\pi r$ and $(1 + m/32\pi r) = \psi$, (7.4) correctly reproduces (7.3). Equation (7.2b) then determines the total mass m to be the $\epsilon \to 0$ limit of (7.1):

$$m = \lim_{\epsilon \to 0} 16\pi\{-\epsilon + [\epsilon^2 + (e/8\pi)^2 + m_0\epsilon/8\pi]^{1/2}\} \,. \tag{7.5}$$

In (7.5) the parameter ϵ has been introduced by setting $\delta^3(\mathbf{r})/r$ equal to $\delta^3(\mathbf{r})/\epsilon$, and is thus essentially the "radius" of the δ-function. [This interpretation of the δ-function corresponds to viewing $\delta^3(\mathbf{r})$ as the limit of a shell distribution $\delta(r-\epsilon)/4\pi r^2$ of radius ϵ. In V, it is shown that the results of this section are, in fact, independent of the model chosen for $\delta^3(\mathbf{r})$ in the point limit.] The result (7.5) is that $m = 2|e|$ and hence the total mass is finite and independent of the bare mechanical mass. The gravitationally renormalized electrostatic self-energy is now finite. The analysis also points out that mass only arises if a particle has nongravitational interaction with a field of finite range.[20] For example, an electrically neutral particle coupled to a Yukawa field would acquire a mass by virtue of *this* coupling.

A solution may also be obtained for the case of two particles of like charge. The results are consistent with those obtained above for one particle (see V). The energy is just $2(|e_1| + |e_2|)$, that is, just the sum of the individual masses, independent of both the mechanical masses and the interparticle separation r_{12}. The absence of an interaction term can be traced to the cancellation of the Newtonian attraction of the masses with the Coulomb repulsion of the like charges.[21] This also checks with the result expected at large separation for particles whose mass and charge are related by $m_{1,2} = (e^2{}_{1,2}/4\pi\gamma)^{1/2}$. For opposite charges, such a cancellation should not occur. However, we have not been able to obtain a rigorous solution for a pure particle state of two opposite charges.

All the usual Newtonian theory results may be obtained from (7.5) or from the corresponding equation [(3.8) of V] for two bodies in the appropriate limit. This "dilute" limit consists in regarding

gravitational self-energy $\frac{1}{2}\rho\phi$ as part of the source, *i.e.*, $\nabla^2\phi = \frac{1}{4}\rho \equiv \frac{1}{4}(\rho_0 + \frac{1}{2}\rho\phi)$. Eliminating ρ, one obtains $\nabla^2\phi = \frac{1}{4}\rho_0(1 - \frac{1}{2}\phi)^{-1}$. In terms of $\chi = 1 - \frac{1}{2}\phi$, this is just (7.2a) in the neutral case. It is interesting that this argument yields the rigorous field equations in the frame (4.22) for this situation. For the point electric charge, the same argument with ρ_0 replaced by $\rho_0 + \frac{1}{2}\rho_e\phi_e$ (ϕ_e being the electrostatic potential) leads to an equation for ϕ which, while different from (7.2a), does yield the correct total energy.

[20] It is interesting to note that, even though $m = 0$ for the neutral particle, this does not imply that space is everywhere flat. The metric is indeed flat for $r > \epsilon$, but rises steeply in the interior. For example, $\int_0^\infty d^3r\sqrt{^3g}\;{}^3R = m_0$, which shows that space is curved at the origin.

[21] This cancellation implies that the solution should be static (since there is no initial potential between the particles). This has indeed been shown to be the case for $m_{1,2} = 2|e_{1,2}|$ by Papapetrou (1947).

m_0 and e small compared to ϵ before passing to $\epsilon = 0$, thus preventing the strong (non-Newtonian) gravitational interactions. Alternately, the "dilute" limit represents a perturbation expansion of the theory in powers of the coupling constant γ. From (7.1) one easily sees that

$$E \sim m_0 + (e^2/4\pi - \gamma {m_0}^2)/2\epsilon + O(1/\epsilon^2)\,. \tag{7.6}$$

[For two bodies, the Newtonian and Coulomb interaction terms $(e_1 e_2/4\pi - \gamma (m_0)_1 (m_0)_2)/r_{12}$ also appear.] Since m_0 and e are unrelated, there is no cancellation in the self-energy term of (7.6). In fact, the perturbation expansion consists of an infinite series of more and more divergent terms. The inapplicability of such an expansion is shown by the finite rigorous answer $m = 2|e|$. If one attempted to use standard perturbation renormalization techniques, one would indeed find that the theory was unrenormalizable (see V).

7.3. *Determination of full metric due to a charge. Stability of the particle. Vanishing of self-stress.* We have been able to find the self energy purely from the initial value (constraint) equations. The full solution of the problem consists also in the specification of the N, $N_i(g_{0\mu})$ for our frame as well as the time development of the system (fuller details are in VA). The equations determining N and N_i are obtained by taking the time derivatives of the coordinate conditions (4.22c,d). These are basically linear combinations of the right-hand sides of (3.15a,b) [including, of course, the coupling term from (6.7)]. Fortunately, the initial values and the coordinate conditions for our problem reduce these equations to quite simple form. At the initial time, (3.15a) becomes

$$\partial_0 g_{ij} = N_{i|j} + N_{j|i}\,. \tag{7.7}$$

Taking the time derivative of (4.22c), one obtains a homogeneous equation for N_i which has the solution

$$N_i(\mathbf{r}, 0) = 0\,. \tag{7.8}$$

The relevant part of (3.15b) is now just the $\partial_0 \pi^{ii}$ equation. The coordinate condition (4.22d), the isotropic form of g_{ij}, and the constraint equation (7.2) simplify this equation to

$$\partial_m(\chi^2\, \partial_m N) = \tfrac{1}{4}\, N\{\chi^{-2} \mathcal{E}^{iL} \mathcal{E}^{iL} + m_0\, \delta^3(\mathbf{r})\}\,. \tag{7.9}$$

For the point particle, the solution for N is just (see VA):

$$N = (1 + |e|/8\pi r)^{-1} = \chi^2\,. \tag{7.10}$$

Note that the total solution for $g_{\mu\nu}$, is everywhere *non-singular* (except, of course, at the particle). In VA, it was shown that this coordinate system led to a *completely* non-singular metric $g_{\mu\nu}$ initially, for a spread-out distribution of charge and mass with arbitrary m_0 and ϵ. (This is in contrast to the standard Reissner-Nordstrom metric in isotropic coordinates for which N always has a singularity for small enough ϵ.)

Further, the point charge is a stable object, since the equations (7.8) and (7.10) for $g_{\mu\nu}$ initially, lead to a static solution ($\partial_t g_{\mu\nu} = \partial_t \pi^{ij} = 0$). The cancellation of the (repulsive) electrostatic self-forces by the gravitational ones is made clear by examining the components of $\mathfrak{T}^{ij}$, the total system's spatial stress density. The conservation requirement on the total stress tensor $\mathfrak{T}^{\mu\nu}$ implies that ${\mathfrak{T}^{ij}}_{,j} = -{\mathfrak{T}^{0i}}_{,0} = 0$ for the static case, where $\mathfrak{T}^{ij}$ are the total system's spatial stresses. Thus, in the notation of the orthogonal decomposition, $\mathfrak{T}^{ij}$ reduces to $\mathfrak{T}^{ij} = \mathfrak{T}^{ijTT} + \mathfrak{T}^{ijT}$, since it is transverse. Spherical symmetry means that $\mathfrak{T}^{ijTT}$, vanishes, since no preferred transverse direction can be distinguished. Hence, $\mathfrak{T}^{ij}$ has at most one independent component which may be taken as $\mathfrak{T}^{ii}$. For our static point solutions $\mathfrak{T}^{ii}$ (as calculated from either the Landau-Lifshitz or Einstein pseudotensors) vanishes everywhere for arbitrary m_0 and e. In the rest frame, then, $\mathfrak{T}^{\mu\nu} = \rho\, \delta_0^\mu\, \delta_0^\nu$ where $\int d^3r\, \rho = m$, the clothed mass. In a moving system, therefore, $\mathfrak{T}^{\mu\nu} = \rho(dx^\mu/d\tau)(dx^\nu/d\tau)$ (τ is Lorentz proper time). The vanishing of $\mathfrak{T}^{ij}$ in the rest frame, then, is necessary for the structure of the total stress tensor to be that of a renormalized mass m. (This requirement is stronger than the usual one that $P_\mu = \int d^3r \mathfrak{T}^0{}_\mu$ transform like the energy-momentum of a particle.[22]) Thus the point charge is a completely stable object, without any *ad hoc* pressure (cohesive) terms required, and its mass is completely determined by its field interactions.

7.4. *Comparison with standard Reissner-Nordstrom solutions.* It is interesting to compare our results with the standard discussion of Schwarzschild (neutral) and Riessner-Nordstrom (charged) solutions. In these approaches, no bare matter parameters were ever introduced, so that the self-energy problem could not be formulated. The source term used in the standard treatment is the Maxwell stress tensor, together with a matter source term of a perfect fluid with four-velocity U^μ:

$$T^{\mu\nu} = (\rho_{00} + p_0)U^\mu U^\nu + p_0 g^{\mu\nu}\ . \qquad (7.11)$$

The scalar "proper rest-mass density" ρ_{00} is, as can be seen from this equation, the mass density in the locally inertial rest-frame, while p_0 is the pressure in this frame (see, for example, Møller, 1952). In our treatment, on the other hand, the matter was introduced dynamically rather than as an externally prescribed source. This was accomplished by including a Lagrangian for the particle. Comparing (7.11) with the source terms in our equations,[23] one finds that $\rho_{00} = m_0\chi^{-6}\,\delta^3(\mathbf{r})$, which shows that ρ_{00} includes "clothing" effects through the χ^{-6}-factor. The pressure terms in the usual source tensor (7.11) are introduced expressly in order to make the distribution stable. Since a pressure term summarizes the presence of other forces (not being treated dynamically), these forces would contribute to the clothing of the original mechanical mass and hence the parameter m_0 would now represent this original mechanical mass *plus* the clothing due to the forces giving rise to the pressure term. In our analysis, where no phenomenological pressure terms are invoked, one cannot at the outset ask that a solution be static. In fact, extended initial distributions of matter and charge are not, in general, stable, although the point limit does have this property.

8. Outlook

8.1. *Discussion of quantization.* In this survey, we have seen that general relativity may be viewed as an ordinary classical field, once the meaning of the coordinate invariance has been analyzed. It was then possible to treat the dynamics of the gravitational field according to techniques common to classical field theory. In this way, many physical properties could be directly understood from their counterparts in, say, electrodynamics. For example, the canonical formalism provided a unique definition of energy and gravitational radiation. The gravitational field, of course, also has aspects not found elsewhere. In particular, its sources are the total energies of *all* other fields. The attractive, static interaction, part of the total energy provided the possibility of a compensating effect on the flat-space self-energies of other fields. Thus, a stable classical point electron of finite

[22]The standard static Reissner-Nordstrom solution has $\mathfrak{T}^{ii} \neq 0$ but $\int d^3r \mathfrak{T}^{ii} = 0$. Thus it obeys the weaker condition on P_μ but its total stress tensor is not that of a particle. This is due to the presence of the *phenomenological* pressure terms which were needed there to stabilize the particle.

[23]In VA, the comparison with (7.11) was carried out in terms of a "dust" model of the particle, that is, with a Lagrangian describing the dynamics of a continuous distribution of matter in gravitational interaction, and without phenomenological pressure forces.

mass exists when gravitation is included.

A realistic elementary particle theory must, of course, be formulated in quantum terms to determine whether the quantum self-mass is also finite. A full quantum treatment includes the quantization of the dynamical modes of the gravitational field, and not only of the effects arising due to a q-number source. Since a complete set of Poisson bracket relations among the classical canonical variables has been obtained, a correspondence principle quantization may be performed immediately by transcribing the P.B.'s into commutators. Such a quantization may indeed be valid when appropriate care of the ordering of the canonical operators in the non-linear parts of the Hamiltonian is taken. However, certain consistency criteria proper to quantum theory and extending beyond this possible ambiguity must be examined. The reduced system (involving only the canonical formalism and its two degrees of freedom) does not represent the full statement of general relativity: one also has equations to determine $g_{0\mu}$, for example, which are not part of this canonical theory as such. Furthermore, one encounters these variables as soon as one makes a Lorentz transformation from the initial frame,[24] an operation that must be allowed for any sensible quantum theory. Since the equations defining $g_{0\mu}$ are now quantum ones, one must establish the consistency of the ordering of the Hamiltonian and the $g_{0\mu}$ equations in one Lorentz frame with those in another frame. Finally, we saw classically that there was an infinite number of *a priori* equivalent sets of simple canonical variables with conserved Hamiltonians, each of which could, of course, be used as a basis for such a quantum scheme; quantum mechanically, however, the relation between these sets of variables need no longer be one of a unitary transformation due to the operator character of the variables. Hence these are no longer *a priori* equivalent starting points for quantization. The classical canonical transformations among these sets will cause the coordinates associated with one set to be functions of both the canonical variables and coordinates of the other. This would lead to the phenomenon, in the quantum theory, of the coordinates of one set being q-numbers when expressed in terms of the variables of the other set,[25] as was discussed in II and IVA.

In view of the many ambiguities which could arise in an attempt to quantize consistently at this level, it would seem more fruitful to return to the Lagrangian in four-dimensional form, *i.e.*, $\mathfrak{g}^{\mu\nu} R_{\mu\nu}[\Gamma^{a}{}_{\lambda\rho}]$ and try to repeat our reduction to canonical form within the framework of quantum theory. There, one can use the manifest Lorentz covariance of the original Lagrangian as an aid in proving the Lorentz covariance of the canonical quantum form that should arise. Further, the ordering ambiguities are drastically reduced: since at most cubic terms enter in this Lagrangian, one can show easily that there is a simple three-parameter family of available Hermitian quantum Lagrangians, all of which are generally covariant. These different orderings of the Lagrangian differ from each other only by double commutators, *i.e.*, by effects of order $\hbar^2$. The basic requirement of consistency between the Lagrange and Heisenberg equations of motion should single out one of these forms, since commutators of Bose fields do not contribute to Lagrange equations but presumably affect the Heisenberg ones. In an investigation based on the four-dimensional form of the quantum Lagrangian, one ultimately expects to arrive at canonical forms very similar to those obtained here.

[24]In electrodynamics one similarly has that the gauge functions (*i.e.*, the scalar potential and longitudinal part of the vector potential) mix with the dynamical variables (*i.e.*, the transverse parts of the vector potential) when a Lorentz transformation is made.

[25]A similar situation could arise in particle mechanics with a transformation $\bar{t} = t + f(p,q)$. Here one would not so transform, since a preferred time coordinate t has been decided upon; in gravitation, however, no such single preferred coordinate frame exists in the interior, at least classically. Whether the quantum theory may force a preferred frame through the consistency requirements is not known.

Thus the classical results should represent an excellent guide in formulating the quantum theory.[26]

8.2. *Speculations on quantum self-energy problem.* Leaving aside these technical questions on the rigor of gravitational quantization, however, one may speculate as to the quantum effects on the self-energy problem. The numerical value obtained classically for the mass m of a point particle with electronic charge, $m = e/(4\pi\gamma)^{1/2} \simeq 10^{18} m_e$, is much too large. Of course, one would not expect classical theory to give correct numerical values for masses. Any realistic discussion of self-masses must be based on quantum theory. However, if the effective flat-space cutoff ($a \sim [e/(4\pi)^{1/2}]\gamma^{1/2} \sim 10^{-34}$ cm) obtained in this paper were to hold also in quantum theory,[27] the numerical values for the mass and charge would be quite different. For example, using such a cutoff in Landau's estimate (Landau, 1956) for the renormalized charge, one finds ${e_r}^2/4\pi \approx 10^{-2}$ (essentially independent of the bare charge) as noticed by Landau. Thus, Landau obtains

$$\begin{aligned} {e_r}^2 &= e^2[1 + (2/3\pi)\nu(e^2/4\pi)\ln\{(\hbar/mc)/a\}]^{-1} \\ &\approx [(\nu/12\pi^2)\ln\{(\hbar/mc)/a\}]^{-1} \end{aligned} \tag{8.1}$$

where $\nu \approx 10$ is essentially the number of charged fields. Equation (8.1) was obtained by summing the dominant terms in each self-energy diagram. With an effective cutoff of physical origin, the usual objections to making an estimate of what otherwise is a divergent series need no longer hold. The methods of Landau also yield a formula for the renormalized mass in terms of the bare mass and the charge.[28] In the classical theory, we saw that the bare mass did not enter. To what extent this is maintained in the quantum theory (*e.g.*, to what extent the bare mass distribution remains effectively a δ-function) is not clear. It should also be emphasized that we have assumed in the above discussion the simplest possibility, that the gravitational effects on the quantum theory can be summarized in terms of a cutoff of the type considered by Landau.

Finally, the treatment given here has not touched on several major topics of interest in the classical theory. For example, the problem of motion (Einstein–Infeld–Hoffman analysis),has not been examined from the present viewpoint. Here, it is hoped that the canonical methods will shed further light on the manner in which the coupling of particles with the various components of the metric determines their motion. A particular issue where these techniques might be relevant is the question of radiation by the moving masses. Some further topics involve spaces with non-flat boundary conditions, which arise in cosmology, as well as spaces where the topology is not Euclidian. It remains to be seen whether an approach of this type can be applied to such situations.

[26] It has been shown in I that, if the theory can be quantized at all, it obeys Bose statistics (as intuitively expected).

[27] An effective quantum gravitational cutoff might, on dimensional grounds, be $\sim (\gamma\hbar c^{-3})^{1/2} \sim 10^{-33}$ cm. This differs from the classical a merely by the factor $a^{1/2} \equiv (e^2/4\pi\hbar c)^{1/2}$ and would not affect the discussion in the text.

[28] This formula, $m = m_0(e^2/{e_r}^2)^{9/4\nu}$, does not make clear what relation between m, m_0, and e_r might be expected, since the estimate of (8.1) fails to determine e with sufficient accuracy.

REFERENCES

Arnowitt, R., and S. Deser, 1959, *Phys. Rev.*, **113**, 745 (I).
Arnowitt, R., S. Deser, and C.W. Misner:
1959, *Phys. Rev.*, **116**, 1322 (II).
1960, *Phys. Rev.*, **117**, 1595 (III).
1960, *Nuov. Cim.*, **15**, 487.
1960, *Phys. Rev. Lett.*, **4**, 375.
1960, *Phys. Rev.*, **118**, 1100 (IV).
1960, *J. Math. Phys.*, **1**, 434 (IIIA).
1960, *Phys. Rev.*, **120**, 313 (V).
1960, *Phys. Rev.*, **120**, 321 (VA).
1960, *Ann. Phys.*, **11**, 116 (VB).
1961, *Nuov. Cim.*, **19**, 668 (IVA).
1961, *Phys. Rev.*, **121**, 1556 (IVB).
1961, *Phys. Rev.*, **122**, 997 (IVC).
Brill, D., 1959, *Ann. Phys.*, **7**, 466.
Eisenhart, L.P., 1949, *Riemannian Geometry*, Princeton University Press, Princeton.
Lanczos, C., 1949, *The Variational Principles of Mechanics*, Toronto University Press.
Landau, L.D., 1955, in *Niels Bohr and the Development of Modern Physics*, Pergamon Press, London.
Misner, C.W., and J.A. Wheeler, 1957, *Ann. Phys.*, **2**, 592.
Møller, C., 1952, *The Theory of Relativity*, Oxford University Press.
Papapetrou, A., 1947, *Proc. Roy. Irish Acad.*, **51A**, 191.
Petrov, A.Z., 1954, *Sci. Not. Kazan State Univ.*, **114**, 55.
Pirani, F.A.E., 1957, *Phys. Rev.*, **105**, 1089.
Schrödinger, E., 1950, *Space-Time Structure*, Cambridge University Press.
Schwinger, J., 1951, *Phys. Rev.*, **82**, 914.
——— 1953, *Phys. Rev.*, **91**, 713.

Chapter 4
Aleksey Petrov 1954, algebraic classification of the Weyl tensor

Original paper: Aleksey Petrov, Klassifikacya prostranstv opredelyayushchikh polya tyagoteniya. *Uchenye Zapiski Kazanskogo Gosudarstvennogo Universiteta im. V. I. Ulyanovicha-Lenina [Scientific Proceedings of Kazan State University, named after V.I. Ulyanov-Lenin]* **114** (8), 55-69 (1954)). Jubilee (1804-1954) Collection.

Reprinted with the kind permission of the Department of Relativity and Gravitation, Institute of Physics, Kazan Federal University.

Paper translated by Jacek Jezierski, Institute of Theoretical Physics, University of Warsaw, e-mail: Jacek.Jezierski@fuw.edu.pl, Malcolm A. H. MacCallum, e-mail: M.A.H.MacCallum@qmul.ac.uk, and A. G. Polnarev.

Previously published in *Gen. Relativ. Gravit.* **32** (2000), pages 1661 – 1664 (the editorial note + biography), 1665 – 1685 (the paper).

Petrov's biography printed in the journal was replaced with an extended version newly written by Asya V. Aminova.

4.1 Editorial note to:

Aleksey Petrov
The classification of spaces defining gravitational fields

Malcolm A. H. MacCallum

In the paper translated below Petrov completes and summarizes results which he had been working on for some years. They led to the introduction of the now famous Petrov types classifying the Weyl tensor. The ideas were taken up and developed over the next few years by a number of authors (e.g. [2, 3, 4, 5, 6, 10]): for an exposition of the resulting classification in modern terms see e.g. [12].

Petrov's methods are taken from classical linear and multilinear algebra applied to the action of the Weyl tensor in bivector space, and rest on the classification of linear maps using the notation normally associated with Segre [1] (and used, for example, in the Segre classification of the Ricci tensor; indeed, Petrov's classification, applied to the Plebanski tensor, is a step in the algorithms for Ricci tensor classification [13]). The term symmetrically-double refers to matrices of the form given in the first equation of (15) in the paper. The exposition given by Petrov is otherwise self-contained, except for the use of the notion of m/n orthogonality. This refers to the interrelation of two bivectors defining subspaces with dimensions p and q where $p \leq q \leq 4$. If there are k common directions, the bivectors are said to be k/p-parallel. Similarly if the p-dimensional subspace contains s directions orthogonal to the q-dimensional subspace, the bivectors are called s/p-perpendicular.

Theorems 1 and 2 show, in modern language, that the bivector space can be split into self-dual and anti-self-dual subspaces, and that the Weyl tensor acts on these two parts in complex conjugate ways, so that only one subspace need be considered. This is easily translated into the classification using spinors [6].

One passage of the original text is rather unclear. It comes in the derivation of the canonical forms for the first type, where Petrov considers the effect of a frame transformation. The conclusion is that in general (in the notation of the paper) one can choose a φ and ψ so that $\underset{2}{\widetilde{\nu}} = 0$ but the text does not express this very precisely.

Petrov's classification itself gives three types. In the paper these are not given a shorthand notation but in his later book [7] they are referred to as T_i $(i = 1, 2, 3)$ for vacuum and $\dot{T}_i$ for Einstein spaces, the case he considers here. However, his names have been entirely superseded by the names used by Pirani [3] and Jordan et. al [8], types I, II, III, N, D and 0, though these are invariably called Petrov types. The correspondence is that T_1 is I together with D as a special case, T_2 is II with N as

Malcolm A. H. MacCallum
Department of Mathematics, University of Bristol
Bristol BS8 1TW, U. K.
e-mail.m.a.h.maccallum@qmul.ac.uk

a special case, and T_3 is III (0 is the zero Weyl tensor). Petrov mentions the special cases but does not elaborate.

It is also worth noting that the frame freedom referred to by Petrov in his second and third types can in general be removed by normalization of the remaining arbitrary constants. Petrov types I, II and III fix frames completely, up to discrete groups permuting simple principal directions in cases I and II. It is only in Petrov types N and D (and of course 0) that there is a true residual frame freedom, and in type N it is not the freedom Petrov speaks of for his second type.

The physical and mathematical importance of the classification arises in various ways. One is in the understanding of gravitational radiation. The different parts of the Weyl tensor can be interpreted as longitudinal and transverse effects [11], and the famous peeling theorem [9] shows that in asymptotically flat space the first few terms in an expansion in $1/r$, where r is a suitable radial variable, are of different and increasingly general Petrov type. Another is that many known solutions of Einstein's equations were derived by assuming that, or have the property that, the Weyl tensor is algebraically special: part III of [12] is devoted to such methods and solutions.

There is an earlier translation of this important paper: "Classification of spaces supporting gravitational fields" by M. Karweit, Translation no. 29, Jet Propulsion Laboratory, Pasadena, California, 1963 (I thank Dr. F.B. Estabrook for sending me a copy of this). Substantial parts of the text are also repeated in [7]. These translations provided useful checks of the following new English version.

Acknowledgements. The Editors of the journal are grateful to R. A. Zalaletdinov for suggesting this paper for the "Oldies" series, for providing a copy of the original paper, and for his help in contacting the appropriate persons in Russia. We are also grateful to D. Gal'tsov and A. Gusev for their help in obtaining the permission for the republication and for their active role in preparing the biographical note. The Editors of this book express their gratitude to Alexei Starobinsky and Sergei Sushkov for their help in obtaining the permission for the reprinting here. We are most grateful to Asya V. Aminova for writing the detailed new biography of Aleksey Petrov printed below.

References

1. Segre, C. (1884). Sulla teoria e sulla classificazione delle omografie in uno spazio lineare ad un numero qualcunque di dimensioni. *Memorie della R. Accademia dei Lincei, ser. 3a* **XIX**, 127.
2. Géhéniau, J. (1957). Une classification des espaces einsteiniens. *C. R. Acad. Sci. (Paris)* **244**, 723.
3. Pirani, F.A.E. (1957). Invariant formulation of gravitational radiation theory. *Phys. Rev.* **105**, 1089.
4. Debever, R. (1959). Tenseur de super-énergie, tenseur de Riemann: cas singuliers. *C. R. Acad. Sci. (Paris)* **249**, 1744.
5. Bel, L. (1959). Quelques remarques sur la classification de Petrov. *C. R. Acad. Sci. (Paris)* **248**, 2561.

6. Penrose, R. (1960). A spinor approach to general relativity. *Ann. Phys.* **10**, 171.
7. Petrov, A.Z. (1961). *Einstein spaces*. Gosudarstvennoe Izdatel'stvo Fiziko-Matematicheskoy Literatury, Moscow. (English edition, Pergamon, Oxford, 1969, translated by R.F. Kelleher and J. Woodrow.)
8. Jordan, P., Ehlers, J., and Sachs, R.K. (1961). Beiträge zur Theorie der reinen Gravitationsstrahlung. *Akad. Wiss. Lit. Mainz, Abhandl. Math.-Nat. Kl.*, no.1.
9. Sachs, R.K. (1962). Gravitational waves in general relativity. VIII. Waves in asymptotically flat space-time. *Proc. Roy. Soc. Lond. A* **270**, 103.
10. Debever, R. (1964). Le rayonnement gravitationnel: Le tenseur de Riemann en relativité générale. *Cahiers Phys.* **18**, 303.
11. Szekeres, P. (1965). The gravitational compass. *J. Math. Phys.*, **6**, 1387.
12. D. Kramer, H. Stephani, M.A.H. MacCallum, and E. Herlt (1980). *Exact solutions of Einstein's field equations*. Deutscher Verlag der Wissenschaften, Berlin, and Cambridge University Press, Cambridge. (Russian translation: "Tochnie resheniya uravnenii Einshteina", 418 pp., translated by I.V. Mitskievich, V.D. Zakharov and S.V. Rumyantsev and edited by Yu. S. Vladimirov, Energoisdat, Moscow, 1982).
13. F.M. Paiva, M.J. Rebouças, G.S. Hall, and M.A.H. MacCallum (1998). Limits of the energy-momentum tensor in general relativity. *Class. Quant. Grav.* **15**, 1031–1041.

4.2 Aleksey Zinovjevitch Petrov – a brief biography

By Asya V. Aminova
Department of Relativity and Gravitation
Institute of Physics of Kazan Federal University
Asya.Aminova@ksu.ru

Born in Parish village of Koshki municipality, Samara guberniya, on October 28 (15 in Old Style), 1910, Alexei Zinovievich Petrov was the eleventh of twelve children of the village pastor, Zinovii Vasilievich Georgievskii. His father died of tuberculosis when Alyosha was five years old. Shortly thereafter, his house caught fire and was completely destroyed. The disaster compelled his mother, Zoya Nikolaevna, to give her two youngest sons, Alyosha and his brother Sevir, a year older, to their paternal aunt, Ekaterina Vasilievna Petrova, a village teacher. She adopted the boys and gave them her surname.

Petrov received elementary education in Koshki and in 1926 he finished a seven-year school in the town of Melekes. At the school he was behind the class in mathematics. His teacher had to postpone his mathematics examination until the autumn, but, after studying hard, Petrov coped with all the tasks.

Petrov studied for one year at the pedagogical college of Melekes, but had to leave and find employment due to the retirement of his foster mother and consequent financial hardship. He worked in Saratov as a carpenter. In 1931 he and Sevir found work in Kazan in the building of an electric power station. Once, at a bookstall, Alexei came across a university textbook on mathematics. This decided his fate. In 1932 he entered the department of physics and mathematics at the University of Kazan.

This University was then on the crest of a remarkable wave. On returning from Göttingen, Nikolai Gurievich Chetaev worked out his famous "Kazan program". Nikolai Grigorievich Chebotarev founded an algebra school. Investigations in the theory of Riemann and generalized spaces by Petr Alexeyevich Shirokov again put the University in the foreground in the development of non-Euclidean geometry, following the tradition of Lobachevsky (see the Preface of B. L. Lapev to [1]). These people became mentors for Alexei Petrov. Veniamin Grigorievich Kopp, who was in the same year as Petrov, recalls: "You would often drop in to see him while he was working and try to talk to him, but Alexei answered all the questions with an inaudible mumble, characteristic of him at work. If you heard such a mumble, you should leave – it was impossible to tear him away from his work."

The gifted and persistent student drew the attention of Shirokov, who, at that time, was one of the most prominent Soviet geometers, studying and promoting the heritage of Lobachevsky, and holding international competitions which reflected the latest advances in non-Euclidean geometry.[1] The theme he offered to Petrov would later become the subject of Petrov's doctoral thesis, and its name would become the title of Petrov's classic monograph "Einstein spaces".

After graduating from the University Petrov worked as a teacher in the Kazan institute and at the same time did his PhD work under Shirokov. Sometimes he existed on just a piece of bread and a glass of water. The first research works of Petrov were devoted to solving the classical problem of geodesic mappings of Riemannian spaces with indefinite metrics. In his paper published in 1949 Petrov gives canonical forms of geodesically corresponding 3-dimensional Riemanian metrics and gives a sketch of a solution to the problem for one type of n-dimensional Riemanian space[2]. His studies were interrupted by the war. In December 1941, in a 40-degree frost, Petrov was sent to the front at Moscow to command a mortar detachment.

The defence of his dissertation passed off successfully in January, 1943 when battery commander Petrov received a short leave of absence for it. But in August that year he was badly wounded and returned home as a disabled soldier of the second highest group of disablement. He became ill and irritable. The disability status was cancelled after long treatment. After his demobilization Petrov worked as an Assistant Professor at the Kazan Aviation Institute, where, in 1945, he joined the Communist Party. Later that year he transferred to the Geometry section of the University. By that time Petrov had already married and had a little son Alyosha. Work at the University promised vast opportunities for pursuing science, and material comforts were of little importance to Petrov. Early in the post-war years Petrov completely formed his scientific predilections: the application of mathematical methods (geometry, group theory, algebra) to physical field theory. In 1946 he began research in the theory of Einstein spaces and in 1952–54 proved a remarkable theorem that afterwards brought him world fame. The theorem stated the existence of three types

[1] Among the laureates of these competitions were S. Lie, D. Hilbert, H. Weyl, E. Cartan, W. Killing and V. A. Fock. Lecturers included F. Klein, H. Poincaré and P. A. Shirokov [1].

[2] A general solution to the problem of geodesic mappings, which allows for an uncountable set of basic types of Riemanian spaces of arbitrary signature and any dimension (unlike previous results, which are all restricted to the three basic types) was given by A. V. Aminova in 1987.

of Einstein spaces. This "Petrov Classification" has proved immensely useful in finding and describing exact solutions of Einstein's field equations and in the study of gravitational radiation. It became the basis of Petrov's doctoral thesis and was defended with huge success in 1957 at the State University of Moscow (MSU).

Upon returning to Kazan, after the completion of his doctoral research, he headed the University admissions commission. But the overwrought heart could not put up with the new anxiety. On the diagnosis of cardiac infarction Petrov was placed in the old hospital just opposite the University. Hardly was the first danger past when he demanded to be discharged from the hospital. His appeal was refused. Then he persuaded his wife to bring him pen and paper and began to work in the hospital. He also consulted post-graduate students.

In October 1956 Petrov became Professor at the Geometry Chair of the University of Kazan and in 1960 he headed the Chair of Relativity Theory and Gravitation, which he helped to found at the Physical Faculty. He expended much effort on organizational and methodical work. He was a brilliant lecturer. He delivered a large number of special courses on gravitation, relativity theory and related disciplines. He devoted much time to working with students, post-graduates and Chair's assistants, headed several regular seminars, and was an editor of the periodic collection of works "Gravitation and the Relativity Theory" published by the University of Kazan. In a short time he created a scientific school that became famous.

In 1960 Petrov was elected Chairman of the gravity section in the Scientific and Technical Council of the Ministry of Higher and Secondary Education of the USSR and Chairman of the Soviet commission in the International Committee for General Relativity and Gravitation[3]. He occupied these posts till the end of his life and played a significant role in the organization and development of investigations in the field of gravitation. In the 1960s his two monographs, "Einstein spaces" and "New methods in the general theory of relativity", were published, summing up the results of his investigations of many years. They were translated into several foreign languages.

Petrov also showed keen interest in experimental confirmation of the general theory of relativity. For that purpose he organized an experimental laboratory at his department and later, during his work at the Institute of Theoretical Physics of the Ukrainian Academy of Sciences, he led together with V. B. Braginsky a series of experiments on observation of gravitational radiation. He was an active popularizer of relativity. His pamphlet "Space, time and matter" (1961) is characterized by a clear style easily understandable by the ordinary reader. It ran into several editions and was translated into Japanese.

On December 26, 1969 Petrov was elected a member of the Ukrainian Academy of Sciences (UAS) and in July 1970 he became the Head of the department of relativity theory and gravitation in the Institute of Theoretical Physics of UAS. At the time of his departure to Kiev his final illness had begun. In Kiev, although he had been ill for a long time, he continued to work in hospital. Every day, according to

[3] The records of the International Society on General Relativity and Gravitation, and its forerunner the International Committee, contain no mention of such a body within it as the "Soviet Commission". Petrov was co-opted to the Committee in 1962 [note by Malcolm A. H. MacCallum].

a special schedule, he met in his ward with post-graduate students and department workers and directed and consulted them.

In April 1972 Petrov was awarded the Lenin award for the series of works on "invariant group methods for investigations in the relativity theory". On May 9 that year he died in the hospital following an operation.

Not long before his death, late in 1971, Petrov came to Kazan. He was taciturn and estranged. During his conversation with V. G. Kopp he said "I live now in a beautiful room looking onto a garden. There is silence all around. That's the place where it is so wonderful to work". These words reveal completely Petrov's character as a man whose aspiration to knowledge was irrepressible.

References

1. Shirokov, P. A., *Izbrannye raboti po geometrii* ("Selected works on geometry"). Kazan University Press, Kazan 1966.
2. Aminova, A. V., *Alexei Zinovievich Petrov*. In the book: *Rasskazy o kazanskikh uchenykh*. Ed. V. V. Kuzmin. Tatarskoe knizhnoe izdatelstvo [Tatar Book Press] 1983.
3. Aminova, A. V., *A. Z. Petrov*. Lecture Notes in Theoretical and Mathematical Physics, Vol. 1, Part 1. Ed. A. V. Aminova. Kazan State University, "B.O.G.", Kazan 1996.
4. Aminova, A. V., Hall, G. S., *Alexei Zinovievich Petrov: Life and activity, 1910 – 1972*. Kazan University press, Kazan 2010, 32 pp.
5. See also:
   http://www.ksu.ru/petrov_school
   http://petrov2010.ksu.ru

General Relativity and Gravitation, Vol. 32, No. 8, 2000

The Classification of Spaces Defining Gravitational Fields[1]

A. Z. Petrov
(Kazan University, Chair of geometry.)

This paper gives the improved proofs of results obtained previously by the author and published first in 1951 [1]. More precisely, it is shown that for a V_4 defining a gravitational field and given in the form

$$ds^2 = g_{ij}dx^k dx^j \tag{1}$$

of a fundamental tensor, satisfying the field equations

$$R_{ij} = \kappa g_{ij} \tag{2}$$

(we will call such manifolds T_4), a classification can be set up by examining the algebraic structure of the curvature tensor.

1. BIVECTOR SPACE

Let us consider a given point P of our manifold T_4 and let us assign to it a local centered-affine geometry E_4. In this E_4 we pick out all the tensors which satisfy the conditions: 1) the number of covariant indices, as well as the number

[1] Original title: *Klassifikacya prostranstv opredelyayushchikh polya tyagoteniya. Uchenye Zapiski Kazanskogo Gosudarstvennogo Universiteta im. V. I. Ulyanova-Lenina [Scientific Proceedings of Kazan State University, named after V.I. Ulyanov-Lenin]*, **114** (8), 55–69 (1954). Jubilee (1804–1954) Collection. Reprinted with the kind permission of the Izdatelstvo Kazanskogo Gosudarstvennogo Universiteta [The Publishing House of the Kazan State University]. Translated by J. Jezierski, Institute of Theoretical Physics, University of Warsaw, and M. A. H. MacCallum, School of Mathematical Sciences, Queen Mary and Westfield College, Mile End Road, London E1 4NS, U.K. E-mail: M.A.H.MacCallum@qmw.ac.uk. The translators thank Dr. A. G. Polnarev for checking the final draft.

0001-7701/00/0800-1665$18.00/0 © 2000 Plenum Publishing Corporation

of contravariant indices, should be even, and 2) the co- and contra-variant indices can be grouped in disjoint antisymmetric pairs. We will consider each such pair as one collective index and we shall denote it with a Greek letter, in contrast to the indices for T_4 and E_4 for which we stick to Latin letters. In this way, from the number of possible values which the collective index can take, we obtain a manifold of dimension $N = n(n-1)/2$ (6 dimensions for $n = 4$), it being known that the tensors of E_4 satisfying the above conditions define in this space tensors with half the rank.

One can say that to each point in T_4 is assigned, in this way, a local 6-dimensional centered-affine geometry with the group[2]

$$\begin{aligned} \eta^{\alpha'} &= A^{\alpha'}_{\alpha}\eta^{\alpha}, & \eta^{\alpha} &= A^{\alpha}_{\alpha'}\eta^{\alpha'}, \\ |A^{\alpha'}_{\alpha}| &\neq 0, & A^{\alpha}_{\beta'}A^{\beta'}_{\gamma} &= \delta^{\alpha}_{\gamma}, \end{aligned} \tag{3}$$

In fact, if we order the collective indices (choosing only one pair from the two possibilities ij and ji) we obtain the six possible collective indices. Let us stick to, for example, the following enumeration

$$1-14, \quad 2-24, \quad 3-34, \quad 4-23, \quad 5-31, \quad 6-12.$$

Now let us consider the transformation of the components T^{ij} of some in general nonsimple bivector,

$$T^{i'j'} = A^{i'j'}_{ij}T^{ij},$$

setting $A^{\alpha'}_{\alpha} = 2A^{[i'j']}_{ij}$ (where $A^{i'}_{i} = (\partial x^{i'}/\partial x^{i})_P$) we obtain in collective indices the transformation

$$T^{\alpha'} = A^{\alpha'}_{\alpha}T^{\alpha},$$

i.e., the set of bivectors of T_n (for this question the dimension is not important) gives in E_N a set of contravariant vectors under the condition that the relations (3) are satisfied. As for these relations, they can be checked directly by changing to Latin indices.

We shall call the manifold obtained *bivector space*. In the sequel, the curvature tensor of T_4 will be of special interest. Corresponding to it in bivector space is a symmetric tensor of rank 2

$$R_{ijkl} \longrightarrow R_{\alpha\beta} = R_{\beta\alpha}.$$

[2] Translators' note: Here and elsewhere, minor misprints in the original have been corrected.

In each local E_6 we can introduce a metric, using for this purpose any tensor in T_4 having the properties

$$M_{ijkl} = M_{klij} = -M_{jikl} = -M_{ijlk},$$

on condition that the corresponding rank two tensor in E_6 is nonsingular. Let us choose as such a fundamental tensor in E_6 the tensor

$$g_{ikjl} = g_{ij}g_{kl} - g_{il}g_{kj} \longrightarrow g_{\alpha\beta} = g_{\beta\alpha}. \tag{4}$$

It is easy to check that $g_{\alpha\beta}$ gives a nondegenerate definition of measure, because $|g_{ij}| \neq 0$ and[3]

$$|g_{\alpha\beta}| = p|g_{ij}|^{2n}, \quad p \neq 0.$$

For a definite g_{ij}, $g_{\alpha\beta}$ will also be definite; for an indefinite g_{ij}, the tensor $g_{\alpha\beta}$ will also in general be indefinite. We note that in the sequel we will consider only those gravitational fields which correspond to a real distribution of matter in space; for this it is necessary [2] that at any given point of T_4 the fundamental tensor g_{ij} in a real coordinate system can be reduced to the form:

$$(g_{ij}) = \begin{pmatrix} -1 & & & \\ & -1 & & \\ & & -1 & \\ & & & 1 \end{pmatrix}, \tag{5}$$

i.e. in this way we arrive at the so-called Minkowski spacetime. Then from (4) it follows that for a frame corresponding to the matrix (5) the fundamental tensor in R_6 takes the form

$$(g_{\alpha\beta}) = \begin{pmatrix} -1 & & & & & \\ & -1 & & & & \\ & & -1 & & & \\ & & & 1 & & \\ & & & & 1 & \\ & & & & & 1 \end{pmatrix}, \quad |g_{\alpha\beta}| = -1, \tag{6}$$

i.e. in fact, $g_{\alpha\beta}$ is an indefinite tensor.

[3] This equation is wrong, though the conclusion is correct. We could not see any simple correction. (Translators)

2. CLASSIFICATION OF T_4

A series of the most interesting problems arising in the study of Riemannian manifolds is related with the curvature tensor of a V_n. With the help of this tensor, as is known, one can introduce the notion of the curvature of a V_n in a given two-dimensional section[4] at a given point, or, which is the same, the Gaussian curvature of a two-dimensional surface of geodesics at a given point

$$K = \frac{R_{ijkl}V^{ij}V^{kl}}{g_{pqrs}V^{pq}V^{rs}}, \tag{7}$$

where g_{pqrs} is given by (4), and the two-dimensional section defined by vectors $\underset{1}{V}{}^{i}$, $\underset{2}{V}{}^{j}$, can be characterized by the simple bivector $V^{ij} = \underset{[1}{V}{}^{i}\underset{2]}{V}{}^{j}$. We introduce a *generalized curvature of* V_n which can be obtained if in (7) we relax the requirement of the simplicity of the bivector V^{ij}. The generalized invariant K at some point of V_n will be a homogeneous function of degree zero of the bivector V^{ij} (in general, not simple) and obviously, it makes sense in bivector space, where it can be written in the form

$$K = \frac{R_{\alpha\beta}V^{\alpha}V^{\beta}}{g_{\alpha\beta}V^{\alpha}V^{\beta}}. \tag{8}$$

We consider the problem of finding the critical values of K or, equivalently, those vectors V^{α} in R_N which correspond to the critical values of K. We shall call these critical values of K *stationary curvatures of* V_n and the corresponding bivectors V^{α}, *stationary sections* of V_n. Hence the problem is reduced to the determination of *unconstrainedly stationary vectors* V^{α} in bivector space based on the necessary and sufficient conditions for stationarity:

$$\frac{\partial K}{\partial V^{\alpha}} = 0. \tag{9}$$

It is necessary to take into account that for indefinite tensors g_{ij}, $g_{\alpha\beta}$ is also indefinite, and therefore the appearance of null[5] stationary sections is possible

$$g_{\alpha\beta}V^{\alpha}V^{\beta} = 0. \tag{10}$$

Initially we exclude this case, but return to it below.

[4] Translators' note: literally, 'direction'.

[5] Translators' note: literally, "isotropic", a term no longer much used with this meaning.

If (10) does not hold, then (9) leads to the conclusion that

$$(R_{\alpha\beta} - K g_{\alpha\beta}) V^{\beta} = 0, \tag{11}$$

i.e. the stationary directions of the V_n will be the principal axes of the tensor $R_{\alpha\beta}$ in bivector space, and the stationary curvatures of the V_n will be the characteristic values of the secular equation

$$|R_{\alpha\beta} - K g_{\alpha\beta}| = 0. \tag{12}$$

Let us assume now that (10) holds for stationary V^{α}. Because we are interested only in the K such that (9) is satisfied, K is a continuous function of V^{α} and it is therefore necessary that the following condition should be fulfilled

$$R_{\alpha\beta} V^{\alpha} V^{\beta} = 0.$$

So it is possible to calculate the value of K for a null stationary direction V^{α}, assuming continuity of K with respect to the variable V^{α}, in the following manner:

$$K(V^{\alpha}) = \lim_{\mathrm{d}V^{\alpha} \to 0} K(V^{\alpha} + \mathrm{d}V^{\alpha}).$$

If for a given V^{α} we write

$$\varphi = g_{\alpha\beta} V^{\alpha} V^{\beta}, \quad \psi = R_{\alpha\beta} V^{\alpha} V^{\beta}, \tag{13}$$

then for the null stationary V^{α}

$$K(V^{\alpha}) = \lim_{\mathrm{d}V^{\alpha} \to 0} \frac{\psi(V^{\alpha} + \mathrm{d}V^{\alpha}) - \psi(V^{\alpha})}{\varphi(V^{\alpha} + \mathrm{d}V^{\alpha}) - \varphi(V^{\alpha})} = \lim \frac{\sum_{\sigma} \frac{\partial}{\partial V^{\sigma}} \psi \, \mathrm{d}V^{\sigma} + \cdots}{\sum_{\sigma} \frac{\partial}{\partial V^{\sigma}} \varphi \, \mathrm{d}V^{\sigma} + \cdots}$$

and, since the limit must be independent of how $\mathrm{d}V^{\alpha}$ changes,

$$K(V^{\alpha}) = \frac{\frac{\partial}{\partial V^{\sigma}} \psi}{\frac{\partial}{\partial V^{\sigma}} \phi} = \frac{R_{\sigma} V^{\alpha}}{g_{\sigma} V^{\alpha}},$$

i.e. we again obtain (11).

The determination of stationary curvatures and directions in R_N leads to the investigation of the pair of quadratic forms (13). Hence reduction of this pair of forms in real space to canonical form gives a classification for the curvature tensor of the V_n at a given point and in a neighbourhood which contains this point and in which the *characteristics of the K-matrix*

$$\|R_{\alpha\beta} - K g_{\alpha\beta}\| \tag{14}$$

remain unchanged. To each type of characteristic of the matrix (14) corresponds a particular type of gravitational field. This defines the desired classification of the T_4.

Using a real transformation, the matrix $\|g_{\alpha\beta}\|$ can always be reduced to the form (6), and it remains to simplify the matrix $\|R_{\alpha\beta}\|$ by applying a real orthogonal transformation.

Theorem 1. The matrix $\|R_{\alpha\beta}\|$ for an orthogonal frame (6) will be symmetrically-double.

For the frame (5), the field equations take the form:

$$\sum_k e_k R_{ikjk} = \kappa g_{ij}; \quad e_k = \pm 1,$$

i.e. if $i = j$

$$\sum_k e_k R_{ikik} = \kappa e_i$$

and if $i \neq j$

$$e_k R_{ikjk} + e_l R_{iljl} = 0 \quad (i, j, k, l \neq).$$

Expressing all these relations in terms of collective indices in bivector space and taking into account the enumeration introduced in §1, we obtain for the matrix the expressions:

$$\|R_{\alpha\beta}\| = \left\| \begin{array}{c|c} M & N \\ \hline N & -M \end{array} \right\|$$

$$M = \left\| \begin{matrix} m_{11} & m_{12} & m_{13} \\ m_{21} & m_{22} & m_{23} \\ m_{31} & m_{32} & m_{33} \end{matrix} \right\|$$

$$N = \left\| \begin{matrix} n_{11} & n_{12} & n_{13} \\ n_{21} & n_{22} & n_{23} \\ n_{31} & n_{32} & n_{33} \end{matrix} \right\|, \qquad \begin{matrix} m_{\alpha\beta} = m_{\beta\alpha} \\ n_{\alpha\beta} = n_{\beta\alpha} \\ (\alpha, \beta = 1, 2, 3) \end{matrix} \tag{15}$$

where $\sum_{i=1}^{3} m_{ii} = \kappa$, and $\sum_{i=1}^{3} n_{ii} = 0$, due to the Ricci identity, which completes the proof. We note that the same type of matrix, but under the additional assumption of orthogonality, was obtained by V.F. Kagan by considering the group of Lorentz transformations [3]. The same type of matrix was considered, again under the orthogonality condition, by Ya.S. Dubnov [4] and A.M. Lopshitz [5]. The fact proved in the above theorem holds for any orthogonal frame, and therefore, taking into account that an orthogonal frame is determined for $n = 4$ only up to 6 degrees of freedom, one can consider further simplifications of the matrix by choice of six rotations.

First we prove a theorem which at first glance strongly constrains the number of possible types of characteristic of the matrix (14).

Theorem 2. The characteristics of the K-matrix (14) always consist of two identical parts.

We reduce the matrix (14) to a simpler form by so-called elementary transformations which, as is well known, do not change the elementary divisors of a matrix or, therefore, its characteristics. Let us express this matrix in the form

$$\left\| \begin{array}{c|c} m_{\alpha\beta} + K\delta_{\alpha\beta} & n_{\alpha\beta} \\ \hline n_{\alpha\beta} & -m_{\alpha\beta} - K\delta_{\alpha\beta} \end{array} \right\|$$

where $\delta_{\alpha\beta}$ is the Kronecker symbol. Adding to each of the first three columns the corresponding columns among the last three, multiplied by i, we derive the equivalent matrix

$$\left\| \begin{array}{c|c} m_{\alpha\beta} + in_{\alpha\beta} + K\delta_{\alpha\beta} & n_{\alpha\beta} \\ \hline -i(m_{\alpha\beta} + in_{\alpha\beta} + K\delta_{\alpha\beta}) & -m_{\alpha\beta} - K\delta_{\alpha\beta} \end{array} \right\| ;$$

adding to each of the last three rows the corresponding row among the first three, multiplied by i, we reduce the matrix to the form:

$$\left\| \begin{array}{c|c} m_{\alpha\beta} + in_{\alpha\beta} + K\delta_{\alpha\beta} & n_{\alpha\beta} \\ \hline 0 & -m_{\alpha\beta} + in_{\alpha\beta} - K\delta_{\alpha\beta} \end{array} \right\| .$$

Finally, multiplying the first three columns by $i/2$ and adding them to the corresponding last three columns, and doing the same with the last three rows, we end up with the matrix[6]

$$\left\| \begin{array}{c|c} m_{\alpha\beta} + in_{\alpha\beta} + K\delta_{\alpha\beta} & 0 \\ \hline 0 & -m_{\alpha\beta} + in_{\alpha\beta} - K\delta_{\alpha\beta} \end{array} \right\| \equiv \left\| \begin{array}{c|c} P(K) & 0 \\ \hline 0 & -\overline{P}(K) \end{array} \right\| ,$$

equivalent to the K-matrix (14). Thus the problem becomes the investigation of two three-dimensional matrices $P(K)$ and $\overline{P}(K)$, whose corresponding elements are complex conjugate. Thence we deduce that the elementary divisors of these two matrices are also complex conjugate, and therefore their characteristics have the same forms. In this way the characteristics of our K-matrix divide into two parts which repeat each other—the theorem holds.

We note that the principal directions and invariant subspaces[7] of the K-matrix should also be complex conjugate in pairs.

Now we can carry out the classification of gravitational fields, which is expressed by

Theorem 3. There exist three and only three types of gravitational field.

The three-dimensional matrix $P(K)$ can have only one of three possible types of characteristic: [1 1 1], [2 1], [3], if we omit the cases where some of the elementary divisors have the same roots and therefore some of the numbers inside the square brackets should be enclosed in round brackets (for example, [(1 1)1], [(2 1)], etc.).

The characteristics of $\overline{P}(K)$ must take a similar form. Thus the characteristics of the K-matrix will be written

$$1.[1\overline{1}, 1\overline{1}, 1\overline{1}];\ 2 : [2\overline{2}, 1\overline{1}];\ 3 : [3\overline{3}], \tag{16}$$

where an overlined number denotes the degree of the elementary divisor whose root is complex conjugate to the root of the elementary divisor whose degree is given by the previous number.

Each of these types of gravitational field has to be considered separately below, and, which is most important, one has to get the canonical form of the matrix $\|R_{\alpha\beta}\|$ for each of these types.

[6] In the lower right corner on both sides of the equation the sign was opposite in the original text. Corrected by the translator.

[7] Translator's note: literally, 'bundles'.

3. THE CANONICAL FORM OF THE MATRIX $\|R_{\alpha\beta}\|$

Let us consider the first type, with characteristic $[1\bar{1}, 1\bar{1}, 1\bar{1}]$. Because in this case the characteristics are of simple type, the tensor $\|R_{\alpha\beta}\|$ has 6 non-null mutually orthogonal principal directions [6]. These directions in bivector space at a given point of T_4 give bivectors of specific structure, as will be shown.

Let us denote the components of the vectors of a real orthogonal frame at a point of T_4 by $\underset{k}{\xi}^i$ $(k, i = 1, 2, 3, 4)$; the simple bivectors $\underset{[k}{\xi}^i \underset{l]}{\xi}^j (k \neq l)$, describing the 2-dimensional planes[8] given by vectors of the frame, we will, for simplicity, denote $\underset{kl}{\xi}^{ij}$. In bivector space these simple bivectors describe 6 independent non-null mutually-orthogonal coordinate vectors $\underset{\sigma}{\xi}^\alpha = \delta^\alpha_\sigma$, and any vector in R_6, in particular the vectors of the principal directions of $R_{\alpha\beta}$, can be expressed in terms of these vectors.

We shall show that as vectors of the principal directions (these being uniquely defined only in the case when the roots of the characteristic polynomial (12) are distinct) one can take vectors of the form:

$$W^\alpha = \lambda(\underset{1}{\xi}^\alpha \pm i \underset{4}{\xi}^\alpha) + \mu(\underset{2}{\xi}^\alpha \pm i \underset{5}{\xi}^\alpha) + \nu(\underset{3}{\xi}^\alpha \pm i \underset{6}{\xi}^\alpha). \tag{17}$$

In fact the condition that W^α determines a principal direction of the tensor $R_{\alpha\beta}$ can be written:

$$(R_{\alpha\beta} - K g_{\alpha\beta}) W^\beta = 0. \tag{18}$$

However this system of 6 equations, by the double symmetry of the K-matrix, reduces to the three equations:

$$(m_{s1} \pm i n_{s1} + k)\lambda + (m_{s2} \pm i n_{s2})\mu + (m_{s3} \pm i n_{s3})\nu = 0 \ (s = 1, 2, 3).$$

In order that λ, μ, ν be a nonzero solution of this system it is necessary and sufficient that K should be a root of one of the equations

$$|P(K)| = 0, \qquad |\overline{P}(K)| = 0, \tag{19}$$

i.e. a root of the characteristic polynomial (12), which proves the theorem.

To a vector W^α (17) of the manifold R_6 at a given point of T_4 corresponds the bivector of maximal rank

[8] Translators' note: literally, "area elements".

$$W^{ij} = \lambda(\underset{14}{\xi}^{ij} \pm i\underset{23}{\xi}^{ij}) + \mu(\underset{24}{\xi}^{ij} \pm i\underset{31}{\xi}^{ij}) + \nu(\underset{34}{\xi}^{ij} \pm i\underset{12}{\xi}^{ij}). \tag{20}$$

It is not hard to see that under a real orthogonal transformation W^{ij} is mapped into a bivector of the same type, while $\lambda, \mu, \nu \longrightarrow \overset{*}{\lambda}, \overset{*}{\mu}, \overset{*}{\nu}$, so that the norm of the bivector is invariant

$$\lambda^2 + \mu^2 + \nu^2 = \overset{*}{\lambda}{}^2 + \overset{*}{\mu}{}^2 + \overset{*}{\nu}{}^2.$$

Let the roots of (12), $\underset{s}{K}$ $(s = 1, 2, 3)$ correspond to the vectors of principal directions $\underset{s}{W}^{\alpha}$; then the roots $\underset{s+3}{K}$, by the previous considerations, must correspond to $\overline{\underset{s}{W}}{}^{\alpha}$ with a suitable numbering of roots. To the root $\underset{1}{K}$ corresponds the bivector

$$\underset{1}{W}^{pq} = \underset{1}{\lambda}(\underset{14}{\xi}^{pq} + i\underset{23}{\xi}^{pq}) + \underset{1}{\mu}(\underset{24}{\xi}^{pq} + i\underset{31}{\xi}^{pq}) + \underset{1}{\nu}(\underset{34}{\xi}^{pq} + i\underset{12}{\xi}^{pq}),$$

and to $\underset{4}{K}$ the bivector

$$\underset{4}{W}^{pq} = \overline{\underset{1}{\lambda}}(\underset{14}{\xi}^{pq} - i\underset{23}{\xi}^{pq}) + \overline{\underset{1}{\mu}}(\underset{24}{\xi}^{pq} - i\underset{31}{\xi}^{pq}) + \overline{\underset{1}{\nu}}(\underset{34}{\xi}^{pq} - i\underset{12}{\xi}^{pq}).$$

Let us decompose the bivector $\underset{1}{W}^{pq}$ in the form of a sum of two real bivectors $\underset{1}{V}^{pq} + i\underset{1}{\overset{*}{V}}{}^{pq}$; then

$$\underset{4}{W}^{pq} = \underset{1}{V}^{pq} - i\underset{1}{\overset{*}{V}}{}^{pq}.$$

Let

$$\lambda = \underset{1}{a} + i\underset{1}{b}, \quad \mu = \underset{2}{a} + i\underset{2}{b}, \quad \nu = \underset{3}{a} + i\underset{3}{b},$$

where $\underset{s}{a}$, $\underset{s}{b}$ are real numbers $(s = 1, 2, 3)$ and therefore

$$\underset{1}{V}^{pq} = \underset{1}{a}\,\underset{14}{\xi}^{pq} + \underset{2}{a}\,\underset{24}{\xi}^{pq} + \underset{3}{a}\,\underset{34}{\xi}^{pq} - \underset{1}{b}\,\underset{23}{\xi}^{pq} - \underset{2}{b}\,\underset{31}{\xi}^{pq} - \underset{3}{b}\,\underset{12}{\xi}^{pq};$$

$$\overset{*}{\underset{1}{V}}{}^{pq} = \underset{1}{b}\,\underset{14}{\xi}^{pq} + \underset{2}{b}\,\underset{24}{\xi}^{pq} + \underset{3}{b}\,\underset{34}{\xi}^{pq} + \underset{1}{a}\,\underset{23}{\xi}^{pq} + \underset{2}{a}\,\underset{31}{\xi}^{pq} + \underset{3}{a}\,\underset{12}{\xi}^{pq}.$$

Because $\underset{1}{W}^{\alpha}$ is a nonnull vector in R_6, we can always assume that it is a unit vector

$$g_{\alpha\beta}\underset{1}{W}^{\alpha}\underset{1}{W}^{\beta} = 1,$$

and we conclude that:

$$\sum_{s=1}^{3} \underset{s}{a}\underset{s}{b} = 0, \tag{21}$$

$$\sum_{s=1}^{3} \underset{s}{b}^{2} - \underset{s}{a}^{2} > 0. \tag{22}$$

Now one can assert that: 1) the real bivectors $\underset{1}{V}^{pq}$ and $\overset{*}{\underset{1}{V}}{}^{pq}$ are singly foliated.

In fact, writing down the condition for simplicity, we get (21).

2) they are 0-parallel.

They cannot be $\frac{2}{2}$-parallel, because this is possible only under the condition that the coefficients of the same $\underset{ij}{\xi}^{pq}$ are proportional; in our case they would have to be zero. For example:

$$\frac{\underset{1}{a}}{\underset{1}{b}} = -\frac{\underset{1}{b}}{\underset{1}{a}}, \quad \underset{1}{a}^{2} + \underset{1}{b}^{2} = 0;$$

they cannot be $\frac{1}{2}$-parallel, because then $\underset{1}{W}^{\alpha}$ would be a singly foliated complex bivector, but, writing down the simplicity condition, we end up with a contradiction with (21) and (22). Thus we are left only with the above possibility;

3) These bivectors are $\frac{2}{2}$-*perpendicular*. For this it is necessary and sufficient, for arbitrary i, j, to satisfy the equations

$$\underset{1}{V}_{is}\overset{*}{\underset{1}{V}}{}^{sj} = 0.$$

It is easy to see that they lead to (21) and therefore hold.

Let us consider the simple bivector $\underset{1}{V}^{pq}$. Its norm, from (22), is

$$g_{\alpha\beta}\underset{1}{V}^{\alpha}\underset{1}{V}^{\beta} = \sum_{s=1}^{3} \underset{s}{b}^2 - \underset{s}{a}^2 > 0.$$

In the plane of this simple bivector we can always choose two real orthogonal and nonnull vectors η^p, ν^p. Then the norm of our bivector may be expressed in the form

$$2\eta_p\eta^p \cdot \nu_q\nu^q,$$

and therefore these two vectors are both spacelike or both timelike. Their norms cannot be > 0, because if we take these two orthogonal real vectors as basis vectors, we reach a contradiction with the law of inertia of quadratic forms. Consequently, these two vectors have negative norms. Because of that, after renormalizing we can take them as the vectors $\overset{*}{\underset{2}{\xi}}{}^i$, $\overset{*}{\underset{3}{\xi}}{}^i$ of a new real orthogonal frame.

In precisely the same way, we can define two real nonnull orthogonal vectors (orthogonal to each other and to $\overset{*}{\underset{2}{\xi}}{}^i$, $\overset{*}{\underset{3}{\xi}}{}^i$) in the plane of $\overset{*}{\underset{1}{V}}{}^{pq}$, but now with opposite signs of the norms, so that

$$g_{\alpha\beta}\overset{*}{\underset{1}{V}}{}^{\alpha}\overset{*}{\underset{1}{V}}{}^{\beta} < 0;$$

we call these vectors $\overset{*}{\underset{1}{\xi}}{}^i$, $\overset{*}{\underset{4}{\xi}}{}^i$. In this coordinate system

$$\overset{*}{\underset{1}{W}}{}^{pq} = \underset{14}{\xi}{}^{pq} + i\,\underset{23}{\xi}{}^{pq},$$

$$\overset{*}{\underset{4}{W}}{}^{pq} = \underset{14}{\xi}{}^{pq} - i\,\underset{23}{\xi}{}^{pq}.$$

We note that the frame $\{\overset{*}{\xi}\}$ is chosen up to a rotation in the plane $\{\overset{*}{\underset{2}{\xi}}\overset{*}{\underset{3}{\xi}}\}$ and a Lorentz rotation in the plane $\{\overset{*}{\underset{1}{\xi}}\overset{*}{\underset{4}{\xi}}\}$. We are, of course, only interested in the bivectors $\underset{\sigma}{W}^{pq}$ up to a scalar factor.

Now, writing down the orthogonality condition for $\overset{*}{\underset{1}{W}}{}^{pq}$ and $\overset{*}{\underset{2}{W}}{}^{pq}$, we obtain, obviously, that the bivector of the second principal direction must have the form:

$$\overset{*}{\underset{2}{W}}{}^{pq} = \overset{*}{\underset{2}{\mu}}(\overset{*}{\underset{24}{\xi}}{}^{pq} + i\overset{*}{\underset{31}{\xi}}{}^{pq}) + \overset{*}{\underset{2}{\nu}}(\overset{*}{\underset{34}{\xi}}{}^{pq} + i\overset{*}{\underset{12}{\xi}}{}^{pq}).$$

Using the freedom of the frame shown above we perform the rotations:[9]

$$\underset{1}{\xi}{}^{p} = \cosh\varphi\overset{*}{\underset{1}{\xi}}{}^{p} + \sinh\varphi\overset{*}{\underset{4}{\xi}}{}^{p},$$

$$\underset{4}{\xi}{}^{p} = \sinh\varphi\overset{*}{\underset{1}{\xi}}{}^{p} + \cosh\varphi\overset{*}{\underset{4}{\xi}}{}^{p};$$

$$\underset{2}{\xi}{}^{p} = \cos\psi\overset{*}{\underset{2}{\xi}}{}^{p} + \sin\psi\overset{*}{\underset{3}{\xi}}{}^{p}$$

$$\underset{3}{\xi}{}^{p} = -\sin\psi\overset{*}{\underset{2}{\xi}} + \cos\psi\underset{3}{\xi}{}^{p}.$$

After these transformations $\underset{1}{W}$ will have the same form as before, and consequently $\underset{2}{W}$ too appears as

$$\underset{2}{\tilde{W}}{}^{pq} = \underset{2}{\tilde{\mu}}(\underset{24}{\tilde{\xi}}{}^{pq} + i\underset{31}{\tilde{\xi}}{}^{pq}) + \underset{2}{\tilde{\nu}}(\underset{34}{\tilde{\xi}}{}^{pq} + i\underset{12}{\tilde{\xi}}{}^{pq}).$$

where

$$\underset{2}{\tilde{\nu}} = \sin\psi\cosh\varphi + p\cos\psi\cosh\varphi + q\sin\psi\sinh\varphi$$
$$+ i(\cos\psi\sinh\varphi + q\cos\psi\cosh\varphi - p\sin\psi\sinh\varphi),$$

$$p + iq = \frac{\overset{*}{\underset{2}{\nu}}}{\overset{*}{\underset{2}{\mu}}}$$

and where $\overset{*}{\underset{2}{\mu}}$ can be assumed different from zero, since otherwise we would satisfy $\varphi = \psi = 0$. One can find real φ and ψ for every $\underset{2}{\tilde{\nu}} = 0$. Now the frame is uniquely determined, and in this frame, if we take account of the orthogonality of $\underset{1}{W}, \underset{2}{W}, \underset{3}{W}$, these bivectors will take the form (up to a scalar factor):

[9] The notation for the hyperbolic functions in the original was ch and sh rather than the cosh and sinh usual in English (Translator).

$$\underset{1}{W}^{pq} = \underset{14}{\xi}^{pq} + i\,\underset{23}{\xi}^{pq},$$
$$\underset{2}{W}^{pq} = \underset{24}{\xi}^{pq} + i\,\underset{31}{\xi}^{pq},$$
$$\underset{3}{W}^{pq} = \underset{34}{\xi}^{pq} + i\,\underset{12}{\xi}^{pq}$$

and, from the complex conjugacy shown above,

$$\underset{4}{W}^{pq} = \overline{\underset{1}{W}^{pq}}, \quad \underset{5}{W}^{pq} = \overline{\underset{2}{W}^{pq}}, \quad \underset{6}{W}^{pq} = \overline{\underset{3}{W}^{pq}}.$$

Now, writing out the condition (18) for each of these bivectors and taking into account that

$$\underset{\alpha}{\xi}^{\sigma} = \delta^{\sigma}_{\alpha},$$

we easily obtain

$$m_{ii} = -\underset{i}{\alpha}, \quad m_{ij} = 0, \quad n_{ii} = -\underset{i}{\beta}, \quad n_{ij} = 0, \quad (i = 1, 2, 3;\ i \neq j)$$

and, therefore, for the first type of T_4 we obtain the following canonical form of the matrix:

$$(R_{\alpha\beta}) = \left\|\begin{array}{ccc|ccc} -\underset{1}{\alpha} & & & -\underset{1}{\beta} & & \\ & -\underset{2}{\alpha} & & & -\underset{2}{\beta} & \\ & & -\underset{3}{\alpha} & & & -\underset{3}{\beta} \\ \hline -\underset{1}{\beta} & & & \underset{1}{\alpha} & & \\ & -\underset{2}{\beta} & & & \underset{2}{\alpha} & \\ & & -\underset{3}{\beta} & & & \underset{3}{\alpha} \end{array}\right\| \tag{23}$$

and the real parts of the stationary curvatures are related by the conditions

$$\sum_{1}^{3} \underset{s}{\alpha} = \kappa \tag{24}$$

and the imaginary parts, because of the Ricci identity

$$R_{1423} + R_{1234} + R_{1342} = 0,$$

obey the conditions

$$\sum_{1}^{3} \underset{s}{\beta} = 0. \tag{25}$$

Let us now consider a T_4 with a characteristic of the second type $[21, \overline{21}]$. As was shown above (§2), for the principal directions and invariant subspaces of the K-matrix, we can take the principal directions and invariant subspaces of the matrices $P(K)$ and $\overline{P}(K)$. From this it follows that it is sufficient to consider, for example, a matrix $P(K)$ with the characteristic [21].

With such a characteristic, the tensor $P_{\alpha\beta} = -m_{\alpha\beta} + in_{\alpha\beta}$ of the three-space has [6] one non-null principal direction

$$(P_{\alpha\beta} - \underset{1}{K}\,\delta_{\alpha\beta})\underset{1}{W}^{\beta} = 0, \tag{26}$$

and one null principal direction orthogonal to $\underset{1}{W}$

$$(P_{\alpha\beta} - \underset{2}{K}\delta_{\alpha\beta})\underset{2}{W}^{\beta} = 0. \tag{27}$$

Besides these, there is a null vector $\underset{3}{W}^{\beta}$, orthogonal to $\underset{1}{W}^{\beta}$ but not orthogonal to $\underset{2}{W}^{\beta}$, that together with the latter forms an invariant plane $\{\underset{2}{W}, \underset{3}{W}\}$ of the tensor $P_{\alpha\beta}$, which can be expressed by the condition:

$$(P_{\alpha\beta} - \underset{2}{K}\delta_{\alpha\beta})\underset{3}{W}^{\beta} = \sigma \underset{2}{W}_{\alpha}, \tag{28}$$

where σ is any nonzero scalar; we can choose its value. This arbitrariness is a consequence of the result that $\overline{\underset{2}{W}}$, $\overline{\underset{3}{W}}$, being null, can be multiplied by any number without changing their norms.

Every principal direction or subspace of $P_{\alpha\beta}$ will determine a corresponding principal direction or subspace of the tensor $R_{\alpha\beta}$; all of them will be determined by bivectors of the type (17).

Let the root $\underset{1}{K}$ correspond to a simple elementary divisor $(K - \underset{1}{K})$ of the K-matrix field and to a principal direction given by $\underset{1}{W}^{\alpha}$. Because this bivector is nonnull one can apply all the arguments given for the $\underset{1}{W}^{\alpha}$ considered in the previous case. Consequently one can choose a real frame such that referred to it

$$\underset{1}{W}^{pq} = \underset{14}{\xi}^{pq} + i\underset{23}{\xi}^{pq}.$$

This frame is fixed up to a rotation in the plane $\{\underset{2}{\bar{\xi}}\underset{3}{\bar{\xi}}\}$ and a Lorentz rotation in the plane $\{\underset{1}{\bar{\xi}}\underset{4}{\bar{\xi}}\}$. Since the bivectors $\underset{2}{W}^{pq}$ and $\underset{3}{W}^{pq}$ must be orthogonal to $\underset{1}{W}^{pq}$, they must take the form:

$$\underset{2}{W}^{pq} = \underset{2}{\mu}(\underset{24}{\xi}^{pq} + i\underset{31}{\xi}^{pq}) + \underset{2}{\nu}(\underset{34}{\xi}^{pq} + i\underset{12}{\xi}^{pq}),$$
$$\underset{3}{W}^{pq} = \underset{3}{\mu}(\underset{24}{\xi}^{pq} + i\underset{31}{\xi}^{pq}) + \underset{3}{\nu}(\underset{34}{\xi}^{pq} + i\underset{12}{\xi}^{pq}).$$

The nullity condition for these bivectors leads to the relations

$$\underset{2}{\mu}^2 + \underset{2}{\nu}^2 = 0, \quad \underset{3}{\mu}^2 + \underset{3}{\nu}^2 = 0,$$

i.e.

$$\underset{2}{\nu} = e_1 i\underset{2}{\mu}, \quad \underset{3}{\nu} = e_2 i\underset{3}{\mu},$$

where e_1 and e_2 are ± 1. Finally, writing down the fact that the bivectors cannot be orthogonal, we obtain that $e_1 = -e_2$. Consequently one can, for example, assume that

$$\underset{2}{W}^{pq} = \underset{24}{\xi}^{pq} + i\underset{31}{\xi}^{pq} + i(\underset{34}{\xi}^{pq} + i\underset{12}{\xi}^{pq}),$$
$$\underset{3}{W}^{pq} = \lambda\{\underset{24}{\xi}^{pq} + i\underset{31}{\xi}^{pq} - i(\underset{34}{\xi}^{pq} + i\underset{12}{\xi}^{pq}), \}$$

where λ is an arbitrary nonzero scalar factor.

Now it remains for us to write down conditions analogous to (26), (27) and (28) for the tensor $R_{\alpha\beta}$, taking into account again that, as in the previous case, $\underset{\nu}{\xi}^{\alpha} = \delta^{\alpha}_{\nu}$. Those conditions will take the form:

$$(R_{\alpha\beta} - \underset{1}{K} g_{\alpha\beta})\underset{1}{W^{\beta}} = 0,$$
$$(R_{\alpha\beta} - \underset{2}{K} g_{\alpha\beta})\underset{2}{W^{\beta}} = 0,$$
$$(R_{\alpha\beta} - \underset{2}{K} g_{\alpha\beta})\underset{3}{W^{\beta}} = \sigma g_{\alpha\beta}\underset{2}{W^{\beta}}.$$

The tensor $g_{\alpha\beta}$ is given by the matrix (6). Taking here $\alpha = 1, 2, \ldots, 6$, we easily obtain that the *matrix* $R_{\alpha\beta}$ (11) will take the form:

$$(R_{\alpha\beta}) = \left\| \begin{array}{ccc|ccc} -\underset{1}{\alpha} & 0 & 0 & -\underset{1}{\beta} & 0 & 0 \\ 0 & -\underset{2}{\alpha}+\sigma & 0 & 0 & -\underset{2}{\beta} & \sigma \\ 0 & 0 & -\underset{2}{\alpha}-\sigma & 0 & \sigma & -\underset{2}{\beta} \\ \hline -\underset{1}{\beta} & 0 & 0 & \underset{1}{\alpha} & 0 & 0 \\ 0 & -\underset{2}{\beta} & \sigma & 0 & \underset{2}{\alpha}-\sigma & 0 \\ 0 & \sigma & -\underset{2}{\beta} & 0 & 0 & \underset{2}{\alpha}+\sigma \end{array} \right\|, \quad \sigma \neq 0. \tag{29}$$

Here σ may be chosen freely, but $\neq 0$; $\underset{s}{\alpha}$ and $\underset{s}{\beta}$, as in the first case, are related by conditions:

$$\underset{1}{\alpha} + 2\underset{2}{\alpha} = \kappa, \quad \underset{1}{\beta} + 2\underset{2}{\beta} = 0. \tag{30}$$

The frame is fixed up to a rotation in the plane $\{\underset{2}{\xi}\,\underset{3}{\bar{\xi}}\}$ and a Lorentz rotation in the plane $\{\underset{1}{\bar{\xi}}\,\underset{4}{\xi}\}$.

It remains to consider the third type with characteristic $[3, \bar{3}]$. For such characteristics [6] of the tensor $P_{\alpha\beta}$ one can find a single principal null direction $\underset{1}{W^{\beta}}$, and moreover, two vectors $\underset{2}{W^{\beta}}$ and $\underset{3}{W^{\beta}}$ possessing the properties:

$$(P_{\alpha\beta} - \underset{1}{K}\delta_{\alpha\beta})\underset{1}{W^{\beta}} = 0,$$
$$(P_{\alpha\beta} - \underset{1}{K}\delta_{\alpha\beta})\underset{2}{W^{\beta}} = \sigma\delta_{\alpha\beta}\underset{1}{W^{\beta}},$$
$$(P_{\alpha\beta} - \underset{1}{K}\delta_{\alpha\beta})\underset{3}{W^{\beta}} = \tau\delta_{\alpha\beta}\underset{2}{W^{\beta}}, \tag{31}$$

where σ and τ are arbitrary numbers $\neq 0$. The vector $\underset{2}{W^{\beta}}$ is nonnull, but $\underset{3}{W^{\beta}}$

is null. Moreover, $\underset{1}{W}^{\beta}$ is orthogonal to $\underset{2}{W}^{\beta}$ but not orthogonal to $\underset{3}{W}^{\beta}$, and the vector $\underset{2}{W}^{\beta}$ is orthogonal to $\underset{3}{W}^{\beta}$.

Because $\underset{2}{W}^{pq}$ is a nonnull vector, we can, as in the two previous cases, make a corresponding choice of frame (with two degrees of freedom) and write this bivector in the form:

$$\underset{2}{W}^{pq} = \underset{24}{\xi}^{pq} + i\underset{31}{\xi}^{pq}.$$

Then for the bivectors $\underset{1}{W}$ and $\underset{3}{W}$, if we take into account the conditions of nullity and orthogonality shown above, we obtain the expressions:

$$\underset{1}{W}^{pq} = \underset{14}{\xi}^{pq} + i\underset{23}{\xi}^{pq} + i(\underset{34}{\xi}^{pq} + i\underset{12}{\xi}^{pq}),$$
$$\underset{3}{W}^{pq} = \lambda\{\underset{14}{\xi}^{pq} + i\underset{23}{\xi}^{pq} - i(\underset{34}{\xi}^{pq} + i\underset{12}{\xi}^{pq})\},$$

where λ is an arbitrary number $\neq 0$. Further following the same arguments as for the preceding type of characteristic we write down conditions (30) for the tensor $R_{\alpha\beta}$ that guarantee that $\underset{1}{W}^{\alpha}$ is a principal direction (in bivector space), and the vectors $\underset{1}{W}^{\alpha}$, $\underset{2}{W}^{\alpha}$ and $\underset{3}{W}^{\alpha}$ determine the invariant planes of the tensor $R_{\alpha\beta}$. These conditions are:

$$(R_{\alpha\beta} - Kg_{\alpha\beta})\underset{1}{W}^{\beta} = 0,$$
$$(R_{\alpha\beta} - Kg_{\alpha\beta})\underset{2}{W}^{\beta} = \sigma g_{\alpha\beta}\underset{1}{W}^{\beta},$$
$$(R_{\alpha\beta} - Kg_{\alpha\beta})\underset{3}{W}^{\beta} = \tau g_{\alpha\beta}\underset{2}{W}^{\beta}. \tag{32}$$

where σ and τ are numbers other than zero.

Taking into account that a bivector W^{pq} at a given point of T_4 corresponds in the local metric bivector space to a vector $\underset{nt}{W}^{pq} \longrightarrow \underset{\sigma}{W}^{\alpha}$ and having in mind that for a coordinate frame

$$\underset{nt}{\xi}^{pq} \longrightarrow \underset{\sigma}{\xi}^{\alpha} = \delta^{\alpha}_{\sigma},$$

it is easy to verify that the system of equations (32) is reduced to the following 9 independent equations[10]

[10] In the 6th equation below, the right hand side was corrected from σ_i to $i\sigma$ (Translator).

$$\begin{aligned}
m_{11} + in_{11} + im_{13} - n_{13} &= -K,\\
m_{12} + in_{12} + im_{23} - n_{23} &= 0,\\
m_{13} + in_{13} + im_{33} - n_{33} &= -iK,\\
m_{12} + in_{12} &= -\sigma,\\
m_{22} + in_{22} &= -K,\\
m_{23} + in_{23} &= -i\sigma,\\
m_{11} + in_{11} - im_{13} + n_{13} &= -K,\\
m_{12} + in_{12} - im_{23} + n_{23} &= -\tau,\\
m_{13} + in_{13} - im_{33} + n_{33} &= iK,
\end{aligned}$$

where $K = \alpha + i\beta$ is one of the two triply repeated roots of the characteristic polynomial

$$|R_{\alpha\beta} - K g_{\alpha\beta}| = 0,$$

and the numbers σ and τ are different from zero but otherwise arbitrary. This arbitrariness arises from the arbitrariness of the number λ and it is a consequence of the nullity of the vectors $\underset{1}{W}^{\alpha}$ and $\underset{3}{W}^{\alpha}$. One can for example assume that σ and τ are real numbers.

Solving this system and taking into account the conditions

$$\sum_{s=1}^{3} e_s m_{ss} = \kappa, \quad \sum_{s=1}^{3} e_s n_{ss} = 0,$$

it is easy to convince ourselves that $\tau = 2\sigma$, $\beta = 0$, $\alpha = \kappa/3$, so that the matrix $\|R_{\alpha\beta}\|$ takes the form:

$$(R_{\alpha\beta}) = \left\| \begin{array}{ccc|ccc}
-\frac{\kappa}{3} & -\sigma & 0 & 0 & 0 & 0\\
-\sigma & -\frac{\kappa}{3} & 0 & 0 & 0 & -\sigma\\
0 & 0 & -\frac{\kappa}{3} & 0 & -\sigma & 0\\
\hline
0 & 0 & 0 & \frac{\kappa}{3} & \sigma & 0\\
0 & 0 & -\sigma & \sigma & \frac{\kappa}{3} & 0\\
0 & -\sigma & 0 & 0 & 0 & \frac{\kappa}{3}
\end{array} \right\|, \qquad (33)$$

where σ is an arbitrary real number different from zero: the frame is fixed up to a rotation in the plane $\{\underset{1}{\xi}\underset{3}{\xi}\}$ and a Lorentz rotation in the plane $\{\underset{2}{\xi}\underset{4}{\xi}\}$.

Hence, as the final result, we obtain the following

Theorem. There exist three fundamentally distinct types of gravitational field:

The 1-st type with a characteristic of the K-matrix of the simple type $[111, \overline{111}]$; for this type a real orthogonal frame is fixed at each point of the T_4 and with respect to it the matrix $\|R_{\alpha\beta}\|$ takes the form (23) under the conditions (24) and (25).

The 2-nd type with a characteristic of the non-simple type $[21, \overline{21}]$; for this type the frame is fixed only up to two degrees of freedom and with respect to it the matrix $\|R_{\alpha\beta}\|$ takes the form (29) under the conditions (30).

The 3-rd type with a characteristic of the non-simple type $[3, \overline{3}]$; for this type the frame also has two degrees of freedom and with respect to it the matrix $\|R_{\alpha\beta}\|$ takes the form (33).

Here the overlined numbers in the characteristics denote the exponents of those elementary divisors which have roots complex conjugate to those of the corresponding non-overlined numbers.

The three types indicated allow, of course, further more detailed classification. For example, one can pick out the cases of multiple and real roots, as we considered earlier. This result, obtained by me in 1950, was first published in 1951 [1]. In that article the formulation is not precise enough. A proof of the third theorem in §2 was also given by A.P. Norden in 1952 (unpublished), who considered bi-affine spaces. The proof given here is a third approach and it is obviously the simplest one.

Regarding the work presented in §3, i.e. the determination of the canonical form of the matrix $(R_{\alpha\beta})$ in an orthogonal anholonomic frame, it is necessary to make the following remark. At first glance, one could treat the problem as follows: because we know the characteristic of the matrix $\|R_{\alpha\beta} - Kg_{\alpha\beta}\|$, one might consider it possible to write down at once the canonical form of this matrix using general algebraic theory [6]. However, it is not possible because the coefficients of the real linear transformations can only be numbers of the form

$$A_{\alpha}^{\alpha'} = 2A_{ij}^{[i'j']},$$

where $A_i^{i'} = (\partial x^{i'}/\partial x^i)_P$ are the coefficients of some real orthogonal transformation at a given point P in the manifold T_4. That is, we can only use transformations from some subgroup of the group of all real orthogonal transformations in 6-dimensional space.

This fact, which made necessary the arguments in §3, is in our case obvious and represents a concrete application of the general theorem proved by G.B. Gurevich [7].

REFERENCES

1. A. Z. Petrov. On spaces defining gravitational fields. *Dokl. Akad. Nauk SSSR*, **XXXI,** 149–152 (1951).
2. L. Landau and E. Lifshitz. *Theory of fields*. Moskva-Leningrad (1941), pp. 263–268.
3. V. F. Kagan. On some systems of numbers…, *Izdatelstvo Moskovskogo Gosudartsvennogo Universiteta* (1926–27), pp. 1–24.
4. Ya.S. Dubnov. On symmetrically-double orthogonal matrices. *Izdatelstvo Moskovskogo Gosudartsvennogo Universiteta* (1926–27), pp. 33–54.
5. A. M. Lopshitz. Vector solutions of the problem of symmetrically-double matrices. Proceedings of the All-Russian Mathematical Congress, Moscow, 1927, Moskva-Leningrad (1928), 186–187.
6. A. Z. Petrov. Theorem on the principal axes of a tensor. *Isvestiya Kazanskogo Fiziko-Matematicheskogo Obshchestra [Bulletin of the Kazan Physico-Mathematical Society]*, **14,** 43 (1949).
7. G. B. Gurevich. On some linear transformations of symmetric tensors or polyvectors. *Mat. Sbornik*, **26,** 463–469 (1950).

Chapter 5
Wolfgang Kundt's seminar notes (1957), the Bianchi classification of 3-D spatial symmetry groups by the method of Schücking

The original text had never been published. It existed only in the form of handwritten seminar notes, and was circulated among personal acquaintances of members of the Hamburg relativity seminar. It was published in the Golden Oldies series for the first time.

Reprinted with the kind permission of Wolfgang Kundt and of all the surviving authors of the editorial note (H. Wahlquist died in 2008, C. Behr could no longer be located).

Previously published in *Gen. Relativ. Gravit.* **35** (2003), pages 475 – 489 (the editorial note + Schücking's biography), 491 – 498 (Kundt's seminar notes).

The editorial note was re-edited and updated for this reprinting.

5.1 Editorial note to:

Wolfgang Kundt
The spatially homogeneous cosmological models

Andrzej Krasiński · Christoph G. Behr · Engelbert Schücking · Frank B. Estabrook · Hugo D. Wahlquist · George F. R. Ellis · Robert Jantzen · Wolfgang Kundt

Part 1: In pursuit of the original source

By Andrzej Krasiński

Today, the Bianchi classification of 3-dimensional Lie algebras is no longer presented by the original Bianchi method [1]. The now-common approach is usually credited to C. G. Behr, but one never sees any reference to a paper by him. Tracing the references back in time one is most often led either to the paper by Ellis and

Andrzej Krasiński
N. Copernicus Astronomical Center, Polish Academy of Sciences, Bartycka 18, 00 716 Warsaw, Poland
e-mail: akr@camk.edu.pl

Christoph G. Behr
Eduard-Steinle-Straße 19, 70 619 Stuttgart, Germany (current address unknown)

Engelbert Schücking
29 Washington Square West, New York, NY 10011, USA
e-mail: elschucking@msn.com

Frank B. Estabrook
Jet Propulsion Laboratory, Mail Stop 169-327, 4800 Oak Grove Drive, Pasadena, California 91109, USA
e-mail: frank.b.estabrook@jpl.nasa.gov

Hugo D. Wahlquist (deceased 2008)

George F. R. Ellis
Department of Mathematics and Applied Mathematics, University of Cape Town, Rondebosch 7700, Cape Town, South Africa
e-mail: George.Ellis@uct.ac.za

Robert Jantzen
Department of Mathematical Sciences, Villanova University, Villanova, Pennsylvania 19085, USA
e-mail: jantzen@villanova.edu

Wolfgang Kundt
Institut für Astrophysik der Universität Bonn, Auf dem Hügel 71, D-53121 Bonn, Germany
e-mail: wkundt@astro.uni-bonn.de

MacCallum [2] or to the paper by Estabrook, Wahlquist and Behr [3]. The latter, along with a brief summary of the (then-) new method, contains a promise that a separate paper by Behr would be published, where the approach would be presented in more detail. However, a thorough search through the author indexes in *Physics Abstracts* for 1968 and the following years has revealed no entries by Behr.

At this point, this Editor (A. K.) launched an investigation into the mystery. It turned out that Christoph Behr left research in physics soon after 1968 and had never published the planned paper. Back in 2000/2001, when the investigation was under way, he was retired and lived in Stuttgart, Germany,[1] but had not preserved any notes and was not able to reconstruct the original account. George Ellis recalled that he first learned the approach from notes taken by Wolfgang Kundt at the relativity seminar in Hamburg. The actual inventor of the approach turned out to be Engelbert Schücking.

Several e-mails followed. W. Kundt had found his old notes and had kindly agreed to translate them to English and write them up for publication. The text reprinted here was published in 2003 in the *Gen. Relativ. Gravit.* journal for the first time [4] (and, like every newly published text, was approved for publication by two referees). This is the closest approximation to the original source that it was possible to reconstruct today. It is preceded by short personal recollections on this subject of the main players in this story: C. Behr, E. Schücking, F. B. Estabrook with H. D. Wahlquist, G. Ellis and W. Kundt, and by editorial comments written by Robert Jantzen.

Part 2: The unfinished road to publication

By Christoph G. Behr

The approach to the Bianchi classification in question goes back to 1966 and 1967, when I was working as Postdoctoral Resident Research Associate at the Physics Section of the Jet Propulsion Laboratory at Pasadena, California. I belonged to a small group of relativists headed by Frank B. Estabrook. During that time I carried out research in relativistic cosmology. Emphasis was put on spatially homogeneous world models (i.e. on space-times admitting of groups of motions that are transitive on space-like 3-dimensional orbits). In addition to this I considered solutions of Einstein's field equations that allow for groups of motions which are transitive on the whole space-time manifold. As suggested by and frequently discussed with Professor Engelbert L. Schücking, then at the University of Texas at Austin, I investigated transitions between cosmologies with non-isometric groups of motions by means of singular transformations. This research was motivated by the wish to get insight into the physical and geometrical properties of, and relations between, known solutions of the field equations. Further on I hoped that new solutions could be constructed by this method.

The "Bianchi – Behr classification" was the first result of this research. The then-new approach is presented to some extent in the paper entitled "Dyadic analysis of

[1] In 2012, when this book was being prepared, C. Behr could no longer be located.

spatially homogeneous world models", published by Frank B. Estabrook, Hugo D. Wahlquist and myself [3] in 1968.

About 1967 I had in mind to publish a paper containing the then-new approach to Bianchi classification in detail, together with some results in cosmology I hoped to obtain. Later on I gave up this plan after having encountered considerable difficulties, which prevented me from coming to the desired cosmological results.

In December 1966 I gave a seminar on "Degenerations of Lie Algebras" at the University of Texas at Austin. Further on, I gave a talk entitled "Contractions of Lie groups and general relativity" at the 1967 Winter Meeting of the American Physical Society in Pasadena, California. In 1968 I left the Jet Propulsion Laboratory – and relativistic cosmology.

Since about 1972 I have been no longer engaged in relativity.

At Austin, Texas, in the summer of 1966, Engelbert Schücking informed me about his original idea for the classification of 3-dimensional Lie algebras. If my memory serves me right, Engelbert Schücking did this in the way as it is stated by Wolfgang Kundt in the present document (*Schücking's Program, step* 2). Then I left Austin for Pasadena. I certainly agree that my approach to the Bianchi classification [3] should not be called "Bianchi – Behr classification" – but "Bianchi – Schücking – Behr classification", as suggested by Robert Jantzen in his editorial comments (section 6 of the present text).

Part 3: The new approach to the Bianchi classification and its background

By Engelbert Schücking

In 1955 I was hired by Otto Heckmann, director of the Hamburg Observatory, to work on a new edition of his book [5]. Heckmann had generalized the Newtonian cosmology of Edward Arthur Milne and William McCrea [6] by considering also linear anisotropic flows of the substratum with velocity v_j, Euclidean coordinates x_k and time t given by

$$v_j = A_{jk}x_k, \qquad A_{jk} = A_{kj}(t); \qquad j,k = 1,2,3.$$

I knew about Kurt Gödel's model of a rotating universe in Einstein's theory of gravitation [7] and showed him the exact Newtonian analog of Gödel's model. This made it clear that a global Newtonian flow could also show rotation and that the restriction to symmetric A_{jk} was not warranted. The existence of a Newtonian analog to Gödel's relativistic model suggested the possible existence of relativistic analogs to Heckmann's anisotropic Newtonian models and he allowed me to work on their construction. This led at once to the integration of the Bianchi type I models with Euclidean space sections. I did not know of Edward Kasner's [8] solution and Abraham Taub's [9] model for the vacuum case. But it was clear that there could be other models with homogeneous 3-spaces which were not flat.

In Howard Robertson's article on cosmology [10] I found a reference to Luigi Bianchi and got hold of Bianchi's text [11] which discusses also homogeneous 3-dimensional Riemannian manifolds. The original paper of 1898 I saw only many years later when Remo Ruffini gave me a copy of it. I knew, possibly from the book by Sophus Lie and Georg Scheffers [12], a classification of complex 3-dimensional Lie algebras, but that was not much help for the real case.

The derivation of Bianchi's classification was a necessary step in attempting the construction of the anisotropic dust models with homogeneous space sections. Only a few results of these calculations were published [13], [14]. The basic technique was using the automorphism group of each Lie algebra, i.e. the homogeneous linear transformations which leave the third rank tensor of the structure constants fixed, in a similar way as one employs the Galois group for the solution of algebraic equations.

Gödel showed me in 1961 that he had done all models for each Bianchi type separately. When I asked him when he was going to publish his results, he said: "not in the next ten years". He never did and this full drawer in one of his filing cabinets still waits for an editor. I expect that one will find there the most elegant version of the Bianchi classification. He was one of the most ingenious mathematicians.

Wolfgang Kundt's paper certainly gives the gist of my seminar talk in 1957 but emphasis and telegraphic style of proofs are certainly his. It is possible that my own notes are still extant, but they would be in boxes that have not been looked into for forty years and retrieval might take some time. Anyhow, I am grateful to Wolfgang for his reconstruction.

I had helped Christoph Behr with his "Diplomarbeit" (Thesis, Hamburg 1960). In 1965 and 66 Christoph spent some time at the Relativity Center in Austin, Texas. He was looking for a problem and I suggested we derive the homogeneous models from a simplified variational problem. This had been done by Hermann Weyl for the Schwarzschild-Droste solution in his book [15][2]. The idea is to integrate over a right- or left-invariant volume of the homogeneous submanifolds and vary the Lagrangean without disturbing the symmetry. One should then get a one-dimensional variational problem, exactly what Weyl got.

If I remember correctly, this worked only for vanishing vector of the structure constants (in Ellis and MacCallum's classification: class A). I was not able to understand why not in the other cases and, therefore, did not publish the result.[3] I suppose it would have been at that occasion that I familiarized Christoph Behr with the classification of the three-dimensional structure constants of real Lie algebras. He then went to Caltech to work with Frank Estabrook and Hugo Wahlquist.

By the way, the problem of simplified variational problems is not without interest, especially for theories with internal symmetry spaces. Alex Harvey once showed me a paper by the Brandeis mathematician Richard Palais who called it the "principle of minimal criticality". His theorem says that it sometimes works and sometimes not, but I was unable to follow his higher mathematics.

The appendix A of MacCallum's magisterial paper on the classification of 4-dimensional Lie algebras [18] should be reprinted here. It is, among other things, a remarkable piece of historical detective work.

[2] Droste's original paper [16] was reprinted in the Golden Oldies series. Droste himself also used the variational approach (editor).

[3] The reason is nicely explained by Malcolm A. H. MacCallum in Ref. [17] (editor).

Part 4: Applying first order tetrad formalism to the new approach

By Frank B. Estabrook and Hugo D. Wahlquist

Directly motivated by the successes of the Newman–Penrose tetrad formalism in problems where there are preferred null congruences, Estabrook and Wahlquist in 1963 developed a complete 3+1 formalism for orthonormal tetrad fields aligned with a preferred timelike congruence. We adopted symbols for vectors and tensors in the local orthogonal 3-space using the terminology customary in fluid dynamics, and called the result "dyadic analysis", a la Gibbs. We were able to discuss this approach with participants at the First (1963) and Second (1964) Texas Symposiums, among whom were some of the contributors to this Golden Oldie. Our paper on dyadics appeared in print in J.Math.Phys. in 1964 [19].

As a result of a conversation there with Dennis Sciama, we exchanged letters with and sent a preprint to G. F. R. Ellis, who shortly afterwards sent us a copy of his Cambridge thesis, in which he noted some of our results. It was immediately obvious that we had been treading parallel paths, while using very different formalisms. In the last section of his thesis, Ellis applied his formulation to homogeneous dust cosmologies, not, however, employing the Bianchi–Behr classification scheme of which he was unaware at the time. As Ellis points out in his comment here, it is quite surprising that Engelbert Schücking, who apparently was in possession of this superior classification in 1957, never published it, nor utilized it in his subsequent contributions to cosmology.

Schücking himself proposed that Christoph Behr join us at JPL as a NAS / NRC postdoc, and when he arrived in September, 1966, Christoph showed us the group theoretic approach to homogeneous cosmologies he was pursuing. Since a preferred timelike congruence is available, and the three spacelike orthonormal triad vectors can be invariant under the isometry group action (so their Ricci rotation coefficients are only time dependent), it was for us a natural application of the 3+1 dyadic formalism. We were able to treat these cosmologies using 1st order equations for the dyadic variables, rather than 2nd order equations for the metric coefficients, as had been the customary approach. Doing so accomplished a great simplification of the evolution equations, and we also discovered that the Jacobi identities appeared as the simple eigenvector equation, $\mathbf{n}.F = 0$, where the symmetric dyadic F and the 3-vector $\mathbf{n}$ comprise the 9 "structure constants" of the algebra (but here treated as time-dependent Ricci rotation coefficients, or affinity fields).

Orienting the spacelike orthonormal triads to diagonalize F, one is immediately led to a classification of sub-cases somewhat different than Bianchi's, simpler and more natural. We grouped them into two classes, vanishing $\mathbf{n}$ and non-vanishing $\mathbf{n}$, later designated types A and B, respectively, by Ellis and MacCallum [2]. We introduced the subscript h notation into the continuous types VI and VII; Bianchi type III became type VI with $h = -1$. Christoph was already familiar with this purely algebraic structure, which appeared quite naturally as a constraint in the dyadic differential equations, so we agreed to call the resulting classification of types Bianchi–Behr types. The most significant consequence of combining this (to us new) algebraic classification with the 1st order dyadic equations was the quick proof of the general conservation law for Bianchi–Behr types, including h. It was published in the joint paper with Behr [3].

We also worked out the consequences for a few of the simplest cases. Type IX had been carefully treated by Behr [20] using the 2nd order metric equations, but in the diagonalized orthonormal frame the symmetric sub-type studied by Gödel [21] was particularly elegant. For vacuum, or with cosmological constant, a quartic Hamiltonian emerged which anticipated Misner's mixmaster models.

Preprints of our cosmology paper were circulated at the Third Texas Symposium (in New York, January 1967). We remember a long interaction there with Wolfgang Kundt, who was most interested in our local proof that, regardless of the matter content and its behavior (excluding singularities), Bianchi–Behr type is conserved. After his return to Hamburg he wrote that, working with Bernd Schmidt, he now understood this group-theoretically.

At the London GRG meeting in the summer of 1968 after our paper appeared, we learned that, again, closely parallel work had been underway at Cambridge by Ellis and MacCallum [2], and that they had found the same type classification to be of great value. As is frequently the case in relativity, new developments may have several independent sources. In any event, to Chris Behr belongs the credit for initiating and motivating our contribution with him, where the Bianchi–Behr classification was first published in a refereed journal article and used to derive significant new results.

Part 5: How the new approach began to be known

By George F. R. Ellis

The present streamlined understanding of the relationship between the Lie algebra structure constants and the Bianchi group types, needed to study systematically the dynamics of spatially homogeneous universe models, was developed by Engelbert Schücking in Hamburg in the years 1956 to 1957, and became general knowledge through the papers of Estabrook, Wahlquist, and Behr (1968) [3] on the one hand and of Ellis and MacCallum (1969) [2] on the other. Both sets of authors derived their understanding of this topic from Schücking. The implicit or explicit use of Lie groups to characterise spatially homogeneous geometries invariant under a G3 of isometries, from now on referred to as Bianchi geometries, actually dates from much earlier, but the method of classification did not arrive at its present concise form until the work of Schücking.

Already in 1921 E Kasner [8] looked at the Bianchi I (abelian) spatially homogeneous anisotropic models in the vacuum case and in 1933 G Lemaitre [22] examined the matter case, but neither did so from a group theory viewpoint – these are obvious anisotropic generalisations of the Robertson–Walker models that did not require group theory in their derivation. Study of their dynamics was later picked up by Schücking, B. B. Robinson, A. K. Raychaudhuri, K. S. Thorne, and others, but usually without specific group theory characterisation.

In 1935–1936, H. P. Robertson [23] and A. G. Walker [24] in discussing the Robertson–Walker models gave generic Lie group prescriptions for obtaining these solutions, in effect treating them as Bianchi geometries, but apparently never used specific Lie algebra types in their calculations. In particular they did not identify the group types that applied to these models, nor explain how the simply transitive and multiply transitive subgroups were related to each other. Their actual derivation of the metrics rather relied directly on their spatial homogeneity and isotropy, cf. [10].

The first systematic use of Lie group theory to study Bianchi geometries was by Kurt Gödel, who developed the study of the geometry and dynamics of Bianchi IX cosmological models in 1952 [21] as a follow up to his stationary space-time homogeneous model (the Gödel universe of 1949 [7]). Thus he was the first to introduce these anisotropic models explicitly characterised by their group symmetries, but did so in an elusive and enigmatic way, to a considerable degree relying on the special properties of Bianchi IX symmetries, i.e. the SO(3) symmetry group. However according to Schücking (who visited him in Princeton) he carried out many calculations for more general group types, but never published them (details are probably still extant in his notes)[4]. Then Taub (1951) [9] derived the dynamic equations for the generic vacuum Bianchi geometries, explaining the techniques needed to develop these equations in a non-holonomic and non-orthogonal basis, thus giving the first easily available systematic exposition of methods usable for all the spatially homogeneous Bianchi models. Somewhat later Schücking developed a similar version suitable for cosmology (he derived the equations with a fluid source term) and published a brief note on it in the 1958 Solvay conference proceedings [13], and a detailed account in the Appendix to the Heckmann and Schücking article in the Witten volume (1962) [14]. He emphasized the role of the automorphism group in conjunction with use of a non-orthogonal basis in simplifying the equations, without giving details. This was taken up much later by Jantzen and others. The Heckmann and Schücking article however did not give the specific structure constant decomposition that is used today. It should be mentioned that A. Z. Petrov in Russia also carried out an independent systematic examination of space-times invariant under Lie groups, resulting in a major book [25], but he did not focus specifically on cosmological models.

When working on my PhD thesis in Cambridge from 1961 to 1964, on advice from Dennis Sciama I investigated first homogeneous (steady state) anisotropic space-times and then Bianchi models. In doing so I developed a generic 1+3 tetrad approach, working from geometric ideas in Helgason's text on Differential Geometry and Symmetric Spaces, supplemented by knowledge of Lie algebras from Cohn's book on the topic and the detailed study of Continuous Groups of Transformations in Eisenhart's book of that name. I was also given helpful advice by Schücking when he visited Hermann Bondi and Felix Pirani at King's College, London, in about 1961; this gave me access to a typewritten version of the crucial 1962 article in the Witten book. The orthonormal tetrad formalism I developed was utilised in my PhD thesis in 1964 and published in 1967 in a paper [26] that systematically examined

[4] For more on Gödel's contribution to relativistic cosmology, see G. F. R. Ellis, the editorial note preceding Refs. [7] and [21].

all Locally Rotationally Symmetric ('LRS') dust spaces (invariant under multiply transitive groups of isometries). At about the same time C. Behr was working on Bianchi IX models in Hamburg and L. Shepley in Princeton was developing somewhat similar methods for all the Bianchi models, under advice from John Wheeler. The dynamics of these models were then studied by others, including P. T. Saunders, C. W. Misner (who introduced differential form methods), S. W. Hawking, and C. B. Collins. But none of these workers used the classification that is now in use. The 3-dimensional Lie algebras were rather classified in terms of the reduction to the canonical forms for their structure constants determined by Bianchi [1], made accessible by Taub's (1951) paper [9].

For me a key event was a visit that I and Graham Dixon made to the 1st Institute of Theoretical Physics in Hamburg in about 1968. We stayed with Wolfgang Kundt and his wife in Hoheneichen and took part in seminars with the Hamburg group, which did not then include Schücking, Ehlers, or Sachs (they were all in Texas by that time). One evening Wolfgang showed me his notes on the Schücking seminar that are reproduced here, and I spent a night poring over them. I believe I must have copied down a large part of them by hand, as was my custom in those days, which was before Xeroxes were available (and long before either word processors or Latex were on the horizon). The key element that I had not seen anywhere else was the representation of the structure constants given in the Lemma in Step 2 of Kundt's notes, its importance arising firstly from the fact that this immediately solved the Jacobi identities as shown in that Lemma, and secondly that it consequently gave such a simple classification of the Lie Algebras on using the obvious diagonal bases (in contrast to the numerous pages of projective geometry used by Lie and Bianchi). Engelbert now comments that the representation used was obvious as it was just a reduction to irreducible parts, but this step is only 'obvious' once one has raised the lower two indices on the structure constants by use of the antisymmetric 3-symbol to give a quantity with two upstairs indices.

On my return to Cambridge I immediately adapted my 1+3 tetrad methods to this formalism and worked out the dynamical consequences for Bianchi models with the help of Malcolm MacCallum, who became my PhD student. Thus in effect Ellis and MacCallum (1969) [2] was a development of Ellis (1967) [26] in the light of Schücking's seminar paper, as reported by Kundt. This work was developed independently of the Estabrook–Wahlquist–Behr 1968 paper [3], which appeared shortly before ours did and contained much overlapping material in a completely different formalism. Our paper included some other aspects however, giving the non-diagonal form that is useful in some cases, giving a complete classification of the relation of the Bianchi models to the multiply transitive spatially homogeneous models and in particular showing that $k = -1$ RW models corresponded to Bianchi VII (and not Bianchi VIII as many had believed up to that time)[5], and showing the exceptional status of the Bianchi VI_h models with $h = -1/9$. We also introduced the Class A / Class B notation that is now in common use. This orthonormal formalism for Bianchi models later became much more powerful through introduction of

[5] This had been independently discussed by Grishchuk [27], in a paper unknown to us at the time.

an expansion-normalised version of the variables by John Wainright, providing the basis for systematic use of dynamical-systems methods to illuminate the dynamics of these models (see [28] for an exposition).

The puzzle is that the Schücking seminar recorded in the note following, containing the crucial representation and resulting Lemma together with the classification of the 3-d Lie algebras and details of how this classification corresponds to that given by Bianchi, took place in January 1957. But Schücking did not include this material in the 1962 Witten article, nor mention it to me when I visited him in London at about that time. It seems strange that such an important feature was not included in that article, and so was not for example mentioned in either my own PhD thesis in 1964 nor in Shepley's thesis in 1965. We have here an enigmatic aspect of Engelbert – he had the key to the best representation of the Bianchi structure constants already in 1957, but chose not to publish it even when writing in detail about these spaces.

He has explained to me now that he was planning to present these results in a book which never got written for two reasons: first he did not attain the deeper insight into the mechanism of Einstein's gravitation he had hoped for by using these methods, and second he believed that what he had done was already contained in unpublished work of Gödel's. Personally I doubt that – Engelbert's work was developed in the conducive environment of the Hamburg groups consisting of P. Jordan, O. Heckmann, E. Schücking, J. Ehlers, W. Kundt, and latterly R. Sachs and M. Trümper, where elegant covariant methods were communally developed and treasured. The development of Schücking's classification method was perhaps natural in that context, but it was not obvious to other workers in the field at the time, and I am not aware of any evidence that Gödel developed this particular insight. In any case I was deeply indebted to Wolfgang Kundt for showing me the seminar notes on Engelbert's work, for they made my paper with Malcolm MacCallum possible, which then became a useful source for many other workers.

Part 6: Relations between different approaches to the Bianchi classification

By Robert Jantzen

To give the document some context as briefly summarized in [3] and its reference [14], Gödel had first used Bianchi type VIII and IX dust models in cosmology in 1949 and 1952 [7], [21], while Taub had methodically investigated all the vacuum Bianchi models in 1951 [9], giving the form of the spacetime metric and the Killing vector fields for all Bianchi types explicitly. This was followed up by work on dust models by Heckmann and Schücking [14], reported here as Schücking's program then in progress in 1957. In this present document, the diagonal form of the structure constants introduced by Ellis and MacCallum [2] (in which the symmetric part of the structure constant matrix is diagonal and the vector is aligned with one of the basis vectors) is suggested but not implemented. It is this more convenient choice of representatives which was described in the Behr article and later universally used in Bianchi cosmology which has come to be known as the Bianchi–Behr classification, but should clearly now be called the Bianchi–Schücking–Behr classification.

The one remaining detail not discussed in the literature is the actual relation between Bianchi's canonical choice of structure constants appearing in the article below and the logical diagonal form choices in the Ellis–MacCallum scheme. In their notation

$$C^\alpha{}_{\beta\gamma} = \varepsilon_{\beta\gamma\delta} n^{\beta\delta} + a_\delta \delta^{\delta\alpha}_{\beta\gamma} , \quad [e_\beta, e_\gamma] = C^\alpha{}_{\beta\gamma} e_\alpha ,$$
$$C^{\alpha\delta} = C^\alpha{}_{\beta\gamma} \varepsilon^{\delta\beta\gamma} = C^{(\alpha\delta)} + C^{[\alpha\delta]} = n^{\alpha\delta} + \varepsilon^{\alpha\delta\gamma} a_\gamma ,$$

where

$$n^{\alpha\delta} = C^{(\alpha}{}_{\beta\gamma} \varepsilon^{\delta)\beta\gamma} , \; a_\gamma = \tfrac{1}{2} C^\alpha{}_{\gamma\alpha} , \; n^{\alpha\delta} a_\delta = 0 , \; a_\alpha a_\beta = \tfrac{1}{2} h \varepsilon_{\alpha\gamma\mu} \varepsilon_{\beta\delta\nu} n^{(\gamma\delta)} n^{(\mu\nu)} ,$$

the last relation first given by Collins and Hawking [29], apparently due to an observation of Hawking. The invariant constant h which appears in it was first introduced by Behr and is related in the following way to the constant h which appears below in the Bianchi classification scheme for Bianchi types VI and VII, call it h_B for Bianchi's parameter: $h = -(h_B+1)^2/(h_B-1)^2$ for type VI ($h \le 0$, $-1 \le h_B < 1$, $h_B \neq 0$) and $h = h_B^2/(4-h_B^2)$ for type VII ($h \ge 0$, $0 \le h_B^2 < 2$). For type VI the parameter values $h_B = 0$ ($h = -1$) for Bianchi type III and $h_B = 1$ ($h \to -\infty$) for Bianchi type IV are excluded. The class A type VI corresponds to $h_B = -1$ ($h = 0$), while the class A type VII corresponds to $h_B = 0$ ($h = 0$). The modern designation for these types is in fact VI_h ($h \le 0$) and VII_h ($h \ge 0$) with III = VI_{-1}, and VI_0 and VII_0 denoting the class A subcases.

The diagonal form closest to Bianchi's scheme is $(n^{\alpha\beta}) = \mathrm{diag}(n^{(1)}, n^{(2)}, n^{(3)})$, $(a_b) = (0,0,a)$, $an^{(3)} = 0$, so that $a^2 = hn^{(1)}n^{(2)}$ (Behr's definition of h) and

$$(C^{\alpha\beta}) = \begin{pmatrix} n^{(1)} & -a & 0 \\ a & n^{(2)} & 0 \\ 0 & 0 & n^{(3)} \end{pmatrix} \leftrightarrow \begin{array}{l} [e_2, e_3] = n^{(1)} e_1 - a e_2 \\ [e_3, e_1] = a e_1 + n^{(2)} e_2 \; . \\ [e_1, e_2] = n^{(3)} e_3 \end{array}$$

The ordered diagonal values $n^{(\alpha)}$ can be taken as $1, \ldots, -1, \ldots, 0, \ldots$ as suggested by Schücking, while a is either zero (class A or unimodular) or not zero (class B or nonunimodular), in which case it can be made 1 except for Bianchi types VI and VII. The Bianchi canonical choices for VI and VII instead make the first 2x2 subblock of the matrix $(C^{\alpha\beta})$ respectively offdiagonal and upper-triangular. The transformation between the choices is mediated by

$$X_\alpha = e_\beta A^{-1\beta}{}_\alpha , \; \bar{n}^{\alpha\beta} = \det(A^{-1}) A^\alpha{}_\gamma n^{\gamma\delta} A^\beta{}_\delta , \; \bar{a}_\alpha = a_\beta A^{-1\beta}{}_\alpha ,$$

where the bar indicates a Bianchi basis X_α and its structure constants $\bar{C}^\alpha{}_{\beta\gamma}$.

For type VI with $q = n^{(1)} = -n^{(2)} = 1$, Ellis and MacCallum [2] introduced a second canonical choice for which A is a rotation by angle $\pi/4$ about the third direction leading to the nonzero 2x2 block of $[C]$ of the matrix $(C^{\alpha\beta})$ taking the form

$$[\bar{C}] = \begin{pmatrix} 0 & q-a \\ q+a & 0 \end{pmatrix} = \begin{pmatrix} 0 & h_B \\ -1 & 0 \end{pmatrix}, \qquad [A] = 2^{-1/2} \begin{pmatrix} 1 & -1 \\ 1 & 1 \end{pmatrix},$$

leading to the above relationship between the parameters. The same transformation about the first basis vector can be used in the type VIII case with $q = n^{(1)} = n^{(2)} = -n^{(3)} = 1$ to get a Bianchi basis apart from a factor of 2 and a permutation. (These lead to null bases on the dual space for the indefinite quadratic form associated with $n^{\alpha\beta}$ in this 2x2 block.)

For Bianchi type VII with $q = n^{(1)} = n^{(2)} = 1$, the transformation is instead a scaled Lorentz boost in the 2x2 block

$$[\bar{C}] = \begin{pmatrix} -q & qh_B \\ 0 & -q \end{pmatrix}, \qquad A = \begin{pmatrix} -1 & \tanh 2\theta & 0 \\ \tanh 2\theta & -1 & 0 \\ 0 & 0 & -\cosh 2\theta \end{pmatrix},$$

where $\sinh 2\theta = h^{1/2} = q/a$ and $\tanh 2\theta = \sqrt{h/(1+h)} = h_B/2$.

Apart from permutations and possible scale factors and signs, this explains the relationship between the canonical choices made by the modern classification and those of Bianchi, which were really those of Lie who used the first seven Roman numerals in reverse order compared to Bianchi to describe the equivalence classes of 3-dimensional complex Lie algebras [30, 31]. Bianchi gives his own derivation of Lie's classification with his refinement to the real case in his own book [11], using the dimension of the derived Lie algebra (equivalent to the rank of $(C^{\alpha\beta})$) and its structure to arrive at canonical bases for each type (see [32] for an English translation and commentary). The same paper of Bianchi Remo Ruffini had given Schücking we translated together in 1973 when I was a Princeton undergraduate, just after MacCallum had been a postdoc with Taub explaining in [33] the variational problems referred to above by Schücking, and I then had the pleasure of being Taub's final graduate student before his retirement in 1978. A single parametrized formula giving a coordinate representation for all the diagonal form Lie algebra vector field bases as a function of the four parameters $(n^{(1)}, n^{(2)}, n^{(3)}, a)$ came out of that experience [34].

Part 7: The memorable seminar

By Wolfgang Kundt

In 1956/7, Engelbert Schücking was a young collaborator of Otto Heckmann at Hamburg's Observatory in Bergedorf, and at the same time a member of Pascual Jordan's Seminar on General Theory of Relativity at Hamburg's I. Institute for Theoretical Physics, together with Jürgen Ehlers and myself as the two junior members. I had finished my diploma in October 1955 and still tried to get a better understanding of Einstein's Theory of Gravitation, of Quantization, and of various branches of Mathematics. Engelbert occasionally confronted Jürgen and me with his most recent studies – from all disciplines of physics – in *spontaneous seminar* presentations. One of those is the subsequent approach to a determination, in more or less closed form, of all spatially homogeneous cosmological models, of which I took notes which I edited in polished form at home, in January 1957, mostly in (German) handwriting because of the many special symbols and equations whose typing would have had to be very incomplete anyway. It was my tacit understanding that Engelbert, or one of his younger associates – like his later student Christoph Behr – would take the matter into print, i.e. that my writeup had no other purpose than to help me remember this interesting work. Apparently, no such documentation can be easily accessed today.

Schücking's idea was that the class of known exact solutions of Einstein's field equations, applicable to Cosmology, could be vastly extended (over Friedmann and Lemaître's isotropic class) under the assumption of *spatial homogeneity* because the assumed translational symmetry of spacetime guarantees the existence of 3-d subspaces which are homeomorphic to the respective Lie group of isometric mappings. Spatial homogeneity is, of course, not an exact property of our world, but should be a rather realistic approximation to the Universe on length scales above those of the largest clusters of galaxies, some 10^2Mpc and beyond. A large candidate class of analytic models for our Universe would help understand its large-scale properties and expected evolution in time. How to achieve this goal?

Acknowledgements. The Editor (A. K.) is grateful to all the contributors for their work, for their patience in negotiating the final version acceptable to everyone, and in addition to:

1. George Ellis for important startup information.
2. Hugo D. Wahlquist for directing me to the right track in pursuit of Christoph Behr.
3. Ms. Jutta Gonska, the webmaster of Mathematics Department at the University of Mainz, Germany, for providing Behr's postal address.
4. Wolfgang Kundt for translating his notes to English and preparing them for publication.
5. Bob Jantzen for editorial comments and footnotes.
6. Malcolm MacCallum for several improvements in the final text.

References

1. L. Bianchi, *Memorie di Matematica e di Fisica della Societa Italiana delle Scienze*, serie III, Tomo **XI**, 267 – 352 (1898) ['On three dimensional spaces which admit a group of motions', English translation: *Gen. Rel. Grav.* **33**, 2157 – 2253 (2001)].
2. G. F. R. Ellis and M. A. H. MacCallum, *Commun. Math. Phys.* **12**, 108 (1969).
3. F. B. Estabrook, H. D. Wahlquist and C. G. Behr, *J. Math. Phys.* **9**, 497 (1968).
4. W. Kundt, *Gen. Relativ. Gravit.* **35**, 491 (2003).
5. O. Heckmann, *Theorien der Kosmologie*, Springer, 1942.
6. E. A. Milne, *Quart. J. Math. Oxford* **5**, 64 (1934); W. McCrea and E. A. Milne, *Quart. J. Math. Oxford*, **5**, 73 (1934); reprinted in *Gen. Rel. Grav.* **32**, 1939 and 1949 (2000).
7. K. Gödel, *Rev. Mod. Phys.* **21**, 447 (1949); reprinted in *Gen. Rel. Grav.* **32**, 1409 (2000).
8. E. Kasner, *Am. J. Math.* **43**, 217 (1921); reprinted in *Gen. Rel. Grav.* **40**, 865 (2008).

9. A. Taub, *Ann. Math.* **53**, 472 (1951); reprinted in *Gen. Rel. Grav.* **36**, 2689 (2004).
10. H. P. Robertson, *Rev. Mod. Phys.* **5**, 62 (1933); to be reprinted in *Gen. Rel. Grav.* as a Golden Oldie in due time.
11. Luigi Bianchi, *Lezioni sulla teoria dei gruppi continui finiti di trasformazioni* (Lectures on the theory of finite continuous transformation groups (1902-1903)), Spoerri, Pisa, 1918, see sections 198–199, pp. 550–557.
12. S. Lie and G. Scheffers, *Geometrie der Berührungstransformationen*, Leipzig 1896.
13. O. Heckmann and E. Schücking, World Models, in: *La structure et l'evolution de l'univers. Institut International de Physique Solvay, Onzième Conseil de Physique [Proceedings of the 11th Solvay conference on physics.]*. Editions Stoops, Brussels 1958.
14. O. Heckmann, E. Schücking, Relativistic Cosmology, in *Gravitation, an Introduction to Current Research*, Ed. L. Witten, Wiley, New York 1962, p. 438.
15. H. Weyl, *Space-Time-Matter*, Dover, 1952.
16. J. Droste, *Koninklijke Nederlandsche Akademie van Wetenschappen Proceedings* **19**, 197 (1917); reprinted in *Gen. Rel. Grav.*, **34**, 1541 (2002).
17. M. A. H. Mac Callum, in *General Relativity, an Einstein Centenary Survey*, Eds. S. W. Hawking and W. Israel, Cambridge University Press 1979, p. 552–553.
18. M. A. H. Mac Callum, in *On Einstein's Path*, Ed. A. Harvey, Springer 1999, p. 299–317.
19. F. B. Estabrook and H. D. Wahlquist, *J. Math. Phys.* **5**, 1629 (1964).
20. C.G.Behr, *Astronomische Abhandlungen der Hamburger Sternwarte* Bd **VII** Nr. 5. Hamburg (1965).
21. K. Gödel, in *Proceedings of the International Congress of Mathematicians*, Eds. L. M. Graves et al. Cambridge, Mass. 1952, vol. 1, p. 175; reprinted in *Gen. Rel. Grav.* **32**, 1419 (2000); see also *Kurt Gödel, Collected Works*. Edited by S. Feferman, J. W. Dawson Jr., S. C. Kleene, G. H. Moore, R. M. Solvay and J. van Heijenoort, Oxford University Press 1986, 3 volumes.
22. G. Lemaître, *Ann. Soc. Sci. Bruxelles* **A53**, 51 (1933); reprinted in *Gen. Relativ. Gravit.* **29**, 641 (1997).
23. H. P. Robertson, *Astrophys. J.* **82**, 248 (1935).
24. A. G. Walker, *Proc. London Math. Soc.* **42**, 90 (1936).
25. A. Z. Petrov, Einstein-Räume. Akademie-Verlag, Berlin, 1964.
26. G. F. R. Ellis, Dynamics of Pressure-Free Matter in General Relativity, *J. Math. Phys.* **8**, 1171 (1967).
27. L. P. Grishchuk, *Astron. Zh.* **44**, 1097 (1967) [*Sov. Astr. A. J.* **11**, 881 (1968)].
28. J. Wainwright J, G. F. R. Ellis (Eds.), Dynamical Systems in Cosmology. Cambridge: Cambridge University Press, 1997.
29. C. B. Collins and S. W. Hawking, *Ap. J.* **180**, 317 (1973).
30. S. Lie (with F. Engel), *Theorie der Transformationsgruppen* (Theory of Transformation Groups), Volumes 1–3, Liepzig, Germany, 1888,1890,1893 [reprinted by Chelsea Pub Co, Bronx, NY 1970], see chapter 28, section 136, pp. 713–722.
31. S. Lie and G. Scheffers, *Vorlesungen über Continuierliche Gruppen mit Geometrischen und Anderen Anwendungen* (Lectures on Continuous Groups with Geometric and Other Applications), Leipzig, Germany 1893; reprinted by Chelsea Pub Co, Bronx, NY 1971, see chapter 20, section 2, pp. 565–572.
32. R. T. Jantzen, "Editorial Commentary on the Lie–Bianchi Classification of 3-Dimensional Lie Algebras" accompanying the English translation of sections 198–199 of [11], to be published also in the gr-qc e-print archive.
33. M. A. H. MacCallum and A. H. Taub, *Commun. Math. Phys.* **25**, 173 (1972).
34. R. T. Jantzen, *Commun. Math. Phys.* **64**, 211 (1979).

5.2 Engelbert Schücking – a brief biography

By Engelbert Schücking

When teaching history of astronomy (and astrology) on television I have been advertized as a "Westphalian Gemini" since I was born in that part of the European Union under this sign on May 23, 1926. By age 15, I had contributed 368 days of sunspot observations [1] to the Zürich Sunspot-statistics and realized that there are more intelligent ways to do science. In fact, when I was 12 and decided I needed to learn calculus to determine the orbit of comets, Fräulein Killing, daughter of the Münster mathematician, loaned me a book. The work of her father who first classified semi-simple Lie algebras and discussed symmetries of Riemannian manifolds has dominated my life in science.

I studied at the universities in Münster, Göttingen and Hamburg and got a Dr. of Science degree in mathematics from Hamburg University (a physics degree required a lab in analytic chemistry which I arrogantly refused to take). In Hamburg I worked for Pascual Jordan who had just become interested in relativity and taught me that people with irresponsible ideas in politics can be decent personally. I had taught myself relativity from books by Arthur Eddington and Hermann Weyl, helped Jordan develop his variation of Theodor Kaluza's theory, and investigated with Otto Heckmann the structure and solutions of Newtonian cosmology. I then started the investigation of anisotropic, spatially homogeneous dust models in relativistic cosmology. That led, in collaboration with István Ozsváth, to counter-examples of some versions of Ernst Mach's principle in Einstein's theory of gravitation.

I spent five years at the University of Texas at Austin and started in 1963 with Alfred Schild and Ivor Robinson the Texas Symposia on Relativistic Astrophysics which continue every other year, the next one will be in Florence, Italy in December 2002.

In 1967 I moved to New York University and supervised some twenty theses in relativity (mostly cosmology). More recent work is a book on homogeneous gravitational fields with Eugene Surowitz, the study of the Cabibbo–Kobayashi–Maskawa manifold with István Ozsváth and Jerome Epstein, and an attempt to geometrize the electro-weak interaction, called the "sub-standard theory". I am also interested in the history of science, and I help in translating Einstein's papers into English.

References

1. Astronomische Mitteilungen, ed. by W. Brunner, 140, 141, 142, Zürich, Schulthess and Co. 1941,1942,1943.

Editor's comment: A brief autobiography of E. Schücking may be found in *On Einstein's Path*, Ed. A. Harvey, Springer 1999, pp. 1 – 14. Further historical remarks are in the biographical notes by A Harvey (pp. vii–viii) and the article by I. Ozsváth (pp. 339–352) in the same volume.

General Relativity and Gravitation, Vol. 35, No. 3, March 2003 (© 2003)

The Spatially Homogeneous Cosmological Models[1]

Wolfgang Kundt[2]

Received July 15, 2002

Schücking's *program* of calculating all spatially homogeneous dust-filled spacetimes is presented.

KEY WORDS: Cosmological models; dust; Lie algebra; Einstein field equation.

SCHÜCKING'S PROGRAM

Schücking's *program* of calculating all spatially homogeneous dust-filled spacetimes runs like this: Wanted are all normal-hyperbolic 4-d (Lorentzian) cosmological metrics which are invariant under a 3-d semi-group of isometries, i.e. which are mapped onto themselves by each element of the group (of 'motions'), together with a normalized timelike geodetic vector field—the 4-velocity of the cosmic substratum—and which satisfy Einstein's field equations for pressure-free matter (dust). This program will be achieved by the following *five steps*:

Step 1 is a short, elementary introduction into the theory of local representations of an r-d semi-group by permutations of an n-d manifold; needed in the sequel are only *Lie's second and third theorem*. *Step 2* determines a system of *representatives* (affine tensors of 'structure constants' $c^{\gamma}{}_{\alpha\beta}$) of all *3-d Lie algebras* (over the real numbers), as solutions of Lie's 3rd theorem. *Step 3* integrates the vectors $\xi^i_\alpha(x)$ of *infinitesimal transformations* for the representatives (in a 4-d manifold with suitable initial conditions; Lie's 2nd theorem). *Step 4* determines necessary conditions on all the sought-after spatially homogeneous metrics $g_{ij}(x^k)$, by *integrating* the differential equations of the *group of motions* in the normalization

[1]For the historical background see ref. [1].

[2]Institut für Astrophysik der Universität Bonn Auf dem Hügel 71, D-53121 Bonn, Germany; e-mail: wkundt@astro.uni-bonn.de

chosen in step 3 (for the representatives of all the groups). *Step 5 finally* restricts the freedom left in all the metrics of step 4 by additionally imposing Einstein's field equations.

STEP 1: LIE GROUPS AS PERMUTATION GROUPS

A set $\mathcal{G}$ containing the elements σ, τ is called a *semi-group* iff (1) for each two $\sigma, \tau \in \mathcal{G}$, $\mathcal{G}$ also contains their *product* $\sigma\tau$, (2) this product is *associative*, and (3) $\mathcal{G}$ contains a *unit* element ε (with $\varepsilon\sigma = \sigma\varepsilon$ for all σ).

We talk of a *representation* of an r-d semi-group $\mathcal{G}$ as a local, sufficiently differentiable permutation group (described by coordinates $a := (a^\kappa)$) of an n-d manifold $\mathcal{W}$ (described by coordinates $x := (x^i)$) iff

$$\bar{x}^i = \bar{x}^i(x^j, a^\kappa), \quad 1 \le i, j \le n, 1 \le \kappa \le r, \tag{1}$$

holds with (i):

$$\det(\bar{x}^i_{,j}) \neq 0 \text{ in } \mathcal{G}, \tag{2}$$

i.e., if for every $\sigma \in \mathcal{G}$ there exists a local inverse map, further (ii):

$$\chi^\alpha(a)\ \bar{x}^i_{,\alpha}(x, a) = 0 \text{ in } \mathcal{G} \Rightarrow \chi^\alpha(a) = 0 \text{ in } \mathcal{G}, \tag{3}$$

i.e., the representation is r-d, all parameters a^κ are essential, and (iii):

$$x^i = \bar{x}^i(x^j, 0) \tag{4}$$

for formal convenience. (A comma denotes partial differentiation). As a first result we get:

Lemma 1. *Defining the functions* φ^ν *through*

$$\bar{x}^i\big(\bar{x}^j\big(x^k, \underset{1}{a}{}^\lambda\big), \underset{2}{a}{}^\mu\big) =: \bar{x}^i\big(x^k, \varphi^\nu\big(\underset{1}{a}{}^\alpha, \underset{2}{a}{}^\beta\big)\big), \tag{5}$$

we have: det $\varphi^\nu_{,k_1} \neq 0 \neq \det(\varphi^\nu_{,\lambda_2})$.

For a proof, differentiate eq. (5) partially w.r.t. $\underset{1}{a}{}^\lambda$ and $\underset{2}{a}{}^\mu$, and refer to properties (2), (3). Next:

Theorem 1. *(of Lie): For* $\bar{x}^i = \bar{x}^i(x, a)$ *we have*

$$\bar{x}^i_{,\alpha} = \xi^i_\beta(\bar{x})\ A^\beta_\alpha(a) \,. \tag{6}$$

Proof: $\underset{2}{a}{}^\alpha = \underset{2}{a}{}^\alpha(\underset{1}{a}{}^\beta, \varphi^\gamma)$ because of lemma 1. Define the inverse matrix ψ^k_i by $\bar{x}^i_{,j}(\bar{x}, \underset{2}{a})\ \psi^k_i := \delta^k_j$; it exists because of condition (2). Lie's first theorem is obtained by total differentiation of eq. (5) w.r.t. $\underset{1}{a}{}^\alpha$ at fixed φ^γ at the position $\underset{2}{a}{}^\alpha =$

0, and transvection[3] with $\psi^k_i(\bar{x}, 0)$, whereby $\xi^k_\beta(\bar{x}(x, \underset{1}{a})) = -\psi^k_i(\bar{x}, 0)\, \bar{x}^i{}_{,\beta}(\bar{x}, 0)$, $A^\beta_\alpha(\underset{1}{a}) = \underset{2}{a}{}^\beta{}_{,\underset{1}{\alpha}}(\underset{1}{a})$.

Lemma 2. $\det(A^\beta_\alpha) \neq 0$.

Proof: If this determinant would vanish, theorem 1 would contradict condition (3).

Lemma 3. *The inverse matrix B^β_γ of A^α_β yields the equations*

$$\xi^i_\alpha(\bar{x}) = \bar{x}^i{}_{,\beta}\, B^\beta_\alpha\,, \tag{7}$$

Proof: Transvect eqs. (6) with B^β_γ.

Lemma 4. *The r infinitesimal transformations $\xi^i_\alpha(\bar{x})$ are pointwise linearly independent: An equation $c^\alpha \xi^i_\alpha(\bar{x}) = 0$ for all i at some point $\bar{x}$ implies $c^\alpha = 0$.*

Proof: Follows from lemma 3 with condition (3).

Theorem 2. *(of Lie): The integrability conditions of theorem 1, $\bar{x}^i_{[,\alpha,\beta]} = 0$ — in which (square) brackets denote forming the skew part — take the explicit form*

$$2\xi^k_{[\alpha}\, \xi^i_{\beta],k} = 2\, B^\kappa_\alpha\, B^\lambda_\beta\, A^\gamma_{[\kappa,\lambda]}\, \xi^i_\gamma \;=:\; c^\gamma{}_{\alpha\beta}\, \xi^i_\gamma\,, \tag{8}$$

with constant arrays $c^\gamma{}_{\alpha\beta}(a) = c^\gamma{}_{[\alpha\beta]}(a)$. Or shorter:

$$[X_\alpha, X_\beta] = c^\gamma{}_{\alpha\beta} X_\gamma \tag{9}$$

holds for the directional derivatives $X_\alpha := \xi^i_\alpha\, \partial_i$, where large square brackets denote forming twice the skew part (Lie product). The $c^\gamma{}_{\alpha\beta}$ are called the 'structure constants' of the group; they form an affine tensor under linear transformations of the basis elements X_α.

Proof: When eq. (8) is differentiated w.r.t. a^κ, one arrives at a linear dependence as in lemma 4, which must be trivial; i.e. the $c^\gamma{}_{\alpha\beta}$ have vanishing derivatives w.r.t. a^κ.

Lemma 5. *(of Maurer-Cartan): The structure constants $c^\gamma{}_{\alpha\beta}$ introduced in theorem 2 satisfy*

$$2\, A^\gamma_{[\mu,\nu]} = A^\alpha_\mu\, A^\beta_\nu\, c^\gamma{}_{\alpha\beta}. \tag{10}$$

Proof: follows from the defining equation of the $c^\gamma{}_{\alpha\beta}$ in eq. (8), by transvection with the matrices $A^\alpha_\mu\, A^\beta_\nu$.

[3] Transvection is the operation of multiplication of two tensors and subsequent contraction over one or several pairs of (upper and lower) indices.

Theorem 3. *(of Lie): Any affine tensor of structure constants $c^{\gamma}{}_{\alpha\beta}$ gives rise to a (Lie) group iff these constants satisfy the compatibility and integrability conditions*

$$c^{\varepsilon}{}_{(\alpha\beta)} = 0 = c^{\delta}{}_{\{\alpha\beta}c^{\varepsilon}{}_{\gamma\}\delta}, \tag{11}$$

i.e., if they are skew in their two lower indices, and if they satisfy the (quadratic) Jacobi identity, in which curly brackets denote forming the cyclic part.

Proof: In his ad-hoc presentation, Schücking left out the proof of this basic result from Lie theory. The result can be alternatively expressed by demanding that the square brackets in eqs. (9) define a Lie product. Eqs. (9) clearly imply the skewness of $c^{\gamma}{}_{\alpha\beta}$ in its two lower indices, and also imply the Jacobi identity by one further differentiation. The non-trivial part of the proof consists in showing that these two conditions form, at the same time, a sufficient set for integrability. It can be reduced to the key result from differential forms that every closed form—i.e., every form whose exterior derivative vanishes—can be integrated in a simply connected domain.

STEP 2: REPRESENTATIVES FOR ALL 3-D LIE ALGEBRAS

Program: Search for all solutions of eqs. (11) by real arrays $c^{\gamma}{}_{\alpha\beta}$ for $r = 3$. The arrays $c^{\gamma}{}_{\alpha\beta}$ transform as $X^{\gamma}X_{\alpha}X_{\beta}$ under arbitrary affine transformations of the X_{α}, i.e., are affine tensors (where upper and lower indices transform cogrediently); consequently, all affinely inequivalent solutions of eqs. (11) can be obtained by presenting a complete set of affinely inequivalent normal forms (representatives). These representatives will be chosen with a maximum number of vanishing coefficients.

Lemma 6. *For $r = 3$, define the affine density $a^{\iota\kappa}$ by*

$$a^{\iota\mu} := \frac{1}{2}\delta^{\iota\kappa\lambda}c^{\mu}{}_{\kappa\lambda}, \tag{12}$$

whose inverse reads: $c^{\mu}{}_{[\kappa\lambda]} = \delta_{\kappa\lambda\iota}a^{\iota\mu}$, and the vector b_{ι} by

$$b_{\iota} := \frac{1}{2}\delta_{\iota\kappa\lambda}a^{\kappa\lambda}, \tag{13}$$

with inverse: $a^{[\kappa\lambda]} = \delta^{\kappa\lambda\mu}b_{\mu}$. Theorem 3 (of Lie) is then equivalent with

$$a^{(\kappa\lambda)}b_{\lambda} = 0. \tag{14}$$

Proof: The skew symmetry of the $c^{\mu}{}_{\kappa\lambda}$ in $\kappa\lambda$ is absorbed into the definition of the density $a^{\iota\kappa}$. Eq. (14) is equivalent with the Jacobi identity—the second of eqs. (11)—because of $a^{[\kappa\lambda]}\, b_{\lambda} = \delta^{\kappa\lambda\mu}\, b_{\mu}\, b_{\lambda} = 0$, as is seen by evaluating $a^{\kappa\lambda}\, b_{\lambda}$ for a suitable sequence of contractions of the $\delta_{\iota\kappa\lambda}$.

Consequences: As is well known, the matrices $(\delta_{\iota\kappa}\, d_\kappa)$ with $d_\kappa = \pm 1, 0$ form a complete system of representatives of affinely inequivalent symmetric tensors $a^{(\kappa\lambda)}$. The corresponding system of representatives for the densities $a^{(\kappa\lambda)}$ is obtained from them by calculating modulo $\pm(\delta_{\iota\kappa})$, i.e. by invoking reflections. For every representative $a^{(\kappa\lambda)}$, all permitted b_λ can be obtained by solving eqs. (14), from which $a^{[\kappa\lambda]}$ and $a^{\kappa\lambda}$ follow via the inverse of (13). The thus obtained $a^{\kappa\lambda}$ are arranged according to their (affine) rank—which is invariant, because of $r - \text{rank}(a^{\kappa\lambda}) = \dim\{b_\lambda \mid a^{\kappa\lambda}$ $\text{b}_\lambda = 0\}$—and subsequently further simplified, if possible, by suitable affine transformations. Calculations worth mentioning occur only for $\text{rank}(a^{(\kappa\lambda)}) = 2$. Here the representatives

$$(a^{(\kappa\lambda)}) = \begin{pmatrix} 1 & 0 & 0 \\ 0 & \pm 1 & 0 \\ 0 & 0 & 0 \end{pmatrix}$$

lead to:

$$(a^{\kappa\lambda}) = \begin{pmatrix} 1 & a & 0 \\ -a & \pm 1 & 0 \\ 0 & 0 & 0 \end{pmatrix},$$

whence:

$$\begin{pmatrix} 1 & 1 & 0 \\ -1 & k & 0 \\ 0 & 0 & 0 \end{pmatrix}$$

with $k \gtrless 0$ for $\pm$; in the lower case, the form:

$$\begin{pmatrix} 0 & h & 0 \\ -1 & 0 & 0 \\ 0 & 0 & 0 \end{pmatrix}$$

can be achieved, with $h \neq 0, 1$, in the upper case the form:

$$\begin{pmatrix} -1 & h & 0 \\ 0 & -1 & 0 \\ 0 & 0 & 0 \end{pmatrix}$$

with $h^2 < 4$. The following table summarizes the result for the matrices $(a^{\kappa\lambda})$ with occasional minor (affine) modifications in order to make them agree with Bianchi's choice:

$\mathrm{char}(a^{(\kappa\lambda)})$	$repr(a^{(\kappa\lambda)})$	$r=1$	$r=2$	$r=3$
$c=0$	0		$\begin{matrix} 0 & 1 & 0 \\ -1 & 0 & 0 \\ 0 & 0 & 0 \end{matrix}$	
$c=1$	$\begin{matrix} 1 & 0 & 0 \\ 0 & 0 & 0 \\ 0 & 0 & 0 \end{matrix}$	$\begin{matrix} 1 & 0 & 0 \\ 0 & 0 & 0 \\ 0 & 0 & 0 \end{matrix}$	$\begin{matrix} 1 & 1 & 0 \\ -1 & 0 & 0 \\ 0 & 0 & 0 \end{matrix}$	
$c=2-$	$\begin{matrix} 1 & 0 & 0 \\ 0 & -1 & 0 \\ 0 & 0 & 0 \end{matrix}$	$\begin{matrix} 0 & 0 & 0 \\ -1 & 0 & 0 \\ 0 & 0 & 0 \end{matrix}$	$\begin{matrix} 0 & h & 0 \\ -1 & 0 & 0 \\ 0 & 0 & 0 \end{matrix}, h \neq 0, 1$	
$c=2+$	$\begin{matrix} 1 & 0 & 0 \\ 0 & 1 & 0 \\ 0 & 0 & 0 \end{matrix}$		$\begin{matrix} -1 & h & 0 \\ 0 & -1 & 0 \\ 0 & 0 & 0 \end{matrix}, h^2 < 4$	
$c=3-$	$\begin{matrix} 1 & 0 & 0 \\ 0 & -1 & 0 \\ 0 & 0 & -1 \end{matrix}$			$\begin{matrix} 0 & 0 & 1 \\ 0 & -2 & 0 \\ 1 & 0 & 0 \end{matrix}$
$c=3+$	$\begin{matrix} 1 & 0 & 0 \\ 0 & 1 & 0 \\ 0 & 0 & 1 \end{matrix}$			$\begin{matrix} 1 & 0 & 0 \\ 0 & 1 & 0 \\ 0 & 0 & 1 \end{matrix}$

.

Here $c := \mathrm{char}(a^{(\kappa\lambda)})$ stands for the rank and signs of the diagonal matrix $a^{(\kappa\lambda)}$, and r for the rank of $a^{\kappa\lambda}$. These two characteristic 'numbers' together, (c, r), are related to Bianchi's 9 types I through IX by:

Bianchi	I	II	III	IV	V	VI	VII	VIII	IX
(c,r)	$(0,0)$	$(1,1)$	$(2-,1)$	$(1,2)$	$(0,2)$	$(2-,2)$	$(2+,2)$	$(3-,3)$	$(3+,3)$

.

Note that Bianchi types VI and VII are not single types, but one-parameter classes of types, described by a parameter h.

Definitions (12) and (13) can now be used to calculate Lie's structure tensors $c^{\mu}{}_{\kappa\lambda}$; they read:

type	$[X_1, X_2]$	$[X_1, X_3]$	$[X_2, X_3]$
I	0	0	0
II	0	0	X_1
III	0	X_1	0
IV	0	X_1	$X_1 + X_2$
V	0	X_1	X_2
VI	0	X_1	hX_2
VII	0	X_2	$-X_1 + hX_2$
VIII	X_1	$2X_2$	X_3
IX	X_3	$-X_2$	X_1

. (15)

STEP 3: INTEGRATION OF INFINITESIMAL TRANSFORMATIONS

Table (15) and Lie's theorem 2 can now be used to calculate the infinitesimal transformations $\xi^i_\alpha(x)$, by integrating eqs. (9) in suitable coordinate systems. For all types, such a coordinate choice can be made that realizes $\xi^i_1(x) = \delta^i_1$ globally, in a suitable (simply connected) environment. In addition, for types I through VII, $\xi^i_2(x) = \delta^i_2$ can be achieved, which makes the integration trivial. For the other, slightly more difficult types, here ends Schücking's presentation; but feasibility is well in sight.

STEP 4: INTEGRATION OF EQUATIONS OF MOTION

In order to find a representative system of spacetimes admitting the above Lie groups as transitive spacelike isometries, metrics $g_{ij}(x^k)$ must be found which satisfy Killing's equation

$$0 = 2\xi_{(i;j)} = g_{ij,k}\xi^k + g_{kj}\xi^k_{,i} + g_{ki}\xi^k_{,j} \tag{16}$$

for the representative infinitesimal transformations $\xi^i(x)$ found in step 3— $\xi_i := g_{ij}\xi^j$, and a semicolon standing for covariant differentiation—together with the corresponding equations for the (geodetic) timelike unit 4-vector field $u^i(x)$ that describes the motion of the cosmic substratum:

$$0 = u_{j,i}\xi^i + u_i\xi^i_{,j} \; ; \tag{17}$$

which equation expresses the vanishing Lie derivative of $u^i(x)$ along the group trajectories; $u^i\, u_i = -1$, $u_{i;k}\, u^k = 0$. Eqs. (16, 17) have to be solved simultaneously, for the components $g_{ij}(x)$ of the metric tensor and the components $u^i(x)$ of the 4-velocity vector. To this end, (at least) two different pathways offer themselves: (i) $u^i(x)$ can be chosen to define the local system of rest, whence—without loss of generality—$u^i(x) = \delta^i_0$, $g_{00} = 1$, $g_{0\iota,0} = 0$. Or: (ii) one chooses $g_{0\iota} = 0$, and gets the other components of $g_{ij}(x)$ and $u^i(x)$ via integration.

STEP 5: INTEGRATION OF FIELD EQUATIONS

It remains to satisfy Einstein's field equations (with a cosmological constant Λ):

$$G_{ij} = (\Lambda + \kappa\mu/2)g_{ij} - \kappa\mu\, u_i u_j, \tag{18}$$

in which $\kappa := 8\pi G/c^2$ is the (modified) constant of gravity, and μ the mass density of the cosmic substratum. Eqs. (18) are equivalent to the 10 scalar equations obtainable from them by transvection with the tensors $u^i u^j$, $u^i \xi^j_\alpha$, and $\xi^i_\alpha\ \xi^j_\beta$. These transvections can be re-expressed through the defining equations of Ricci's curvature tensor:

$$R_{ij} v^i w^j = 2v^i_{[;i;k]} w^k = \left(v^i_{;i}\right)_{,k} w^k - \left(v^i_{;k} w^k\right)_{;i} + v_{(i;k)} w^{(i;k)} - v_{[i,k]} w^{[i,k]}, \tag{19}$$

whose evaluation, for the vectors u and ξ, can profit from the equations $u^i_{;k} u^k = 0 = u^i u_{i;k}$ and $\xi_{(i;k)} = 0$, which are consequences of eqs. (16) and the geodesy of $u^i(x)$. In this way, satisfying Einstein's field equations for the spatially homogeneous cosmological models has been reduced to calculations which can be performed by hand. This ends Schücking's 1957 presentation.

REFERENCES

[1] Krasiński, A., Christoph G. Behr, Engelbert Schücking, Frank B., Estabrook, Hugo D., Wahlquist, George F. R., Ellis, Robert Jantzen, and Wolfgang Kundt (2003). The Bianchi classification in the Schücking-Behr approach, *Gen. Rel. Grav.* **35**, 475.

Part II
Discussion of physical effects

Chapter 6
Hans Thirring and Josef Lense 1918 – 1921, Machian effects in rotating matter

Original papers:

1. Hans Thirring, Über die Wirkung rotierender ferner Massen in der Einsteinschen Gravitationstheorie. *Physikalische Zeitschrift* **19**, 33 (1918).

2. Hans Thirring, Berichtigung zu meiner Arbeit: "Über die Wirkung rotierender Massen in der Einsteinschen Gravitationstheorie". *Physikalische Zeitschrift* **22**, 29 (1921).

3. Josef Lense and Hans Thirring, Über den Einfluß der Eigenrotation der Zentralkörper auf die Bewegung der Planeten und Monde nach der Einsteinschen Gravitationstheorie. *Physikalische Zeitschrift* **19**, 156 (1918).

Papers translated by Bahram Mashhoon, 223 Physics Building, University of Missouri, Columbia, MO 65211-7010, USA, email: mashhoonb@missouri.edu; Friedrich W. Hehl, Institut für Theoretische Physik, Universität zu Köln, Zülpicher Straße 77, D-50937 Köln, email: hehl@thp.uni-koeln.de; and Dietmar S. Theiß.

The previous publication in the *Gen. Relativ. Gravit.* journal combined the English translations of these papers, the editorial comments and partial biographies of the authors in a single article:

B. Mashhoon, F.W. Hehl and D.S. Theiß, "On the gravitational effects of rotating masses: The Thirring-Lense papers", *Gen. Relativ. Gravit.* **16**, 711 – 750 (1984).

This combined text is copied here with the kind permissions of Hirzel Verlag, Bahram Mashhoon and Friedrich Hehl. For this reprinting, a supplement to the editorial note was added and the biographies of the authors were newly supplied.

6.1 Supplement to the editorial note to

The Thirring – Lense papers: Gravitomagnetism, GP-B and Mach's principle

Bahram Mashhoon

The pioneering work of Hans Thirring [1] and Josef Lense [2] that was highlighted in collaboration with Friedrich Hehl and Dietmar Theiss some three decades ago has in the meantime flourished into the mature subject of gravitoelectromagnetism ("GEM"). Reviews of GEM as well as detailed treatments of its various aspects can be found in [3, 4, 5] and the references cited therein. Among the various gravitomagnetic effects that have been elucidated thus far, one should mention, in particular, the *gravitomagnetic clock effect* [6]. On the experimental side, the exterior gravitomagnetic field of the Earth has now been directly measured via Gravity Probe B ("GP-B") [7, 8, 9]. This magnificent space experiment involved four superconducting gyroscopes and a telescope that were launched on 20 April 2004 into a polar orbit of radius 642 km aboard a drag-free satellite.

The general relativistic precession of a gyroscope in orbit about the Earth would mainly consist of a geodetic effect and a frame-dragging effect due to the rotation of the Earth. For the GP-B orbit, Einstein's theory predicts the respective precession rates of -6606.1 mas/yr and -39.2 mas/yr; the two effects occur in orthogonal directions, and are measured relative to the telescope that tracks the 5.9 magnitude guide star IM Pegasi. Here "mas" stands for milliarcsecond and 1 mas $= 4.848 \times 10^{-9}$ rad. GP-B data collection started in August 2004 and ended in August 2005. The preliminary results of the analysis of 155 days of data were published in 2009 [10]. The complete analysis of GP-B data from all of the four gyroscopes results in a measured geodetic drift rate of -6601.8 ± 18.3 mas/yr and a frame-dragging drift rate of -37.2 ± 7.2 mas/yr [11].

Regarding the observational aspects of the precession of the orbit of a test particle in the field of a rotating source, namely, the *Lense-Thirring effect*, there has also been progress in the analysis of laser ranging to the LAGEOS and LAGEOS II spacecraft [12, 13].

Are inertial forces of gravitational origin? In the early days of *general relativity*, it was thought, in accordance with Mach's principle, that this new relativistic theory of the gravitational field could perhaps explain inertial effects as being due to the gravitational influence of distant masses in the universe. Indeed, this fundamental issue provided the original motivation for Thirring's work on gravitomagnetism. In

Bahram Mashhoon
Department of Physics and Astronomy, University of Missouri
223 Physics Building, Columbia, MO 65211-7010, USA
e-mail: mashhoonb@missouri.edu

our commentary on Thirring's paper that is republished here, we argued against that interpretation on the basis of the modern geometric formulation of Einstein's theory. Meanwhile, our viewpoint has been further developed and expounded in a number of publications [14, 15, 16, 17]; in this approach, the issue of the origin of inertia in general relativity remains essentially the same as in Newtonian mechanics, except that the global inertial frames of Newtonian theory are now replaced by local inertial frames in conformity with Einstein's local principle of equivalence.

Various other viewpoints regarding Mach's principle do still exist, as would become evident from a perusal of [18]. For instance, in Barbour's approach to Mach's principle, the universe as a whole determines the local inertial reference frames of general relativity. In this connection, let us note that the universality of the gravitational interaction implies that every physics experiment in the laboratory is gravitationally influenced by the whole mass-energy content of the universe. In the traditional approach to physics, one starts from the conception of the universe in terms of the Newtonian absolute space and time and works up toward a relativistic framework for observational cosmology. However, Barbour starts from a Machian conception of the universe as a whole. Barbour's theoretical scheme for a fully Machian view of general relativity has so far been implemented for a finite spatially closed universe model [19, 20].

Acknowledgment: I am grateful to Friedrich Hehl for many helpful discussions and for suggesting the original project back in 1981.

References

1. http://en.wikipedia.org/wiki/Hans_Thirring
2. http://en.wikipedia.org/wiki/Josef_Lense
3. J. F. Pascual-Sánchez, L. Floría, A. San Miguel and F. Vicente (eds.), Reference Frames and Gravitomagnetism (World Scientific, Singapore, 2001).
4. R. Ruffini and C. Sigismondi (eds.), Nonlinear Gravitodynamics (World Scientific, Singapore, 2003).
5. L. Iorio (ed.), The Measurement of Gravitomagnetism: A Challenging Enterprise (NOVA Science, Hauppauge, N.Y., 2006).
6. B. Mashhoon, F. Gronwald and H. I. M. Lichtenegger, Lect. Notes Phys. **562**, 83 (2001). arXiv: gr-qc/9912027.
7. http://einstein.stanford.edu
   http://www.gravityprobeb.com
8. C. W. F. Everitt et al., in Near Zero: Festschrift for William M. Fairbank, C. W. F. Everitt , ed. (Freeman, San Francisco, 1986).
9. C. Lämmerzahl, C. W. F. Everitt and F. W. Hehl (eds.), Gyros, Clocks, Interferometers...: Testing Relativistic Gravity in Space, Lect. Notes Phys. **562** (Springer, Berlin, 2001).
10. C. W. F. Everitt et al., Space Sci. Rev. **148**, 53 (2009).
11. C. W. F. Everitt et al., Phys. Rev. Lett. **106**, 221101 (2011).
12. I. Ciufolini and E. C. Pavlis, Nature (London) **431**, 958 (2004).
13. L. Iorio, Space Sci. Rev. **148**, 363 (2008).
14. B. Mashhoon, Phys. Lett. A **126**, 393 (1988).
15. H. Lichtenegger and B. Mashhoon, "Mach's Principle", in The Measurement of Gravitomagnetism: A Challenging Enterprise, L. Iorio, ed. (NOVA Science, Hauppage, N.Y., 2007), chapter 2. arXiv: physics/0407078 [physics.hist-ph].

16. B. Mashhoon and H. Kaiser, "Inertia of Intrinsic Spin", Physica B **385-386**, 1381 (2006). arXiv: quantum-ph/0508182.
17. B. Mashhoon and P.S. Wesson, "Mach's Principle and Higher-Dimensional Dynamics", Ann. Phys. (Berlin) **524**, 63 (2012). arXiv: 1106.6036 [gr-qc].
18. J. Barbour and H. Pfister (eds.), Mach's Principle: From Newton's Bucket to Quantum Gravity, Einstein Studies **6** (Birkhäuser, Boston, 1995).
19. J. Barbour, Ann. Phys. (Berlin) **524**, A39 (2012). arXiv: 1108.3057 [gr-qc].
20. B. Mashhoon and P.S. Wesson, Ann. Phys. (Berlin) **524**, A44 (2012). arXiv: 1108.3059 [gr-qc].

6.2 Hans Thirring – a brief biography

By Andrzej Krasiński, translated and abstracted from Ref. [1]

Hans Thirring was born on 23 March 1888 in Vienna. In 1907 he passed the examination for the secondary school certificate at the Sophiengymnasium and began to study mathematics, physics and physical exercise at the Vienna University. Fascinated by the lectures of Fritz Hasenöhrl he made physics his main subject, but managed, in his spare time, to pass the state examination for gym teachers in 1910.

Hasenöhrl employed Thirring as an assistant at the Institute of Theoretical Physics even before he obtained his PhD degree, in October 1910. He held this post unitl 1921. Thirring's promotion to PhD occurred in 1911, on the basis of the thesis entitled "On some thermodynamical relations in the neighbourhood of the critical point and the triple point". One year later he passed the state examination in mathematics and physics. In 1915 he completed his habilitation, on the basis of theoretical investigations of specific heat of crystals. He became interested in this subject under the influence of his friend Erwin Schrödinger. After Hasenöhrl was killed on the front on 7 October 1915, Thirring took over his mechanics lectures for that academic year.

Soon after, Thirring volunteered to work in the military technical committee on the development of electrical military devices. He became involved in this subject in 1914, with the work on selenium photo-cells.

In 1918, two articles (those reprinted in the following pages) made him one of the most famous champions of the still heavily controversial theory of relativity. In consequence, in 1919 he was employed at the Technical University as a lecturer on physics. In 1920 he gave a course of lectures on mechanics at Vienna University. Then, in 1921 he became Associate Professor of Theoretical Physics at Vienna University and a member of the faculty at the Institute of Theoretical Physics. He became full Professor in 1927.

He had developed his anti-war views already during World War I, and in the 1930s he became increasingly involved in world peace movements. After Austria was incorporated into the German "Reich", Thirring was forcibly furloughed from the University in April 1938, because of "endangering the defence willingness of the German youth" ("Gefährdung des Wehrwillens der deutschen Jugend"), and also in consequence of his personal relations with Einstein and Freud. Shortly thereafter he left his Institute, and in November 1938, at his own request, was retired.

Subsequently, he became scientific advisor in the Elin AG company in Vienna, then from 1942 to 1945 he worked at Siemens & Halske in Vienna. Nevertheless, thanks to the friendly support of his colleague Prof. Ludwig Flamm, his office at the University was at his disposal throughout the Nazi period, and he could work there without being disturbed.

He returned to teaching in the fall of 1945 as a guest professor at the University of Innsbruck, with lectures on "World peace as a psychological problem". In the academic year 1946/47 he became the dean of the Philosophical Faculty at Vienna University. He also took up his post at the Institute of Theoretical Physics. In 1946 he became a corresponding member of the Austrian Academy of Sciences.

With his real retirement approaching, in 1957, Thirring turned away from the usual bureaucratic routine in the search for his successor. Following the US model, he influenced the dean to survey the twelve most important theoreticians of the German-speaking countries (four Nobel laureates among them) to name the best-qualified candidate for this post among the younger Austrian physicists. Eleven of them named Thirring's son Walter.

From 1957 to 1963 Thirring was a member of the higher chamber of the Austrian Parliament (Bundesrat) as a representative of the Austrian Socialist Party (SPÖ), although he was not a member of it. While there, he formulated the plan of one-sided disarmament of Austria, which became known as the Thirring Plan. It included the dissolution of the army; the borders of the country were to be protected by UN troops. Throughout his life he felt connected to the Pugwash movement. He organised the first Pugwash Conference in Vienna in 1957. For his peace activities Thirring was twice nominated for the Nobel peace prize.

Apart from his professional duties, Thirring was engaged in several other activities. He was a member of, among others, the German-Austrian Alpine Club (D-ÖAV), the Ghibellineo Choir, the German Physical Society, the Chemical-Physical Society, the German Society for Technical Physics, the Paneuropa Union, the Union of German High School Teachers, the German Ski Club, the Biophysical Society for Shortwave Research, the Electrotechnical Union and the American Physical Society.

He died in Vienna on 22 March 1976, one day before his 88th birthday.

In addition to being a scientist, teacher and politician, Thirring was also an inventor. In 1927/28 he constructed a sound-recording movie camera and projector that worked with use of the "Thirring selenium cells". These apparatuses were used until 1938 to produce the Austrian weekly film chronicle and also short movies. He had several patents for his other inventions, 2 in Austria and one in each of Germany, Spain, France and Italy. In 1937 he organised the "First International Congress on Shortwave Research" in Vienna. Being a dedicated skier he constructed the "Thirring coat", famous in its time, a kind of sail spanned between the arms and legs of a skier that gave the illusion of gliding. To accompany this, in 1939 he wrote the book "Der Schwebelauf" ("The gliderun").

Thirring was active as a popularizer of the theory of relativity and of the quantum theory. He himself considered as his most important achievement the book "Homo sapiens. Psychology of human relations", written in 1938 – 1945 (published in 1947, and in short form in 1953, under the title "The art of human coexistence", by the Austrian UNESCO committee). Its subjects were the most common behavioural errors in private life and in politics, and the possibilities of avoiding or correcting them.

Selected publications by Hans Thirring (apart from those mentioned above and those reprinted further on):

• Die Idee der Relativitätstheorie (1921), 3rd edition, corrected and extended: Springer, Vienna 1948; translated among others to English, French, Japanese and Swedish.
• (with Otto Halpern) Grundgedanken der neueren Quantentheorie, Teil 1 u. 2, Springer 1928/29.
• Alte Probleme – neue Lösungen in den exakten Wissenschaften (1934).
• Geschichte der Atombombe. Mit einer elementaren Einführung in die Atomphysik (1946).
• Der Weltfriede als psychologisches Problem (1946).
• Atomkrieg und Weltpolitik (1948).
• Wissenschaftliches Denken im Alltag und in der Weltpolitik (1951).
• Power Production. The Practical Application of World Energy (1956).
• Der Weg der theoretischen Physik von Newton bis Schrödinger. Springer, Vienna 1962.
• Kernenergie gestern, heute und morgen (1963, with H. Grümm).

The source for this text [1] credits Ref. [2] as the origin of all information reported here. Detailed biographies of Hans Thirring may be found in Refs. [3, 4].

References

1. http://austria-lexikon.at/af/Wissenssammlungen/Biographien/Thirring,_Hans
2. Walter Höflechner, *Materialien zur Entwicklung der Physik und ihrer 'Randfächer' Astronomie und Meteorologie an den österreichischen Universitäten 1752–1938 [Materials for the development of physics and its 'peripheral professions', astronomy and meteorology, at the Austrian universities 1752–1938]*. 1992 – 2002, unpublished.
3. Gabriele Kerber, Ulrike Smola, Brigitte Zimmel (eds.), *Hans Thirring - ein Homo sapiens: Zitate, Bilder und Dokumente anlässlich der 101. Wiederkehr seines Geburtstages [Hans Thirring - Homo sapiens: quotations, pictures and documents on the occasion of the 101st anniversary of his birthday]*. Fassbaender, Wien 1989 (catalogue of an exhibition).
4. Brigitte Zimmel and Gabriele Kerber (eds.), *Hans Thirring: ein Leben für Physik und Frieden [Hans Thirring: a life for physics and peace]*. Böhlau, Wien 1992.

6.3 Josef Lense – a brief biography

By Andrzej Krasiński, translated and compiled from Refs. [1], [2] and [3].

Josef Lense was born in Vienna on 28 October 1890. He studied mathematics, astronomy and theoretical physics at the University of Vienna in 1909 – 1913. He obtained his PhD degree on 13 February 1914 on the basis of the work entitled "The jovicentral movement of the small planets" [4]; his adviser was Samuel Oppenheim. His habilitation work, done with Wilhelm Wirtinger and completed in 1921, was "On the integration of a p-fold differential expression depending on n independent variables".

In April 1927 Lense was appointed associate professor for applied mathematics at the Technische Hochschule (later renamed Technische Universität) in Munich. In 1928 he became full professor there, and in 1946 the director of the Institute of Mathematics and Analytical Mechanics. He held this post until 1961, even though he retired in 1958. Several generations of students of mathematics, physics, chemistry, engineering and architecture at the TU owe their mathematical education to Lense. The lecture room could accommodate 860 persons, and Lense's audiences indeed included hundreds of students.

His first area of research activity was astronomy, to which he would return from time to time in later years. In mathematics, he worked, among other things, on properties of special functions (the Bessel functions, their conformal transformations and zeros, the zeros of the Bernoulli polynomials, asymptotic expansions of the Hankel functions) and on complex differential geometry (isotropic manifolds). He also worked on several problems of theoretical physics, among them Newton's law in non-Euclidean spaces, kinetic gas theory and a function-theoretical discussion of the three body problem; plus, of course, the paper reprinted here. Several textbooks written by him ([5], which had its 3rd edition before 1980, and [6] – [10]) are characteristic for their precision and concise presentation. Ref. [6] contains a complete presentation of advanced mathematics on just 260 pages. He is also remembered for his contributions to the "Handbuch für Physik" (1928 and 1955).

Music played an important role in the life of Josef Lense. He had education in singing and cultivated it throughout his life, most often accompanying himself on a piano. His lectures at the TU on "Mathematics and music" were famous. He would illustrate his considerations about symmetry of a piece of music with his own performance. The author of Ref. [1] fondly recalls how Lense's perfect command of the piano allowed the two of them to play together violin sonatas by Händel, Haydn, Mozart, Beethoven and Schubert. Lense was also an opera enthusiast. He owned the opera scores by Wagner and Verdi and used them to play the pieces himself.

Lense's work in science earned him several honours. In 1948 he became a member of the Bavarian Academy of Sciences, from 1950 to 1961 he was a member of Circolo Matematico di Palermo, and from 1968 on he was a corresponding member of the Austrian Academy of Sciences. From 1950 to 1952 he was the Dean and deputy Dean of the Faculty of Science at the TH München. In 1966 he received the Bavarian Order of Merit.

Josef Lense died on 28 December 1985 in Munich. He was aware of the Thirring – Lense papers being prepared for reprinting in the *General Relativity and Gravitation* journal. He obtained a preliminary typescript of the freshly translated articles from Friedrich Hehl, to which he responded with gratitude and appreciation, and with a few suggestions of corrections. He also supplied a few details of his CV, which are included in the above.[1]

References

1. Josef Heinhold, Prof. Lense 90.
   http://homepage.mac.com/titurel/MathApprObit/LenseHeinhold.pdf [a copy from *TUM Mitteilungen 6/80*] [TUM = Technische Universität München].
2. Peter Vachenauer, Josef Lense 1890 – 1985.
   http://www-hm.ma.tum.de/geschichte/node22.html
3. Peter Vachenauer, A brief CV of Josef Lense. *Int. J. Mod. Phys.* **D14**, 1973 – 1975 (2005).
4. J. Lense, *Astronomische Nachrichten* **196**, 341 (1914).
5. J. Lense, *Reihenentwicklungen in der Mathematischen Physik [Series expansions in mathematical physics]*. De Gruyter, Berlin, 1933.
6. J. Lense, *Vorlesungen über Höhere Mathematik [Lectures on advanced mathematics]*. Oldenbourg, Munich, 1948.
7. J. Lense, *Vom Wesen der Mathematik und ihren Grundlagen [On the gist of mathematics and its foundations]* (1949).
8. J. Lense, *Kugelfunktionen [Spherical functions]*. Akademische Verlagsgesellschaft, Leipzig 1950. Second edition 1954.
9. J. Lense, *Kurze Einführung in die Differentialgeometrie des Raumes [A brief introduction to differential geometry of space]* (1953).
10. J. Lense, *Analytische projektive Geometrie [Analytic projective geometry]*. Oldenbourg, Munich 1965.

[1] Copy of a letter sent on 28 Oct. 1981 by Josef Lense to Friedrich Hehl, kindly supplied to A. K. by Bahram Mashhoon.

General Relativity and Gravitation, Vol. 16, No. 8, 1984

On the Gravitational Effects of Rotating Masses: The Thirring–Lense Papers[1]

BAHRAM MASHHOON, FRIEDRICH W. HEHL, and DIETMAR S. THEISS

Institut für Theoretische Physik, Universität zu Köln, D-5000 Cologne 41, Federal Republic of Germany

Received May 30, 1982

Abstract

The purpose of this work is to provide a critical analysis of the classical papers of H. Thirring [*Phys. Z.*, **19**, 33 (1918); *Phys. Z.*, **22**, 29 (1921)] and J. Lense and H. Thirring [*Phys. Z.*, **19**, 156 (1918)] on rotating masses in the relativistic theory of gravitation and to render them accessible to a wider circle of scholars. An English translation of these papers is presented which follows the original German text as closely as possible. This is followed by a concise account of the significance of the results of these papers as well as the possibility of measuring the gravitational effects of rotating masses.

§(1): *Introduction*

Hans Thirring (1888–1976) was born and educated in Vienna. He studied mathematics and physics at the University of Vienna and received his Ph.D. in 1911. His habilitation thesis of 1915 was on the theory of the specific heat of solids. His major fields of interest were the theory of relativity and applied physics; moreover, he had wide-ranging interests in psychology, politics, and in issues related to world energy resources. He became professor of physics in Vienna in 1921. For political reasons he had to leave the University in 1938. In 1945 he was reinstated and became the director of the Institute for Theoretical Physics. Among the books he authored there is one entitled *Die Idee der Relativitätstheorie*.

Josef Lense (b. 1890) was born and educated in Vienna. He studied astronomy at the University, obtaining his Ph.D. in 1914 [cf. Die jovizentrische Bewe-

[1] Dedicated to Professor Josef Lense on the occasion of his 92nd birthday.

0001-7701/84/0800-0711$03.50/0 © 1984 Plenum Publishing Corporation

gung der kleinen Planeten, *Astron. Nachrichten*, **196**, 341–346 (1913)]. His habilitation thesis of 1921 on the integration of differential expressions, as well as his subsequent work on quadratic forms, was motivated by the geometric aspects of the relativity theory. He became a lecturer in Vienna in 1921, and a professor of mathematics at the Technical University of Munich in 1927. In addition, from 1946 until his retirement, he was the director of the Mathematical Institute in Munich. He has authored many books in mathematics, a subject in which his major fields of interest have been differential geometry and mathematical physics. He presently makes his home in Munich.

The collaboration of Lense and Thirring began in Vienna after Thirring, aware of Lense's background in astronomy, inquired whether Lense would be willing to carry through, in perturbation theory, the integration of the equations of motion of a test body in the field of a rotating mass together with the application of the results to the orbits of the planets and moons [1]. This resulted in a joint publication in the *Physikalische Zeitschrift* of 1918.

§(2): *English Translation of the Thirring-Lense Papers*

On the Effect of Rotating Distant Masses in Einstein's Theory of Gravitation

HANS THIRRING

The considerations of this study can best be made clear by a quotation from Einstein's fundamental paper of 1914.[1] He says the following in the introduction:

"At first it seems that such an extension of the theory of relativity has to be rejected for physical reasons. Namely, let K be a permissible coordinate system in the Galilei-Newtonian sense, K' a uniformly rotating coordinate system with respect to K. Then centrifugal forces act on masses which are at rest relative to K', while not on masses at rest relative to K. Already Newton considered this as proof that the rotation of K' has to be interpreted as "absolute," and that one cannot thus treat K' as "at rest" with the same justification as K. This argument is not valid, however, as explained by E. Mach in particular. That is, we need not attribute the existence of those centrifugal forces to the motion of K'; rather, we can attribute them as well to the average rotational motion of the ponderable distant masses in the surrounding relative to K', whereby we treat K' as at rest. If the Newtonian laws of mechanics do not admit such a conception, this could well be caused by the defects of this theory. . . ."

Since Einstein's theory seems to have been brought to completion in the

[1] A. Einstein, *Berl. Ber.*, 1914, p. 1030; see also *Ann. Phys. (Leipzig)*, **49**, p. 769 (1916).

publications of 1915, the question suggests itself: Is the new theory devoid of the defects of the Newtonian theory insofar as the rotation of distant masses, according to their equations, indeed produces a gravitational field equivalent to a "centrifugal field"? Perhaps one could regard a discussion of this question futile by stating that the required equivalence is guaranteed by the general covariance of the field equations. But things are not quite as simple since the boundary conditions for $g_{\mu\nu}$ at spatial infinity play a role as well. These questions of principal interest have been treated in the papers of De Sitter[1] and Einstein.[2] We will not go into these general questions, rather we will calculate and study the field of rotating distant masses using a specific and concrete example. For this purpose, the method of approximate integration of the field equations given by Einstein[3] is perfectly appropriate. This method will serve as the foundation for the following calculations. We choose as an example the field inside a uniformly rotating infinitely thin hollow sphere endowed with a constant mass density.

In the first section of this paper (which can be skipped without affecting the understanding of the rest) the $g_{\mu\nu}$ inside the spherical shell will be calculated approximately; in the second section the motion of a point mass in this field will be discussed.

A. Calculational Part: Calculation of the $g_{\mu\nu}$ for the Neighborhood of the Center of the Rotating Hollow Sphere

Notation:

a, radius of the hollow sphere;
M, its mass;
ω, its angular velocity;
x, y, z, rectangular coordinates of a point on the surface of the sphere;
x_0, y_0, z_0, coordinates of the point under consideration;
χ, gravitational constant;
ρ_0, naturally measured space density of matter.

In connection with the approximations used for calculating the field, the following should be mentioned in advance: The field in the neighborhood of the center of the sphere will be considered to be so weak that in the field equations only terms of first order with respect to $\gamma_{\mu\nu}$ are taken into account ($\gamma_{\mu\nu}$ is defined by $g_{\mu\nu} = -\delta_{\mu\nu} + \gamma_{\mu\nu}$). Since the higher-order terms are neglected, it is possible to use Einstein's method for the approximate integration of the field

[1] W. de Sitter, *Amsterdam Proc.*, **19**, 527, 1917.
[2] A. Einstein, *Berl. Ber.*, 1917, p. 142.
[3] A. Einstein, *Berl. Ber.*, 1916, p. 688.

equations. According to the second type of approximation often used, the components of the velocity of the ponderable matter will be considered as small with respect to unity (velocity of light), so that in this crudest approximation, which yields Newton's theory, already the first powers can be neglected. Let us apply this approximation—which is totally independent of the first one—insofar as we drop the terms of third and higher orders in the velocities in comparison to 1. Finally, our calculations relate to the neighborhood of the center of the sphere; let r be the distance between the point under consideration and the center of the sphere and let R be the distance between the point under consideration and the integration element. Then we will expand $1/R$ in a power series in r/a up to second order.

Einstein's method of approximate integration yields for the computation of the $g_{\mu\nu}$ the following formulas:

$$g_{\mu\nu} = -\delta_{\mu\nu} + \gamma_{\mu\nu}, \qquad \delta_{\mu\nu} = 1, \quad \mu = \nu \qquad (1)$$

$$= 0, \quad \mu \neq \nu$$

$$\gamma_{\mu\nu} = \gamma_{\mu\nu}' - \frac{1}{2} \sum_{\alpha} \gamma_{\alpha\alpha}' \qquad (2)$$

$$\gamma_{\mu\nu}' = -\frac{\chi}{2\pi} \int \frac{T_{\mu\nu}(x, y, z, t - r)}{R} \, dV_0 \qquad (3)$$

Here $T_{\mu\nu}$ denotes the covariant energy tensor of matter and dV_0 the spatial volume element of integration space,

$$R^2 = (x - x_0)^2 + (y - y_0)^2 + (z - z_0)^2$$

The coefficients $g_{\mu\nu}$ of the line element refer to the coordinates

$$x_1 = x, \quad x_2 = y, \quad x_3 = z, \quad x_4 = it$$

According to the first type of approximation, we are allowed to substitute the covariant energy tensor by the contravariant one. Neglecting the stresses, it is given by

$$T_{\mu\nu} = T^{\mu\nu} = \rho_0 \frac{dx_\mu}{ds} \frac{dx_\nu}{ds}$$

$$= \rho_0 \frac{dx_\mu}{dx_4} \frac{dx_\nu}{dx_4} \left(\frac{dx_4}{ds} \right)^2 \qquad (4)$$

Let the hollow sphere rotate with angular velocity ω about the z axis, then for one of its points with polar coordinates, a, ϑ, φ, we have

$$\frac{dx_1}{dx_4} = -i\frac{dx}{dt} = ia\omega \sin\vartheta \sin\varphi$$

$$\frac{dx_2}{dx_4} = -i\frac{dy}{dt} = -ia\omega \sin\vartheta \cos\varphi \tag{5}$$

$$\frac{dx_3}{dx_4} = 0$$

These values upon substitution into (4) yields the following scheme for $T_{\mu\nu}$:

$$T_{\mu\nu} = \rho_0 \left(\frac{dx_4}{ds}\right)^2 \cdot \left\{\begin{matrix} -a^2\omega^2 \sin^2\vartheta \sin^2\varphi & +a^2\omega^2 \sin^2\vartheta \sin\varphi\cos\varphi & 0 & ia\omega \sin\vartheta \sin\varphi \\ +a^2\omega^2 \sin^2\vartheta \sin\varphi\cos\varphi & -a^2\omega^2 \sin^2\vartheta \cos^2\varphi & 0 & -ia\omega \sin\vartheta \cos\varphi \\ 0 & 0 & 0 & 0 \\ ia\omega \sin\vartheta \sin\varphi & -ia\omega \sin\vartheta \cos\varphi & 0 & 1 \end{matrix}\right\} \tag{6}$$

Since we take ρ_0 to be the naturally measured matter density, one has to insert the naturally measured spatial volume element for dV_0 as well in order to ensure the tensorial character of the integral (3). For this the following formula[1] holds:

$$dV_0 = \sqrt{g}\, i \frac{dx_4}{ds} dV \tag{7}$$

For the integration we introduce polar coordinates. Thus

$$\sqrt{g}\, dV = a^2 da \sin\vartheta\, d\vartheta\, d\varphi \tag{8}$$

Finally, one still has to express $1/R$ in terms of the integration variables. We choose the coordinate system such that the point under consideration is situated in the Z-X plane. Its coordinates are then

$$x_0 = r\sin\vartheta_0, \quad y_0 = 0, \quad z_0 = r\cos\vartheta_0$$

Then we have

$$\begin{aligned} R^2 &= (a\sin\vartheta\cos\vartheta - r\sin\vartheta_0)^2 + a^2\sin^2\vartheta\sin^2\varphi + (a\cos\vartheta - r\cos\vartheta_0)^2 \\ &= a^2\left[1 - \frac{2r}{a}(\sin\vartheta\cos\varphi\sin\vartheta_0 + \cos\vartheta\cos\vartheta_0) + \frac{r^2}{a^2}\right] \end{aligned}$$

Neglecting the terms mentioned at the beginning, the expansion in a binomial series yields

[1] See A. Einstein, *Berl. Ber.*, 1914, 1058, equation (47a).

$$\frac{1}{R}=\frac{1}{a}\left\{1+\frac{r}{a}(\sin\vartheta\cos\varphi\sin\vartheta_0+\cos\vartheta\cos\vartheta_0)-\frac{1}{2}\frac{r^2}{a^2}\right.$$
$$\left.+\frac{3}{2}\frac{r^2}{a^2}(\sin\vartheta\cos\varphi\sin\vartheta_0+\cos\vartheta\cos\vartheta_0)^2\right\} \tag{9}$$

We denote the expression inside the curly bracket by K and write

$$\frac{1}{R}=\frac{K}{a} \tag{9a}$$

Substitution of (6), (7), (8), and (9a) into (3) yields

$$\gamma_{11}{}'=\frac{i\chi}{2\pi}\rho_0 a^3\omega^2\,da\int_0^{2\pi}d\varphi\int_0^{\pi}d\vartheta\left(\frac{dx_4}{ds}\right)^3\sin^3\vartheta\sin^2\varphi K$$

$$\gamma_{22}{}'=\frac{i\chi}{2\pi}\rho_0 a^3\omega^2\,da\int_0^{2\pi}d\varphi\int_0^{\pi}d\vartheta\left(\frac{dx_4}{ds}\right)^3\sin^3\vartheta\cos^2\varphi K$$

$$\gamma_{44}{}'=-\frac{i\chi}{2\pi}\rho_0 a\,da\int_0^{2\pi}d\varphi\int_0^{\pi}d\vartheta\left(\frac{dx_4}{ds}\right)^3\sin\vartheta K$$

$$\gamma_{12}{}'=-\frac{i\chi}{2\pi}\rho_0 a^3\omega^2\,da\int_0^{2\pi}d\varphi\int_0^{\pi}d\vartheta\left(\frac{dx_4}{ds}\right)^3\sin^3\vartheta\sin\varphi\cos\varphi K$$

$$\gamma_{14}{}'=\frac{\chi}{2\pi}\rho_0 a^2\omega\,da\int_0^{2\pi}d\varphi\int_0^{\pi}d\vartheta\left(\frac{dx_4}{ds}\right)^3\sin^2\vartheta\sin\varphi K$$

$$\gamma_{24}{}'=-\frac{\chi}{2\pi}\rho_0 a^2\omega\,da\int_0^{2\pi}d\varphi\int_0^{\pi}d\vartheta\left(\frac{dx_4}{ds}\right)^3\sin^2\vartheta\cos\varphi K$$

$$\gamma_{13}{}'=\gamma_{23}{}'=\gamma_{33}{}'=\gamma_{43}{}'=0 \tag{10}$$

The absolute value of the quantity dx_4/ds differs from unity only by terms of order $\omega^2 a^2$; it appears as a factor in the small first-order terms $\gamma_{\mu\nu}{}'$; hence it is sufficient to calculate them starting with the expression for the line element in the "zeroth" approximation:

$$ds^2=-dx_1{}^2-dx_2{}^2-dx_3{}^2-dx_4{}^2$$
$$\frac{ds^2}{dx_4{}^2}=-1-\frac{dx_1{}^2+dx_2{}^2+dx_3{}^2}{dx_4{}^2}$$
$$=-1+\omega^2a^2\sin^2\vartheta$$

$$\frac{ds}{dx_4} = i\left(1 - \frac{\omega^2 a^2}{2}\sin^2\vartheta\right)$$

$$\left(\frac{dx_4}{ds}\right)^3 = i\left(1 + \frac{3}{2}\omega^2 a^2 \sin^2\vartheta\right) \tag{11}$$

Because the accuracy of our calculation extends only to terms of order $\omega^2 a^2$, we can put $(dx_4/ds)^3 = i$ for all those $\gamma_{\mu\nu}'$ which already contain the factor ωa; merely for γ_{44}' will we use the expression (11). Furthermore, we put

$$\rho_0 da = \sigma$$

and then equations (10) become

$$\gamma_{11}' = -\frac{\chi}{2\pi}\sigma a^3\omega^2 \int_0^{2\pi} d\varphi \int_0^{\pi} d\vartheta \sin^3\vartheta \sin^2\varphi\, K$$

$$\gamma_{22}' = -\frac{\chi}{2\pi}\sigma a^3\omega^2 \int_0^{2\pi} d\varphi \int_0^{\pi} d\vartheta \sin^3\vartheta \cos^2\varphi\, K$$

$$\gamma_{44}' = \frac{\chi}{2\pi}\sigma a \int_0^{2\pi} d\varphi \int_0^{\pi} d\vartheta \sin\vartheta\, K \left(1 + \frac{3}{2}\omega^2 a^2 \sin^2\vartheta\right)$$

$$\gamma_{12}' = \frac{\chi}{2\pi}\sigma a^3\omega^2 \int_0^{2\pi} d\varphi \int_0^{\pi} d\vartheta \sin^3\vartheta \sin\varphi\cos\varphi\, K$$

$$\gamma_{14}' = \frac{i\chi}{2\pi}\sigma a^2\omega \int_0^{2\pi} d\varphi \int_0^{\pi} d\vartheta \sin^2\vartheta \sin\varphi\, K$$

$$\gamma_{24}' = -\frac{i\chi}{2\pi}\sigma a^2\omega \int_0^{2\pi} d\varphi \int_0^{\pi} d\vartheta \sin^2\vartheta \cos\varphi\, K$$

$$\gamma_{13}' = \gamma_{23}' = \gamma_{33}' = \gamma_{34}' = 0 \tag{12}$$

If one substitutes the value of K from (9) into these expressions and calculates the integrals, one obtains

$$\gamma_{11}' = -\frac{\chi}{2\pi}\frac{M}{3a}a^2\omega^2\left(1 - \frac{r^2}{5a^2}\right)$$

$$\gamma_{22}' = -\frac{\chi}{2\pi}\frac{M}{3a}a^2\omega^2\left\{1 - \frac{r^2}{5a^2}\left(1 - 3\sin^2\vartheta_0\right)\right\}$$

$$\gamma_{44}' = \frac{\chi}{2\pi}\frac{M}{a}\left\{1 + a^2\omega^2\left[1 - \frac{r^2}{5a^2}\left(1 - \frac{3}{2}\sin^2\vartheta_0\right)\right]\right\}$$

$$\gamma_{24}' = -\frac{i\chi}{2\pi}\frac{M}{3a}\omega r\sin\vartheta_0$$

$$\gamma_{12}' = \gamma_{14}' = \gamma_{13}' = \gamma_{23}' = \gamma_{33}' = \gamma_{43}' = 0 \tag{13}$$

From these expressions one obtains the $\gamma_{\mu\nu}$ and subsequently the $g_{\mu\nu}$ by using the equations (1) and (2). Furthermore, we substitute in place of the polar coordinates r and ϑ_0 of the point under consideration its rectangular coordinates and replace the Einstein gravitational constant χ by the conventional one: $k = \chi/8\pi$ (velocity of light = 1). Then one gets

$$g_{11} = -1 - \frac{2kM}{a}\left\{1 + a^2\omega^2 - \frac{\omega^2}{10}(2z_0{}^2 + x_0{}^2)\right\}$$

$$g_{22} = -1 - \frac{2kM}{a}\left\{1 + a^2\omega^2 - \frac{\omega^2}{10}(2z_0{}^2 - 3x_0{}^2)\right\}$$

$$g_{33} = -1$$

$$g_{44} = -1 + \frac{2kM}{a}\left\{1 + \frac{5a^2\omega^2}{3} - \frac{\omega^2}{6}(2z_0{}^2 - x_0{}^2)\right\}$$

$$g_{24} = -i\frac{4kM}{3a}\omega x_0 \tag{14}$$

all remaining $g_{\mu\nu}$ vanish.

Now we want to free ourselves also from the special choice of the coordinate system. (We had placed the point under consideration in the Z–X plane.) For this purpose, we perform the transformation:

$$x_1' = x_1\cos\alpha + x_2\sin\alpha$$

$$x_2' = -x_1\sin\alpha + x_2\cos\alpha$$

$$x_3' = x_3$$

$$x_4' = x_4 \tag{15}$$

Then by means of the transformation law of a covariant tensor of second rank

$$g_{\sigma\tau}' = \frac{\partial x_\mu}{\partial x_\sigma'}\frac{\partial x_\nu}{\partial x_\tau'}g_{\mu\nu}$$

the coefficient scheme becomes

$$g_{\mu\nu} = \begin{Bmatrix} -1-\dfrac{2kM}{a}\left[1+a^2\omega^2-\dfrac{\omega^2}{10}(2z^2+x^2-3y^2)\right], & +\dfrac{2kM}{a}\dfrac{\omega^2}{5}xy, & 0, & +i\dfrac{4kM}{3a}\omega y \\ +\dfrac{2kM}{a}\dfrac{\omega^2}{5}xy, & -1-\dfrac{2kM}{a}\left[1+a^2\omega^2-\dfrac{\omega^2}{10}(2z^2-3x^2+y^2)\right], & 0, & -i\dfrac{4kM}{3a}\omega x \\ 0, & 0, & -1, & 0, \\ +i\dfrac{4kM}{3a}\omega y, & -i\dfrac{4kM}{3a}\omega x, & 0, & -1+\dfrac{2kM}{a}\left[1+\dfrac{5a^2\omega^2}{3}+\dfrac{\omega^2}{6}(2z^2-x^2-y^2)\right] \end{Bmatrix} \tag{16}$$

The index 0 in the coordinates is dropped here; in the following x, y, z denote the coordinates of the point under consideration.

B. Physical Part: The Motion of a Point Mass inside the Rotating Hollow Sphere. We want to set up the equations of motion of a point mass situated near the center of our rotating hollow sphere. The field in this neighborhood is characterized by the coefficient scheme of the $g_{\mu\nu}$ [equation (16) of the first section].

As is well known, the law of motion for a point mass in Einstein's theory is given by the condition $\delta \int ds = 0$, or, if one carries through the variation,[1]

$$\frac{d^2x_\tau}{ds^2} = \Gamma^\tau_{\mu\nu}\frac{dx_\mu}{ds}\frac{dx_\nu}{ds}, \qquad \tau = 1\ldots4 \tag{17}$$

We have, according to the first type of approximation, for the "field components" $\Gamma^\tau_{\mu\nu}$

$$\Gamma^\tau_{\mu\nu} = -\begin{Bmatrix}\mu\nu\\ \tau\end{Bmatrix} = \begin{bmatrix}\mu\nu\\ \tau\end{bmatrix} = \frac{1}{2}\left(\frac{\partial g_{\tau\nu}}{\partial x_\mu} + \frac{\partial g_{\mu\tau}}{\partial x_\nu} - \frac{\partial g_{\mu\nu}}{\partial x_\tau}\right) \tag{18}$$

We merely want to consider motions of the point mass which are small with respect to the velocity of light, so that we can neglect the squares and products of the components of the velocity. Then one can cancel all those terms on the right-hand side of the equations (17) in which the index 4 does not appear, and, besides, we can substitute the derivatives with respect to s by the derivatives with respect to t. Taking into account that $dx_4/dt = i$, the equations (17) become

$$\frac{d^2x_\tau}{dt^2} = 2i\left(\Gamma^\tau_{14}\frac{dx_1}{dt} + \Gamma^\tau_{24}\frac{dx_2}{dt} + \Gamma^\tau_{34}\frac{dx_3}{dt}\right) - \Gamma^\tau_{44} \tag{19}$$

Hence in what follows, only those of the field components $\Gamma^\tau_{\mu\nu}$ are taken into

[1] Indices appearing twice are to be summed over from 1 to 4.

consideration which contain the index 4 at least once. These are 16 quantities, which (though they are not tensorial components !) can be arranged in our case according to the scheme of an antisymmetric tensor of second rank. Since the partial derivatives with respect to x_4 vanish altogether in the stationary field, the quantities $\Gamma^{\tau}_{\sigma 4}$ can be written in the following way:[2]

$$\Gamma^1_{14} = 0, \qquad \Gamma^1_{24} = \frac{1}{2}\left(\frac{\partial g_{14}}{\partial x_2} - \frac{\partial g_{24}}{\partial x_1}\right),$$

$$\Gamma^1_{34} = \frac{1}{2}\left(\frac{\partial g_{14}}{\partial x_3} - \frac{\partial g_{34}}{\partial x_1}\right), \qquad \Gamma^1_{44} = -\frac{1}{2}\frac{\partial g_{44}}{\partial x_1}$$

$$\Gamma^2_{14} = \frac{1}{2}\left(\frac{\partial g_{24}}{\partial x_1} - \frac{\partial g_{14}}{\partial x_2}\right), \qquad \Gamma^2_{24} = 0,$$

$$\Gamma^2_{34} = \frac{1}{2}\left(\frac{\partial g_{24}}{\partial x_3} - \frac{\partial g_{34}}{\partial x_2}\right), \qquad \Gamma^2_{44} = -\frac{1}{2}\frac{\partial g_{44}}{\partial x_2} \tag{20}$$

$$\Gamma^3_{14} = \frac{1}{2}\left(\frac{\partial g_{34}}{\partial x_1} - \frac{\partial g_{14}}{\partial x_3}\right), \qquad \Gamma^3_{24} = \frac{1}{2}\left(\frac{\partial g_{34}}{\partial x_2} - \frac{\partial g_{24}}{\partial x_3}\right),$$

$$\Gamma^3_{34} = 0, \qquad \Gamma^3_{44} = -\frac{1}{2}\frac{\partial g_{44}}{\partial x_3}$$

$$\Gamma^4_{14} = \frac{1}{2}\frac{\partial g_{44}}{\partial x_1}, \qquad \Gamma^4_{24} = \frac{1}{2}\frac{\partial g_{44}}{\partial x_2},$$

$$\Gamma^4_{34} = \frac{1}{2}\frac{\partial g_{44}}{\partial x_3}, \qquad \Gamma^4_{44} = 0$$

If one substitutes here the special values of the $g_{\mu\nu}$ from (16), one gets the following scheme:

[1]Note that this scheme corresponds to the six-vector $\mathfrak{M}$ of the electromagnetic field. The analogy between electrodynamics and (approximated) gravitational theory goes still further when one bears in mind that by the approximate integration the quantities $g_{14}, g_{24}, g_{34}, g_{44}$ are calculated in the same way in terms of the density and velocity of matter as the potentials $\mathfrak{A}_x$, $\mathfrak{A}_y$, $\mathfrak{A}_x$, Φ in terms of the electric four-current, and furthermore, that in our case the right-hand sides of equations (19) correspond altogether, except for numerical factors, to the components of the ponderomotive force $\mathfrak{E} + [\mathfrak{v}\mathfrak{H}]$!

$$\begin{matrix} 0 & i\dfrac{4kM}{3a}\omega & 0 & -\dfrac{kM}{3a}\omega^2 x \\ -i\dfrac{4kM}{3a}\omega & 0 & 0 & -\dfrac{kM}{3a}\omega^2 y \\ 0 & 0 & 0 & \dfrac{2kM}{3a}\omega^2 z \\ \dfrac{kM}{3a}\omega^2 x & \dfrac{kM}{3a}\omega^2 y & -\dfrac{2kM}{3a}\omega^2 z & 0 \end{matrix} \qquad (21)$$

Then, we obtain the equations of motion for our specific problem from (19) to (21):

$$\begin{aligned} \ddot{x} &= -\frac{8kM}{3a}\omega\dot{y} + \frac{kM}{3a}\omega^2 x \\ \ddot{y} &= \frac{8kM}{3a}\omega\dot{x} + \frac{kM}{3a}\omega^2 y \\ \ddot{z} &= -\frac{2kM}{3a}\omega^2 z \end{aligned} \qquad (22)$$

The right-hand sides of the equations represent the components of the force which our field exerts on the point mass with mass 1. As can be seen, the first terms of the X and Y components correspond to the Coriolis force and the second terms to the centrifugal force. At first sight, the third equation yields the surprising result that this "centrifugal force" possesses an axial component as well. Its occurrence in the field of the rotating sphere can be elucidated in the following way: From the viewpoint of the observer at rest, those surface elements of the hollow sphere situated near the equator have a larger velocity, and consequently a larger apparent (inertial and gravitating) mass, than those close to the poles. Hence the field of a rotating hollow sphere with a constant surface density corresponds to the field of a hollow sphere at rest with a surface density which increases with the polar angle ϑ. In the latter case, it is readily understandable that points lying outside the equatorial plane are dragged into it.

(Incidentally, one can also easily imagine that forces analogous to the centrifugal force occur in the interior of a hollow sphere endowed with such a nonuniform mass density. As is well known, in potential theory one can demonstrate the vanishing of the force field inside a hollow sphere with constant surface charge density as follows: The attractive force of the surface elements, which are seen from P under the angle $d\omega$ (see figure 1), is equal and opposite

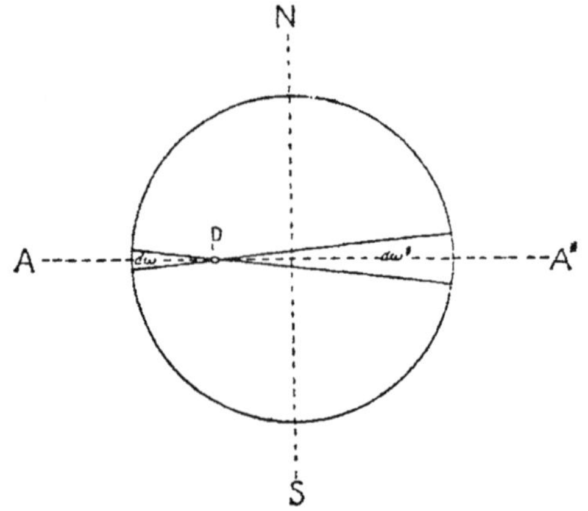

Fig. 1.

to the force exerted by those surface elements which lie within the opposite angle $d\omega'$. Of course, this no longer applies in the case of nonuniform surface density. Let AA' be the equatorial plane; then, with respect to the position of the point P as drawn in the figure, the surface elements lying inside the angle $d\omega$ are closer to the equator on the average, and are therefore specifically heavier than those of $d\omega'$. Hence there results a force in the direction of $A'A$, i.e., a tension that is perpendicular to the rotation axis, points away from it, and becomes smaller the closer the point P is shifted to the center.)

That we merely recognize a radial but never an axial component of the centrifugal force in nature could be reconciled with the results found here as follows: The approximation of the firmament by means of an infinitely thin hollow sphere is just incorrect. But even if we want to improve our approximation (say, by means of a spatial distribution of masses) we would never obtain a field equivalent to a real centrifugal field when using this method of integration. We would obtain such a field if we imagined all masses in the outer space (galaxies, etc.) as rotating and calculated their gravitational effect. But the solution for the retarded potentials [equation (3)] assumes the boundary condition $\lim \gamma_{\mu\nu} = 0$ for spatial infinity. As Einstein has shown in his cosmological paper,[1] these boundary conditions are approximately fulfilled for a coordinate system with respect to which the fixed stars are at rest on the average. Therefore, our solution (16) does not represent the field of a single rotating hollow sphere representing the Universe, rather the field inside of such a hollow sphere outside of which there are masses at still larger distances from the origin that are at rest on the average with respect to the chosen system of reference. For ex-

[1] *Berl. Ber.*, 1917, p. 142.

ample, the field as represented by equation (16) is that which would hold at the center of the Sun if instead of the Sun and all the planets a large hollow sphere existed, say, with the radius of Neptune's orbit and which rotated with an angular velocity ω with respect to the fixed stars. If there were observers at the center of this sphere, situated on a heavenly body, the gravitational field of which can be neglected and which rotates about the same axis with the angular velocity ω', the observers would perceive centrifugal and Coriolis forces consisting of the effect of their own rotation and the effect of the rotation of the hollow sphere. In the following, the influence of the field of the hollow sphere on the centrifugal force caused by its own rotation will be studied.

For this purpose, we introduce a coordinate system tied to the reference body rotating with the angular velocity ω'. This is performed by means of the transformation

$$\left.\begin{aligned} x' &= x\cos\omega'\frac{x_4}{i} + y\sin\omega'\frac{x_4}{i}, \qquad z' = z \\ y' &= -x\sin\omega'\frac{x_4}{i} + y\cos\omega'\frac{x_4}{i}, \qquad x_4' = x_4 \end{aligned}\right\} \tag{23}$$

By this transformation, the quantities of interest $g_{\mu 4}$ become

$$\begin{aligned} g_{14}' &= -iy'\left[\omega'\left(1+\frac{2kM}{3a}\right) - \omega\frac{4kM}{3a}\right] \\ g_{24}' &= ix'\left[\omega'\left(1+\frac{2kM}{3a}\right) - \omega\frac{4kM}{3a}\right] \\ g_{44}' &= -1 + \frac{2kM}{a}\left[1 + \frac{5a^2\omega^2}{3} - \frac{\omega^2}{3}z^2\right] + (x'^2 + y'^2) \\ &\quad \cdot \left\{\omega'^2\left(1+\frac{2kM}{a}\right) - \omega\omega'\frac{4kM}{3a} + \omega^2\frac{kM}{3a}\right\} \end{aligned} \tag{24}$$

If one constructs the equations of motion out of these quantities according to (19) and (20), one will get

$$\begin{aligned} \ddot{x} &= 2\left[\omega'\left(1+\frac{2kM}{a}\right) - \omega\frac{4kM}{3a}\right]\dot{y} + \left\{\omega'^2\left(1+\frac{2kM}{a}\right) - \omega\omega'\frac{4kM}{3a} + \omega^2\frac{kM}{3a}\right\}x \\ \ddot{y} &= -2\left[\omega'\left(1+\frac{2kM}{a}\right) - \omega\frac{4kM}{3a}\right]\dot{x} + \left\{\omega'^2\left(1+\frac{2kM}{a}\right) - \omega\omega'\frac{4kM}{3a} + \omega^2\frac{kM}{3a}\right\}y \\ \ddot{z} &= -\frac{2kM}{3a}\omega^2 z \end{aligned} \tag{25}$$

After setting $M = 0$ here, one obtains the usual centrifugal-Coriolis field:

$$\begin{aligned} \ddot{x} &= 2\omega'\dot{y} + \omega'^2 x \\ \ddot{y} &= -2\omega'\dot{x} + \omega'^2 y \\ \ddot{z} &= 0 \end{aligned} \tag{26}$$

After setting $M \neq 0$, $\omega = 0$, it becomes

$$\begin{aligned} \ddot{x} &= 2\omega'\left(1 + \frac{2kM}{a}\right)\dot{y} + \omega'^2\left(1 + \frac{2kM}{a}\right)x \\ \ddot{y} &= -2\omega'\left(1 + \frac{2kM}{a}\right)\dot{x} + \omega'^2\left(1 + \frac{2kM}{a}\right)y \\ \ddot{z} &= 0 \end{aligned} \tag{27}$$

wherefrom one can see how the inertial effects are influenced by the presence of the surrounding masses M. The centrifugal and Coriolis forces are multiplied by the factor $(1 + 2kM/a)$.

Finally, one can see from equation (25) that if the hollow sphere rotates in the same sense, this results in diminishing the centrifugal and Coriolis forces. When one sets

$$\omega' = \omega \frac{4kM}{3(2kM + a)} \tag{28}$$

then the Coriolis force vanishes. One could call the quantity $4kM/3(2kM + a)$ the dragging coefficient of the hollow sphere with respect to the Coriolis force. The centrifugal force cannot be made to vanish, since the expressions inside the curly brackets of equation (25), when set equal to zero, do not yield real roots for ω. In the reference system "at rest" ($\omega' = 0$), the expression for the centrifugal force was

$$\frac{kM}{3a}\omega^2(x^2 + y^2)^{1/2}$$

If one now allows the reference system to rotate in the same sense as the hollow sphere, the centrifugal force will at first decrease for small values of ω' and will reach a minimum when ω'/ω is equal to half the value of the "dragging coefficient.[1] Thereupon, it increases again and reaches the original value again which it had for $\omega' = 0$ as soon as ω'/ω is equal to the dragging coefficient. Then it will continue to increase along with ω', and for large ω' will reach a value hardly

[1] One can immediately convince oneself by differentiating the term inside the parenthesis.

different from the value it had in the absence of the hollow sphere [namely, $\omega'^2(x^2+y^2)^{1/2}$], since according to our assumptions $2kM/a$ is small compared to 1.

At first sight, it seems contradictory to the spirit of a theory of relativity that the right-hand sides of the equations of motion (25) do not depend on the difference $\omega - \omega'$ alone. One should not forget, however, that we do not deal with only two bodies (point mass and hollow sphere) in the problem treated here; rather, owing to the boundary conditions $\lim \gamma_{\mu\nu} = 0$, still further distant masses at rest relative to the original reference system are introduced as a third element determining the field.

Summary

Using a concrete example, it is shown that forces occur in the (Einsteinian) gravitational field of distant rotating masses which are analogous to the centrifugal and the Coriolis forces, respectively. The peculiarities associated with this special case are discussed.

Vienna, December 1917, Institute for Theoretical Physics of the University.
(Received 21 December 1917.)

Correction to My Paper: "On the Effect of Rotating Distant Masses in Einstein's Theory of Gravitation"[1]

HANS THIRRING

Herr Laue and Herr W. Pauli kindly brought to my attention the following errors in my above-mentioned paper: In the approximate integration of the Einsteinian field equations, the quantity dV in the formulas for the retarded potentials $\gamma_{\mu\nu}$ [equation (3) *l.c.*] simply denotes the usual spatial volume element of integration space (in polar coordinates: $r^2 dr \sin\vartheta\, d\vartheta\, d\varphi$) and not, as erroneously claimed by me, the naturally measured volume element

$$i\frac{dx_4}{ds} r^2 dr \sin\vartheta\, d\vartheta\, d\varphi$$

[1] This journal [*Phys. Z.*], 19, 33, 1918.

Therefore, one has to drop a factor $i(dx_4/ds)$ everywhere in the equations (10) and (12). Furthermore, in going from (12) to (13) we incorrectly set $\int \rho_0 dV = M$. Instead, the correct formula is:

$$\int \rho_0 dV_0 = i \int \rho_0 \frac{dx_4}{ds} dV = M$$

or taking (11) into account,

$$M = 4\pi\sigma a^2 \left(1 + \frac{\omega^2 a^2}{3}\right)$$

By eliminating these errors, one obtains for the coefficients $g_{\mu\nu}$ of the line element, in place of (10), the scheme presented below [equation (16)] *:

$$g_{\mu\nu} = \begin{Bmatrix} -1 - \frac{2kM}{a}\left[1 + \frac{a^2\omega^2}{3} - \frac{2\omega^2}{15}(z^2 + x^2 - 2y^2)\right], & +\frac{2kM}{a}\frac{\omega^2}{5}xy, & 0, & +i\frac{4kM}{3a}\omega y \\ +\frac{2kM}{a}\frac{\omega^2}{5}xy, & -1 - \frac{2kM}{a}\left[1 + \frac{a^2\omega^2}{3} - \frac{2\omega^2}{15}(z^2 + y^2 - 2x^2)\right], & 0, & -i\frac{4kM}{3a}\omega x \\ 0, & 0, & 0, & 0 \\ +i\frac{4kM}{3a}\omega y, & -i\frac{4kM}{3a}\omega x, & 0, & -1 + \frac{2kM}{a}\left[1 + a^2\omega^2 - \frac{2\omega^2}{15}(2z^2 - x^2 - y^2)\right] \end{Bmatrix} \quad (16)$$

The equations of motion for the point mass then read as follows:

$$\begin{aligned} \ddot{x} &= -\frac{8kM}{3a}\omega\dot{y} + \frac{4kM}{15a}\omega^2 x \\ \ddot{y} &= +\frac{8kM}{3a}\omega\dot{x} + \frac{4kM}{15a}\omega^2 y \\ \ddot{z} &= -\frac{8kM}{15a}\omega^2 z \end{aligned} \quad (22)$$

Hence the Coriolis force remains unchanged as compared to my original formula (22); however, for the centrifugal term one needs an additional factor of $\frac{4}{5}$.

Furthermore, the equations of motion with respect to the reference system rotating with angular velocity ω' read [see equation (25)]:

$$\ddot{x} = 2\left[\omega'\left(1 + \frac{2kM}{a}\right) - \omega\frac{4kM}{3a}\right]\dot{y} + \left\{\omega'^2\left(1 + \frac{2kM}{a}\right) - \omega\omega'\frac{8kM}{3a} + \omega^2\frac{4kM}{15a}\right\}x$$

*Translator's note: A typographical error should be corrected in equation (16): The entry for g_{33} must be "–1" instead of "0."

$$\ddot{y} = -2\left[\omega'\left(1+\frac{2kM}{a}\right) - \omega\frac{4kM}{3a}\right]\dot{x} + \left\{\omega'^2\left(1+\frac{2kM}{a}\right) - \omega\omega'\frac{8kM}{3a} + \omega^2\frac{4kM}{15a}\right\}y$$

$$\ddot{z} = -\frac{8kM}{3a}\omega^2 z \qquad (25)$$

Also here the correction of the error simply yields an additional factor of $\frac{4}{5}$ in the terms with ω^2. Because of an additional mistake in the transformation to the rotating coordinate system, there originally occurred a factor of $\frac{4}{3}$ instead of $\frac{8}{3}$ in the term with $\omega\omega'$. Thus, two sentences in the next-to-last paragraph of my paper have still to be corrected as follows: "If one now allows the reference system to rotate in the same sense as the hollow sphere, the centrifugal force will at first decrease for small values of ω' and will reach a minimum when ω'/ω is equal to the 'dragging coefficient'. Thereupon, it increases again and reaches the original value which it had for $\omega' = 0$ as soon as ω'/ω is equal to twice the dragging coefficient."

The principal result of my paper (occurrence of centrifugal and Coriolis forces in the gravitational field of distant rotating masses) remains completely unchanged.

Vienna, 15 October 1920.

(Received 25 October 1980.)

On the Influence of the Proper Rotation of Central Bodies on the Motions of Planets and Moons According to Einstein's Theory of Gravitation

J. LENSE and H. THIRRING

In a paper published recently,[1] one of us has calculated approximately the field inside a rotating hollow sphere according to Einstein's theory of gravitation. This example seemed to be of principal interest in answering the question whether the rotation of distant masses according to Einstein's theory really produces a gravitational field equivalent to a "centrifugal field." Now it appears interesting, from another point of view, to consider the integration of the field

[1] Hans Thirring; this journal [*Phys. Z.*], **19**, 33, 1918; hereafter referred to as paper I.

equations for a rotating solid sphere, which can be simply carried out using the same methods. For, so long as one stood on the basis of Newton's theory, one could exactly substitute the field outside a sphere charged with uniform volume density (at rest or rotating) by the field of a material point of the same mass. Likewise, according to Einstein's theory, the field of an incompressible fluid sphere *at rest* is equivalent to that of a point mass,[1] but this is no longer valid for rotating spheres. As will be shown in what follows, additional terms appear here which correspond to centrifugal and Coriolis forces. Now since the planets move in the field of the Sun, which itself rotates, and the moons in turn in the field of the rotating planets, it does not seem excluded from the outset that one can obtain a novel confirmation of Einstein's theory by observing the perturbations caused by these additional terms. In the following, we calculate numerically the perturbations in the planetary orbital elements, which turn out to be below observational limits. For the moons of Jupiter, however, one obtains relatively large secular perturbations which should still lie within the errors of measurement.

§1. *The Calculation of the $g_{\mu\nu}$ for the Field of the Rotating Solid Sphere*

Notation:

- l, radius of the sphere;
- M, its mass;
- ω, its angular velocity;
- x', y', z', rectangular coordinates of a point of the integration space;
- x, y, z, coordinates of the point under consideration;
- k, gravitational constant;
- ρ_0, naturally measured space density of matter.

The calculation is performed in analogy with that in the paper cited at the beginning: Einstein's method of approximate integration[2] is used, but this time, when forming the energy tensor of matter, the velocity of the masses producing the field is regarded as so small compared to 1 (velocity of light) that one can neglect the squares and products of the components of the velocity. (This is in contrast to the example treated in the previous paper. The terms of the centrifugal force, which are proportional to* w^2, are omitted and only the Coriolis terms appear.) Neglecting such terms is completely justified when one considers that $l\omega$ for the Sun and all the planets is very small in a system of units in which the velocity of light = 1. Besides, in the case treated here we consider the field at a larger distance from the surface of the sphere. Let r denote the distance be-

[1] K. Schwarzschild, *Berl. Ber.*, 1916, p. 424.
[2] A. Einstein, *Berl. Ber.*, 1916, p. 688.
*Translator's note: A typographical error should be corrected here. Read "ω^2" instead of "w^2."

tween the point under consideration and the center of the sphere, r' the distance between the center and the integration element, and R the distance between the point under consideration and the integration element, then we expand $1/R$ in a series in r'/r up to the second order.

Then, exactly as in paper I, we start from the approximate solution given by Einstein[1]:

$$g_{\mu\nu} = -\delta_{\mu\nu} + \gamma_{\mu\nu}, \qquad \delta_{\mu\nu} \begin{matrix} =1, & \mu = \nu \\ =0, & \mu \neq \nu \end{matrix}$$

$$\gamma_{\mu\nu} = \gamma'_{\mu\nu} - \frac{1}{2}\delta_{\mu\nu}\sum_{\alpha}\gamma'_{\alpha\alpha} \tag{1}$$

$$\gamma'_{\mu\nu} = -\frac{k}{2\pi}\int \frac{T_{\mu\nu}(x', y', z', t-R)}{R}\, dV_0$$

Thereafter, we form the energy tensor of stress-free matter

$$T_{\mu\nu} = T^{\mu\nu} = \rho_0 \frac{dx_\mu}{ds}\frac{dx_\nu}{ds} = \sigma_0 \frac{dx_\mu}{dx_4}\frac{dx_\nu}{dx_4}\left(\frac{dx_4}{ds}\right)^2 \tag{2}$$

with the expressions for the components of velocity:

$$\begin{aligned} \frac{dx_1}{dx_4} &= -i\frac{dx'}{dt} = ir'\omega \sin\vartheta' \sin\varphi' \\ \frac{dx_2}{dx_4} &= -i\frac{dy'}{dt} = -ir'\omega \sin\vartheta' \cos\varphi' \\ \frac{dx_3}{dx_4} &= 0 \end{aligned} \tag{3}$$

(r', ϑ', φ' polar coordinates of a point of the sphere; the rotation is performed about the Z axis), and by neglecting the terms with ω^2 we obtain

$$T_{\mu\nu} = \rho_0 \left(\frac{dx_4}{ds}\right)^2 \cdot \left\{ \begin{matrix} 0, & 0, & 0, & ir'\omega\sin\vartheta'\sin\varphi' \\ 0, & 0, & 0, & -ir'\omega\sin\vartheta'\cos\varphi' \\ 0, & 0, & 0, & 0 \\ ir'\omega\sin\vartheta'\sin\varphi', & -ir'\omega\sin\vartheta'\cos\varphi', & 0, & 1 \end{matrix} \right. \tag{4}$$

[1] In the corresponding equation (2) of paper I, the factor $\delta_{\mu\nu}$ has been inadvertently omitted.

According to equations (7) and (8) of paper I, we have to set for dV_0

$$dV_0 = i\,\frac{dx_4}{ds}\,r'^2\,dr'\,\sin\vartheta'\,d\vartheta'\,d\varphi' \tag{5}$$

In order to express $1/R$ in terms of the integration variables, we choose the coordinate system again such that the point under consideration is placed in the Z-X plane. Then, by introducing polar coordinates,

$$x = r\sin\vartheta, \qquad y = 0, \qquad z = r\cos\vartheta$$

hold and one has

$$R^2 = (r'\sin\vartheta'\cos\varphi' - r\sin\vartheta)^2 + r'^2\sin^2\vartheta'\sin^2\varphi' + (r'\cos\vartheta' - r\cos\vartheta)^2$$
$$= r^2\left[1 - \frac{2r'}{r}(\sin\vartheta'\cos\varphi'\sin\vartheta + \cos\vartheta'\cos\vartheta) + \frac{r'^2}{r^2}\right]$$

Upon expanding in binomial series up to second term,

$$\frac{1}{R} = \frac{1}{r}\left\{1 + \frac{r'}{r}(\sin\vartheta'\cos\varphi'\sin\vartheta + \cos\vartheta'\cos\vartheta) - \frac{1}{2}\frac{r'^2}{r^2}\right.$$
$$\left. + \frac{3}{2}\frac{r'^2}{r^2}(\sin\vartheta'\cos\varphi'\sin\vartheta + \cos\vartheta'\cos\vartheta)^2\right\} \tag{6}$$

We again denote the expression inside the curly bracket by K and write

$$\frac{1}{R} = \frac{K}{r} \tag{6a}$$

If we now substitute (4), (5), and (6a) into the last equation (1), then we obtain

$$\gamma'_{44} = -\frac{i\chi}{2\pi}\frac{\rho_0}{r}\int_0^l r'^2\,dr'\int_0^{2\pi} d\varphi'\int_0^{\pi} d\vartheta'\left(\frac{dx_4}{ds}\right)^3 \sin\vartheta'\,K$$

$$\gamma'_{14} = \frac{\chi}{2\pi}\frac{\rho_0}{r}\,\omega\int_0^l r'^3\,dr'\int_0^{2\pi} d\varphi'\int_0^{\pi} d\vartheta'\left(\frac{dx_4}{ds}\right)^3 \sin^2\vartheta'\sin\varphi'\,K$$

$$\gamma'_{24} = -\frac{\chi}{2\pi}\frac{\rho_0}{r}\,\omega\int_0^l r'^3\,dr'\int_0^{2\pi} d\varphi'\int_0^{\pi} d\vartheta'\left(\frac{dx_4}{ds}\right)^3 \sin^2\vartheta'\cos\varphi'\,K$$

$$\gamma'_{11} = \gamma'_{22} = \gamma'_{33} = \gamma'_{12} = \gamma'_{13} = \gamma'_{23} = \gamma'_{34} = 0 \tag{7}$$

By neglecting the ω^2 terms and using the first type of approximation, we have

$$\left(\frac{dx_4}{ds}\right)^3 = i$$

[See equation (11) of paper I.] When one substitutes this value for $(dx_4/ds)^3$ as well as the expression for K from (6) and (6a) into (7), one obtains after evaluation of the integrals

$$\begin{aligned}
\gamma'_{44} &= \frac{\chi}{2\pi}\frac{M}{r} \\
\gamma'_{14} &= 0 \\
\gamma'_{24} &= -i\,\frac{\chi}{2\pi}\frac{M}{r}\frac{l}{5r}\,\omega l \sin\vartheta \\
\gamma'_{11} &= \gamma'_{22} = \gamma'_{33} = \gamma'_{12} = \gamma'_{13} = \gamma'_{23} = \gamma'_{34} = 0
\end{aligned} \tag{8}$$

Moreover, it follows from (1) that once rectangular coordinates are introduced and the Einsteinian gravitational constant is replaced by the Newtonian one $k = \chi/8\pi$,

$$\begin{aligned}
g_{11} &= g_{22} = g_{33} = -1 - \frac{2kM}{r} \\
g_{44} &= -1 + \frac{2kM}{r} \\
g_{24} &= -i\,\frac{4kM}{r}\frac{lx}{5r^2}\,\omega l \\
g_{12} &= g_{13} = g_{23} = g_{14} = g_{34} = 0
\end{aligned} \tag{9}$$

If now by a rotation of the system one frees oneself from the special choice of the coordinates in which the point under consideration is situated in the Z-X plane, one obtains the final scheme for the coefficients:

$$g_{\mu\nu} = \left\{\begin{matrix}
-1 - \frac{2kM}{r}, & 0, & 0, & i\,\frac{4kM}{5r}\frac{ly}{r^2}\,\omega l \\
0, & -1 - \frac{2kM}{r}, & 0, & -i\,\frac{4kM}{5r}\frac{lx}{r^2}\,\omega l \\
0, & 0, & -1 - \frac{2kM}{r}, & 0 \\
i\,\frac{4kM}{5r}\frac{ly}{r^2}\,\omega l, & -i\,\frac{4kM}{5r}\frac{lx}{r^2}\,\omega l, & 0, & -1 + \frac{2kM}{r}
\end{matrix}\right. \tag{10}$$

§2. *The Equations of Motion of a Point Mass in the Field of a Rotating Solid Sphere*

In the following, the equations of motion of a point mass in the field of a rotating solid sphere will be set up and we will assume that its velocity is so small that we can neglect the squares and the products of the components of velocity with respect to 1. Thereby, we emphasize from the outset that we are only interested in finding those *perturbation terms* of the planetary motion which are due to the rotation of the central bodies. In order to obtain a sufficiently exact solution of the planetary problems in accordance with Einstein's theory, one still has to add to the perturbation terms calculated here those which lead to the known perihelion precession.[1] The terms due to the proper rotation of the central bodies already emerge from the first approximation of Einstein's theory, while the perihelion perturbation just mentioned was only obtained in the second approximation. Nevertheless, it is still not permissible to take the former into account and to neglect the latter. The reason why one is not allowed to do that results from the following considerations: The ansatz for the force developed below and the Newtonian one differ by additional terms which are proportional to $\omega l v$, where v represents the velocity of the planets or the moons, respectively, while ωl represents the velocity of a point on the equator of the central body. But now the inequality

$$v > \omega l \tag{11}$$

is valid for both the Sun–planets system as well as the planets–moons systems that will be considered. If we take the terms with $\omega l v$ into account, we have all the more reason to consider those terms in the equations of motion which contain the squares and products of the components of velocity of the point mass. But if we do that, then we are not at all allowed to calculate with the first approximation alone, for those terms which are added in the second approximation to the Newtonian terms are in ratio to the former as $\alpha/r:1$ ($\alpha = 2kM$). But the square of the velocity of a planet is also of the same order as α/r; thus taking the quadratic terms in the velocities into account logically requires the consideration of terms emerging from the second approximation. It follows therefrom that the calculations performed here would not actually make sense owing to the validity of the inequality (11). Nevertheless, they can be used in practice if one keeps in mind that all perturbations considered here are so small as to allow us to superpose them linearly. One bases the computation on the equations of motion as given by Einstein in his Mercury paper and adds the perturbation terms calculated in what follows. In this way, one arrives at the desired result which is the calculation of an orbit including relativistic effects.

[1]A. Einstein, *Berl. Ber.*, 1915, p. 831.

As shown in paper I, by neglecting the above-mentioned terms and with the coordinates $x_1 = x, x_2 = y, x_3 = z, x_4 = it$ the general equations of motion

$$\frac{d^2 x_\tau}{ds^2} = \Gamma^\tau_{\mu\nu} \frac{dx_\mu}{ds} \frac{dx_\nu}{ds}$$

become

$$\frac{d^2 x_\tau}{dt^2} = 2i\left(\Gamma^\tau_{14} \frac{dx_1}{dt} + \Gamma^\tau_{24} \frac{dx_2}{dt} + \Gamma^\tau_{34} \frac{dx_3}{dt}\right) - \Gamma^\tau_{44} \tag{12}$$

The 16 quantities $\Gamma^\tau_{\sigma 4}$ that appear here are, according to the first type of approximation for a stationary field, given by

$$\Gamma^1_{14} = 0, \qquad \Gamma^1_{24} = \frac{1}{2}\left(\frac{\partial g_{14}}{\partial x_2} - \frac{\partial g_{24}}{\partial x_1}\right),$$

$$\Gamma^1_{34} = \frac{1}{2}\left(\frac{\partial g_{14}}{\partial x_3} - \frac{\partial g_{24}}{\partial x_1}\right), \qquad \Gamma^1_{44} = -\frac{1}{2}\frac{\partial g_{44}}{\partial x_1}$$

$$\Gamma^2_{14} = \frac{1}{2}\left(\frac{\partial g_{24}}{\partial x_1} - \frac{\partial g_{14}}{\partial x_2}\right), \qquad \Gamma^2_{24} = 0,$$

$$\Gamma^2_{34} = \frac{1}{2}\left(\frac{\partial g_{24}}{\partial x_3} - \frac{\partial g_{34}}{\partial x_2}\right), \qquad \Gamma^2_{44} = -\frac{1}{2}\frac{\partial g_{44}}{\partial x_2}$$

$$\Gamma^3_{14} = \frac{1}{2}\left(\frac{\partial g_{34}}{\partial x_1} - \frac{\partial g_{14}}{\partial x_3}\right), \qquad \Gamma^3_{24} = \frac{1}{2}\left(\frac{\partial g_{34}}{\partial x_2} - \frac{\partial g_{24}}{\partial x_3}\right),$$

$$\Gamma^3_{34} = 0, \qquad \Gamma^3_{44} = -\frac{1}{2}\frac{\partial g_{44}}{\partial x_3}$$

$$\Gamma^4_{14} = \frac{1}{2}\frac{\partial g_{44}}{\partial x_1}, \qquad \Gamma^4_{24} = \frac{1}{2}\frac{\partial x_{44}}{\partial x_2},$$

$$\Gamma^4_{34} = \frac{1}{2}\frac{\partial g_{44}}{\partial x_3}, \qquad \Gamma^4_{44} = 0 \tag{13}$$

For our field, as given by the equations (10), this scheme becomes*

*Translator's note: The entry for Γ^3_{14} in (14) must have "r^2" in the denominator instead of "z^2."

$$
\begin{matrix}
0, & -i\dfrac{2kM}{5r^2}\dfrac{\omega l^2}{r}\dfrac{x^2+y^2-2z^2}{r^2}, & -i\dfrac{6kM}{5r^2}\dfrac{\omega l^2}{r}\dfrac{yz}{r^2}, & \dfrac{kM}{r^2}\dfrac{x}{r} \\
+i\dfrac{2kM}{5r^2}\dfrac{\omega l^2}{r}\dfrac{x^2+y^2-2z^2}{r^2}, & 0, & +i\dfrac{6kM}{5r^2}\dfrac{\omega l^2}{r}\dfrac{xz}{r^2}, & \dfrac{kM}{r^2}\dfrac{y}{r} \\
+i\dfrac{6kM}{5r^2}\dfrac{\omega l^2}{r}\dfrac{yz}{z^2}, & -i\dfrac{6kM}{5r^2}\dfrac{\omega l^2}{r}\dfrac{xz}{r^2}, & 0, & \dfrac{kM}{r^2}\dfrac{z}{r} \\
-\dfrac{kM}{r^2}\dfrac{x}{r}, & -\dfrac{kM}{r^2}\dfrac{y}{r}, & -\dfrac{kM}{r^2}\dfrac{z}{r}, & 0
\end{matrix}
\tag{14}
$$

By substituting these values for the $\Gamma^{\tau}_{\sigma 4}$ into (12), we obtain the desired equations of motion*:

$$
\begin{aligned}
\ddot{x} &= \frac{kM}{r^2}\frac{\omega l^2}{r}\left[\frac{4}{5}\frac{x^2+y^2-2z^2}{r^2}\cdot+\frac{12}{5}\frac{yz}{r^2}\dot{z}\right]-\frac{kM}{r^2}\frac{x}{r} \\
\ddot{y} &= -\frac{kM}{r^2}\frac{\omega l^2}{r}\left[\frac{4}{5}\frac{x^2+y^2-2z^2}{r^2}\dot{x}+\frac{12}{5}\frac{z}{r^2}\dot{z}\right]-\frac{kM}{r^2}\frac{y}{r} \\
\ddot{z} &= \frac{kM}{r^2}\frac{\omega l^2}{r}\frac{12}{5}\frac{\underline{\underline{z}}}{r}\frac{x\dot{y}-y\ddot{x}}{r}-\frac{kM}{r^2}\frac{z}{r}
\end{aligned}
\tag{15}
$$

The last terms on the right-hand side represent the Newtonian force; as discussed above, one has to substitute in its place the force components according to Einstein's Mercury paper. The first terms on the right-hand side are the perturbation terms of interest caused by the proper rotation of the central bodies.

§3. *Calculation of the Perturbations due to the Proper Rotation of the Central Body*

The perturbation terms appearing in the equations (15) have to be regarded as components X, Y, Z of the perturbing force caused by the proper rotation of the central bodies. We decompose them into three other mutually perpendicular components S, T, W, where S denotes the radial one, T the transversal one, and W the orthogonal one (i.e., normal to the plane of the planetary orbit) and we introduce the following conventional astronomical notation:

*Translator's note: In the first of equations (15), a "y" is missing in "$\dot{y}$", in the second, an "x" is missing in "xz/r^2", and in the last one, "$z/\!/r$" and "$\ddot{x}$" must be replaced by "z/r" and "$\dot{x}$," respectively.

a, semimajor axis;
e, eccentricity;
$p = a(1 - e^2)$, semilatus rectum;
$i = \sphericalangle\, y\Omega\Pi$, inclination;
$\Omega = \sphericalangle\, XO\Omega$, longitude of the node;
$\tilde{\omega}$ = broken $\sphericalangle\, XO\Pi$, longitude of the pericenter;
L_0, mean longitude of the epoch, i.e., mean longitude of the planet or satellite at time $t = 0$ (also a broken angle, measured from the X axis);
$v = \sphericalangle\, \Pi OP$, true anomaly;
$u = \sphericalangle\, \Omega OP = v + \tilde{\omega} - \Omega$, argument of the latitude;
U, period of revolution of the planet or satellite in days;
$n = 2\pi/U = (kM/a^3)^{1/2}$, mean daily motion;
$C = r^2\dot{v} = na^2(1 - e^2)^{1/2}$, twice the areal velocity.

Furthermore, we abbreviate the constant appearing in the equations (15) by setting $4kM\omega l^2/5 = K$.

Now we have*

$$x = r(\cos u \cos \Omega - \sin u \sin \Omega \cos i)$$

$$y = r(\cos u \sin \Omega + \sin u \cos \Omega \cos i)$$

$$z = r \sin u \sin i$$

$$r = \frac{p}{1 + e \cos v}$$

$$x\dot{y} - y\ddot{x} = C \cos i$$

Moreover,

$$S = X(\cos u \cos \Omega - \sin u \sin \Omega \cos i) + Y(\cos u \sin \Omega + \sin u \cos \Omega \cos i) + Z \sin u \sin i$$

$$T = -X(\sin u \cos \Omega + \cos u \sin \Omega \cos i) - Y(\sin u \sin \Omega - \cos u \cos \Omega \cos i) + Z \cos u \sin i$$

$$W = X \sin \Omega \sin i - Y \cos \Omega \sin i + Z \cos i$$

If one substitutes the values of X, Y, Z obtained from equations (15) into these formulas for S, T, W by using the given relations and notations, one finds after a longer intermediary calculation

*Translator's note: In the fifth formula below, read "$\dot{x}$" instead of "$\ddot{x}$."

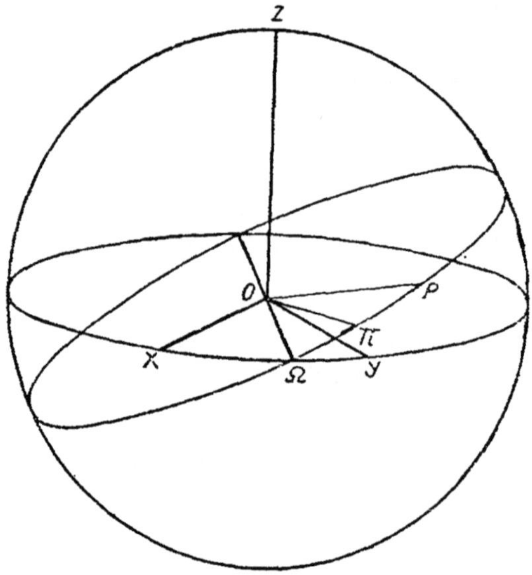

Fig. 2. Π and P denote the positions of the pericenter and of the planet or satellite, respectively, which are projected from the middle point O of the central body onto the sphere.

$$
\begin{aligned}
S &= \frac{KC\cos i}{r^4} \\
T &= -\frac{K\dot{r}\cos i}{r^3} = -\frac{KCe\cos i\sin v}{pr^3} \\
W &= \frac{K\sin i}{r^4}(2C\sin u + r\dot{r}\cos u) \\
&= \frac{KC\sin i}{r^4}\left(\frac{re\sin v\cos u}{p} + 2\sin u\right)
\end{aligned}
\tag{16}
$$

The variation of the orbital elements caused by the perturbing force is given by the equations

$$
\begin{aligned}
\frac{da}{dt} &= \frac{2}{n(1-e^2)^{1/2}}\left(Se\sin v + T\frac{p}{r}\right) \\
\frac{de}{dt} &= \frac{(1-e^2)^{1/2}}{na}\left[S\sin v + T\left(e + \frac{r+a}{a}\cos v\right)\right] \\
\frac{di}{dt} &= \frac{1}{C}Wr\cos u
\end{aligned}
$$

$$\frac{d\Omega}{dt}=\frac{1}{C\sin i}Wr\sin u$$

$$\frac{d\tilde{\omega}}{dt}=\frac{(1-e^2)^{1/2}}{nae}\left[-S\cos v+T\left(1+\frac{r}{p}\right)\sin v\right]+2\sin^2\frac{i}{2}\,\frac{d\Omega}{dt}$$

$$\frac{dL_0}{dt}=-\frac{2}{na^2}Sr+\frac{e^2}{1+(1-e^2)^{1/2}}\frac{d\tilde{\omega}}{dt}+2(1-e^2)^{1/2}\sin^2\frac{i}{2}\,\frac{d\Omega}{dt}$$

which, by substituting the values (16), can be represented in the following form:

$$\frac{da}{dt}=0$$

$$\frac{de}{dt}=\frac{K\cos i}{Ca}\sin v\cdot\dot{v}$$

$$\frac{di}{dt}=\frac{K\sin i}{Cp}\cos u\;[e\sin v\cos u+2(1+e\cos v)\sin u]\;\dot{v}$$

$$\frac{d\Omega}{dt}=\frac{K}{Cp}\sin u[e\sin v\cos u+2(1+e\cos v)\sin u]\;\dot{v}$$

$$\frac{d\tilde{\omega}}{dt}=-\frac{K\cos i}{Ca}\left(2+\frac{1+e^2}{e}\cos v\right)\dot{v}+2\sin^2\frac{i}{2}\,\frac{d\Omega}{dt}$$

$$\frac{dL_0}{dt}=-\frac{2K\cos i}{na^2p}(1+e\cos v)\,\dot{v}+\frac{e^2}{1+(1-e^2)^{1/2}}\frac{d\tilde{\omega}}{dt}+2(1-e^2)^{1/2}\sin^2\frac{i}{2}\,\frac{d\Omega}{dt}$$

In the sense of perturbation theory, we regard the orbital elements on the right-hand side as constants due to the extraordinarily small factors K. We only integrate over v, i.e., we calculate the perturbations to first order in K, and observe that $u=v+\tilde{\omega}-\Omega$. Moreover, if we introduce $K_1=K/na^3$, we obtain

$$\Delta a=0$$

$$\Delta e=-\frac{K_1\cos i}{(1-e^2)^{1/2}}\cos v$$

$$\Delta i=-\frac{K_1\sin i}{2(1-e^2)^{3/2}}(\cos 2u+2e\cos v\cos^2 u)$$

$$\Delta\Omega=\frac{K_1}{(1-e^2)^{3/2}}\left[v-\frac{1}{2}\sin 2u+e\left(\sin v-\frac{1}{2}\sin 2u\cos v\right)\right]$$

$$\Delta\tilde{\omega}=-\frac{K_1\cos i}{(1-e^2)^{3/2}}\left(2v+\frac{1+e^2}{e}\sin v\right)+2\sin^2\frac{i}{2}\,\Delta\Omega$$

$$\Delta L_0 = -\frac{2K_1 \cos i}{1-e^2}(v + e \sin v) + \frac{e^2}{1+(1-e^2)^{1/2}} \Delta\tilde{\omega} + 2(1-e^2)^{1/2} \sin^2 \frac{i}{2} \Delta\Omega$$

There follows the interesting result that the perturbations in the semimajor axis vanish exactly. While only periodic terms appear in Δe and Δi, additional secular terms occur in the perturbation of the other elements; namely, because of $v = nt$ + periodic terms,

$$\Delta\Omega = \frac{K_1}{(1-e^2)^{3/2}} nt$$
$$\Delta\tilde{\omega} = \Delta L_0 = -\frac{2K_1}{(1-e^2)^{3/2}} \left(1 - 3 \sin^2 \frac{i}{2}\right) nt \tag{17}$$

§4. *Numerical Results*

The numerical evaluation shows that these secular perturbations on the Sun-planets system even for the period of one century are beyond any observational possibility; for they reach a maximum of 0.01″ (for the perihelion of Mercury). Things are different in the planet-moons systems. Here somewhat larger numbers are encountered. It is better for the numerical calculation to transform the formulas (17). Let us introduce the following notation:

l, radius of the planet in centimeters;
τ, period of rotation of the planet in days;
a, semimajor axis of the satellite in centimeters;
a_1, semimajor axis of the planetary orbit in centimeters;
U, orbital period of the satellite in days;
U_1, orbital period of the planet in days;
J, number of days in a year;
c, velocity of light in cm $\sec^{-1}$.

Then the formulas resulting from (17)

$$2\Delta\Omega = -\Delta\tilde{\omega} = -\Delta L_0 = \frac{\pi^2 J}{9c^2} \frac{l^2}{\tau U^2} \tag{18}$$

yield, in units of arc seconds per century, the perturbations in the elements of the satellite orbit due to the rotation of the planet. We have set $e^2 = i^2 = 0$, which is allowed for the moons involved within the desired accuracy.

In addition, the perturbations are superposed additively in the sense of Section 2 with those discussed by Einstein in his Mercury paper, which are partly due to the direct effect of the planet and partly due to the perturbing effect of the Sun. The former are given by

$$\Delta\Omega = 0, \qquad \Delta\tilde{\omega} = \Delta L_0 = \frac{5\pi^2 J}{24c^2} \frac{a^2}{U^3(1 - e^2)} \tag{19}$$

and the latter[1] by*

$$4\Delta\Omega = \Delta\tilde{\omega} = \Delta L_0 = \frac{5\pi^2 J}{12c^2} \frac{a_1{}^2}{U_1{}^3} \tag{20}$$

both in arc seconds per century. In equation (20), the eccentricity and the inclination of the planetary and satellite orbits were neglected again due to the fact that these terms are extremely small as shown in Table I. They are considerably smaller for all remaining moons.

The perturbations due to the proper rotation of the planet are contained in Table II. They remain below 0.5″ for all other satellites.

The largest terms are analogous to the Einsteinian perihelion motion of Mercury [formulas (19)], as emerges from Table III.† They are smaller than 0.5″ for the moons that are left out.

If we now add the three types of terms in order to obtain all relativistic effects, the following has to be taken into account: The correction of Newton's

Table I

	$\Delta\Omega$	$\Delta\tilde{\omega} = \Delta L_0$
Moon	+1.9″	+7.7″
Both moons of Mars	+0.7″	+2.7″

Table II

	Jupiter			Saturn				
	V	I	II	1	2	3	4	5
$\Delta\Omega$	+1′53″	+ 9″	+2″	+20″	+10″	+ 5″	+2″	+1″
$\Delta\tilde{\omega} = \Delta L_0$	−3′46″	−18″	−4″	−41″	−19″	−10″	−5″	−2″

[1] W. de Sitter, Planetary motion and the motion of the moon according to Einstein's theory, *Amsterdam Proc.*, **19**, 1916. In the formulas (20), XY plane denotes the plane of planetary orbit. In de Sitter's paper, on p. 379 formula (38), the factor $\frac{1}{4}$ is missing in front of $\delta\tilde{\omega}$.

*Translator's note: The result of de Sitter cited in footnote 1 is, in fact, correct. Hence equation (20) gives values of $\Delta\tilde{\omega}$ and ΔL_0 too large by a factor of 4. Tables I and IV must be corrected accordingly.

†Translator's note: The existence of a satellite of Saturn, mentioned as "Saturn 10" or "Themis" in Tables III and IV, was announced by Pickering in 1905. This has not been confirmed by subsequent observations.

Table III[a]

		$\Delta\tilde{\omega} = \Delta L_0$		$\Delta\tilde{\omega} = \Delta L_0$
Mars	1	22″	Jupiter I	4′28
	2	2	II	1 24
Saturn	1	5′46″	III	26
	2	3 03	IV	6
	3	1 47	V	36 37
	4	59	Uranus 1	22
	5	25	2	10
	6	3	3	3
	7	2	4	1
	10	2	Neptune's moon	5

[a] $\Delta\Omega = 0$.

law treated in Einstein's Mercury paper is caused by a perturbing force acting along the radius vector. Its components, according to the cited paper, amount to

$$S = -\frac{3n^2a^3C}{2c^2}\frac{\dot{v}}{r^2}, \qquad T = W = 0$$

so they are independent of the choice of the coordinate system. Consequently, the corresponding perturbations [formulas (19) and Table III] can be related to an arbitrary XY plane. However, the variations of the elements contained in the formulas (20) and caused by the perturbing effect of the Sun (representing the deviation from the classical form) are referred to the orbital plane of the planet as already mentioned. The same holds for the numbers in Table I, whereas in Table II, which contains the perturbation terms caused by the rotation of the planets, everything refers to the equatorial plane of the central body in accordance with the choice of the coordinate system in the present paper. Therefore, the following is valid for Table IV, which summarizes all the relativistic effects: Terms (19) and (20) only appear in connection with the Moon and the two moons of Mars; hence, the plane of reference is the orbital plane of the planet. In the case of the satellites of Jupiter and Saturn, however, the reference plane is the equatorial plane of the corresponding central body since here again only terms (18) and (19) appear. The perturbations of the moons of Uranus and the moon of Neptune contain only terms (19); consequently, the plane of reference can be chosen arbitrarily.

The following can be stated about the column "Δt": The secular perturbation in the mean longitude causes a variation in the mean daily motion. That is, in the absence of relativistic effects a certain correction should be made for the time that elapses between two definite events, such as the darkenings of the moons of Jupiter. This correction for a time span of a hundred years is given in

Table IV

	$\Delta\Omega$	$\Delta\tilde{\omega} = \Delta L_0$	Δt
Moon	2″	8″	13.9s
Mars 1. Phobos	1	25	0.5
2. Deimos	1	5	0.4
Jupiter I	9	4′ 10″	29.5
II	2	1 20	18.9
III	0	26	12.5
IV	0	6	7.1
V	1′ 53″	32 51	1m 5.4s
Saturn 1. Mimas	20	5 05	19.2
2. Enceladus	10	2 44	15.0
3. Tethys	5	1 37	12.2
4. Dione	2	54	9.2
5. Rhea	1	23	6.9
6. Titan	0	3	3.3
7. Hyperion	0	2	2.7
10. Themis	0	2	2.9
Uranus 1. Ariel	0	22	3.7
2. Umbriel	0	10	2.7
3. Titania	0	3	1.5
4. Oberon	0	1	1.0
Neptune's moon	0	5	2.1

the last column of Table IV and can be obtained by means of the formula

$$\Delta t = \frac{1}{15} U \Delta L_0$$

Summary

The perturbation terms in the orbits of the planets and moons which, according to Einstein's theory, are caused by the proper rotation of the central body turn out to be smaller than those which emerge from the second approximation of the theory and lead to the perihelion motion of Mercury. They need not at all be taken into consideration in comparison with the latter terms in the case of the planetary orbits; however, for the orbits of the moons of Jupiter and Saturn they have to be taken into account. The secular perturbations due to all relativistic effects involved were calculated for the moons of the exterior planets. Though in some cases–for example, for the 5th moon of Jupiter–they reach a considerable magnitude, the present observations should not be precise enough as to allow a test of the theory using the perturbations of the orbits of the moons.

Vienna, February 1918. Institute for Theoretical Physics of the University.

(Received 21 February 1918.)

§(3): *Discussion of Thirring's Papers*

Hans Thirring investigated the gravitational "magnetic" field generated by a rotating hollow sphere. However, Mach's ideas on the origin of inertia provided the main impetus for his work. Mach criticized Newton's formulation of the laws of mechanics since the motion of matter was referred to an absolute space. This notion is related to an ensemble of global inertial frames each in uniform motion with respect to the others. A body in uniform motion persists in that state and offers "resistance" when accelerated with respect to absolute space. This characteristic property of matter is embodied in the notion of inertial mass. An observer can determine if matter is accelerated with respect to inertial frames since extra inertial forces (proportional to the inertial mass) appear in an accelerated frame. The success of Newtonian mechanics has to be reconciled with the fact that only the motion of matter with respect to other "matter" is physically detectable.

Newton considered the introduction of the abstract notion of absolute space necessary for the mathematical formulation of the laws of mechanics. On the contrary, Mach considered all motion to be relative. In rejecting the notion of absolute space Mach had predecessors in Leibniz and Berkeley, among others. If only relative motion has significance, the inertial frames must be determined by matter.

To give these vague ideas a more definite formulation, one may extend the principle of relativity to accelerated motion and postulate that inertial forces are due to the gravitational *field* generated by all the matter in the universe. According to Einstein's relativistic theory of gravitation (which has observational support for macroscopic phenomena), however, these notions must be rejected since they imply the *global equivalence* of inertial and certain gravitational forces in contrast to Einstein's principle of equivalence which is purely local. To illustrate this point, consider a variant of Newton's bucket experiment in which the bucket is uniformly accelerated. Other than forces of electromagnetic origin (such as viscosity), the fluid in the bucket is also subject to a *uniform* inertial force field (relative to the bucket). A contradiction arises, however, if the bucket is now treated as freely falling in the gravitational field generated by all the matter in the universe in accelerated motion, since according to Einstein's theory the only external gravitational forces that affect the motion of the fluid relative to the bucket are *tidal* forces.[1] This example also illustrates why the case of rotational acceleration is interesting: The (first-order) tidal and centrifugal forces have similar functional forms. It is clear that these arguments are based on the tensor character of the gravitational field in Einstein's theory and are completely independent of any boundary conditions on the space-time metric at spatial infinity.

[1] This argument against the relativity of accelerated motion can be formulated equally well within the framework of Newton's theory of gravitation.

Moreover, the gravitational interaction of the distant matter in the universe cannot be responsible for the inertial mass if Einstein's principle of equivalence is valid [2].[2]

Let us return to Mach's criticism of Newton's absolute space: Matter moves with respect to absolute space but cannot affect it. Mach's solution was that the configuration of distant matter in the universe determines the characteristics of this inertial space. The local metrical properties of space-time are related to the distribution of matter according to Einstein's theory of gravitation. Furthermore, this theory transforms the question of the inertial properties of our local space-time into a problem of modern cosmology. An important requirement for our local frame to be inertial is the local isotropy of space. This has been tested by several null experiments of great precision [4, 5]. The precise relationship between our local inertial frame and the rest frame determined by distant matter is a difficult problem of observational astronomy. It follows from the study of the motions of the inner planets of the solar system that the two frames do not rotate with respect to each other with an uncertainty of ~1 second of arc per century [6–8]. The discoveries of Hubble expansion and the cosmic microwave background radiation have opened up new possibilities for the discussion of this problem. Severe restrictions may be placed on the rotation of the universe from the recent studies of the isotropy of the microwave background radiation. Moreover, the dipole temperature anisotropy of this radiation may be used to determine the linear motion of our local frame with respect to the rest frame of the radiation [9].

The major contribution of the work of Thirring (and of Lense and Thirring) was the demonstration of how the local inertial frames are affected by the motion of nearby matter ("dragging of the inertial frames"). It is interesting to elucidate this issue in a coordinate invariant manner in relation to the work of Thirring. Consider first a freely falling "laboratory" frame in an external field and assume that a Fermi frame [10] is set up along the geodesic path of a representative point in the laboratory ("center of mass"). Let τ be the proper time along this path and let $\mathbf{\Omega}_D(\tau)$ be the rate of rotation of the locally nonrotating frame with respect to a certain local fiducial frame (e.g., the rest frame of the laboratory). If it is established that this latter frame does not rotate with respect to the rest frame of inertial observers at spatial infinity, $\mathbf{\Omega}_D$ corresponds to the dragging of the local inertial frames with respect to those of infinity. The motion of free particles with respect to the Fermi frame may be described to first order by

[2] In the theories of Newton and Einstein, the strength of the gravitational interaction is constant in space and time. On the basis of Mach's ideas regarding the origin of inertia, a decrease in the inertial mass may, however, be expected as a result of the Hubble expansion of the universe. This may be interpreted as a decrease in the "constant" of gravitation on a cosmological time scale. There is no firm evidence at present for such a variation [3].

$$\frac{d^2 X^i}{d\tau^2} + K^i{}_j(\tau)\, X^j = 0$$

where X^i are the (Fermi) coordinates of a particle in the neighborhood of the representative point. Here $K_{ij}(\tau)$ is a symmetric traceless matrix denoting certain ("electric") components of the Riemann tensor evaluated in the Fermi frame. To express this equation of motion with respect to the fiducial frame which rotates with frequency $\mathbf{\Omega} = -\mathbf{\Omega}_D(\tau)$ with respect to the Fermi frame, let $x_i = M_{ij}(\tau) X^j$ be the relevant coordinate transformation and

$$\frac{dM_{ij}}{d\tau} = e_{ikl}\Omega_l M_{kj}$$

where e_{ijk} is the alternating symbol with $e_{123} = 1$. The equation of motion of a free particle with respect to the fiducial frame is then

$$\frac{d^2 x_i}{d\tau^2} + 2e_{ijk}\,\Omega_j \frac{dx_k}{d\tau} + \left(\Omega_i \Omega_j - \delta_{ij}\Omega^2 + e_{ilj}\frac{d\Omega_l}{d\tau} + k_{ij}\right) x_j = 0$$

where $k_{ij} = M_{im}M_{jn}K_{mn}$.

Thirring considers a hollow sphere of mass M and radius a rotating uniformly with angular frequency ω. The shell has uniform mass density so that its angular momentum is $J = \frac{2}{3} Ma^2 \omega$. The effect of rotation on the gravitational field in the interior of the sphere is considered to first order in the Newtonian constant k and to second order in ω. A test particle at rest at the center of the sphere follows a geodesic C in this field. The Fermi frame along C rotates uniformly in the same sense as the hollow sphere with a frequency

$$\Omega_D = \frac{2kJ}{a^3} = \frac{4kM}{3a}\,\omega$$

with respect to the rest frame of the inertial observers at infinity. In a local frame that rotates with frequency $\mathbf{\Omega} = -\mathbf{\Omega}_D$ with respect to the Fermi frame, the equation of motion of a free test particle near the center of the hollow sphere can be written as above with k_{ij} determined simply by

$$k_{ij}(\tau) = -\frac{1}{2}\left(\frac{\partial^2 g_{44}}{\partial x_i \partial x_j}\right)_C$$

Thirring's calculation of the metric tensor to second order in ω implies that k_{ij} is diagonal with $k_{11} = k_{22} = -(4kM/15a)\,\omega^2$ and $k_{33} = -2k_{11}$. The equation of motion comprising the dragging of the inertial frames and the gravitational tidal force of the rotating mass is the same as that given by Thirring, as expected.[3] Thirring neglected any stresses in the shell which, however, is not in general permissible due to the conservation laws of energy and momentum [11, 12]. When due account is taken of elastic stresses in the rotating shell, only terms in

[3]Hence Thirring's interpretation of the ω^2-terms as analogous to centrifugal forces cannot be maintained, even apart from the fact that gravitational forces can only be locally identified with inertial forces.

the metric tensor proportional to ω^2 turn out to be different from those given by Thirring; consequently, the proper tidal matrix is in fact just $\frac{1}{2}$ of that obtained from Thirring's results.

The weak-field result of Thirring for the dragging of the inertial frames is valid in the limit $kM \ll a$; its generalization to strong fields has been discussed by a number of authors [13–17]. These studies provide further insight into the dependence of Ω_D/ω on the nature of the matter distribution.

To summarize, Mach's ideas provided strong stimuli for the development of the relativistic theory of gravitation by subordinating the existence of inertia to the matter distribution and by suggesting the possible equivalence of inertia and gravitation. Following Einstein [18], attempts have been made to use extensions of Mach's ideas to solve the problem of boundary conditions in relativistic cosmology. The gravitational field equations are local differential equations for the metrical properties of space-time. Once the matter content has been specified, boundary conditions are in general necessary to provide a unique solution for the cosmic gravitational field. To avoid this problem, it is tempting to postulate, for instance, that the universe is spatially closed. It appears, however, that a much more extensive knowledge of the matter and radiation content throughout the universe is necessary before the problem of the cosmic field can be adequately tackled.[4]

§(4): *Discussion of the Paper of Lense and Thirring*

Josef Lense and Hans Thirring investigated the influence of the gravitational "magnetic" field of a rotating body on the motion of test masses in orbit around it. The gravitational field of a rotating body is characterized by a "vector potential," which in the first approximation corresponds to the gravitational "magnetic" dipole due to the mass current. The vector potential is given by[5,6]

$$\mathbf{G} \approx -2k \frac{\mathbf{J} \times \hat{\mathbf{r}}}{r^2}$$

for $r \gg 2kM$ and $r \gg A$, where $A = J/M$ is the angular momentum per unit mass of the body. The precession of a test gyroscope (i.e., a test gravitational "magnetic" dipole) at rest outside a matter distribution indicates the dragging of the inertial frames or, equivalently, the presence of the gravitational "magnetic"

[4] Review papers on Mach's ideas [19–21] may be consulted for more extensive discussions and further references to the literature.

[5] For a localized matter distribution, the "vector" potential is defined by $G_i = -g_{i4}/g_{44}$ in a system of Schwarzschild-like coordinates.

[6] Lense and Thirring calculated the gravitational "vector" potential for a *rotating solid sphere* in the weak-field approximation. In this procedure, deviations of the source from spherical symmetry are naturally reflected in the vector potential in a form similar to the multipole expansions of electrodynamics [22, 23].

field $-2\boldsymbol{\Omega}_D$,

$$\boldsymbol{\Omega}_D = -\frac{1}{2}\nabla \times \mathbf{G} = \frac{kJ}{r^3}[3(\hat{\mathbf{r}}\cdot\hat{\mathbf{J}})\hat{\mathbf{r}} - \hat{\mathbf{J}}]$$

The weak-field analogy with electrodynamics provides insight into the gravitational field of a rotating body.[7] This analogy is incomplete, however, in contrast to the notion of the dragging of the inertial frames which is of general validity. To illustrate this point, consider a test gyroscope with its center of mass at rest in the stationary field of a Kerr black hole [26]. Let $\mathbf{S}$ be the spin vector of the gyroscope with respect to a local orthonormal tetrad frame whose spatial axes correspond to the spherical coordinate axes of static observers at infinity. The gyroscope precesses according to the formula

$$\frac{d\mathbf{S}}{d\tau} = \boldsymbol{\Omega}_D \times \mathbf{S}$$

where the precession frequency is given by

$$\boldsymbol{\Omega}_D = f(r,\vartheta)\,\nabla\times\left(\frac{k\mathbf{J}\times\hat{\mathbf{r}}}{r^2}\right) + g(r,\vartheta)\,\frac{kM}{r}\cdot\frac{(k\mathbf{J}\times\hat{\mathbf{r}})\times\hat{\mathbf{r}}}{r^3}$$

in Boyer-Lindquist coordinates [27]. The functions f and g are determined from

$$\Sigma^{3/2}(\Sigma - 2kMr)f = r^4\Delta^{1/2}$$

$$kMg = (\Delta^{-1/2}\Lambda - r)f$$

where $\Sigma = r^2 + A^2\cos^2\vartheta$, $\Lambda = r^2 - A^2\cos^2\vartheta$, and $\Delta = r^2 + A^2 - 2kMr$. The radial coordinate ranges from the static limit

$$r = kM + (k^2M^2 - A^2\cos^2\vartheta)^{1/2}$$

to infinity. The weak-field results are obtained for $r \gg 2kM$,

$$f(r,\vartheta) = 1 + \frac{kM}{r} + \cdots$$

$$g(r,\vartheta) = 1 + \frac{5}{2}\left[1 - \frac{1}{5}\frac{A^2}{k^2M^2}(1 + 2\cos^2\vartheta)\right]\frac{kM}{r} + \cdots$$

while f and g both diverge as the static limit surface is approached. At this surface, the assumption that the gyroscope's center of mass is at rest as determined by the inertial observers at infinity breaks down. Despite the dramatic increase

[7]It is interesting to note that the formal analogy with electrodynamics, pointed out by Einstein [24], was explored by Thirring [25], who confirmed that in the weak-field limit the geodesic equation may be written in the form of the Lorentz force law in terms of certain gravitational "electric" and "magnetic" fields.

in the magnitude of precession frequency close to the static limit surface, the general direction of this vector is similar to its weak-field limit: At the poles, $\mathbf{\Omega}_D$ is parallel to $\mathbf{J}$ and is given by $2kr^{-3}f(r, 0)\,\mathbf{J}$, but as ϑ varies, its direction gradually deviates away from $\mathbf{J}$ until it reverses at the equator, where $\mathbf{\Omega}_D = -kr^{-3}(1 - 2kM/r)^{-1}\,\mathbf{J}$. Thus, even in the strong-field region, the gravitational influence of a rotating mass is characterized by the dragging of the inertial frames.

In the weak-field approximation a test body follows a perturbed Keplerian orbit about a central rotating mass. To characterize the motion, one may use instead of the position and velocity of the test mass at any given moment of time the six orbital parameters of the osculating ellipse which the test particle would follow if the perturbing force were turned off at that instant. In terms of these parameters, the equations of motion then reduce to the Lagrange "planetary" equations. Lense and Thirring found that the perturbing effect of the gravitational "magnetic" field produced no change in the semimajor axis (a) of the osculating ellipse, while the eccentricity (e) and the inclination (i) of the orbit were affected only in a periodic manner, so that they remained unchanged when averaged over time. The only secular effects appeared in the longitude of the ascending node, the longitude of the pericenter, and the mean longitude. They discovered that the average rate of advance of the ascending node was given by

$$\mathbf{\Omega}_{\mathrm{LT}} = \frac{2k\mathbf{J}}{a^3(1 - e^2)^{3/2}}$$

while the longitude of the pericenter, as well as the mean longitude, varied on the average at a rate $(1 - 3\cos i)\,\Omega_{\mathrm{LT}}$. It follows from these results that the average rate of change of the longitude of the pericenter relative to the longitude of the node is given by

$$-3\Omega_{\mathrm{LT}}(\hat{\mathbf{J}} \cdot \hat{\mathbf{L}})\,\hat{\mathbf{L}}$$

where $\mathbf{L}$ is the orbital angular momentum and $\hat{\mathbf{J}} \cdot \hat{\mathbf{L}} = \cos i$. Thus the frequency of pericenter precession is on the average given by

$$\boldsymbol{\omega}_{\mathrm{LT}} = \Omega_{\mathrm{LT}}[\hat{\mathbf{J}} - 3(\hat{\mathbf{J}} \cdot \hat{\mathbf{L}})\,\hat{\mathbf{L}}]$$

which also characterizes the precession of the orbit as a whole. This result may be expressed as follows [28]: For an unperturbed orbit, the orbital angular momentum and the Lenz vectors are constants of the motion that characterize the orientation of the orbit. The rotation of the central mass produces on the average a precessional motion in these vectors with a frequency $\boldsymbol{\omega}_{\mathrm{LT}}$, which therefore corresponds to the average precession frequency of the elliptical orbit.[8]

The pericenter precession due to the rotation of the central mass was first

[8]Note that the average frequency of precession of the orbital angular momentum is given, equally well, by $\mathbf{\Omega}_{\mathrm{LT}}$.

considered for an equatorial orbit by de Sitter [29]. Lense and Thirring provided a general treatment valid for any orbital inclination.[9] This precession effect has been utilized in the theoretical determination of the angular momentum of the Kerr black hole [31, 32].

The observation of the Lense–Thirring precession in the orbits of the planets and moons could provide a new significant test of Einstein's theory in the solar system. This possibility was investigated in considerable detail by Lense and Thirring, who pointed out that the effects were too small to be measurable. This conclusion is equally valid today, although new possibilities have been opened up by the advent of space exploration.

To measure the Lense–Thirring precession, an orbit must have considerable eccentricity or inclination; furthermore, the observations must be so accurate as to allow the extraction of this small effect from the competing–and generally more significant–Newtonian and relativistic precessions. The multipole moments of the central masses and many-body effects account for the main Newtonian component, while the relativistic part is due to the purely spherical post-Newtonian fields of the Sun and the planets. The main effects of this post-Newtonian field are Einstein's pericenter precession and de Sitter's three-body effect due to the tidal influence of the Sun on a planet–moon orbit. Lense and Thirring discussed the relativistic perturbations in detail and provided estimates of the known significant effects, namely, Einstein's pericenter precession, the Lense–Thirring orbital precession, and de Sitter's three-body effect.[10] Moreover, they drew attention to the fact that for a given period of time (e.g., a century), the most significant relativistic orbital precession effect in the solar system is the Einstein precession for the moons of Jupiter and Saturn. The prospects for the observation of this effect have been reexamined recently [34]. A long-range observing program with improved accuracy is, in any case, necessary. The recent discovery of a new satellite of Jupiter and several new inner moons of Saturn by the Voyager spacecraft [35, 36] is encouraging, since the relativistic effects are expected to be even greater for these satellites. It is a cherished hope that, as space exploration continues, the various relativistic orbital effects can become measurable in the not-too-distant future.

The artificial satellites provide a new tool in attempts to test the predictions of Einstein's theory. Ginzburg [37] considered the possibility of measuring the perigee shift of an artificial satellite due to the rotation of the Earth, thereby providing a test of the Lense–Thirring pericenter precession. On the other hand, an interesting experiment has been suggested by Van Patten and Everitt [38, 39]

[9] In spite of a claim to the contrary [30], the Lense–Thirring results correctly give the frequency of the pericenter motion for any inclination angle i, including $i = \pi$.

[10] A summary of the various relativistic effects in the orbits of the planetary satellites has been given by Lense [33]. In terms of the general order of magnitude, the Einstein precession is the largest for these orbits, followed by the Lense–Thirring and de Sitter effects, respectively.

for the measurement of the Lense–Thirring precession of the orbital angular momentum. In this experiment, the dragging of the line of nodes is measured by placing two counter-orbiting drag-free spacecraft in polar orbits about the Earth.[11]

An alternative approach to the problem of determining the gravitational "magnetic" field of a rotating mass was considered by Schiff [41] and independently by Pugh. It consists in measuring the precession rate of a gyroscope in Earth orbit due to the coupling of the gyroscope's angular momentum to the rotation of the Earth. An experiment of this type using a superconducting gyroscope has been under active development at Stanford University by Fairbank, Everitt and their collaborators [42].

The Stanford gyroscope experiment and the Van Patten–Everitt counter-orbiting satellites experiment are the only two significant "laboratory" experiments thus far proposed to measure the dragging of the inertial frames due to a rotating mass [3].

Note Added in Proof

In a recent paper, V. B. Braginsky and A. G. Polnarev [*JETP Lett.*, **31**, 415 (1980)] have suggested a new experiment to measure the gravitational influence of a rotating mass. In this experiment, the contribution of the gravitational "magnetic" field of the Earth to the relative tidal acceleration experienced by masses at the ends of a spring in orbit around the Earth is measured. This idea has been discussed in detail by B. Mashhoon and D. S. Theiss [*Phys. Rev. Lett.*, **49**, 1542 (1982)], and the various general relativistic effects have been calculated. To distinguish the effect of the gravitational magnetic field from inertial effects, a local inertial frame along the orbit must be precisely defined. In fact, local gyroscopes must satisfy essentially the same performance criteria as in the Stanford gyroscope experiment.

Furthermore, a new gravitational effect of a rotating mass has recently been found by B. Mashhoon and D. S. Theiss. See B. Mashhoon, *Gen. Rel. Grav.*, March, 1984.

Acknowledgments

It is a pleasure to thank Professor J. Lense for helpful comments. The work of one of the authors (B.M.) was supported by the Alexander von Humboldt Foundation.

[11] In addition, a preliminary analysis [40] indicates that the Van Patten–Everitt experiment may also be used, by means of satellite-to-satellite range-rate measurements, to search for gravitational waves that are incident on the Earth and have a frequency in the range 10^{-4} Hz $\lesssim \nu \ll 10$ Hz.

References

1. Lense, J. (1981). Private communication.
2. Brans, C. H. (1977). *Phys. Rev. Lett.*, **39**, 856.
3. Shapiro, I. I. (1980). In *General Relativity and Gravitation: One Hundred Years After the Birth of Albert Einstein*, Vol. 2, Held, A., ed. (Plenum Press, New York), p. 469.
4. Hughes, V. W., Robinson, H. G., and Beltran-Lopez, V. (1960). *Phys. Rev. Lett.*, **4**, 342.
5. Drever, R. W. P. (1961). *Philos. Mag.*, **6**, 683.
6. Clemence, G. M. (1947). *Rev. Mod. Phys.*, **19**, 361.
7. Clemence, G. M. (1957). *Rev. Mod. Phys.*, **29**, 2.
8. Schiff, L. I. (1964). *Rev. Mod. Phys.*, **36**, 510.
9. Weiss, R. (1980). *Ann. Rev. Astron. Astrophys.*, **18**, 489.
10. Mashhoon, B. (1977). *Astrophys. J.*, **216**, 591.
11. Bass, L., and Pirani, F. A. E. (1955). *Philos. Mag.*, **46**, 850.
12. Hönl, H., and Maue, A. W. (1956). *Z. Phys.*, **144**, 152.
13. Brill, D. R., and Cohen, J. M. (1966). *Phys. Rev.*, **143**, 1011.
14. Cohen, J. M., and Brill, D. R. (1968). *Nuovo Cimento*, **B56**, 209.
15. De la Cruz, V., and Israel, W. (1968). *Phys. Rev.*, **170**, 1187.
16. Lindblom, L., and Brill, D. R. (1974). *Phys. Rev. D*, **10**, 3151.
17. Orwig, L. P. (1978). *Phys. Rev. D*, **18**, 1757.
18. Einstein, A. (1955). *The Meaning of Relativity* (Princeton University Press, Princeton, New Jersey).
19. Goenner, H. (1970). In *Ernst Mach: Physicist and Philosopher*, Boston Studies in the Philosphy of Science, Vol. VI, Cohen, R. S., and Seeger, R. J., eds. (Reidel, Dordrecht).
20. Reinhardt, M. (1973). *Z. Naturforsch.*, **28a**, 529.
21. Raine, D. J. (1981). *Rep. Prog. Phys.*, **44**, 1151.
22. Teyssandier, P. (1977). *Phys. Rev. D*, **16**, 946.
23. Teyssandier, P. (1978). *Phys. Rev. D*, **18**, 1037.
24. Einstein, A. (1913). *Phys. Z.*, **14**, 1261.
25. Thirring, H. (1918). *Phys. Z.*, **19**, 204.
26. Mashhoon, B. (1971). *J. Math. Phys.*, **12**, 1075.
27. Boyer, R. H., and Lindquist, R. W. (1967). *J. Math. Phys.*, **8**, 265.
28. Landau, L. D., and Lifshitz, E. M. (1971). *The Classical Theory of Fields*, 3rd ed. (Pergamon Press, Oxford).
29. De Sitter, W. (1916). *Mon. Not. Roy. Astron. Soc.*, **76**, 699.
30. Kalitzin, N. St. (1958). *Nuovo Cimento*, **9**, 365.
31. Kerr, R. P. (1963). *Phys. Rev. Letters*, **11**, 237.
32. Boyer, R. H., and Price, T. G. (1965). *Proc. Cambridge Philos. Soc.*, **61**, 531.
33. Lense, J. (1918). *Astron. Nachr.*, **206**, 117.
34. Hiscock, W. A., and Lindblom, L. (1979). *Astrophys. J.*, **231**, 224.
35. Voyager 2 Jupiter results: (1979). *Science*, **206**, 925.
36. Voyager 2 Saturn results: (1982). *Science*, **215**, 499.
37. Ginzburg, V. L. (1959). *Sci. Am.*, **200**, 149.
38. Van Patten, R. A. and Everitt, C. W. F. (1976). *Celest. Mech.*, **13**, 429.
39. Van Patten, R. A., and Everitt, C. W. F. (1976). *Phys. Rev. Letters*, **36**, 629.
40. Mashhoon, B. (1979). *Astrophys. J.*, **227**, 1019.
41. Schiff, L. I. (1960). *Phys. Rev. Lett.*, **4**, 215.
42. Everitt, C. W. F. (1973). In *Experimental Gravitation*, Bertotti, B., ed. (Academic Press, New York), p. 331.

Chapter 7
Myron Mathisson 1931 + 1937, equations of motion of particles with spin

Original papers:

1. Myron Mathisson, Die Mechanik des Materieteilchens in der allgemeinen Relativitätstheorie [The mechanics of a matter particle in general relativity].[1] *Zeitschrift für Physik* **67**, 826 – 844 (1931).

2. Myron Mathisson, Neue Mechanik materieller Systeme. *Acta Physica Polonica* **6**, 163 – 200 (1937).

Reprinted with the kind permissions of Springer Verlag, the current owner of publication rights to *Zeitschrift für Physik*, of the editors of *Acta Physica Polonica* and of Andrzej Trautman.

Papers translated by Anita Ehlers, email: Anita.Ehlers@t-online.de.

Previously published in *Gen. Relativ. Gravit.* **42** (2010), pages 985 – 987 (the editorial note + biography), 989 – 1010 (paper 1), 1011 – 1048 (paper 2).

[1] In the *Gen. Relativ. Gravit.* re-publication the title was erroneously translated as "...mechanics of matter particles...".

Gen Relativ Gravit (2010) 42:985–987
DOI 10.1007/s10714-010-0937-0

GOLDEN OLDIE EDITORIAL

Editorial note to: Myron Mathisson, The mechanics of matter particles in general relativity

and to: New mechanics of material systems

Andrzej Trautman

Published online: 19 February 2010

Keywords Mathisson–Papapetrou equations · Equations of motion in general relativity · Golden Oldie

The two papers by Mathisson that appear here as Golden Oldies [1,2], contain an implicit polemic with Einstein and his approach to the problem of motion. In the first paper [1], Mathisson shows that nonlinearity of the field equations is not essential for obtaining from them the equations of motion. After giving a neat description of the Liénard–Wiechert potential, Mathisson treats the analogous problem for a weak gravitational field $h_{\mu\nu}$. He introduces, following Einstein, the tensor $\psi^{\nu}_{\mu} = h^{\nu}_{\mu} - \frac{1}{2}\delta^{\nu}_{\mu}h^{\rho}_{\rho}$ so that the linearized field equations in empty space are equivalent to $\Box\psi^{\nu}_{\mu} = 0$ and $\psi^{\nu}_{\mu,\nu} = 0$. The tensor field $\psi^{\mu\nu} = m^{\mu\nu}(u)/r$—where u is a suitably defined retarded time—satisfies the wave equation and the Einstein condition $\psi^{\nu}_{\mu,\nu} = 0$ implies $m^{\mu\nu} = m\dot{z}^{\mu}\dot{z}^{\nu}$, $\dot{m} = 0$ and the equation of motion $\ddot{z}^{\mu} = 0$.

A few months after returning from Kazan (see his biography below), Mathisson sent for publication his most important paper [2] where he introduced the notion of a 'gravitational skeleton' and gave a derivation of the coupling between spin and curvature. In Mathisson's definition of the gravitational skeleton one can see the germ of the idea of a distribution, as later introduced by Laurent Schwartz: Mathisson uses 'test functions' $p_{\mu\nu}$ and the equation

The republications of the original Mathisson papers can be found in this issue following the editorial note and online via doi:10.1007/s10714-010-0938-z (The mechanics of matter particles in general relativity) and via doi:10.1007/s10714-010-0939-y (New mechanics of material systems).

A. Trautman (✉)
Instytut Fizyki Teoretycznej, Uniwersytet Warszawski, Hoża 69, Warszawa, Poland
e-mail: andrzej.trautman@fuw.edu.pl

$$\int_D T^{\mu\nu} p_{\mu\nu} d^4x = \int_L \left(m^{\mu\nu} p_{\mu\nu} + m^{\mu\nu\rho} p_{\mu\nu;\rho} + \cdots\right) ds$$

to replace the continuous energy-momentum tensor $T^{\mu\nu}$ filling a world-tube D by an equivalent—as far as the external field is concerned—distribution with support on a time-like world-line $L \subset D$. Taking $p_{\mu\nu} = \xi_{\mu;\nu} + \xi_{\nu;\mu}$ and using the conservation law $T^{\mu\nu}{}_{;\nu} = 0$, Mathisson proves that the above integral over L vanishes for arbitrary ξ vanishing at both ends of the world-line ('Mathisson's variational principle'). He then uses this principle to derive the equations of motion, which in modern form are

$$\dot{p}_\mu + \frac{1}{2} R_{\mu\nu\rho\sigma} u^\nu s^{\rho\sigma} = 0,$$
$$\dot{s}_{\mu\nu} + u_\mu p_\nu - u_\nu p_\mu = 0,$$

where p^μ and $s^{\rho\sigma}$ are, respectively, the momentum and spin (intrinsic angular momentum) of a body moving in space–time with curvature described by the Riemann tensor $R_{\mu\nu\rho\sigma}$; dots denote covariant derivatives in the direction of the 4-velocity u^μ.

In October of 2007, there was held, at the Stefan Banach International Mathematical Center in Warsaw, a conference devoted to Myron Mathisson, his life, work, and influence on current research. The proceedings of the conference were published, as a special issue of *Acta Physica Polonica B: Proceedings Supplement* **1** No.1 (2008) and are freely available at the address http://th-www.if.uj.edu.pl/acta/sup1/t1.htm. In particular, the article by T. Sauer & A. Trautman contains a biography of Mathisson and a list of all his scientific publications. W. G. Dixon gave a report on *Mathisson's New Mechanics: Its Aims and Realisation* and S.R. Czapor & R.G. McLenaghan on *Hadamard's Problem of Diffusion of Waves* and Mathisson's contribution to its solution. These and other papers presented at the conference contain a fairly thorough discussion of Mathisson's work and its subsequent developments. For this reason, the Editors of the Golden Oldies felt that there was no need to present here a detailed analysis of Myron Mathisson's contributions.

Myron Mathisson: a brief biography

By Andrzej Trautman

Myron Mathisson (1897–1940) was a Polish Jew, born and educated in Warsaw, where he obtained, in 1931, the Ph. D. degree under the supervision of Czesław Białobrzeski, a professor of theoretical physics at the University.

In 1929 Mathisson wrote, in French, a letter to Einstein, criticizing, in rather strong words, Einstein's approach to the problem of equations of motion in general relativity and outlining his own ideas on the subject. Einstein answered by inviting Mathisson to Berlin. Mathisson declined the invitation on the ground that he was not yet prepared to collaborate with Einstein; in the following years Einstein and Mathisson exchanged about 20 letters that are accessible at the Albert Einstein Archives at the Hebrew University of Jerusalem. On several occasions, Einstein supported Mathisson to obtain fellowships to pay scientific visits abroad.

In a series of six papers published in the 1930s, Mathisson outlined a new method of deriving equations of motion of bodies in general relativity. He associated with such bodies a distribution with support on a time-like world-line, representing the motion of their center of mass. In particular, he showed that for a body with an intrinsic angular momentum (spin), this world-line is no more a geodesic: Mathisson discovered a new interaction between spin and Riemannian curvature; the resulting equation is now associated with the names of Mathisson and Papapetrou. The latter published in 1951 another derivation of the coupling between curvature and spin. Mathisson's papers contained an early explicit use of the geometry of null (optical) elements to describe fields in relativity. He gave a new derivation of the equation of motion of a radiating electron.

In 1932, Mathisson obtained a habilitation that allowed him to use the title of 'docent' (analog of the German *Privat-Dozent*) and give lectures at the University of Warsaw, but did not imply a permanent, salaried position. In spite of Einstein's support, he never obtained an academic post in Poland.

His visits to Paris (1935 and 1939) and a collaboration with Jacques Hadamard were particularly fruitful. Mathisson developed a new method to analyze the properties of the fundamental solution of linear hyperbolic differential equations and used it to resolve Hadamard's conjecture on the class of equations for which the Huygens principle in four space–time dimensions is valid. Mathisson's work and his paper of 1939 so impressed Hadamard that, after Mathisson's death, he dedicated to him his paper on *The Problem of the Diffusion of Waves*, Ann. Math. **42** (1942) 510–522.

Mathisson spent the academic year 1936–37 lecturing at the University of Kazan. From the fall of 1937 to the spring of 1939 he lived in Cracow and collaborated there with Jan Weyssenhoff, a professor of theoretical physics at the Jagiellonian University.

Mathisson's ideas were further developed, during and after World War II, by Weyssenhoff and his students. In the spring of 1939 Mathisson went to Cambridge and continued there his work on the problem of motion. He impressed the physicists there so that, when in 1940 he died of tuberculosis, Dirac edited and presented for publication the notes left by Mathisson; he also wrote Mathisson's obituary notice that appeared in Nature.

References

1. Die Mechanik des Materieteilchens in der allgemeinen Relativitätstheorie. Z. Phys. **67**, 826–844 (1931)
2. Neue Mechanik materieller Systeme. Acta Physica Polonica **6**, 163–200 (1937)

Gen Relativ Gravit (2010) 42:989–1010
DOI 10.1007/s10714-010-0938-z

GOLDEN OLDIE

Republication of: The mechanics of matter particles in general relativity

Myron Mathisson

Published online: 9 March 2010

An editorial note to this paper can be found in this issue preceding this Golden Oldie and online via doi:10.1007/s10714-010-0937-0. The companion paper to this one can be found in this issue after this Golden Oldie and online via doi:10.1007/s10714-010-0939-y.

Original paper: Myron Mathisson, "Die Mechanik des Materieteilchens in der allgemeinen Relativitätstheorie", *Z. Phys.* **67**, 826–844 (1931). Reprinted with the kind permission of the Editors of European Physical Journal D, the current owner of the publication rights. Translated by Anita Ehlers, Anita.Ehlers@t-online.de.

Editorial responsibility: A. Krasiński, e-mail: akr@camk.edu.pl.

M. Mathisson (Deceased September 13, 1940)
Warsaw, Poland
e-mail for correspondence: akr@camk.edu.pl

The Mechanics of Matter Particles in General Relativity[1]

By **Myron Mathisson** in Warsaw

With 1 figure. (Received on December 23, 1930)

General remarks. – Linearisation of the Maxwell-Einstein equations on the Euclidean background – Retarded quantities as line functions. – Use of Dirac's δ-function for the calculation of divergences. – The method of calculation. – Gravitational potentials of an arbitrarily moving charged matter particle in first approximation. – The mechanical equations with radiation damping. – The problem of motion in General Relativity Theory: constancy of charge and mass, uniqueness of potentials, conditions at infinity. – Summary.

§ 1. In this paper the fundamental equations of mechanics of a charged matter particle will be deduced from Einstein's equations for gravitation. This is done without introducing restrictions concerning the particle's radiation and acceleration. For this purpose not only differential equations, but also their solutions – the potentials – will have to be considered, which supposes a mathematically transparent method of approximation. It goes without saying that no quasistatic generalisations of rest potentials may be used: quasistatic considerations are meaningful only if radiation is neglected.

If, in this paper, we talk about line singularities of potentials as lines of matter we consider the mathematical form of potentials outside of certain world tubes, in the same way as one talks about Newtons's potential of a homogeneous sphere that is represented on the outside of the sphere by a function that is singular at the center of the sphere. Some physicists prefer the concept of matter as a singularity since it answers the question concerning the spatial structure of particles in the simplest and most radical way, by robbing the particle of all structure, and since the singularity which is beyond the field is apparently the carrier of the particle's individuality and of the constancy of charge and matter. It is seen, in view of possible quantum possibilities, as a rather welcome danger that an assumption about the kind of

[1]Original title: Die Mechanik des Materieteilchens in der allgemeinen Relativitätstheorie, *Z. Phys.* **67**, 826–844 (1931). Reprinted with the kind permission of the Editors of European Physical Journal D, the current owner of the publication rights. Translated by Anita Ehlers, Anita.Ehlers@t-online.de.

singularity can lessen the particle's ability to adapt to the outer surrounding field.

These points of view are rather vague. A precise formulation of the problem of motion in General Relativity Theory can be gained comfortably only when we, at the end of this paper, discuss our results.

§ 2. We introduce the fundamental concept of *background* in the following way. Let it be possible to represent the components of the metric tensor for all points of a world domain as a sum $g_{ik}+\gamma_{ik}$, where (γ_{ik}) is assumed to be small and g_{ik} represents the metric tensor of some Riemannian world region W_4. We will call the metric manifold W_4 the background and regard (γ_{ik}) as well as the electromagnetic potentials φ_i as *tensor fields* in W_4. The lowering and raising of indices will refer to the metric of the background. We will expand the Einstein-Maxwell equations with respect to the γ_{ik}, and from that we will get a sequence of approximate equations of increasing order with respect to the γ_{ik}. In a close neighbourhood of the particles the equations and their solutions will anyway lose their meaning, since it may well be assumed that our expansions will no longer converge in that neighbourhood. Therefore we assume that single particles are embedded in tight, separate tubes which extend infinitely in a timelike direction and are in this way excluded from our world domain W_4, which otherwise encompasses the whole infinite world.

The solutions of our equations are supposed to be absolutely smooth in the world domain W_4, where there is no matter, i.e. to be continuous with continuous derivatives of arbitrary order. Our solutions may be assumed invalid for the interior of matter tubes. Even so, we will get some restricting equations for some world lines within the tubes. We will denote these world lines as lines of singularity since our potentials of finite order of approximation will be determined, if seen purely analytically, within the tubes, too, but will have poles (and other peculiarities) on the singularity lines. We presuppose about the singularity lines that the tube regions are separate from each other and that in all their points a small (but finite) chord (Δx_i) can be expanded with respect to the arc $\Delta s = \sqrt{g_{ik}\Delta x^i \Delta x^k}$.[2]

[2]Editor's footnote: The last equality was corrected, in the original the square root was missing. The exact meaning of the sentence is not clear, presumably it was a strange method of saying that the singularity lines are smooth.

It may seem strange that a certain well defined line inside a tube could be influenced by conditions outside the tube. One should notice, however, that by writing down electrodynamic potentials, with the help of which we achieve the approximation of order zero, a "centralisation" is achieved, since these potentials have poles on well determined timelike world lines. In this paper we will discuss the case of a *Euclidean background* and use equations of lowest order with regard to the γ_{ik}. We assume about the coordinate system that the g_{ik} are constant in it, whereas the γ_{ik} are to vanish in spatial infinity. Our system of equations consists of the usual Maxwell equations (of order zero) and Einstein's equations of gravitation in first approximation

$$\Box\psi_{ik} + \frac{\partial z_i}{\partial x^k} + \frac{\partial z_k}{\partial x^i} - g_{ik}\frac{\partial z^r}{\partial x^r} = -2\kappa T_{ik}, \qquad z_i = \frac{\partial \psi_i^k}{\partial x^k}, \quad \psi_i^k = \gamma_i^k - \frac{1}{2}\delta_i^k\gamma_r^r.$$

Also in the higher orders of approximation on the Euclidean background the left-hand sides of the equations agree with the left-hand sides of the zeroth or first order. To the right hand sides will be added expressions which follow from the development of the electromagnetic and gravitational equations with respect to the γ_{ik}. A linear equation has to be solved at each step of the approximation, since the right hand side will prove to be determined by the preceding approximations.

The introduction of the Euclidean background does not imply a return to Special Relativity Theory. Rather this background plays the role of an auxiliary metric which is introduced for the coordinate system. Any part of a Riemannian world can, with respect to an arbitrary background, under rather broad assumptions always be represented by series expansion. For our background however it is essential that it comprises the whole world (with the exclusion of the tubes) and – above all – that it represents a homogeneous solution of the gravitational equations for the empty world, $R_{ik} = 0$.

§ 3. We will pass to the formulae now. For a Riemannian manifold the variation formula needs no explanation:

$$\frac{1}{2}\delta\int_1^2 Q dt = [q_i\delta x^i]_1^2 - \int_1^2 \left(\frac{dq_i}{dt} - \frac{1}{2}\frac{\partial g_{\alpha\beta}}{\partial x^i}q^\alpha q^\beta\right)\delta x^i dt, \quad q^i = \frac{dx^i}{dt}, \quad Q = g_{ik}q^i q^k.$$

Here t denotes any parameter of the initial curve, (δx^i) is the small displacement which deforms the initial curve into the varied neighbouring one. The

δx^i do not vanish at the endpoints, which gives rise to the expression

$$[q_i \delta x^i]_1^2 = (q_i \delta x^i)_2 - (q_i \delta x^i)_1.$$

If the initial curve as well as the varied one is a geodesic null line, our variation formula leads to

$$[q_i \delta x^i]_1^2 = 0. \tag{1}$$

The vectors (q^i) at points 1 and 2 are transformed into each other by parallel translation along the null line.

With respect to tensor calculus we now base ourselves on the ground of Special Relativity and consider the world as a Euclidean one with the metric fundamental form $g_{ik} dx^i dx^k$. We assume our coordinate system only to be linear, i.e., that it can be normalised by a linear transformation, so that the non-vanishing g_{ik} have the values

$$g_{00} = 1, \qquad g_{ii} = -1, \qquad i = 1, 2, 3.$$

The direction past – future may be given for all timelike directions by a future-directed vector. Let a timelike world line be given and on it an initial point P_0. The proper time s for any point P of the world line is determined with respect to P_0 by the length of the arc P_0P, and considered positive or negative depending on whether the vector $\overrightarrow{P_0P}$ is future- or past-directed. At a point O not on the world line we construct the past-directed null cone $\mathfrak{K}(0)$. Our world line is intersected by the cone at a point P. This construction ascribes to the world point $O(x)$ a scalar s and a vector $\overrightarrow{OP}$. The former is determined by the length of the arc P_0P with the corresponding sign, the latter by the generator of our null cone that is directed toward P. We assume this construction being done for all points in our world, which determines a scalar field $s(x)$ and a vector field with the components $l^i(x)$, which have their origin on a world line (l^i are the components of $\overrightarrow{OP}$). We are now dealing with the differential quotients $\frac{\partial s}{\partial x^i}, \frac{\partial l_i}{\partial x^k}$.

In order to get these we will use formula (1). In our Euclidian world let $O(x)$ be point 1, and the point associated with it by the cone construction on the world line P point 2. Applying the same cone construction to the point O' which arises from O by an arbitrary infinitesimal translation (δx^i), we find on our world line a point P' arising from P by a translation $(u^i \delta s)$, with (u^i) the velocity in point P. If we take in formula (1) the components l^i of the null vector $\overrightarrow{P_0 P}$ as q^i, then

$$l_i \delta x^i - l_i u^i \delta s = 0.$$

Since, however, $\delta s = \frac{\partial s}{\partial x^i} \delta x^i$ we get, with $l_r u^r = n$,

$$\frac{\partial s}{\partial x^i} = \frac{l_i}{n}. \tag{2}$$

The $(l^i + \delta l^i)$ associated with point O' can easily be expressed by (l^i) and (δx^i):

$$l^i + \delta l^i = l^i - \delta x^i + u^i \delta s, \qquad \text{i.e.} \qquad \delta l^i = \left(u^i \frac{\partial s}{\partial x^r} - \delta^i_r \right) \delta x^r.$$

This implies

$$\frac{\partial l^i}{\partial x^k} = -\delta^i_k + \frac{u^i l_k}{n}. \tag{3}$$

Let $f(s)$ be a function of the arc-length defined on the world line. With the cone construction we will derive from such functions of the arc-length functions of the world point, by giving the world point $O(x)$ the value of the function at the assigned point P. We will denote the original and the derived function with the same symbol $f(s)$. Both these functions have the same value on the world line. Denoting derivatives by the arc length by overdots, we have

$$\frac{\partial f}{\partial x^i} = \dot{f} \frac{l_i}{n}.$$

If $f(s)$ is equal to a component of the velocity, we get

$$\frac{\partial u_i}{\partial x^k} = \frac{\dot{u}_i l_k}{n},$$

and therefore

$$\frac{\partial n}{\partial x^i} = -u_i + \frac{l_i}{n}(1 + p), \qquad p = l_r \dot{u}^r. \tag{4}$$

§ 4. In our thus derived rules for differentiation the four coordinates are equivalent. These formulae are easy to use. We consider some examples which will be of importance later.

From a vector $e_i(s)$, given on the world line, we build the world vector

$$\varphi_i(x) = \frac{e_i}{n}.$$

The divergence

$$z = \frac{\partial \varphi^i}{\partial x^i}$$

is found to be

$$z = \frac{\dot{e}_i l^i}{n^2} + \frac{e_i u^i}{n^2} - \frac{e_i l^i}{n^3}(1+p).$$

If in all world points $z = 0$, we get for the term of order $1/n^2$

$$\frac{e_i}{n^2}\left(u^i - \frac{l^i}{n}\right) = 0.$$

The vector in parentheses is orthogonal to (u^i). For different l^i it is equal to any vector in the space orthogonal to (u^i). Since (e^i) is thus orthogonal to all vectors of this space, it does not have a component orthogonal to (u^i), and thus has the form (eu^i). Then we have

$$z = \frac{\dot{e}}{n}.$$

Therefore we must have $e =$ const. The condition $z = 0$ did not lead to any restriction on the world line.

Things are different, however, with respect to the relation

$$z_i \equiv \frac{\partial \psi_i^k}{\partial x^k} = 0,$$

where ψ_{ik} is defined in terms of the symmetrical tensor on the worldline by the formula

$$\psi_{ik}(x) = \frac{m_{ik}(s)}{n}.$$

We then get

$$z_i \equiv \frac{1}{n^3}[l^\alpha l^r(\dot{m}_{i\alpha} u_r - m_{i\alpha}\dot{u}_r + l^r(m_{i\alpha} u^\alpha u^r - m_{ir})] = 0.$$

This equation decomposes into the two following ones:

$$m_{ir} - m_{i\alpha}u^{\alpha}u^{r} = 0, \qquad \dot{m}_{i\alpha}u_r - m_{i\alpha}\dot{u}_r = 0.$$

The first tensor equation, transformed to rest with respect to (u^i), gives $m_{ir} = 0$ if $r \neq 0$. Since m_{ik} is symmetric, this implies $m_{ik} = Mu_iu_k$, with M at first a function of the arc length. The second tensor equation gives, because of $(u\dot{u}) = 0$, $\dot{m}_{i\alpha} = 0$, which implies

$$M\dot{u}_i + \dot{M}u_i = 0.$$

Multiplication with u^i gives

$$\dot{M} = 0, \qquad M = \text{const}.$$

This implies $M\dot{u}_i = 0$, i.e., $\dot{u}_i = 0$. The condition $z_i = 0$ thus determined the possible world line to be a straight line, and ascribed the form const u_iu_k to the function m_{ik}. Obviously, through these restrictions the equation $z_i = 0$ in fact becomes an identity.

It may be noted that the function

$$\psi(x) = \frac{f(s)}{n}$$

fulfills the wave equation

$$\Box\psi = 0 \qquad \left(\Box = g^{ik}\frac{\partial^2}{\partial x^i \partial x^k}\right)$$

identically, as can be verified by direct calculation with use of our formulae for differentiation. It is seen even easier by noticing that ψ is a retarded potential.

§ 5. The equation

$$\Box\psi = T \tag{5}$$

is solved by a retarded potential:

$$\psi = \frac{1}{4\pi}\int \frac{T_{t-r}}{r}dV \tag{6}$$

($x^0 = t$ and $x^1 x^2 x^3$ are orthogonal world coordinates; $dV = dx^1 dx^2 dx^3$). We will now transform this solution for our purposes.

Let there be given, in an m-dimensional Euclidean space, p contravariant vectors with the components[3]

$$\underset{1}{v}^{\lambda}, \underset{2}{v}^{\lambda}, \ldots, \underset{p}{v}^{\lambda}, \qquad 1 \leq p \leq m, \qquad \lambda = 1, 2, ..., m.$$

The tensor quantity built from these vectors, skew symmetric in all pairs of indices, with the components[4]

$$v^{\lambda_1 \ldots \lambda_p} = \frac{1}{p!} \begin{vmatrix} \underset{1}{v}^{\lambda_1} & \cdots & \underset{1}{v}^{\lambda_p} \\ \vdots & \cdots & \vdots \\ \underset{p}{v}^{\lambda_1} & \cdots & \underset{p}{v}^{\lambda_p} \end{vmatrix} = \frac{1}{p!} \left(\underset{1}{v}^{\lambda_1} \underset{2}{v}^{\lambda_2} \ldots \underset{p}{v}^{\lambda_p} - \underset{2}{v}^{\lambda_1} \underset{1}{v}^{\lambda_2} \ldots \underset{p}{v}^{\lambda_p} + \ldots \right)$$

we will call the p-vector spanned by the given vectors. The connection between $v^{\lambda_1 \ldots \lambda_p}$ and the generating vectors will be denoted by square brackets:

$$v^{\lambda_1 \ldots \lambda_p} = \underset{1}{v}^{[\lambda_1} \underset{2}{v}^{\lambda_2} \ldots \underset{p}{v}^{\lambda_p]}.$$

For arbitrary tensors we put

$$v^{\ldots [\lambda_1 \ldots \lambda_p] \ldots}_{\ldots} = \frac{1}{p!} \left(v^{\ldots \lambda_1 \lambda_2 \ldots \lambda_p \ldots}_{\ldots} - v^{\ldots \lambda_2 \lambda_1 \ldots \lambda_p \ldots}_{\ldots} + \ldots \right).$$

All quantities generated by the $p!$ different permutations of the indices in brackets are summed up, with plus- or minus signs according to whether the permutation is even or odd. The dots indicate that there can be other indices, too. We get the following development:

$$\underset{1}{v}^{[\lambda_1} \ldots \underset{p}{v}^{\lambda_p]}$$

$$= \frac{1}{p} \underset{1}{v}^{\lambda_1} \underset{2}{v}^{[\lambda_2} \ldots \underset{p}{v}^{\lambda_p]} - \frac{1}{p} \underset{1}{v}^{\lambda_2} \underset{1}{v}^{[\lambda_1} \underset{3}{v}^{\lambda_3} \ldots \underset{p}{v}^{\lambda_p]} + \frac{1}{p} \underset{1}{v}^{\lambda_3} \underset{2}{v}^{[\lambda_1} \underset{3}{v}^{\lambda_2} \underset{4}{v}^{\lambda_4} \ldots \underset{p}{v}^{\lambda_p]} - \ldots . \quad (7)$$

[3]Editor's footnote: The next formula had a typo in the original, the v in the first factor was missing, only its indices were present. Corrected by the editor.

[4]Editor's footnote: In the original paper, there were several typos inside the determinant: three of the v-s were missing, and so were some of the underset indices. Corrected by the editor.

If the infinitesimal vectors with components $d_1x^\nu, ..., d_px^\nu$ form the p-vector, we get an element. We introduce the notation

$$f^{\lambda_1 \ldots \lambda_p} d\tau_p = d_1 x^{[\lambda_1} \ldots d_p x^{\lambda_p]}.$$

If the element $f^{\lambda_1 \ldots \lambda_p} d\tau_p$ occurs outside an integral we usually omit the $d\tau_p$.

Let now (a^λ) be the unit vector of the time axis in the coordinate system (6), and correspondingly $\left(\underset{i}{h}{}^\lambda\right)$, $i = 1, 2, 3$, the unit vectors of the other axes. We introduce the notation

$$A^{\cdots} = A^{\lambda_1\lambda_2\lambda_3\lambda_4} = a^{[\lambda_1} \underset{1}{h}{}^{\lambda_2} \underset{2}{h}{}^{\lambda_3} \underset{3}{h}{}^{\lambda_4]}.$$

We have

$$A^{\cdots} dV = A^{\cdots} dx^1 dx^2 dx^3 = a^{[\lambda_1} f^{\lambda_2\lambda_3\lambda_4]} d\tau_3.$$

Here $f^{\lambda_1\lambda_2\lambda_3]} d\tau_3$ is the surface element of the past light cone which has its vertex at the point $O(x)$ for which $\psi(x)$ results from (6). If (l^i) is the null vector drawn from $O(x)$ to the integration point, then $r = -l^i a_i$. But we have

$$\frac{a^{[\lambda_1} f^{\lambda_2\lambda_3\lambda_4]}}{l_i a^i} = \frac{u^{[\lambda_1} f^{\lambda_2\lambda_3\lambda_4]}}{l_i u^i},$$

where (u^i) is any velocity vector $(u_i u^i = 1)$. For, with (u^i) decomposed parallel to (a^i) and in the plane of $f^{\lambda_1\lambda_2\lambda_3}$, the second component matters neither for $l_i u^i$ nor for $u^{[\lambda_1} f^{\lambda_2\lambda_3\lambda_4]}$, and the component in direction of (a^i) can be replaced by (u^i) in the numerator as well as is the denominator. Therefore we can replace formula (6) by the invariant relation

$$-4\pi A^{\cdots} \psi(x) = \int u^{[\lambda_1} f^{\lambda_2\lambda_3\lambda_4]} \frac{T}{n} d\tau_3, \qquad n = (lu). \tag{8}$$

$A^{\cdots}$ and $u^{[\lambda_1} f^{\lambda_2\lambda_3\lambda_4]}$ are supposed to have the same orientation. Each four-vector built from normalised vectors that are orthogonal to each other can be taken as $A_{\cdots}$. All such tetrads transform into each other by a Lorentz transformation with determinant + 1. All components $A^{\lambda_1\lambda_2\lambda_3\lambda_4}$ remain unchanged thereby.

We will use the arbitrariness of the (u^i)-field by taking it as a velocity field of a family of timelike world lines. The world lines generating the walls of the matter tubes are to belong to this family. Only one line of the family is to pass through any one world point.

The correct form of the Gaussian formula in the four-dimensional Euclidean world is the following:

$$\int_{\tau_4} f^{\lambda_1\lambda_2\lambda_3\lambda_4}\frac{\partial w^\mu}{\partial x^\mu}d\tau_4 = -\int_{\tau_3} f^{[\lambda_1\,\lambda_2\lambda_3}w^{\lambda_4]}d\tau_3. \tag{9}$$

Here a suitable relation between the orientation of $f^{\lambda_1\lambda_2\lambda_3\lambda_4}$ and $f^{\lambda_1\lambda_2\lambda_3}$ is assumed, which does not concern us here. This formula will provide the value of

$$z_i = \frac{\partial \omega_i^k}{\partial x^k}$$

if ω_{ik} is given by the formulae

$$4\pi A^{\cdots}\omega_{ik}(x) = \int u^{[\lambda_1}\,f^{\lambda_2\lambda_3\,\lambda_4]}\frac{T_{ik}}{n}d\tau_3, \qquad (T_{ik} = T_{ki}), \tag{10}$$

which is to be interpreted in the sense of (8). The integration comprises the whole past light cone with the exception of those areas which are contained in the matter tubes. We presuppose in all world points outside the tubes[5]

$$\frac{\partial T_i^k}{\partial x^k} = 0.$$

To transform z_i into an integral over the intersection of the tubes with the past light cone we define an auxiliary function φ. Through each world point $A(x)$ we draw a world direction parallel to the vector (a_i) $(a_i a^i = 1)$ which intersects the cone of $O(x_0)$ at the point A_0 (Fig. 1). Let σ be the length of the timelike segment A_0A, which is positive or negative according to whether $\overrightarrow{A_0A}$ is past- or future-directed. We put

$$\varphi = \frac{\delta(\sigma)}{l_i a^i},$$

the (improper) function $\delta(x)$ being defined by the equations

$$\int_{-\infty}^{+\infty} f(x)\delta(x)dx = f(0), \qquad \delta(x) = 0 \;\; (x \neq 0)$$

[5]Editor's footnote: The words "outside the tubes" did not exist in the original, and were added by the editors – as dictated by the logic of the construction.

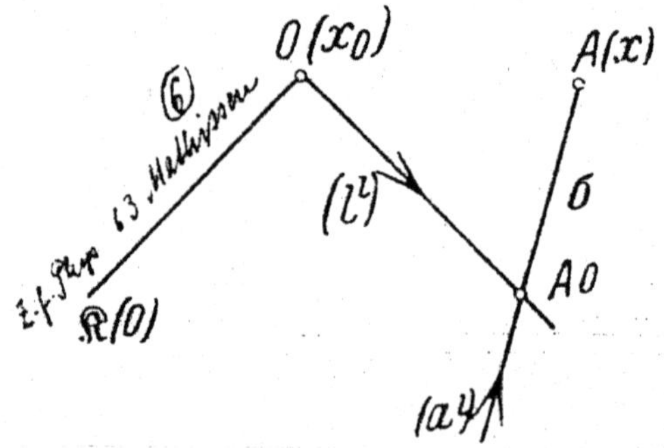

Figure 1.

(introduced into quantum mechanics by Dirac). The δ-function can be obtained from a proper function by a limit process, and it can be used in the calculations without any worry as if it were a continuous function. We replace now in (10) (u^i) by (a^i) and express ω_i^k, with the help of the δ-function, as an integral over the whole world (with exception of the tubes). Since only the relative position of $O(x_0)$ and $A(x)$ matters for the value of φ, this implies

$$\frac{\partial \varphi}{\partial x_0^i} + \frac{\partial \varphi}{\partial x^i} = 0.$$

Therefore it is permitted to replace

$$z_i = \frac{\partial \omega_i^k}{\partial x_0^k}$$

by

$$4\pi A^{\cdots} z_i = \int f^{\lambda_1\lambda_2\lambda_3\lambda_4} \frac{\partial \varphi}{\partial x_0^k} T_i^k d\tau_4 = - \int f^{\lambda_1\lambda_2\lambda_3\lambda_4} \frac{\partial \left(\varphi T_i^k\right)}{\partial x^k} d\tau_4,$$

because of $\frac{\partial T_i^k}{\partial x^k} = 0$. Formula (9) thus implies

$$4\pi A^{\cdots} z_i = \int f^{[\lambda_1\,\lambda_2\lambda_3} T_i^{\lambda_4]} \varphi d\tau_3. \tag{11}$$

The integration is to be taken over the surfaces of the world tubes and over the "infinitely distant tube surface" comprising the world. We presuppose that the integral over the infinitely distant surface vanishes, in consequence of the correspondingly fast decrease of T_{ik}.

Formula (2) implies

$$\frac{\partial\sigma}{\partial x_0^i} = -\frac{l_i}{(la)}$$

(with a minus sign since in that formula the arc length is calculated towards the null cone, but σ is counted from the null cone). If ds denotes the arc element of a worldline on the tube surface we get, because of $dx^i = u^i ds$,

$$d\sigma = \frac{\partial\sigma}{\partial x^i}dx^i = \frac{(lu)}{(la)}ds, \qquad ds = \frac{(la)}{(lu)}d\sigma.$$

The surface element $f^{\lambda_1\lambda_2\lambda_3}$ of the tube wall amounts to $u^{[\lambda_1} f^{\lambda_2\lambda_3]} d\tau_2 ds$; $f^{\lambda_1\lambda_2}$ is the element of the corresponding cross-section of the worldline fibre on the tube surface. Transforming in (11) the integral over s into one over σ and taking into regard the integral property of the δ-function, we finally get

$$4\pi A^{\cdots}z_i = \int u^{[\lambda_1} f^{\lambda_2\lambda_3} T_i^{\cdot\lambda_4]}\frac{1}{(lu)}d\tau_2. \tag{12}$$

The domain of integration is formed by the intersection of the tubes and the null cone.

§ 6. The system of Einstein's equations of gravitation decomposes, as is well known, in first approximation by appropriate coordinate transformation into two systems:

$$\frac{1}{2}\Box\psi_i^k = -\kappa T_{ik}; \qquad \frac{\partial\psi_i^k}{\partial x^k} = 0$$

(comp. § 2). This decomposition determines the coordinate system up to a transformation

$$\overline{x}^i = x^i + \xi^i(x),$$

if any linear transformation is permitted, where the ξ^i are of the same order of magnitude as the γ_{ik} and satisfy the equations $\Box\xi^i = 0$. Since we admit singularities we will get non-zero solutions of the (ξ^i) which vanish at infinity. The restrictions up to now thus determine the coordinate system only up to some transformations of first order, even if linear transformations are admitted. This fact does not, however, affect the mechanical equations which concern quantities of order zero. We will come back to that in a different context.

Solutions which correspond to an uncharged mass point should have, as experience tells, simple poles. The equations $\Box\psi_i^k = 0$ imply then the form $\frac{m_i^k(s)}{n}$ for the ψ_i^k. We saw already (§ 4) that the equations $\frac{\partial\psi_i^k}{\partial x^k} = 0$ lead to constant mass and a straight world line. The last result implies that our coordinate system agrees, up to quantities of first order, with the usual linear systems which are determined uniquely up to linear transformations by the principle of inertia.

Proceeding to the calculations referring to charged mass-points we encounter at once a computational difficulty which arises from the integrals

$$\int u^{[\cdot}f^{\cdots]}\frac{T_i^k}{n}d\tau_3.$$

These become infinite when the world lines contract to their singularity lines since on approaching a singularity line the T_{ik} increase as $1/n^4$. What is one to do, however, if one wants to get a particular solution of Poisson's equation $\Delta U = F$ by evaluating Newton's potential $-\frac{1}{4\pi}\int\frac{F}{R}dV$ when F has a pole of third or higher order? One tries to add to the divergent potential, which is at first calculated for the interior of a small sphere, such solutions of the Laplace equation $\Delta U = 0$ which are defined for an infinitely small sphere around the pole as its Newton potentials and which compensate the infinite growth of the potential. Then the radius of the small sphere can be shrunk to zero. We will proceed in a similar way in the case of the equations of gravitation.

The obstacle arising from the infinity of the integrand can be avoided also by substituting in the equation

$$\frac{1}{2}\Box\psi_i^k = -\kappa T_{ik}$$

$\psi_{ik} = \alpha_{ik} + \beta_{ik}$ and trying to remove the poles of third and fourth order from the expression

$$U_{ik} = -\kappa T_{ik} - \frac{1}{2}\Box\beta_{ik}$$

by an appropriate choice of β_{ik}. Then α_{ik} can be found from the equation

$$\frac{1}{2}\Box\alpha_{ik} = U_{ik}.$$

In order to use the relations (12), which is important for the derivation of the mechanical equations, the β_{ik} may be chosen such that the equations

$$\frac{\partial U_i^k}{\partial x^k} = 0$$

hold. That is connected to the fact that the gauge equation

$$\frac{\partial \psi_i^k}{\partial x^k} = 0$$

plays the role of the real source of the mechanical equation. This method enabled the author to actually find a derivation of the mechanical equations.

Both methods sketched here have computational advantages and disadvantages. Since we are dealing here foremost with a formulation of the mechanical problem of General Relativity, we will confine ourselves to listing the results without going into the rather complex calculations.

§ 7. As the zero-th order solution of our equations, the ordinary Maxwell equations, we will assume the Liénard-Wiechert potential vector. With the notations of § 3 and 4 we get for a particle the potential

$$\varphi^i = -\frac{e}{4\pi}\frac{u^i}{n};$$

e is the charge measured in Lorentzian units. The equation $\frac{\partial \varphi^i}{\partial x^i} = 0$ implies e = const. (comp. § 4). One may have objections against the Liénard-Wiechert potential – we will discuss that in our concluding remarks.

We calculate the field tensor of the particle $\varphi_{ik} = \frac{\partial \varphi_i}{\partial x^k} - \frac{\partial \varphi_k}{\partial x^i}$ according to the formulae for differentiation in § 3. The electromagnetic field at a world-point can be decomposed into two parts: the field φ_{ik} of a certain particle considered by us, which gets infinite on a world line S, and into the field F_{ik} of the other particles, which is free of singularities on S. This decomposition is possible because of the linearity of Maxwell's equations. The energy tensor of the whole electromagnetic field is

$$T_{ik} = \frac{1}{4}g_{ik}\varphi_{rs}\varphi^{rs} - \varphi_{ir}\varphi_k{}^r + \frac{1}{2}g_{ik}\varphi_{rs}F^{rs} - (\varphi_{ir}F_k{}^r + \varphi_{rs}F_i{}^r) + \Theta_{ik};$$

Θ_{ik} is the energy tensor derived solely from the F_{ik} and therefore is free of singularities on S. Only those components of T_{ik} are important for mechanics which are on S infinite of at least second order with respect to 1/n. The electromagnetic energy tensor obeys the equation

$$\frac{\partial T_i^k}{\partial x^k} = 0$$

identically. The equation

$$\frac{1}{2}\Box\psi_i^k = -\kappa\, T_i^k$$

has a particular solution ω_i^k, which is given according to formula (8) by

$$4\pi A^{\cdot\cdot}\omega_i^k(x_0) = -2\kappa \int u^{[\cdot} f^{\cdots]} \frac{T_i^k}{(lu)} d\tau_3. \qquad (13)$$

[The origin is $O(x_0)$ instead of $O(x)$]. The interior of the tubes around the singularity lines is to be excluded from integration. We construct the tubes as follows. We choose the point P of the singularity line S as the vertex of a future-directed null cone. Let (u^i) be the velocity at P. On the cone we select a two-dimensional surface which is, seen from the space orthogonal to (u^i) a sphere with center P. Let ε be its radius. If P traverses all its world line S, our surface generates a (three-dimensional) tube surface. We assume ε to be constant. The particular solution (13) depends on ε and is not determined for the interior of the tubes. By performing the calculation sketched in the previous paragraph one gets the following solution independent of ε which has a finite value in all world points outside of the singularity lines:

$$\Omega_{ik}(x_0) = \lim_{\varepsilon\to 0}\left[\omega_{ik} + \frac{\kappa}{8\pi^2}\sum \frac{e^2}{\varepsilon}\cdot\frac{1}{(LU)}\left(\frac{1}{6}g_{ik} - \frac{2}{3}U_iU_k\right)\right]. \qquad (14)$$

The summation is to be done over all singularity lines. The quantities U^i, L^i correspond to the usual u^i, l^i and refer to the singularity lines.

The existence of the integrals occurring in (14) is tied to assumption of sufficiently fast decrease of T_{ik} at spatial infinity.

The formula (12) permits us to calculate the divergences, first as surface integrals over the (two-dimensional) surfaces, in which the tubes intersect the null cone. We pass from the ω_{ik} to the Ω_{ik} and to lim $\varepsilon = 0$. A somewhat long calculation leads to

$$\frac{\partial \Omega_{ik}}{\partial x_0^k} = \sum \frac{1}{4\pi (LU)} \left[-\frac{\kappa e^2}{3\pi} \left(\dddot{U}_i - \dddot{U}_r U^r U_i \right) + 2\kappa e \overset{P}{F}_{ir} U^r \right], \tag{15}$$

with $\overset{P}{F}_{ir}$ denoting the field F_{ik} at point P, which, through the cone-construction, is associated to the world point $O(x_0)$ on the singularity line. The expression (15) is a solution of the wave equation $\Box\psi = 0$ of the form $f(s)/n$.

We build the complete solution ψ_{ik} from Ω_{ik} by adding a solution of the wave equation which vanishes in spatial infinity. We put

$$\psi_{ik} = \Omega_{ik} + \sum \frac{m_{ik}}{(LU)}, \qquad m_{ik} = m_{ki}. \tag{16}$$

The sum is to be over all lines of singularity. The equations $\frac{\partial \psi_i^k}{\partial x_0^k} = 0$ yield now

$$\sum \frac{1}{(LU)^3} \{ L^r L^s (\dot{m}_{is} U_r - m_{ir} U_s) + L^r (m_{is} U^s U_r - m_{ir})$$
$$- \frac{(LU)^2}{4\pi} \left[\frac{\kappa e^2}{3\pi} \left(\dddot{U}_i - \dddot{U}_r U^r U_i \right) - 2\kappa e \overset{P}{F}_i r U^r \right] \Bigg\} = 0.$$

This equation is supposed to hold for any field point which is not on the singularity lines. That is possible only if all the terms vanish, i.e., if the equation holds for each singularity line. Then the vanishing of the term of highest (second) order relative to $1/(LU)$ (comp. § 4) implies:

$$m_{ik} = M U_i U_k.$$

This implies, as a condition for the vanishing of the term of first order relative to 1/(LU),

$$\dot{M} U_i + M \dot{U}_i - \frac{1}{4\pi} \left[\frac{\kappa e^2}{3\pi} \left(\ddot{U}_i - (\ddot{U} U) U_i \right) - 2\kappa e \overset{P}{F}_{ir} U^r \right] = 0.$$

Multiplying with U_i we get

$$\boxed{\dot{M} = 0; \qquad M = \text{const.}} \tag{17}$$

Let $M = \frac{\kappa m_0}{2\pi}$. Going over from the (Lorentz) electrical units used up to now to the electrostatical ones, we have

$$e(\text{Lor.}) = \sqrt{4\pi} e_0 \text{ (elst.)}, \qquad F_{ik}(\text{Lor.}) = \frac{1}{\sqrt{4\pi}} F_{ik} \text{ (elst.)},$$

and finally, by writing F_{ik}, u_i instead of $\overset{P}{F}_{ik}, U_i$ we get the dynamical equations in the form

$$\boxed{m_0 \dot{u}_i + \tfrac{2}{3} e_0^2 \left((\ddot{u} u) u_i - \ddot{u}_i \right) = e_0 F_{ki} u^k.} \tag{18}$$

Both sides of this vector equation are orthogonal to (u^i).

The conformally Euclidean solution[6] for an uncharged particle at rest may be expressed in first approximation for not too small r in the following way:

$$\gamma_{00} = \gamma_{11} = \gamma_{22} = \gamma_{33} = -\frac{2m}{r};$$

the other components are

$$\gamma_{ik} = 0 \quad \left(r^2 = x^{1^2} + x^{2^2} + x^{3^2} \right).$$

The gravitational radius m and the mass measured in g are related by

$$m = \frac{\kappa m_0}{8\pi}.$$

Expressed in (ψ_{ik}) the conformally Euclidean solution is:

$$\psi_{ik} = \frac{4m}{n} u_i u_k = \frac{\kappa m_0}{2\pi} \frac{u_i u_k}{n} \qquad \left(\frac{\partial \psi_i^k}{\partial x^k} = 0 \right).$$

which is thus in agreement with our $\frac{m_{ik}}{(LU)}$.

The electrical part of ψ_{ik} for a particle at rest is contained in Ω_{ik}. We will not consider it further, since for it no comparison with potentials justified experimentally is possible.

We stress again that our method of approximation is independent of the fact that the approximated solutions become unusable in a certain neighbourhood of the matter particles. Even so there exist definite differential equations for the analytically sharp lines of singularity.

[6] The real space then is conformally related to the Cartesian image space; see H. Weyl, Raum, Zeit, Materie, 5. Aufl. Berlin, Springer, 1923, p. 254-255.

§ 8. The equations of motion derived by us agree with those of the older theory of electrons. But we are not so much interested in just incorporating these known equations into the structure of general relativity. The following remarks will clarify the proper meaning of our calculations.

The equations $\frac{\partial\psi_i^k}{\partial x^k} = 0$ led us in the end to conditions which contain on the one hand functions of the position on the world line, but, on the other hand, functions of the arbitrary field point. This leads to the question: Can these conditions be satisfied, for an arbitrary field point, i.e., for a four-dimensional continuum, by assumptions about the path of the one-dimensional world line? Do we not have to be afraid that further approximations could lead to overdetermination? (Or should we, considering quantum possibilities, be glad about that?) *One* source of possible overdetermination is contained in the construction of the vector $z_i = \frac{\partial\psi_i^k}{\partial x^k}$. It decomposes into two terms. The first one is $\frac{\partial\alpha_i^k}{\partial x^k}$, where α_i^k is a (vanishing at infinity) solution of the equation $\Box\alpha_i^k = 0$. The second term is a sum of surface integrals. (The surfaces shrink in the limit to a point.) This is true for further approximations, too, since the equations of gravitation decompose at every step of approximation into the systems

$$\Box\psi_i^k = U_i^k, \qquad \frac{\partial\psi_i^k}{\partial x^k} = 0 \qquad \left(\frac{\partial U_i^k}{\partial x^k} = 0\right)$$

(if a Euclidean background is introduced). We are free to choose the α_i^k, the second term however depends on the given U_{ik}. One might guess that the last fact could lead to the introduction of complicated functions of the field point $O(x_0)$ in z_i, and that the equation $z_i = 0$ could lead to too many, even contradictory, equations for the singularity line. But this case cannot occur as an *inevitable* mathematical necessity. For we have $\Box z_i = 0$, and therefore we have, in any case, the expansion

$$z_i = \frac{Z_i}{n} + \sum \frac{\partial^m}{\partial x^{\lambda_1}\ldots\partial x^{\lambda_m}}\left(\frac{Z_i^{\lambda_1\ldots\lambda_m}}{n}\right), \tag{19}$$

where the $Z_i, Z_i^{\lambda_1\ldots\lambda_m}$ depend only on the position on the line of singularity. The same holds for the second term of z_i which emerges from a surface integral. In our calculations it contained only the first term of (19), Z_i/n. But the other terms can, if they occur, be made to vanish by choosing appropriate particular solutions of the equation $\Box\psi_{ik} = 0$. If, e.g., a term

$$X_i = \frac{\partial^m}{\partial x^{\lambda_1} \ldots \partial x^{\lambda_m}} \left(\frac{Z_i^{\lambda_1 \ldots \lambda_m}}{n} \right)$$

shows up, it will be canceled by adding to ψ_{ik}

$$\alpha_{ik} = \frac{\partial^m}{\partial x^{\lambda_1} \ldots \partial x^{\lambda_{m-1}}} \left(\frac{-Z_{ik}^{\lambda_1 \ldots \lambda_{m-1}}}{n} \right),$$

for we have

$$\frac{\partial \alpha_i^k}{\partial x^k} + X_i = 0.$$

Corresponding relations hold for the electromagnetic Maxwell-Einstein equations which decompose also for the higher approximations on Euclidean background into the systems

$$\Box\varphi_i = s_i, \qquad \frac{\partial \varphi^i}{\partial x^i} = 0 \qquad \left(\frac{\partial s^i}{\partial x^i} = 0 \right).$$

Here, too, those terms which emerge from the equation $\frac{\partial \varphi^i}{\partial x^i} = 0$ and do not have the form Z/n may be canceled by additions to (φ_i).

Does this method determine the potentials ψ_{ik}, φ uniquely? We touched the problem of nonuniqueness already in the context of coordinate determination (§ 6). The solution ψ_{ik} on which we based our derivation of the dynamical equations is actually only *one* solution. It is obviously possible to add to it a solution of the wave equation which vanishes on the infinitely far rim of the past light cone, if this solution has the same singularity lines and fulfills the equations $\frac{\partial \psi_i^k}{\partial x^k} = 0$ identically. The requirement that for vanishing exterior fields our solutions should reduce to those for the matter particle at rest does not remove the non-uniqueness. The same is true for the electrodynamic equations. It is possible, e.g., to add to the potential (φ^i) the vector (α^i) given by

$$\alpha^i = \frac{\partial}{\partial x^k} \left(\frac{u^i \dot{u}^k - u^k \dot{u}^i}{n} \right),$$

since $\Box \alpha^i = 0$, $\frac{\partial \alpha^i}{\partial x^i} = 0$ identically. For a straight world line we have $\alpha_i = 0$. In general[7] $\frac{\partial \alpha_i}{\partial x^k} - \frac{\partial \alpha_k}{\partial x^i} \neq 0$.

If one assigns to matter a structure of its own which is not completely determined by the laws of the field, that is, if one does not imagine matter as lines of singularity of the field, then the knowledge of the structure of matter is necessary for a unique determination of the potentials. If, in addition, the distribution of matter is free of singularities, then there cannot be additional mechanical equations. The prospect of getting hold of this structure theoretically, in a not too arbitrary way seems to us rather unfavorable. From the point of epistemology, too, such a position seems rather unsatisfying.

The calculations from § 7 reveal a further danger which exists also if the matter distribution is free of singularities. For we have been led to the equation $\frac{dM}{ds} = 0$, which is based on the fact that for $\frac{d}{ds}(Mu^i)$ a vector orthogonal to (u^i) cancels.[8] It is, however, quite possible that in further approximations the relation $\frac{dM}{ds} = f(s)$ could appear instead [for the electromagnetic equations this would be $\frac{de}{ds} = \varphi(s)$, comp. § 4]. If now $f(s)$ [resp. $\varphi(s)$] has the form $\frac{dF}{ds}$, with F vanishing in those world points in which the particle does not find an exterior field, then we can at least be certain that mass (resp. charge) assumes in a vanishing exterior field always the same value. If, however, $f(s), \varphi(s)$ cannot be brought to the form $\frac{dF}{ds}$, then M, e are determined only by the prehistory of the particle, and their constancy is lost.

The independence of the solutions (potentials) of the infinitely far past is usually based on concrete assumptions about the states of the field and the motion of matter of the infinite past. It is certainly far from us to ascribe the origin of atoms to a primary parallel rain of particles, as Epicure does in his cosmogony. (The assumption of static relations in the infinite past, as the electron theory does to justify its retarded potentials, is essentially the same as the particle rain). But we have no second thoughts about assuming that T_{ik} and ψ_{ik} vanish as fast on the infinitely far rim of the past light cone

[7]Editor's footnote: Two corrections were done in this sentence by the editors. The original text said "Dabei ist $\frac{\partial \alpha_i}{\partial x^k} - \frac{\partial x_k}{\partial x^i} \neq 0$." The second term in the equation was an obvious typo. The meaning of the original sentence was unclear, and the translation here is the editors' guess.

[8]Editor's footnote: The last half-phrase was faithfully translated from the original, but in fact it should rather say 'that for $\frac{d}{ds}(Mu^i) = 0$ a vector orthogonal to (u^i) cancels'.

as is necessary to justify our integrals. The fact is above any doubt that the infinite past is blurred by later states of the field in its influence on the future, but needs a concrete explanation.

Further works of the author, in which it will be attempted to solve the problems posed here, will be published soon.

Summary. The solution of the equations of gravitation for an arbitrarily moving charged particle is given in first approximation. The equations of motion show up as conditions for the existence of solutions. On this example, the following problems are formulated: 1. of the uniqueness of the potentials; 2. of constancy of charge and mass of the particle; 3. of the conditions at infinity. The linearisation of the equations on a Euclidean background is discussed. The mathematical means necessary for the computational treatment of the problem are developed.

Warsaw, December 1930.

Gen Relativ Gravit (2010) 42:1011–1048
DOI 10.1007/s10714-010-0939-y

GOLDEN OLDIE

Republication of: New mechanics of material systems

Myron Mathisson

Published online: 9 March 2010

An editorial note to this paper can be found in this issue preceding this Golden Oldie and online via doi:10.1007/s10714-010-0937-0. The companion paper to this one can be found in this issue preceding this Golden Oldie and online via doi:10.1007/s10714-010-0938-z.

Original paper: Myron Mathisson, "Neue Mechanik materieller Systeme", *Acta Physica Polonica* **6**, 163–200 (1937). Reprinted with the kind permission of the Editors of Acta Physica Polonica. Translated by Anita Ehlers, Anita.Ehlers@t-online.de.

Editorial responsibility: A. Krasiński, e-mail: akr@camk.edu.pl.

M. Mathisson (Deceased September 13, 1940)
Warsaw, Poland
e-mail for correspondence: akr@camk.edu.pl

New Mechanics of Material Systems[1]

Nowa mechanika systemów materialnych

By MYRON MATHISSON, Warsaw

(Received on 8. September 1937)

The field equations of the empty world (= electromagnetic and gravitational equations) *hold outside of matter.* We formulate this requirement in the form of a *variation equation* (§§1,2), thus decomposing the material system into a sum of *multipoles.* In the case of the equations of gravitation this leads to the concept of the gravitational skeleton (§2). In a gravitational skeleton, the pole is responsible for the mass, whereas dipole and quadrupole are responsible for the angular momentum (§§ 2,3,4,5). This attribution results in different ways depending on whether we ascribe an active role to the material system (field *generating* masses) or a passive one (masses acted upon by an external field). After having developed a method of dealing with our variational equation we derive from it the dynamical laws which the determining pieces of our gravitational skeleton have to obey. This way we are necessarily led to mechanical equations which, in comparison to classical ones, contain new terms. Apart from terms which play a role only in the exterior gravitational fields, we get one term of greatest physical meaning which is not tied to exterior gravitational fields.

While dealing with the problem of motion it turns out that the angular momentum has to be introduced as an antisymmetric tensor. The concept of a "rotation axis of a rigid body" is then constructed from the angular momentum (§4). Movement of the center of gravity and rotation are coupled. FOKKER's approach for the movement of the axis of a symmetric top cannot be sustained (§4). If the gravitational skeleton is to be a complete counterpart of the classical dynamical

[1]Original title: Neue Mechanik materieller Systeme, *Acta Physica Polonica* **6** 163–200 (1937). Reprinted with the kind permission of the Editors of Acta Physica Polonica. Translated by Anita Ehlers, Anita.Ehlers@t-online.de.

model of a body we have to introduce not only a dipole but also a quadrupole (§5). The resulting equations contain, as a special case, the equations of precession, and this circumstance permits a verification of our assumptions (§5, end). Taking into regard only the dipole terms would have its classical analog in a restriction to the case of the inertial *sphere.*

The new mechanical equations admit an *energy theorem.* But a new kind of energy comes in here, the *energy of acceleration* (§6, end).

§1. *Field laws and conservation laws*

Using the example of the MAXWELL-EINSTEIN field equations we will consider some fundamental ideas. Introducing the potential vector φ_α the system of equations of electrodynamics in a RIEMANNian world can be reduced to 4 equations of second order

$$L_\alpha(\varphi_\beta) \equiv \Box\varphi_\alpha - K^{\nu}_{\alpha}\varphi_\nu = S_\alpha \quad (\Box \equiv g^{\mu\nu}\nabla_\mu\nabla_\nu) \tag{1.1}$$

(β is not a tensor index and indicates simply that the operation refers to a *vector*) and an equation of first order

$$\nabla_\nu\varphi^\nu = 0. \tag{1.2}$$

The equations (1.1) hold for world regions where there are charges; otherwise the current vector S_α on the right hand side would have to be zero (∇_α means covariant differentiation, with K we denote the curvature quantities). The separation of the system of equations into a system (1.1) and a side condition (1.2) is, as is known, no restriction of generality and can always be achieved by adding a suitable gradient to the potentials φ_α.

For the linear operator L_α the identical (i.e., valid for two arbitrary vector fields p_α, φ_α) *conditions of reciprocity* hold:

$$\left.\begin{aligned} p^\alpha L_\alpha(\varphi_\beta) - \varphi^\alpha L_\alpha(p_\beta) = \nabla_\alpha w^\alpha, \quad \text{where} \\ w^\alpha = p^\nu\nabla_\alpha\varphi_\nu - \varphi^\nu\nabla_\alpha p_\nu. \end{aligned}\right\} \tag{1.3}$$

If p_α, considered as potential vector, fulfills thc MAXWELL-EINSTEIN equations

$$\nabla_\nu f_{\alpha}^{\cdot\,\nu} = 0, \quad f_{\alpha\beta} = \nabla_\beta p_\alpha - \nabla_\alpha p_\beta,$$

then $L_\alpha(p_\beta)$ depends only on $\nabla_\nu p^\nu$. [Because $L_\alpha(p_\beta) = 0$ is identical with the MAXWELL-EINSTEIN equations if the gauge condition $\nabla_\nu p^\nu = 0$ holds.] Therefore $L_\alpha(p_\beta)$ depends for the gradient vector of a scalar field ξ,

$$p_\alpha = \nabla_\alpha\xi \tag{1.4}$$

only on $\nabla_\nu p^\nu$, i.e., on $\Box\xi$. We have in fact

$$L_\alpha(\nabla_\beta\xi) = \Box\nabla_\alpha\xi - K^\nu_\alpha\nabla_\nu\xi.$$

Because of the relation

$$\nabla_\alpha\nabla_\beta\varphi_\lambda - \nabla_\beta\nabla_\alpha\varphi_\lambda = K^\nu_{\lambda\beta\alpha}\varphi_\nu, \tag{1.5}$$

we get

$$\Box\nabla_\alpha\xi = g^{\mu\nu}\nabla_\mu\nabla_\alpha\nabla_\nu\xi = \nabla_\alpha\Box\xi + K^\nu_\alpha\nabla_\nu\xi$$

and therefore

$$L_\alpha(\nabla_\beta\xi) = \nabla_\alpha q, \quad q = \Box\xi. \tag{1.6}$$

Let the continuous scalar field ξ be chosen such that it vanishes outside of a world region τ_4. The field $q = \Box\xi$, too, vanishes there (the existence and continuity of the derivatives of ξ are always assumed to the necessary degree). Assuming that the potentials φ_α and their first derivatives are continuous in the region τ_4, we get from (1.3) by integration of the scalar density

$$\sqrt{g}\nabla_\nu w^\nu = \frac{\partial(\sqrt{g}w^\nu)}{\partial x^\nu}$$

over τ_4, taking into account the equations (1.1), (1.6):

$$\int_{\tau_4}\sqrt{g}p^\alpha S_\alpha \mathrm{d}\tau_4 - \int_{\tau_4}\sqrt{g}\varphi^\alpha\nabla_\alpha q \mathrm{d}\tau_4 = 0.$$

Partial integration and application of Gauss's theorem to a field which vanishes on the boundary surface, we get by introducing the gauge (1.2):

$$\int_{\tau_4}\sqrt{g}\varphi^\alpha\nabla_\alpha q \mathrm{d}\tau_4 = 0.$$

Therefore we have

$$\int_{\tau_4}\sqrt{g}p^\alpha S_\alpha \mathrm{d}\tau_4 = 0. \tag{1.7}$$

Now we assume that the current S_α fills a timelike world tube, outside of which it vanishes. We call a tube timelike if its bounding surface can be generated by timelike worldlines. We choose a timelike worldline within the tube as the axis of the tube. Let σ_3 be a 3-dimensional cross-section of the tube. It intersects the axis in a point P. We will perform the integration in (1.7) integrating first over the layer between 2 neighbouring σ_3 (their distance is ds along the world line), and, regarding the result as a function of the arc length $s = P_0P$ counted from a fixed point P_0 of the tube axis, finally integrate over S. We develop the field p_α, in which the tube is immersed, for each cross-section σ_3 into a TAYLOR series, with the respective point P of σ_3 being the origin. We assume that the development remains valid for all points of the cross section within the tube,

$$p_\alpha = (p_\alpha)_0 + \frac{\partial p_\alpha}{\partial x_0^\mu}(x^\mu - x_0^\mu) + \dots, \tag{1.8}$$

and that for integration the function p^α can be replaced by its development (1.8). (The coefficients of the development are functions of s alone.) Then the integral in (1.7) assumes the form

$$\int_S \left[(p_\alpha)_0 E^\alpha + \frac{\partial p_\alpha}{\partial x_0^\mu} E^{\mu\alpha} + \frac{\partial^2 p_\alpha}{\partial x_0^\mu \partial x_0^\nu} E^{\mu\nu\alpha} + \dots \right] \mathrm{d}s.$$

(We sum over identical indices.)

This can always be brought to an invariant form by completing the derivatives to covariant ones. Equation (1.7) then assumes the form

$$\int_S [p_\alpha e^\alpha + (\nabla_\mu p_\alpha) e^{\mu\alpha} + (\nabla_\mu \nabla_\nu p_\alpha) e^{\mu\nu\alpha} + \dots] \, \mathrm{d}s = 0, \tag{1.9}$$

$$(p_\alpha = \nabla_\alpha \xi).$$

The $e^{\cdots}$ are tensor components, functions of s. The values of p_α and their derivatives in (1.9) are to be taken along the curve of integration.

The term in the sum (1.9) containing p_α corresponds to a simple pole which would give a term of type $\frac{E}{r}$ in the calculation of the potentials, the section of the development consisting of terms with p_α and $\nabla_\mu p_\alpha$ corresponds to a singularity of pole and dipole, the section with three terms corresponds to a pole, a dipole, and a quadrupole, etc. More about that at the end of §2.

The equation (1.9) is not bound to the assumption that (1.1) stays valid into the interior of the matter. If the potentials are valid (with absolute or sufficient exactness) in a neighbourhood of the matter tube we extend the respective potentials, purely fictitiously, smoothly into the interior of the tubes (*WEYL's fictive fillings*) and build the quantities

$$f_{\alpha\beta} = \nabla_\beta \varphi_\alpha - \nabla_\alpha \varphi_\beta,$$

$$S_\alpha = \nabla_\beta f_{\alpha}^{\cdot\,\beta} \quad (\nabla_\alpha S^\alpha \equiv 0).$$

If the φ_α do not fulfill the condition (1.2) inside the tube, we get (1.2) by adding an appropriate gradient.

The scalar field ξ is (apart from assumptions about continuity and differentiability) only bound by the condition that it vanishes outside of an arbitrarily chosen region τ_4. Therefore (1.9) is a very special variational equation; ξ is the variable element. If we presuppose that our physical entity is characterized sufficiently by the simple pole (*point charge*), we get the equation

$$\int_S \frac{\partial \xi}{\partial x^\alpha} e^\alpha \mathrm{d}s = 0. \tag{1.10}$$

We deal with it according to the following method. At every point of the world line *we decompose* e^α *relative to the velocity vector* u^α,

$$u^\alpha = \frac{\mathrm{d}x^\alpha}{\mathrm{d}s}, \quad (u^\alpha u_\alpha) = 1,$$

i.e., we decompose e^α in the direction of u^α and in a direction orthogonal to u^α:

$$e^\alpha = E u^\alpha + {}^*e^\alpha, \quad {}^*e^\nu u_\nu = 0.$$

The quantity ξ vanishes at the ends of the path of integration. Therefore we have

$$\int_S \frac{\partial \xi}{\partial x^\alpha} u^\alpha \mathrm{d}s = 0.$$

(1.10) can therefore be written

$$-\int_S \xi \frac{\mathrm{d}E}{\mathrm{d}s} \mathrm{d}s + \int_S {}^*e^\alpha \frac{\partial \xi}{\partial x^\alpha} \mathrm{d}s = 0. \tag{1.11}$$

From this follows the vanishing of ${}^*e^\alpha$, since when we decompose $\nabla^\alpha \xi = g^{\alpha\nu} \frac{\partial \xi}{\partial x^\nu}$ relative to u^α (in every point of the world line), only that component of $\nabla^\alpha \xi$ counts which is orthogonal to u^α, and *this can be varied independently of* ξ *along the path of integration.* [It is possible to construct a field ξ in such a way, that the value of ξ and the 3 quantities $\underset{i}{C}^\alpha \nabla_\alpha \xi$ are prescribed in 3 directions $\underset{i}{C}^\alpha$ $(i = 1, 2, 3)$ independent of u^α and independent of each other, and similarly for higher derivatives, – the only restriction being that the chosen values obey the assumptions about continuity and differentiability and vanish at the ends of an arc of the world line. Then the first integral in

(1.11) vanishes, too. This is, however, compatible with the variability of ξ along the world line only if the condition

$$\dot{E} \equiv \frac{\mathrm{d}E}{\mathrm{d}s} = 0, \quad E = \text{const.} \tag{1.12}$$

is fulfilled.]

Equation (1.12) contains a conservation law: an evolution law for a quantity (charge) which depends only on time (proper time). The essential point is that we obtained a law for the one dimensional world line from laws which hold in the 4-dimensional world.

There is a second method for deriving the conservation laws from field laws. Solve the equations (1.1) and insert the found potentials into the gauge condition (1.2). The divergence $\nabla_\nu \varphi^\nu$ should vanish identically; identically means here independent of the coordinates of the point $O(x_0) \equiv O(x_0^0, x_0^1, x_0^2, x_0^3)$ for which we determined the φ_α. This point can be chosen freely within a 4-dimensional region, and that is analogous to the variability of ξ. In order to draw all accessible conclusions it suffices to be able to choose the point $O(x_0)$ freely within a very small 4-dimensional region. This fact confirms an earlier insight that we can use equations and methods of computation which are sufficiently accurate only at distances from matter which are not too small.

§2. *The variational equation of mechanics*

For dealing with our principal problem, the derivation of equations of motion, EINSTEIN's equations of gravitation are the foundation. The conservation are the foundation laws are to be deduced from these field equations. The gravitational equations are, however, not linear in the gravitational potentials. We will replace them by approximate linear equations. Let it be possible to represent the components of the metric tensor for all points of a world region as a sum $g_{\alpha\beta} + \gamma_{\alpha\beta}$, where the $\gamma_{\alpha\beta}$ are supposed to be small, and the $g_{\alpha\beta}$ represent the metric tensor of a RIEMANNian world region W. We will call the metrical manifold W a *background* and view the $\gamma_{\alpha\beta}$ as a tensor field *in* W. Covariant derivatives, shifting indices up and down – everything refers to the metric of the background. The background is supposed to be a possible world region in the sense of general relativity. We can expand EINSTEIN's equations in powers of $\gamma_{\alpha\beta}$. The terms linear in $\gamma_{\alpha\beta}$ will be the linear equations we are looking for. In this way we get a generalization of the well known approximate EINSTEIN's equations of gravitation in which the world of special relativity plays the role of a background.

Let $K^{\epsilon}_{\lambda\alpha\beta}, K_{\lambda\mu}, K$ be the curvature quantities of the background. As we said, covariant differentiation is to be understood with respect to the metric of the background. The complete metric tensor has, with lower indices, the components

$$g_{\alpha\beta} + \gamma_{\alpha\beta}.$$

For the components with upper indices we easily get in first approximation

$$g^{\alpha\beta} - \gamma^{\alpha\beta}.$$

Thus, up to entities of second order it is, in fact,

$$(g_{\alpha\nu} + \gamma_{\alpha\nu})(g^{\nu\beta} - \gamma^{\beta\nu}) = \delta^{\beta}_{\alpha}$$

(equal to the mixed unit tensor).[2]

Using a coordinate system that at a given point is geodesic with respect to the background, the complete curvature component $R_{\alpha\beta\lambda\mu}$ of the complete metric $g_{\alpha\beta} + \gamma_{\alpha\beta}$ at this point is, up to quantities of second order,

$$R_{\alpha\beta\lambda\mu} = K_{\alpha\beta\lambda\mu} + \frac{1}{2}\left(\frac{\partial^2\gamma_{\alpha\mu}}{\partial x^\beta \partial x^\lambda} + \frac{\partial^2\gamma_{\beta\lambda}}{\partial x^\alpha \partial x^\mu} - \frac{\partial^2\gamma_{\beta\mu}}{\partial x^\alpha \partial x^\lambda} - \frac{\partial^2\gamma_{\alpha\lambda}}{\partial x^\beta \partial x^\mu}\right). \tag{2.1}$$

In the same coordinate system, for the same point, we also get

$$\nabla_\beta \nabla_\lambda \gamma_{\alpha\mu} = \frac{\partial^2\gamma_{\alpha\mu}}{\partial x^\beta \partial x^\lambda} - \gamma_{\nu\mu}\frac{\partial}{\partial x^\beta}\begin{Bmatrix}\alpha\lambda\\ \nu\end{Bmatrix} - \gamma_{\alpha\nu}\frac{\partial}{\partial x^\beta}\begin{Bmatrix}\mu\lambda\\ \nu\end{Bmatrix},$$

$$\frac{\partial}{\partial x^\beta}\begin{Bmatrix}\alpha\lambda\\ \nu\end{Bmatrix} - \frac{\partial}{\partial x^\alpha}\begin{Bmatrix}\beta\lambda\\ \nu\end{Bmatrix} = K^{\nu}_{\lambda\beta\alpha};$$

where the three-index symbols refer to the background. At the point considered, the curvature $R_{\alpha\beta\lambda\mu}$ can be represented as follows:

$$\begin{aligned} R_{\alpha\beta\lambda\mu} = K_{\alpha\beta\lambda\mu} &+ \frac{1}{2}\left(\nabla_\beta\nabla_\lambda\gamma_{\alpha\mu} + \nabla_\mu\nabla_\alpha\gamma_{\beta\lambda} - \nabla_\alpha\nabla_\lambda\gamma_{\beta\mu} - \nabla_\mu\nabla_\beta\gamma_{\alpha\lambda}\right) + \\ &+ \frac{1}{2}\left(K^{\nu}_{\lambda\beta\alpha}\gamma_{\nu\mu} + K^{\nu}_{\lambda\beta\mu}\gamma_{\nu\alpha} + K^{\nu}_{\lambda\mu\alpha}\gamma_{\nu\beta}\right). \end{aligned}$$

[2]Editor's footnote: A typo in the last formula has been corrected; the indices were incorrectly placed in one term in the original text.

Because of its invariant character this relation is valid for every point and any coordinate system. Let

$$\psi_\alpha^\beta = \gamma_\alpha^\beta - \gamma\delta_\alpha^\beta, \quad \gamma = \frac{1}{2}\gamma_\nu^\nu.$$

Then, up to quantities of the second order, $R_{\beta\mu}$, i.e. $(g^{\alpha\lambda} - \gamma^{\alpha\lambda})R_{\alpha\beta\lambda\mu}$ is:

$$R_{\beta\mu} = K_{\beta\mu} - \frac{1}{2}\Box\,\gamma_{\beta\mu} + \frac{1}{2}(K_\beta^\nu\gamma_{\nu\mu} + K_\mu^\nu\gamma_{\nu\beta}) - \gamma_\nu^\alpha K_{\beta\alpha\mu}^\nu$$
$$(\Box \equiv g^{\mu\nu}\nabla_\mu\nabla_\nu);$$

here we assume that the $\gamma_{\alpha\beta}$ fulfill the condition

$$\nabla_\nu\psi_\alpha^\nu = 0. \tag{2.2}$$

We calculate R_μ^β as $(g^{\beta\nu} - \gamma^{\beta\nu})R_{\mu\nu}$:

$$R_\mu^\beta = K_\mu^\beta - \frac{1}{2}\Box\gamma_\mu^\beta + \frac{1}{2}(K_\nu^\beta\gamma_\mu^\nu - K_\mu^\nu\gamma_\nu^\beta) - \gamma_\nu^\alpha K^{\nu\beta}_{\cdot\,\cdot\,\alpha\mu}.$$

This implies

$$R = K - \Box\gamma - \gamma_\nu^\alpha K_\alpha^\nu.$$

Let $L_\lambda^\mu(\psi_{\alpha\beta})$ be a linear differential expression (with respect to the $\psi_{\alpha\beta}$), which we define as follows:

$$L_\lambda^\mu(\psi_{\alpha\beta}) \equiv -\frac{1}{2}\Box\,\psi_\lambda^\mu + \frac{1}{2}(K_\nu^\mu\psi_\lambda^\nu - K_\lambda^\nu\psi_\nu^\mu + \delta_\lambda^\mu K^{\nu\epsilon}\psi_{\nu\epsilon}) - K^{\nu\mu}_{\cdot\,\cdot\,\epsilon\lambda}\psi_\nu^\epsilon + \tag{2.3}$$
$$+\frac{1}{2}(K_\lambda^\mu - \frac{1}{2}\delta_\lambda^\mu K)\psi_\nu^\nu.$$

Then we have

$$R_\lambda^\mu - \frac{1}{2}\delta_\lambda^\mu R = K_\lambda^\mu - \frac{1}{2}\delta_\lambda^\mu K + L_\lambda^\mu. \tag{2.4}$$

(Lowering the index μ transforms L_λ^μ into a tensor which is asymmetric with regard to λ, μ).

Without restriction of generality the gauge equations (2.2) can be taken as valid, since otherwise we can put:

$$\psi_{\alpha\beta} = \overline{\psi}_{\alpha\beta} + \nabla_\alpha\zeta_\beta + \nabla_\beta\zeta_\alpha - g_{\alpha\beta}\nabla_\nu\zeta^\nu$$

$$\text{i.e.,}\quad \gamma_{\alpha\beta} = \overline{\gamma}_{\alpha\beta} + \nabla_\alpha \zeta_\beta + \nabla_\beta \zeta_\alpha$$

(*but this corresponds to an infinitesimal coordinate transformation*) – and realize the conditions (2.2) for the new potentials $\psi_{\alpha\beta}$ by determining the ζ_α from the equations:

$$\Box \zeta_\lambda + \nabla_\nu \nabla_\lambda \zeta^\nu - \nabla_\lambda \nabla_\nu \zeta^\nu = \nabla_\nu \psi^\nu_\lambda.$$

Applying formula (1.5) we get

$$\Box \zeta_\lambda + K^\nu_\lambda \zeta_\nu = \nabla_\nu \psi^\nu_\lambda.$$

Determining the ζ_λ from these equations can always force the validity of (2.2).

The linear differential expressions $L^\mu_\lambda(\psi_{\alpha\beta})$ fulfill identical relations of reciprocity which are analogous to the relations (1.3):

$$\left.\begin{aligned} p^\alpha_\beta L^\beta_\alpha(\psi_{\lambda\mu}) - \psi^\alpha_\beta L^\beta_\alpha(p_{\mu\nu}) = \nabla_\alpha w^\alpha, \text{where} \\ w_\lambda = p^{\alpha\beta} \nabla_\lambda \psi_{\alpha\beta} - \psi^{\alpha\beta} \nabla_\lambda p_{\alpha\beta}; \quad p_{\alpha\beta} = p_{\beta\alpha}. \end{aligned}\right\} \tag{2.5}$$

We shall now, according to (2.3), determine $L^\mu_\lambda(\psi_{\alpha\beta})$ for

$$p_{\alpha\beta} = \nabla_\alpha \xi_\beta + \nabla_\beta \xi_\alpha. \tag{2.6}$$

For this we shall use a formula which is analogous to (1.5) which is valid vor every covariant tensor $A_{\lambda\mu}$:

$$\nabla_\alpha \nabla_\beta A_{\lambda\mu} - \nabla_\beta \nabla_\alpha A_{\lambda\mu} = K^\epsilon_{\lambda\beta\alpha} A_{\epsilon\mu} - K^\epsilon_{\mu\beta\alpha} A_{\lambda\epsilon}. \tag{2.7}$$

From the BIANCHI identity

$$\nabla_\alpha K^\varepsilon_{\mu\lambda\beta} + \nabla_\lambda K^\varepsilon_{\mu\beta\alpha} + \nabla_\beta K^\varepsilon_{\mu\alpha\lambda} = 0$$

follows

$$\nabla_\alpha K_{\varepsilon\mu\lambda}^{\cdot\,\cdot\,\cdot\,\alpha} = \nabla_\mu K_{\lambda\varepsilon} - \nabla_\varepsilon K_{\lambda\mu}. \tag{2.8}$$

From (1.5) we get

$$\Box \nabla_\lambda \xi_\mu = g^{\alpha\beta} \nabla_\alpha (\nabla_\lambda \nabla_\beta \xi_\mu + K^\varepsilon_{\mu\lambda\beta} \xi_\varepsilon)$$

and then from (2.7) and (2.8)

$$\Box \nabla_\lambda \xi_\mu = \nabla_\lambda \Box \xi_\mu + K^\varepsilon_\lambda \nabla_\varepsilon \xi_\mu + 2K^\varepsilon_{\mu\lambda\alpha} \nabla^\alpha \xi_\varepsilon + \xi^\varepsilon \left(\nabla_\mu K_{\lambda\varepsilon} - \nabla_\varepsilon K_{\lambda\mu}\right). \tag{2.9}$$

(We write $\nabla^\alpha \xi_\varepsilon$ instead of $g^{\alpha\nu}\nabla_\nu \xi_\varepsilon$.) Now we introduce the following assumption for the background:

$$\left.\begin{array}{l} K^\mu_\lambda - \frac{1}{2}\delta^\lambda_\mu K = \alpha\delta^\mu_\lambda \quad (\alpha = \text{const.}) \\ \text{i.e.} \quad K = -4\alpha, \quad K^\mu_\lambda = -\alpha\delta^\mu_\lambda. \end{array}\right\} \tag{2.10}$$

(2.3) takes the simple form[3]

$$L_{\lambda\mu}(\psi_{\alpha\beta}) \equiv -\frac{1}{2}\Box\psi_{\lambda\mu} \tag{2.11}$$

$$(\text{thus we get } L_{\lambda\mu} = L_{\mu\lambda}).$$

For the field (2.6) we then have the simple relation which we get according to (2.9):

$$\left.\begin{array}{l} L_{\lambda\mu}(p_{\alpha\beta}) = \nabla_\lambda q_\mu + \nabla_\mu q_\lambda, \text{ with} \\ q_\lambda = -\frac{1}{2}(\Box\xi_\lambda - \alpha\xi_\lambda). \end{array}\right\} \tag{2.12}$$

In the following the background will play the role the *external* gravitational field, on which the field of a body is superposed. (2.10) admits only such external fields which exist *outside of uncharged matter* (the constant α represents the cosmological term). This is certainly a very large class of gravitational fields.

We shall replace EINSTEIN's equation of gravitation

$$(G^\mu_\lambda \equiv)\ R^\mu_\lambda - \frac{1}{2}\delta^\mu_\lambda R - \alpha\delta^\mu_\lambda = -\kappa T^\mu_\lambda \tag{2.13}$$

according to (2.4) and (2.10) by the linear equations

$$L^\mu_\lambda(\psi_{\alpha\beta}) = \mu^\mu_\lambda \quad (\mu_{\lambda\mu} = \mu_{\mu\lambda}), \tag{2.14}$$

$$\nabla_\nu \psi^\nu_\lambda = 0. \tag{2.15}$$

In excluding the electromagnetic field we assume that μ^μ_λ vanishes outside of a timelike world tube. Following the pattern of the electromagnetic equations treated above we may consider the filling μ^μ_λ of the world tube as a *virtual* one arising by a smooth continuation of a solution existing outside of the

[3]Editor's footnote: equation-number (2.11) added by the editor. It does not exist in the original, but a reference to it is present in the paragraph below (4.9).

tube. As a virtual filling we use (following an idea of WEYL's) the tensor G^{μ}_{λ}, defined by the left hand side of (2.13), where we consider only terms of first order in $\gamma_{\alpha\beta}$. This way we get the tensor M^{μ}_{λ} for which the identity

$$\nabla_{\nu} M^{\nu}_{\lambda} \equiv 0 \tag{2.16}$$

holds which corresponds to the known identity for the full tensor G^{μ}_{λ}. (M^{μ}_{λ} differs from L^{μ}_{λ} by terms vanishing simultaneously with $\nabla_{\nu}\psi^{\nu}_{\lambda}$). Continuing the solutions of $L^{\mu}_{\lambda} = 0$ smoothly into the interior of the tube by constructing for them $\mu^{\mu}_{\lambda} = M^{\mu}_{\lambda}$ we get a μ^{μ}_{λ}-field which has the postulated properties and fulfills the equations

$$\nabla_{\nu} \mu^{\nu}_{\lambda} = 0. \tag{2.17}$$

The field ξ^{α} of (2.6), with respect to its role and its properties, is to be completely analogous to the scalar field ξ of §1. The vector field ξ^{α}, too, is supposed to vanish outside of an (arbitrarily chosen) world region τ_4 with a certain number of its differential quotients, otherwise it can be chosen arbitrarily. Multiplying the reciprocity relation (2.5) by $\sqrt{g}$ and integrating over τ_4 one gets, because of[4]

$$\nabla_{\nu} w^{\nu} = \frac{1}{\sqrt{g}} \frac{\partial(\sqrt{g} w^{\nu})}{\partial x^{\nu}}, \qquad \int_{\tau_4} \frac{\partial(\sqrt{g} w^{\nu})}{\partial x^{\nu}} \mathrm{d}\tau_4 = 0,$$

the equation[5]

$$\int_{\tau_4} \sqrt{g} p^{\alpha\beta} \mu_{\alpha\beta} \mathrm{d}\tau_4 - 2 \int_{\tau_4} \sqrt{g} \psi^{\alpha\beta} \nabla_{\alpha} q_{\beta} \mathrm{d}\tau_4 = 0.$$

One proves by partial integration that the second integral vanishes on the bounding surface because of the gauge condition (2.2) for $\psi_{\alpha\beta}$ and the vanishing of q_{α}. Thus

$$\left.\begin{array}{c} \int_{\tau_4} \sqrt{g}\, p_{\alpha\beta} \mu^{\alpha\beta} \mathrm{d}\tau_4 = 0 \text{ for} \\ \\ p_{\alpha\beta} = \nabla_{\alpha}\xi_{\beta} + \nabla_{\beta}\xi_{\alpha}, \end{array}\right\} \tag{2.18}$$

in full analogy to (1.7). The transition to an integral along a world line is achieved as for (1.7), and one thus gets an equation corresponding to equation (1.9):

$$\int_S [p_{\alpha\beta} m^{\alpha\beta} + (\nabla_{\lambda} p_{\alpha\beta}) m^{\lambda\alpha\beta} + \ldots] \mathrm{d}s = 0. \tag{2.19}$$

[4]Editor's footnote: The index ν at the second w in the first equation that follows was raised by the editor.

[5]Editor's footnote: Indices of μ in the equation that follows were lowered by the editor.

This is our *variation equation of mechanics.*

If we assume that our physical entity is characterized sufficiently by the simple pole (*point mass*) we have the equation

$$\int_S m^{\alpha\beta}\nabla_\alpha \xi_\beta \mathrm{d}s = 0. \tag{2.20}$$

At every point of the world line we split $m^{\alpha\beta}$ relative to the respective velocity vector u^α, i,e,, we decompose it according to the formula

$$m^{\alpha\beta} = Mu^\alpha u^\beta + M^\alpha u^\beta + M^\beta u^\alpha + {}^*m^{\alpha\beta}, \text{where} \tag{2.21}$$

$${}^*m^{\alpha\beta}u_\beta = 0, \quad M^\alpha u_\alpha = 0.$$

There will be terms under the integral sign which contain differential quotients of ξ_α in directions orthogonal to u^α. These terms lead to

$$({}^*m^{\alpha\beta} + M^\alpha u^\beta)\frac{\partial \xi_\beta}{\partial x^\alpha}. \tag{2.22}$$

We are free to choose $4 \cdot 3 = 12$ entities

$$\left.\begin{aligned} &\underset{i}{c}^\nu \frac{\partial \xi_\beta}{\partial x^\nu} \quad (i = 1, 2, 3;\ \beta, \nu = 0, 1, 2, 3) \\ &\qquad (\underset{i}{c}^\nu u_\nu = 0) \end{aligned}\right\} \tag{2.23}$$

a choice of which is independent of the choice of the ξ_β along the world line. But then (2.20) implies that the group of terms (2.22), integrated over s, alone gives zero. That implies, however,

$${}^*m^{\alpha\beta} + M^\alpha u^\beta = 0 \tag{2.24}$$

and on multiplication by u^α

$${}^*m^{\alpha\beta} = 0, \quad M^\alpha = 0. \tag{2.25}$$

Now equation (2.20) can be written as follows:

$$\int_S Mu^\alpha u^\beta \nabla_\alpha \xi_\beta \mathrm{d}s = 0. \tag{2.26}$$

But

$$Mu^\alpha u^\beta \nabla_\alpha \xi_\beta = u^\alpha \nabla_\alpha (Mu^\beta \xi_\beta) - u^\alpha \nabla_\alpha (Mu^\beta)\xi_\beta. \tag{2.27}$$

The first term on the right is just the derivative $\frac{\mathrm{d}}{\mathrm{d}s}$ of a scalar. The ξ^α-field vanishes at the end points of the path of integration, but it can be chosen freely along the world line. Therefore we get from (2.26), (2.27):

$$u^\alpha \nabla_\alpha (M u^\beta) = 0 \tag{2.28}$$

or, in developed form,

$$\frac{\mathrm{d}(M u^\lambda)}{\mathrm{d}s} + \begin{Bmatrix} \alpha\beta \\ \lambda \end{Bmatrix} M u^\alpha u^\beta = 0. \tag{2.29}$$

Multiplying (2.28) by u_β and taking into regard the normalization $u_\beta u^\beta = 1$ one gets also

$$\dot{M} = \frac{\mathrm{d}M}{\mathrm{d}s} = 0, \quad M = const. \tag{2.30}$$

In this way the equation (2.20) leads to the law of motion (conservation law) (2.29) with constant mass. Obviously (2.29) is, together with (2.24) and (2.25), the sufficient condition for (2.20).

We derived the equations (1.9) and (2.19) from equations (1.7) and (2.18). These are, however, compatible with the arbitrariness of our fields ξ, ξ_α if the continuity equations

$$\nabla_\nu S^\nu = 0, \quad \nabla_\nu \mu^{\lambda\nu} = 0 \tag{2.31}$$

hold at the same time. Since (1.7) and (2.18) follow from the field equations (1.1) and (2.14) by using the gauge (1.2) and (2.15), one sees that the conditions (2.31) must be fulfilled if the gauge is not to lead to contradictions.[6]

The gravitational skeleton. By breaking off equations (2.19) at the term containing differential quotients of order r in $p_{\lambda\mu}$, we decompose our material system into a sum of multipoles. We say that we replaced the system by its *gravitational skeleton.*

The solutions $\psi_{\alpha\beta}$ of a system (2.14) are (for a CAUCHY-problem) either retarded potentials (the Euclidean world as a background) or solutions of FREDHOLM's integral equations in which the free term (the given function) is a retarded potential. By limiting the functions μ_α^β to a narrow timelike tube and letting this shrink to a world line one gets for the retarded potentials, in the simplest case, a simple pole, i.e. an expression of type $m\varphi(x_0, x)$, which

[6]From equations (2.31) the equations (1.7) and (2.18) follow immediately. This would be the shortest way to our variational equations (1.3) and (2.19). Their connection to the field potentials would then be disregarded.

depends on a field point $O(x_0)$ and on the position $A(x)$ of the pole with strength m. $O(x_0)$ and $A(x)$ belong to the same geodetic null line, φ gets infinite as $\frac{1}{r}$ a quantity for $O(x_0) \to A(x)$; here r is a quantity analogous to the usual $r(x_0, x)$ and invariantly defined which converges to zero for $O(x_0) \to A(x)$.[7] Dipoles (solutions with singularities of the type of a classical dipole, getting infinite as $\frac{1}{r^2}$) can be obtained by letting two pole lines converge against a single world line, and similar for higher multipoles. The singularities go over into the solutions of FREDHOLM's equation.

Let now $\psi_{\alpha\beta}$ be solutions of $L_{\mu\nu} = 0$, which are gauged according to (2.15) and become infinite as $\frac{1}{r}$ when approaching the world line. Let the singularity line be taken out of the region τ_4 by a small tube (a channel). Integrating the reciprocity relation (2.5) over the new, smaller region τ_4' we get, since $L_{\mu\nu} = 0$ in τ_4',

$$\int_{\tau_4'} \left[\frac{\partial(\sqrt{g} w^\nu)}{\partial x^\nu} + 2\sqrt{g}\psi^{\alpha\beta}\nabla_\alpha q_\beta \right] \mathrm{d}\tau_4 = 0.$$

Because of the gauge condition (2.15), there holds

$$\psi^{\alpha\beta}\nabla_\alpha q_\beta = \nabla_\alpha(\psi^{\alpha\beta} q_\beta),$$

and we get

$$\int_{\tau_4'} \frac{\partial}{\partial x^\nu}(\sqrt{g} w^\nu + 2\sqrt{g}\psi^{\nu\beta} q_\beta)\mathrm{d}\tau_4 = 0.$$

The left hand side of the equation transforms according to GAUSS' theorem into a surface integral. Since the fields $\xi_\alpha, \nabla_\alpha \xi_\beta, q_\alpha$ vanish on the boundary surface of τ_4, only the integral over the tube surfaces is present in the surface integral. Since we assume that the $\psi_{\alpha\beta}$ become infinite as $\frac{1}{r}$, whereas the surface of a spacelike cross section of the channel converges to zero like r^2 when the channel contracts, the term with q_β and the term with $\nabla_\lambda p_{\alpha\beta}$ vanish when the channel shrinks. $p_{\alpha\beta}$ can be replaced by the first term of the TAYLOR development [see (1.8)], since the following is of order r. If the $\psi_{\alpha\beta}$ are of order $\frac{1}{r^2}$ the terms with $\nabla_\lambda p_{\alpha\beta}$ and q_β can make a non-vanishing contribution for $r \to 0$. q_β, however, is a linear combination of the $\nabla_\lambda p_{\alpha\beta}$:

$$q_\beta = g^{\mu\nu}\nabla_\mu\nabla_\nu\xi_\beta = g^{\mu\nu}\nabla_\mu p_{\nu\beta} - \frac{1}{2}\nabla_\beta(g^{\mu\nu}p_{\mu\nu}).$$

[7] See the literature at the end of the paper.

Equation (2.19) and equation (1.9), too, can thus be justified without using virtual fillings if we consider solutions with lines of singularity (multipole lines). Our considerations imply that the potentials must become infinite at least as $\frac{1}{r^{n+1}}$ if (2.19) is to contain differential quotients of $p_{\alpha\beta}$ of order n. Considering a classic (i.e., static) multipole, for which an equation (1.9) or (2.19) is easily calculated, one infers that the potentials must become infinite as $\frac{1}{r^{n+1}}$ not *at least* but *exactly.*

If the gravitational potentials are given in the neighborhood of the singularity line the left hand side of (2.19) can be calculated as a surface integral or as a limit of a surface integral.

The characterization of the system by a world line and by gravitational potentials in its neighbourhood is therefore at the same time a determination of its gravitational skeleton.

A potential $\breve{\varphi}_\lambda = \nabla_\lambda \xi$ (ξ being an arbitrary scalar field) certainly satisfies the MAXWELL equations (which we assume to be referred to potentials) of the empty, current-free world, since the corresponding electromagnetic field vanishes identically. If the potentials are to fulfill the gauge (1.2) then the scalar field ξ can be chosen arbitrarily from among the solutions of the equation $\Box\xi = 0$. From this follows that it is always possible to introduce *physically irrelevant multipoles* by adding to the potentials non essential potentials

$$\breve{\varphi}_\lambda = \nabla_\lambda \xi, \qquad \Box\xi = 0 \tag{2.32}$$

whose generating scalar field ξ has multipole singularities on certain world lines. *Equation (1.9), however, is independent of these physically non-essential singularities*, since for their derivation we used the (real or virtual) current which the potentials (2.31) do not have influence upon (*gauge invariance*).

Very similar relations are found with respect to the equation of gravitation. The role of the potentials (2.31) is now taken by the gravitational potentials

$$\breve{\gamma}_{\lambda\mu} = \nabla_\lambda \xi_\mu + \nabla_\mu \xi_\lambda \tag{2.33}$$

(ξ_λ - any vector field) which can be added to the $\gamma_{\lambda\mu}$. They fulfill the equations (2.14) identically for $\mu_{\lambda\mu} = 0$, with $\psi_{\alpha\beta}$ being connected to the $\breve{\gamma}_{\lambda\mu}$ by the formula

$$\psi_\alpha^\beta = \breve{\gamma}_\alpha^\beta - \frac{1}{2}\breve{\gamma}_\nu^\nu \delta_\alpha^\beta$$

and ξ_λ is to be chosen among the solutions of the equation

$$\Box\xi_\lambda + K^\nu_\lambda \xi_\nu = 0.$$

[We derived these equations earlier from the condition (2.2).] Thus physically non-essential multipoles can be introduced by means of the ξ_α. But the multipoles of equation (2.19) are, as in equation (1.9) above, independent of those singularities which are physically non-essential. This follows immediately from the structure of the virtual fillings used which are invariant with respect to infinitesimal coordinate transformations, which, as is known, generate the fields (2.32).

Let it be emphasized in addition that two different virtual fillings, which, however, have the same connection to the walls of the tube lead to equal multipole strengths in equation (2.19).

§3. Equations of motion of a dipole

If equation (2.20) is supplemented by dipole terms one gets

$$\int_S \left[p_{\alpha\beta} m^{\alpha\beta} + (\nabla_\lambda p_{\alpha\beta}) m^{\lambda\alpha\beta}\right] \mathrm{d}s = 0. \tag{3.1}$$

This equation poses, as we will see soon, a problem which is closed in itself and solvable: We may break off equation (2.19) at the dipole term without encountering contradictions. It will turn out further that the results, which we arrived at solely by deduction, have a compelling, unique interpretation and that determining physical parts of a material *system*, e.g., the angular momentum, can be described within the conceptual frame of the gravitational skeleton. Posing the problem of a system in the form of a gravitational skeleton has the advantage of avoiding such extremely difficult questions like the existence of rigid bodies and rotation (in a RIEMANNian world) and turning immediately towards a number of physically important parameters.

We decompose $m^{\lambda\alpha\beta}$ relative to u^λ, i.e. we represent it as a sum the first term of which is a tensor orthogonal to u^α and whose remaining terms are products of components of u^α and of components of tensors orthogonal to u^α (compare the decomposition of $m^{\alpha\beta}$ in the preceding section). The components of the decomposition containing u^λ may be "integrated away", since they are reducible to terms of single poles by partial integration (according to the formula

$$\int_S u^\lambda \nabla_\lambda Q \mathrm{d}s = \int_S \frac{\mathrm{d}Q}{\mathrm{d}s} \mathrm{d}s = 0,$$

under the assumption that the scalar Q vanishes at the ends of the path of integration, what happens by means of the ξ^α-field). They are therefore contained in $m^{\alpha\beta}$, e.g.,

$$\int_S \nabla_\lambda p_{\alpha\beta} u^\lambda A^{\alpha\beta} \mathrm{d}s = -\int_S p_{\alpha\beta}(u^\lambda \nabla_\lambda A^{\alpha\beta})\mathrm{d}s. \tag{3.2}$$

We therefore define ($m^{\lambda\alpha\beta}$ being symmetric in α, β):

$$m^{\lambda\alpha\beta} = {}^*m^{\lambda\alpha\beta} + n^{\alpha\lambda}u^\beta + n^{\beta\lambda}u^\alpha + n^\lambda u^\alpha u^\beta. \tag{3.3}$$

The tensors ${}^*m^{\cdot\cdot\cdot}$, $n^{\cdot\cdot}$ (in all indices) and the vector n^λ are orthogonal to u^λ; ${}^*m^{\lambda\alpha\beta}$ is symmetric in α, β. For second derivatives of ξ_β in directions orthogonal to u^α which can be chosen independently of each other and of other ξ_β- quantities we get for a sum at first

$$\frac{\partial^2 \xi_\beta}{\partial x^\lambda \partial x^\alpha}({}^*m^{\lambda\alpha\beta} + n^{\alpha\lambda}u^\beta) = 0, \tag{3.4}$$

and for single terms

$$(n^{\alpha\lambda} + n^{\lambda\alpha})u^\beta + {}^*m^{\lambda\alpha\beta} + {}^*m^{\alpha\lambda\beta} = 0. \tag{3.5}$$

Multiplication by u_β gives

$$n^{\alpha\lambda} + n^{\lambda\alpha} = 0, \qquad \boxed{n^{\alpha\lambda} = -n^{\lambda\alpha}.} \tag{3.6}$$

But then (3.4) can be written as follows:

$$\frac{\partial}{\partial x^\lambda}\left(\frac{\partial \xi_\beta}{\partial x^\alpha} + \frac{\partial \xi_\alpha}{\partial x^\beta}\right) {}^*m^{\lambda\alpha\beta} = 0.$$

In the sum on the left hand side the factors of $m^{\alpha\lambda\beta}$ can take arbitrary values symmetric in α, β – after a system of spatial coordinates orthogonal to u^α and u^α as time axis had been chosen at the (arbitrary) point under consideration. Therefore

$${}^*m^{\lambda\alpha\beta} = 0. \tag{3.7}$$

Before pursuing our problem deductively further we shall look for the physical meaning of the determining parts n^λ of our gravitational skeleton. Let the background be Euclidean and referred to an orthogonal (LORENTZian) coordinate system. Let $T_{\alpha\beta}$ be the tensor from (2.13) of material filling of a timelike world tube. It will take the role of $\mu_{\alpha\beta}$ [Equation (2.14)]. In the interior of the world tube we choose a smooth timelike world line and cut the tube into elements (layers) by constructing in distances ds three-dimensional parallel planes σ_3 which are orthogonal to the world line. We have in all points of the world line

$$\left.\begin{aligned} &u^0 = 1, \qquad u^i = 0 \qquad (i = 1, 2, 3); \\ &n^0 = 0 \quad (\text{because of } n^\lambda u_\lambda = 0), \quad n^{0i} = 0 \quad (\text{because of } n^{\nu i} u_\nu = 0), \\ &m^{i00} = n^i \quad [\text{because of (3.2) and (3.7)}]. \end{aligned}\right\} \tag{3.8}$$

Latin indices take the values 1, 2, 3.

Comparison of (2.18) and (2.19) yields:

$$\int_S \left\{ \int_{\sigma_3} \frac{\partial p_{00}}{\partial x_0^i} y^i T^{00} \mathrm{d}y^1 \mathrm{d}y^2 \mathrm{d}y^3 \right\} \mathrm{d}s = \int_S m^{i00} \frac{\partial p_{00}}{\partial x^i} \mathrm{d}s \tag{3.9}$$

(y are coordinates relative to the intersection of σ_3 with the world line.) Therefore

$$n^i = \int_{\sigma_3} y^i T^{00} \mathrm{d}y^1 \mathrm{d}y^2 \mathrm{d}y^3. \tag{3.10}$$

For this we used the TAYLOR development:

$$p_{\alpha\beta} = (p_{\alpha\beta})_0 + \frac{\partial p_{\alpha\beta}}{\partial x_0^\lambda} y^\lambda + \frac{1}{2} \frac{\partial^2 p_{\alpha\beta}}{\partial x_0^\lambda \partial x_0^\mu} y^\lambda y^\mu + \ldots \tag{3.11}$$

For $T_{\alpha\beta}$ we shall take the known tensor of incoherent matter, $\mu u_\alpha u_\beta$. For masses at rest distributed with the density $T^{00} = \mu$ the n^i is the static moment relative to the y^i-axis. If the world line itself (the trace of which determines the origin $y^i = 0$) passes through the center of mass of sections σ_3, the n^i vanish. If the representative world line is curved, the σ_3-sections are no longer parallel, and the element of a section will get a factor $f \neq 1$. The coordinate systems $(y^1 y^2 y^3)$ then are *instantaneous* space axes. T^{00} will experience corrections for matter moving with velocity v. We say that the n^λ are *essentially* static moments in instantaneous spaces intersecting the world line orthogonally, also in the general case of moving matter and RIEMANNian background. *We may assume that the n^λ can be always made* *to vanish*, namely by correcting the representative world line of the system which, for this purpose, is to be drawn through the "centers of mass" of the σ_3-sections.

Now we come back to the mathematical treatment of our problem. If we use $v^{\cdots}_{\cdots}$ as an abbreviation for any tensors with upper and lower indices

$$u^\alpha \nabla_\alpha v^{\cdots}_{\cdots} = \dot{v}^{\cdots}_{\cdots} \quad (\text{e.g., } \dot{n}^{\beta\lambda} = u^\alpha \nabla_\alpha n^{\beta\lambda}) \tag{3.12}$$

we get, taking into account (1.5) and (3.6) and making use of partial integrations of the kind of (3.2):

$$\int_S \nabla_\lambda \nabla_\alpha \xi_\beta (n^{\alpha\lambda} u^\beta + n^{\beta\lambda} u^\alpha) \mathrm{d}s = \frac{1}{2} \int_S \xi_\varepsilon K^\varepsilon_{\alpha\beta\lambda} (n^{\alpha\lambda} u^\beta + 2n^{\beta\lambda} u^\alpha) \mathrm{d}s -$$

$$- \int_S (\nabla_\lambda \xi_\beta) \dot{n}^{\beta\lambda} \mathrm{d}s. \tag{3.13}$$

We decompose the antisymmetric tensor $\dot{n}^{\beta\lambda}$ relative to u^λ; i.e., we decompose it as follows (the decomposition is unique):

$$\dot{n}^{\beta\lambda} = {}^*(\dot{n}^{\beta\lambda}) + L^\beta u^\lambda - L^\lambda u^\beta, \tag{3.14}$$

$${}^*(\dot{n}^{\beta\lambda}) u_\lambda = {}^*(\dot{n}^{\beta\lambda}) u_\beta = 0; \quad L^\beta u_\beta = 0.$$

Then we have

$$L^\beta = \dot{n}^{\beta\alpha} u_\alpha \tag{3.15}$$

and for the ensemble of terms of (3.1) the differential quotients of the ξ_α in directions orthogonal to u^α contain [compare (2.21)]:

$$- \int_S \frac{\partial \xi_\beta}{\partial x^\lambda} \left[{}^*(\dot{n}^{\beta\lambda}) - L^\lambda u^\beta - {}^*m^{\beta\lambda} - M^\lambda u^\beta \right] \mathrm{d}s = 0. \tag{3.16}$$

We know that this implies the vanishing of the term in brackets; thus both terms, the symmetric part

$$\frac{1}{2}(L^\lambda u^\beta + L^\beta u^\lambda) + {}^*m^{\beta\lambda} + \frac{1}{2}(M^\lambda u^\beta + M^\beta u^\lambda) = 0, \tag{3.17}$$

as well as the antisymmetric one vanishes. Multiplying (3.16) by u^β one gets

$$L^\lambda + M^\lambda = 0, \quad \text{therefore } {}^*m^{\beta\lambda} = 0. \tag{3.18}$$

Then follows [from the vanishing of the antisymmetric part of the expression in brackets of (3.17)]

$$^*(\dot{n}^{\beta\lambda}) = 0, \tag{3.19}$$

or[8]

$$\boxed{\dot{n}^{\beta\lambda} + \dot{n}^{\lambda\alpha} u_\alpha u^\beta - \dot{n}^{\beta\alpha} u_\alpha u^\lambda = 0.} \tag{3.20}$$

These relations are supplemented by the following which result from the condition $n^{\beta\lambda} u_\lambda = 0$:

$$\dot{n}^{\beta\lambda} u_\lambda + n^{\beta\lambda} \dot{u}_\lambda = 0. \tag{3.21}$$

The two systems may be replaced by the system

$$\boxed{\dot{n}^{\beta\lambda} - n^{\lambda\alpha} \dot{u}_\alpha u^\beta + n^{\beta\alpha} \dot{u}_\alpha u^\lambda = 0.} \tag{3.22}$$

from which first (3.21), then (3.20) can be obtained again. $\nabla_\alpha \xi_\beta$ occur only in terms in which they vanish after partial integration:

$$\int_S \nabla_\lambda \xi_\beta (-L^\beta u^\lambda + M^\beta u^\lambda) \mathrm{d}s = \int_S 2\xi_\beta \dot{L}^\beta \mathrm{d}s \tag{3.23}$$

[We made use of (3.18).]

Equation (3.1) yields now, gathering the terms with ξ_α:

$$-\int \xi_\alpha \left[u^\nu \nabla_\nu (M u^\alpha) - \frac{1}{2} K^\alpha_{\beta\mu\lambda} (n^{\mu\lambda} u^\beta + 2n^{\beta\lambda} u^\mu) - 2\dot{L}^\alpha \right] \mathrm{d}s = 0. \tag{3.24}$$

The expression in square brackets must vanish. But then

$$\dot{M} u^\alpha + M \dot{u}^\alpha - \frac{1}{2} K^\alpha_{\beta\mu\lambda} (n^{\mu\lambda} u^\beta + 2n^{\beta\lambda} u^\mu) - 2\dot{L}^\alpha = 0. \tag{3.25}$$

Now we have, from (3.15), (3.21) and (3.22),

$$\dot{L}^\alpha = -n^{\alpha\nu} \ddot{u}_\nu, \tag{3.26}$$

and at the same time $\dot{L}^\alpha = \ddot{n}^{\alpha\nu} u_\nu$, and, since $\ddot{n}^{\alpha\nu}$ is antisymmetric,

$$u_\alpha \dot{L}^\alpha = 0. \tag{3.27}$$

[8]Editor's footnote: The index λ at the last u in the equation that follows raised by the editor.

$K^{\alpha}_{\beta\mu\lambda} u_\alpha u^\mu$ is easily seen to be symmetric in β, λ. Therefore

$$K^{\alpha}_{\beta\mu\lambda} n^{\beta\lambda} u_\alpha u^\mu = 0,$$

because of (3.6). Multiplying equation (3.25) by u^α one gets

$$\dot{M} = 0, \quad \text{thus} \quad \boxed{M = const.} \tag{3.28}$$

For the antisymmetric tensor $n^{\beta\lambda}$ we may write

$$2K^{\alpha}_{\beta\mu\lambda} n^{\beta\lambda} = (K^{\alpha}_{\beta\mu\lambda} - K^{\alpha}_{\lambda\mu\beta}) n^{\beta\lambda} = K^{\alpha}_{\mu\beta\lambda} n^{\beta\lambda}.$$

Considering in addition (3.26) we get, instead of (3.25),

$$\boxed{\begin{array}{c} M\dot{u}^\alpha - K^{\alpha}_{\beta\mu\lambda} u^\beta n^{\mu\lambda} + 2n^{\alpha\lambda}\, \ddot{u}_\lambda = 0. \\ M = const. \end{array}} \tag{3.29}$$

[We recall the fact that, in consequence of (3.12),

$$\dot{u}^\alpha = \frac{\mathrm{d}u^\alpha}{\mathrm{d}s} + \left\{ \begin{array}{c} \lambda\mu \\ \alpha \end{array} \right\} u^\lambda u^\mu.$$

holds.] *The term containing* $\ddot{u}_\lambda$ *will be present also in the case of the Euclidean background.*

In this way equation (3.1) leads to the laws of motion (3.6), (3.22) and (3.29). It is clear that, together with our intermediate statements about the tensor components of pole and dipole [as e.g. (3.18)] our laws of motion constitute sufficient conditions for the existence of (3.1).

For the 6 functions $n^{\alpha\beta}$ and the 4 functions u^α we have the 6 equations (3.22) and the 4 equations (3.29). One can read off the equations that for our 10 functions the relations

$$n^{\alpha\beta} u_\beta = 0, \quad u^\alpha u_\alpha = 1 \tag{3.30}$$

hold. One can use (3.30) to reduce the 10 functions to $3 + 3 = 6$.

The force

$$D^\alpha = K^{\alpha}_{\beta\mu\lambda} u^\beta n^{\mu\lambda} + 2n^{\alpha\lambda} \ddot{u}_\lambda \tag{3.31}$$

is orthogonal to u^α; the same holds even for every one of the 2 terms of the sum on the right hand side of (3.31).

§4. Dipole and rotation; precession

The dipole strengths $n^{\alpha\beta}$ can be interpreted by the same method which helped us in the preceding section in the understanding of the n^{α}- quantities. Let the representative worldline of the system be again, in our LORENTZian coordinates parallel to the time axis. Matter is supposed to perform slow movements relative to the spatial coordinate system in which the center of mass is at rest. Neglecting squares of velocities we have

$$T^{k0} = \left(\frac{\mu v^k}{1-v^2}\right) \cong \mu v^k \quad (k = 1, 2, 3)$$

[v^k are the space components of the velocity field of matter relative to the center of mass (velocity of light = 1)]. We further have, according to the decomposition (3.3) in which u^{α} refer, of course, to the representative world line ($u^0 = 1, u^k = 0$):

$$m^{ik0} = n^{ki} \quad (i, k = 1, 2, 3).$$

Now the following equation will correspond to equation (3.9):

$$\int_S \left\{ \int_{\sigma_3} \frac{\partial p_{k0}}{\partial x_0^i} y^i T^{k0} \mathrm{d}y^1 \mathrm{d}y^2 \mathrm{d}y^3 \right\} \mathrm{d}s = \int_S \frac{\partial p_{k0}}{\partial x_0^i} m^{ik0} \mathrm{d}s,$$

and therefore

$$n^{ki} = \int_{\sigma_3} \mu y^i v^k \mathrm{d}y^1 \mathrm{d}y^2 \mathrm{d}y^3 \tag{4.1}$$

n^{ik} is antisymmetric, therefore

$$n^{ik} = \frac{1}{2} \int_{\sigma_3} \mu (v^i y^k - v^k y^i) \mathrm{d}y^1 \mathrm{d}y^2 \mathrm{d}y^3,$$

or

$$\boxed{n^{\alpha\beta} = \tfrac{1}{2} \Omega^{\alpha\beta},} \tag{4.2}$$

with $\Omega^{\alpha\beta}$ being the angular momentum. $n^{\alpha\beta}$ are *essentially* the components of the angular momentum if it is calculated relative to a point of the world line of the center of mass for a σ_3-section orthogonal to the world line.

If it is permitted to treat our system as a *rigid body* we have

$$v^i = v^i{}_r y^r \qquad (i, r = 1, 2, 3) \tag{4.3}$$

with v_{ir} being the antisymmetric tensor of angular velocity [in orthogonal systems with equal orientation it may be represented by a vector ω^i:

$$(v_{12}, v_{23}, v_{31}) = (-\omega_x, -\omega_y, -\omega_z)].$$

Then we have

$$n^{ik} = v^i{}_r I^{kr} \tag{4.4}$$

with

$$I^{kr} = \int \mu y^k y^r \mathrm{d}y^1 \mathrm{d}y^2 \mathrm{d}y^3. \tag{4.5}$$

If the inertia ellipsoid is a *sphere*, we have in a cartesian coordinate system

$$I^{kr} = I\delta^{kr} = \begin{cases} I & (k = r) \\ 0 & (k \neq r) \end{cases} \tag{4.6}$$

and correspondingly in a non orthogonal linear system

$$n^{ik} = Iv^{ik} \tag{4.7}$$

In this simple case the right hand side of (4.4) is antisymmetric, and our dipole model is adequate for the mechanical problem. In the general case of rotation the quadrupole has to be introduced as an extension of the gravitational model, as we shall see in the following. The scalar value Ω of the angular momentum and the tensor of angular momentum $\Omega^{\alpha\beta}$ are related by

$$\Omega^{\alpha\beta} = \Omega q^{\alpha\beta}, \quad \text{with} \quad q^{\alpha\beta} q_{\alpha\beta} = 1. \tag{4.8}$$

The equation (3.22) yields:

$$\frac{1}{2} u^\alpha \nabla_\alpha (n^{\alpha\beta} n_{\alpha\beta}) = \Omega\dot{\Omega} = 0, \quad \boxed{\Omega = const.} \tag{4.9}$$

If Ω can be represented as a product of the moment of inertia I and of the angular velocity ω, and I = const., then $\omega = const.$

Let us remark that the calculation of the dipole strengths $n^{\alpha\beta}$ runs parallel to the approximate calculation of the dipole terms of the potential $\psi^{\alpha\beta}$. It is namely [9] ψ^{0i}, when the gravitation theory is done on a Euclidean background,

[9]Editor's footnote: The original text said $\psi^{\alpha\beta}$ here. It was corrected to ψ^{0i} by the editor, as implied by the reasoning.

given in a world point O through a retarded potential [compare equation (2.11)], which, up to a constant factor, equals

$$\int \left(\frac{T^{0i}}{r}\right)_{t-\frac{r}{c}} \mathrm{d}y^1 \mathrm{d}y^2 \mathrm{d}y^3 \quad (i = 1, 2, 3),$$

in a coordinate system in which the point of the world line connected to O by a null geodesic is instantaneously at rest. Taking for $\frac{1}{r}$ the first 2 terms of its TAYLOR development and replacing the integration on the light cone by integration over a section orthogonal to the world line one gets (4.1) for the components of the dipole strength.

In the gravitational skeleton of the rotating body the angular momentum tensor is the primary quantity with respect to the axis of rotation, which is constructed from the quantity $\Omega^{\alpha\beta}$. One has to go back to (4.3). The axis of rotation of a symmetric top is orthogonal to the tensor of the angular momentum. Therefore we define for the *axis of rotation* c^α:

$$\left.\begin{aligned} &(\alpha)\ c_\alpha n^{\alpha\beta} = 0,\\ &(\beta)\ c_\alpha u^\alpha = 0,\\ &(\gamma)\ c_\alpha c^\alpha = -1. \end{aligned}\right\} \tag{4.10}$$

Differentiation yields

$$\left.\begin{aligned} &(\alpha)\ \dot{c}_\alpha n^{\alpha\beta} = -c_\alpha \dot{n}^{\alpha\beta} = 0, \text{ by (3.22) and (4.10)}\\ &(\beta)\ \dot{c}_\alpha u^\alpha \ = -c_\alpha \dot{u}^\alpha,\\ &(\gamma)\ \dot{c}_\alpha c^\alpha \ = 0. \end{aligned}\right\} \tag{4.11}$$

$\dot{c}_\alpha$ is therefore orthogonal to three independent space directions: to the surface element $n^{\alpha\beta}$ and to c^α; it is timelike and given by

$$\boxed{\dot{c}^\alpha = -(c_\nu \dot{u}^\nu) u^\alpha;} \tag{4.12}$$

$\dot{u}_\alpha$ is to be taken from (3.29). *The motion of the center of mass and the shift of the axis of rotation are coupled.*

The law (4.12) is incompatible with FOKKER's rule of precession. FOKKER's formula lacks the influence of the second derivatives of the $g_{\alpha\beta}$ (the curvature tensor: this influence may, at least in theory, be arbitrarily decreased by decreasing the angular momentum). This is obvious; FOKKER's prescription has been constructed under the assumption that the conservation law for the axis is *locally Euclidean*, i.e., has, compared to the Euclidean law, additions which vanish in a geodetic coordinate system at the point considered. On the other hand, in FOKKER's approach[10], a coupling is present between the shift of rotation axis and the non-geodesic motion of the center of mass in a curvature-free world, which is equivalent to the one given in (4.12), as we will see soon. (We assume that forces, if there are any, influence only the motion of the center of mass.) One should notice the following.[11] Among equations (4.11) only equation (α) has dynamical content; it relies on equation (3.19) which says that the part of the rotational *momentum* which is orthogonal to u^α is constant during the motion (or, more generally, undergoes parallel displacements). In a fully analogous way FOKKER's rule states that the part of the rotation *axis* which is orthogonal to u^α undergoes parallel displacements.

If d^α is a vector coinciding at a point of the trajectory with the axis of rotation c^α, then at a neighboring point c^α is, according to FOKKER, a vector proportional to the orthogonal component ${}^*d^\alpha$:

$$c^\alpha = \beta\, {}^*d^\alpha, \quad {}^*d^\alpha = d^\alpha - (d_\nu u^\nu) u^\alpha. \tag{4.13}$$

d^α is parallely propagated along the world line of the center of mass: $\dot{d}^\alpha = 0$. We introduce the factor β in order to be able to normalize c^α to unit length. We get

$$\dot{c}^\alpha = \beta\, {}^*\dot{d}^\alpha + \dot{\beta}\, {}^*d^\alpha = -\beta\left[(d_\nu \dot{u}^\nu)u^\alpha + (d_\nu u^\nu)\dot{u}^\alpha\right] + \frac{\dot{\beta}}{\beta} c^\alpha$$

$$\left[({}^*\dot{d}^\alpha) = u^\nu \nabla_\nu {}^*d^\alpha\right].$$

It is easily seen that, because of $d^\alpha = c^\alpha$, in our initial point we have to put

$$\beta = \frac{1}{1 + (d^\alpha u_\alpha)^2} = 1,$$

$$\dot{\beta} = 0.$$

Then FOKKER's $\dot{c}^\alpha$ agrees with our $\dot{c}^\alpha$ [formula (4.12)].

---

[10] When his assumption, formulated for a geodesic, is appropriately extended.

[11] Editor's footnote: Next sentence corrected; in the original text the author refers to eq. (4.12), which must have been a typo.

§5. *The quadrupole*

Considering within the gravitational skeleton the quadrupole term the variational equation of the problem takes the form:

$$\int_S \left[p_{\alpha\beta}m^{\alpha\beta} + (\nabla_\lambda p_{\alpha\beta})m^{\lambda\alpha\beta} + (\nabla_\lambda\nabla_\mu p_{\alpha\beta})m^{\lambda\mu\alpha\beta}\right] \mathrm{d}s = 0. \tag{5.1}$$

The extended gravitational skeleton frees us from the restrictions which are tied to the assumption (3.6) which require $n^{\alpha\beta}$ to be antisymmetric [compare the remarks made above in the context of equation (4.7)].

$m^{\lambda\mu\alpha\beta}$ is symmetric in α, β, of course; it may be assumed to be symmetric in λ, μ; the antisymmetric part (indices in square brackets) would contribute to the pole only an amount

$$(\nabla_\lambda\nabla_\mu p_{\alpha\beta})m^{[\lambda\mu]\alpha\beta} = (\nabla_{[\lambda}\nabla_{\mu]}p_{\alpha\beta})m^{\lambda\mu\alpha\beta} = p_{\varepsilon\beta}Q^{\varepsilon\beta}$$

with

$$Q^{\varepsilon\beta} = \frac{1}{2}(K^{\varepsilon}_{\alpha\mu\lambda}m^{[\lambda\mu]\alpha\beta} + K^{\varepsilon}_{\nu\mu\lambda}m^{[\lambda\mu]\beta\nu}) = K^{\varepsilon}_{\alpha\mu\lambda}m^{[\lambda\mu]\alpha\beta}.$$

Decomposing with respect to u^α:

$$m^{\lambda\mu\alpha\beta} = {}^*m^{\lambda\mu\alpha\beta} + (b^{\lambda\mu\alpha}u^\beta + b^{\lambda\mu\beta}u^\alpha) + b^{\lambda\mu}u^\alpha u^\beta + \ldots \tag{5.2}$$

($^*m^{\cdots\cdot}$ and $b^{\cdots}$ are orthogonal to u^α in all indices), we get, analogously to (3.4),

$$\int_S \frac{\partial^3\xi_\beta}{\partial x^\lambda\partial x^\mu\partial x^\alpha}({}^*m^{\lambda\mu\alpha\beta} + b^{(\lambda\mu\alpha)}u^\beta)\mathrm{d}s = 0,$$

and therefore

$${}^*m^{(\lambda\mu\alpha)\beta} + b^{(\lambda\mu\alpha)}u^\beta = 0. \tag{5.3}$$

(Round brackets indicate a symmetric sum:

$$a^{(\lambda\mu\nu)} = \frac{1}{3!}\sum a^{\lambda\mu\nu}$$

summed over all permutations.)

Multiplication by u_β gives

$$b^{(\lambda\mu\alpha)} = 0. \tag{5.4}$$

But then (5.3) can be written as follows

$$\frac{\partial^2}{\partial x^\lambda \partial x^\mu}\left(\frac{\partial \xi_\alpha}{\partial x^\beta}+\frac{\partial \xi_\beta}{\partial x^\alpha}\right){}^*m^{\lambda\mu\alpha\beta}=0.$$

In this case, too, the conclusion that led us to (3.7) can be applied and we get

$$^*m^{\lambda\mu\alpha\beta}=0. \tag{5.5}$$

In the decomposition (5.2) we may neglect terms containing u^λ or u^μ, since they, as is easily seen, can be reduced to dipole and pole. We will disregard terms containing $b^{\cdots}$ with 3 indices, since we want to introduce only such determining parts of the mechanical system which have classical counterparts (i.e., can be interpreted as *essentially* known quantities of the mechanics of point systems). The non-classical quantities which have to be attributed to the $b^{\cdots}$ are the moments of second order of the velocity field:

$$b^{rki}=\frac{1}{2}\int \mu y^r y^k v^i \mathrm{d}y^1\mathrm{d}y^2\mathrm{d}y^3. \tag{5.6}$$

The method of interpretation is completely parallel to that leading to (4.1).

Thus only the assumption remains that the quadrupole is brought into equation (5.1) by the term $b^{\lambda\mu}u^\alpha u^\beta$:

$$m^{\lambda\mu\alpha\beta}=b^{\lambda\mu}u^\alpha u^\beta. \tag{5.7}$$

This leads, as we shall see, to a closed, consistent system of mechanical variables of the material system.

By elementary transformations, in which we repeatedly make use of commutation relations for covariant derivatives [formulae (1.5) and (2.7)], we obtain:[12]

$$\begin{aligned} &b^{\lambda\mu}u^\alpha u^\beta \nabla_\lambda\nabla_\mu\nabla_\alpha\xi_\beta=\\ &=-(\nabla_\lambda\nabla_\mu\xi_\beta)u^\alpha\nabla_\alpha(b^{\lambda\mu}u^\beta)+K^\varepsilon_{\beta\alpha\mu,\lambda}\xi_\varepsilon b^{\lambda\mu}u^\alpha u^\beta+\\ &+\nabla_\varepsilon\xi_\beta(K^\varepsilon_{\mu\alpha\lambda}b^{\lambda\mu}u^\alpha u^\beta+2K^\beta_{\nu\alpha\lambda}b^{\lambda\varepsilon}u^\nu u^\alpha). \end{aligned} \tag{5.8}$$

[We simply omit a term $u^\alpha\nabla_\alpha(\dots)$, unimportant for us.] We use the notation

$$K^\varepsilon_{\beta\alpha\mu,\lambda}=\nabla_\lambda K^\varepsilon_{\beta\alpha\mu}. \tag{5.9}$$

[12] Editor's footnote: A typo in the indices has been corrected in the first term in eq. (5.8), we replaced $b^{\beta\mu}$ by $b^{\lambda\mu}$.

Though in (5.8) there are no third derivatives of ξ_α the quadrupole term can *not* be reduced to terms of lower polarity. The term containing second derivatives of ξ_α does not have the form $(\nabla_\lambda p_{\alpha\beta})q^{\alpha\beta}$.

We decompose $\dot{b}^{\beta\mu}$ relative to u^α:

$$\dot{b}^{\lambda\mu} = {}^*(\dot{b}^{\lambda\mu}) + Z^\lambda u^\mu + Z^\mu u^\lambda + Z u^\lambda u^\mu. \tag{5.10}$$

It is easily seen that the $b^{\lambda\mu}$-term, insofar as second derivatives of the ξ_α are concerned, is taken into account if we replace ${}^*m^{\lambda\alpha\beta}$ in equation (3.4) by

$$M^{\lambda\alpha\beta} = {}^*m^{\lambda\alpha\beta} - b^{\lambda\alpha}\dot{u}^\beta \tag{5.11}$$

and $n^{\alpha\lambda}$ by

$$N^{\alpha\lambda} = n^{\alpha\lambda} - {}^*(\dot{b}^{\alpha\lambda}). \tag{5.12}$$

In this way one gets, instead of (3.6)

$$N^{\alpha\lambda} \equiv n^{\alpha\lambda} - {}^*(\dot{b}^{\alpha\lambda}) = -N^{\lambda\alpha}. \tag{5.13}$$

We will now try to interpret (5.13) physically. The usual interpretation gives *essentially* the strength of the quadrupole term in the expansion of a NEWTONian potential in multipoles:

$$b^{\lambda\mu} = \frac{1}{2}\sum \mu y^\lambda y^\mu. \tag{5.14}$$

We replaced the integral by a sum; μ is now the mass of an individual element of matter, y^α its coordinates relative to the center of mass in its *simultaneous* position $(y^\alpha u_\alpha = 0)$. The vector

$$\frac{\mathrm{d}y^\alpha}{\mathrm{d}s} = v^\alpha$$

is *essentially* the velocity relative to the center of mass and orthogonal to u^α. ${}^*(\dot{b}^{\alpha\lambda})$ can be *essentially* identified with $\dot{b}^{\alpha\lambda}$ and we get for $N^{\alpha\lambda}$, according to (5.14) and (4.1):

$$N^{\alpha\lambda} = \frac{1}{2}\sum \mu(v^\alpha y^\lambda - v^\lambda y^\alpha) = \frac{1}{2}\Omega^{\alpha\lambda}; \tag{5.15}$$

it is indeed antisymmetric, without restricting the generality of the system [compare the remarks made in connection with formula (4.7)]; it is equal to half the angular momentum tensor formed with respect to the center of mass. (5.13) further implies that *essentially*

$$n^{\alpha\lambda} - n^{\lambda\alpha} \equiv 2n^{[\alpha\lambda]} = \Omega^{\alpha\lambda}. \tag{5.16}$$

Now, to the equation (3.6), because of (5.13), corresponds the equation

$$M^{\lambda\alpha\beta} + M^{\alpha\lambda\beta} = 0;$$

this implies, because of (5.11)

$$\frac{1}{2}\,{}^{*}m^{\lambda\alpha\beta} + \frac{1}{2}\,{}^{*}m^{\alpha\lambda\beta} - b^{\lambda\alpha}\dot{u}^{\beta} = 0,$$

and therefore, since ${}^{*}m^{\lambda[\alpha\beta]} = 0$,

$$\frac{1}{2}\,{}^{*}m^{[\alpha\beta]\lambda} = b^{\lambda[\alpha}\dot{u}^{\beta]}.$$

The term of (5.8) containing the twice differentiated ξ_β is $-(\nabla_\lambda\nabla_\alpha\xi_\beta)\dot{b}^{\lambda\alpha}u^\beta$. Its terms emerging from the decomposition (5.11)

$$-(\nabla_\lambda\nabla_\alpha\xi_\beta)Z^{\lambda}u^{\alpha}u^{\beta},\ -(\nabla_\lambda\nabla_\alpha\xi_\beta)Zu^{\lambda}u^{\alpha}u^{\beta}$$

belong, because of the symmetric factor $u^\alpha u^\beta$: the first to the n^λ term of the dipole, the second (after partial integration) to the pole term. There remains thus

$$\begin{aligned} &-(\nabla_\lambda\nabla_\alpha\xi_\beta)Z^{\alpha}u^{\lambda}u^{\beta} = \\ &= -(2\nabla_{[\lambda}\nabla_{\alpha]}\xi_\beta)Z^{\alpha}u^{\lambda}u^{\beta} - (\nabla_\alpha\nabla_\lambda\xi_\beta)Z^{\alpha}u^{\lambda}u^{\beta} = \\ &= -\xi_\varepsilon K^{\varepsilon}_{\beta\alpha\lambda}Z^{\alpha}u^{\lambda}u^{\beta} + \text{one } n^{\alpha}\text{-dipole term.} \end{aligned} \tag{5.17}$$

We do of course include terms of smaller polarity emerging from the $b^{\lambda\mu}$-terms as parts of the dipole- and pole terms.

We shall transform also the term $(\nabla_\varepsilon\xi_\beta)K^{\varepsilon}_{\mu\alpha\lambda}b^{\lambda\mu}u^{\alpha}u^{\beta}$ of (5.8). We split it according to the formula:

$$K^{\varepsilon}_{\mu\alpha\lambda}b^{\lambda\mu}u^{\alpha} = \rho^{\varepsilon} + \rho u^{\varepsilon} \quad (\rho_\alpha u^\alpha = 0). \tag{5.18}$$

The coefficient of $\nabla_\varepsilon\xi_\beta$ in (5.8) is then

$$\rho u^{\varepsilon}u^{\beta} + \rho^{\varepsilon}u^{\beta} + \rho^{\varepsilon\beta} \quad (\rho^{\varepsilon\beta}u_\varepsilon = \rho^{\varepsilon\beta}u_\beta = 0)$$

with

$$\rho^{\varepsilon\beta} = 2K^{\beta}_{\nu\alpha\lambda} b^{\lambda\varepsilon} u^{\nu} u^{\alpha}. \tag{5.19}$$

$\rho u^{\varepsilon} u^{\beta}$ gives a simple pole. Since $n^{\beta\lambda}$ is, as formula (5.13) shows, no longer antisymmetric, we split $\dot{n}^{\beta\lambda}$ according to

$$\dot{n}^{\beta\lambda} = {}^{*}(\dot{n}^{\beta\lambda}) + L^{\beta} u^{\lambda} - L'^{\lambda} u^{\beta} + L u^{\beta} u^{\lambda}, \tag{5.20}$$

instead of doing it by (3.14). We have[13]

$$L = \dot{n}^{\beta\lambda} u_{\beta} u_{\lambda},$$

$$\dot{n}^{\beta\lambda} u_{\lambda} = L^{\beta} + L u^{\beta},$$

$$\dot{n}^{\beta\lambda} u_{\beta} = -L'^{\lambda} + L u^{\lambda},$$

$$L^{\beta} + L'^{\beta} = \dot{n}^{\beta\lambda} u_{\lambda} - \dot{n}^{\lambda\beta} u_{\lambda} = 2\dot{n}^{[\beta\lambda]} u_{\lambda} = \dot{\Omega}^{\beta\lambda} u_{\lambda} = -\Omega^{\beta\lambda} \dot{u}_{\lambda}. \tag{5.21}$$

We now count the term $L u^{\beta} u^{\lambda}$ in (5.20) as a pole term. Then we get instead of (3.13)

$$\int_S \nabla_{\lambda} \nabla_{\alpha} \xi_{\beta} (N^{\alpha\lambda} u^{\beta} + n^{\beta\lambda} u^{\alpha}) \mathrm{d}s =$$

$$= \frac{1}{2} \int_S \xi_{\varepsilon} K^{\varepsilon}_{\beta\alpha\lambda} (N^{\alpha\lambda} u^{\beta} + 2n^{\beta\lambda} u^{\alpha}) \mathrm{d}s -$$

$$- \int_S (\nabla_{\lambda} \xi_{\beta}) \left[{}^{*}(\dot{n}^{\beta\lambda}) + L^{\beta} u^{\lambda} - L'^{\lambda} u^{\beta} \right] \mathrm{d}s.$$

Equation (3.16) has to be replaced by

$$\frac{\partial \xi_{\beta}}{\partial x_{\lambda}} \left[{}^{*}(\dot{n}^{\beta\lambda}) - \rho^{\lambda\beta} - (L'^{\lambda} + M^{\lambda} + \rho^{\lambda}) u^{\beta} - {}^{*}m^{\beta\lambda} \right] = 0,$$

and from it follow the equations

$$L'^{\lambda} + M^{\lambda} + \rho^{\lambda} = 0, \tag{5.22}$$

$${}^{*}(\dot{n}^{\beta\lambda}) - \rho^{\lambda\beta} - {}^{*}\dot{m}^{\beta\lambda} = 0. \tag{5.23}$$

Since (5.22) takes the place of (3.18), we now have instead of (3.23)

$$\int_S \nabla_{\lambda} \xi_{\beta} (-L^{\beta} u^{\lambda} + M^{\beta} u^{\lambda}) \mathrm{d}s = \int_S \xi_{\beta} (\dot{L}^{\beta} + \dot{L}'^{\beta} + \dot{\rho}^{\beta}) \mathrm{d}s. \tag{5.24}$$

[13]Editor's footnote: In the last formula of (5.21), a typo has been corrected, the fourth term was $-\dot{n}^{\beta\lambda} u_{\beta}$ instead of $-\dot{n}^{\lambda\beta} u_{\lambda}$.

The $b^{\lambda\mu}$-term under the integral sign gives, all in all, the following expression with the factor ξ_β:

$$\xi_\beta(K^\beta_{\nu\alpha\mu,\lambda}b^{\lambda\mu}u^\alpha u^\nu + \dot{\rho}^\beta - K^\beta_{\nu\alpha\lambda}Z^\alpha u^\lambda u^\nu). \qquad (5.25)$$

The term with Z^α results from (5.17). One gets from (5.10)

$$Z^\alpha = \dot{b}^{\varepsilon\alpha}u_\varepsilon - \dot{b}^{\varepsilon\nu}u_\varepsilon u_\nu u^\alpha$$

and from (5.18)[14]

$$\rho^\beta = K^\beta_{\mu\alpha\lambda}b^{\lambda\mu}u^\alpha - u^\beta K^\nu_{\mu\alpha\lambda}b^{\lambda\mu}u_\nu u^\alpha.$$

Equation (3.23) will get an additional term which corresponds to (5.25). Equating the coefficients of ξ_α to zero, we get the equation of motion of the center of mass:

$$u^\nu\nabla_\nu(Mu^\alpha) + p^\alpha = 0. \qquad (5.26)$$

We shall give the expression for p^α only for the simple case of vanishing curvature $K_{\dots\dots}$. According to (5.24) and (5.21) we have then[15]

$$p^\alpha = \Omega^{\alpha\lambda}\ddot{u}_\lambda + \dot{\Omega}^{\alpha\lambda}\dot{u}_\lambda. \qquad (5.27)$$

Equation (5.23) gives after splitting off its antisymmetric part:

$$^*(\dot{n}^{[\beta\lambda]}) - \rho^{[\lambda\beta]} = 0. \qquad (5.28)$$

Now we have [compare (3.14) and (3.15)]

$$^*(\dot{n}^{[\beta\lambda]}) = \dot{n}^{[\beta\lambda]} + \dot{n}^{[\beta\alpha]}u_\alpha u^\lambda - \dot{n}^{[\lambda\alpha]}u_\alpha u^\beta.$$

Thus, using (5.19) and writing $\frac{1}{2}\Omega^{\beta\lambda}$ instead of $n^{[\beta\lambda]}$ [compare (4.2), where we assumed $n^{\alpha\beta}$ to be antisymmetric, and (3.21)]

$$\boxed{\tfrac{1}{2}(\dot{\Omega}^{\beta\lambda} - \Omega^{\lambda\alpha}\dot{u}_\alpha u^\beta + \Omega^{\beta\alpha}\dot{u}_\alpha u^\lambda) = (K^\beta_{\nu\alpha\varepsilon}b^{\varepsilon\lambda} - K^\lambda_{\nu\alpha\varepsilon}b^{\varepsilon\beta})u^\nu u^\alpha.} \qquad (5.29)$$

[14]Editor's footnote: The index on the l.h.s. in the equation that follows corrected from λ to β.

[15]Editor's footnote: The index at $\ddot{u}$ in the equation that follows lowered by the editor.

$\frac{1}{2}\Omega^{\beta\lambda}$ has simply been substituted for $n^{[\beta\lambda]}$. We know that $\Omega^{\beta\lambda}$ is *essentially* the angular momentum. If $b^{\varepsilon\lambda}$ has the form

$$A(u^{\varepsilon}u^{\lambda} - g^{\varepsilon\lambda})$$

(*sphere* of inertia), then the right hand side of (5.29) vanishes, which, as we will see soon, reflects *essentially* the classical precession.

Probing classical circumstances. If one wants to go to the limit $c = \infty$ one has to substitute in (5.29):

$$u^0 = 1,\ u^i = 0,\ \dot{u}^{\alpha} = 0,\ b^{0\alpha} = 0 \quad (\text{because of } b^{\nu\alpha}u_{\nu} = 0).$$

In this way we get equation (5.29) in its non relativistic form:

$$K^{i}_{00s}b^{sk} - K^{k}_{00s}b^{si} = \frac{1}{2}\frac{\mathrm{d}\Omega^{ik}}{\mathrm{d}t} \tag{5.30}$$

(Latin indices run, as always, through the values 1,2,3; the index 0 corresponds to the time: $x^0 = t$). For the curvature tensor we take the approximation (2.1) by regarding $K_{....}$ as a weak field on Euclidean background. Then we get for the *static* central symmetric field:

$$K_{i00s} = \frac{1}{2}\frac{\partial^2 g_{00}}{\partial x^i \partial x^s} = -m\frac{\partial^2 \frac{1}{r}}{\partial x^i \partial x^s};\ K^{i}_{00s} = -K_{i00s}.$$

(We assumed $g^{00} = 1 - \frac{2m}{r}$). The NEWTONian force is

$$X^i = \frac{\partial}{\partial x^i}\frac{km_0}{r},$$

the momentum., referred to the center of mass:

$$X^{ik} = \sum \mu(X^i y^k - X^k y^i).$$

We have $m = \frac{km_0}{c^2}$, m_0 the mass of the central body in gr, k the usual gravitational constant. Since for the center of mass

$$\sum \mu y^i = 0,$$

we get, up to higher terms of the Taylor development of X^i [compare (5.14)]:

$$X^{ik} = 2km_0\left(\frac{\partial^2 \frac{1}{r}}{\partial x^i \partial x^s}b^{sk} - \frac{\partial^2 \frac{1}{r}}{\partial x^k \partial x^s}b^{si}\right)$$

and (5.30) is nothing else than the classical theorem of the time derivative of the angular momentum:

$$\frac{d\Omega^{ik}}{dt} = X^{ik}.$$

[We divided the right hand side of (5.30) by c^2, since we introduced *sec* as unit of time.]

§6. Important special cases. The energy equation. The theory of special relativity.

If the metric $ds^2 = (dx^0)^2 - \sum(dx^i)^2$ is the background, the right hand side of the equations (5.29) has to be zero. But then the second term on the right hand side in (5.27) vanishes, and we have the following equations for the motion of the center of mass:

$$M\dot{u}^\alpha + \Omega^{\alpha\nu}\,\ddot{u}_\nu = P^\alpha \quad (M = const.), \tag{6.1}$$

We introduce P^α as an external force. For the angular momentum $\Omega^{\lambda\mu}$ we have the equations

$$\dot{\Omega}^{\beta\lambda} - \Omega^{\lambda\alpha}\dot{u}_\alpha u^\beta + \Omega^{\beta\lambda}\dot{u}_\alpha u^\lambda = 0. \tag{6.2}$$

We assume the force P^α to be orthogonal to u^α:

$$P^\alpha u_\alpha = 0. \tag{6.3}$$

This condition follows necessarily from (6.2). In the case of a body with a gravitational skeleton consisting of a single pole one gets the set of equations

$$\frac{d(Mu^\alpha)}{ds} = P^\alpha \quad (P^\alpha u_\alpha = 0),$$

in which P^α consists in the action of the electromagnetic field of other bodies and the reaction of radiation (the body is charged like a simple electrical pole). Assuming that the gravitational dipole and quadrupole do not lead to changes in the electromagnetic actions of the force one sees that the equation

$$\frac{d(Mu^\alpha)}{ds} + \Omega^{\alpha\nu}\ddot{u}_\nu = P^\alpha$$

must hold, if gravitational multipole and electromagnetical action exist together. The last equation implies $M = const.$ and equation (6.1). We assume that the charge does not evoke precession effects, so that (6.2) remains valid.

Equation (6.1) can be brought to the form:

$$\frac{\mathrm{d}}{\mathrm{d}s}(Mu^\alpha - 2L^\alpha) = P^\alpha, \quad L^\alpha = -\frac{1}{2}\Omega^{\alpha\nu}\dot{u}_\nu \tag{6.4}$$

[compare (3.26)]. Because of (6.3) P^0 is, as is well known, the power of the force P^i:

$$P^0 = P^1 v^1 + P^2 v^2 + P^3 v^3.$$

v^i is the velocity of the center of mass:

$$v^i = \frac{\mathrm{d}x^i}{\mathrm{d}t} \quad (t = x^0).$$

We introduce in (6.4) the system time t instead of the proper time:

$$\frac{\mathrm{d}}{\mathrm{d}s} = \frac{1}{\sqrt{1-v^2}}\frac{\mathrm{d}}{\mathrm{d}t}, \quad \left(v^2 = \sum v_i^2\right),$$

$$\frac{\mathrm{d}}{\mathrm{d}t}(Mu^0 - 2L^0 - A) = 0. \tag{6.5}$$

A is the work of P^i; it contains an arbitrary additive constant. From (6.5) we get the energy theorem in the form

$$\frac{M}{\sqrt{1-v^2}} + \Omega^{0i}\dot{u}_i = A.$$

But

$$\Omega^{0i}u_0 = -\Omega^{ki}u_k, \text{ i.e. } \Omega^{0i} = -\Omega^{ki}v_k;$$

$$\Omega^{0i}\dot{u}_i = -\Omega^{ki}\frac{\dot{v}_i v_k}{\sqrt{1-v^2}},$$

thus

$$\boxed{\frac{M}{\sqrt{1-v^2}} + \frac{\Omega^{ik}}{1-v^2}v_i\frac{\mathrm{d}v_k}{\mathrm{d}t} = A.} \tag{6.6}$$

Introducing the usual time unit ($c = 3.10^{10}$ *cm sec*$^{-1}$) and, assuming a LORENTZian coordinate system that is determined up to orthogonal transformations of the *spatial* axes, also vector symbols, then (6.6) takes the form[16]

$$\boxed{\frac{c^2M}{\sqrt{1-\frac{v^2}{c^2}}}+\frac{[\mathfrak{v}\mathfrak{w}]\dot{\mathfrak{v}}}{c^2(1-\frac{v^2}{c^2})}=A.} \tag{6.7}$$

In addition to the work of the external force (or the potential energy: $A = -E_{pot} + const.$, if the force can be derived from a potential) and to the relativistic energy $c^2M + \frac{1}{2}Mv^2 + \ldots$, our energy theorem (6.7) contains *energy of acceleration*, which expresses itself as scalar product of two vectors: the vector product of angular momentum and velocity and of acceleration. It is

$$(\Omega^{12}, \Omega^{23}, \Omega^{31}) = (\mathfrak{w}_z, \mathfrak{w}_x, \mathfrak{w}_y)$$

the angular momentum of the system with respect to the center of mass,

$$(v^1, v^2, v^3) = (\mathfrak{v}_x, \mathfrak{v}_y, \mathfrak{v}_z)$$

the velocity of the center of mass,

$$[\mathfrak{w}\mathfrak{v}]_x = \mathfrak{w}_y\mathfrak{v}_z - \mathfrak{w}_z\mathfrak{v}_y = -(\Omega^{13}v^3 + \Omega^{12}v^2) = -\sum \Omega^{1k}v^k, \text{ etc.}$$

In the non-relativistic version ($\frac{v}{c} \ll 1$) our theorem of energy yields, if the force admits a potential:

$$\boxed{E_{kin} + E_{pot} + \frac{1}{c^2}[\mathfrak{v}\mathfrak{w}]\frac{d\mathfrak{v}}{dt} = E = const.} \tag{6.8}$$

Static gravitational field. If the metric $\mathrm{d}s^2 = f^2\mathrm{d}t^2 + g_{ik}\mathrm{d}x^i\mathrm{d}x^k$ $(i, k = 1, 2, 3)$ of the background is independent of the time $x^0 = t$, then for a covariant tensor A_λ

$$\dot{A}_0 \equiv u^\nu\nabla_\nu A_0 = \frac{\mathrm{d}A^0}{\mathrm{d}s} - \frac{\partial g_{\mu\nu}}{\partial x^0}A^\mu A^\nu = \frac{\mathrm{d}A_0}{\mathrm{d}s}. \tag{6.9}$$

[16] Editor's footnote: A typo in eq. (6.7) was corrected; in the original text the denominator of the first term was $\sqrt{1-\frac{v^2}{v^2}}$.

Since the g_{0i} are equal to zero, all $K^0_{\lambda\mu\nu}$ vanish, and the 0-th component of equation (3.29) simplifies for a charge free body ($P^\alpha = 0$) to

$$M\dot{u}^0 + 2n^{\alpha\nu}\ddot{u}_\nu = 0.$$

For the covariant 0-th component we get from (6.9), because of (3.26) and replacing $n^{\alpha\beta}$ by $\frac{1}{2}\Omega^{\alpha\beta}$:

$$\frac{\mathrm{d}}{\mathrm{d}s}(Mu_0 - 2L_0) = 0,$$

$$L^0 = -\frac{1}{2}\Omega^{0\nu}\dot{u}_\nu = -\frac{1}{2}\Omega^{ki}v_k\dot{u}_i,$$

$$Mu_0 - 2L_0 = const.$$

We use here the relations

$$v^i = \frac{\mathrm{d}x^i}{\mathrm{d}t},\quad u^0 = \frac{\mathrm{d}t}{\mathrm{d}s} = \frac{1}{\sqrt{f^2 - v^2}},\quad u^i = \frac{v^i}{\sqrt{f^2 - v^2}},\quad v^2 = -g_{ik}v^iv^k.$$

We thus arrive at a relation analogous to (6.6):

$$\boxed{\frac{Mf^2}{\sqrt{f^2 - v^2}} + f^2\sqrt{f^2 - v^2}\,\Omega^{ik}u_i\dot{u}_k = E = const.} \tag{6.10}$$

It is assumed that the mechanical system has a dipole skeleton (inertia sphere).

Abstract[17]

We assume the laws of the field (electromagnetic and gravitational) outside matter in the form of the MAXWELL–EINSTEIN equations. (Our deductions are, anyway, based only on certain fundamental attributes of the structure of those equations.) We characterize matter through the potentials of the field that it creates. These potentials can be represented as sums of potentials of multipoles (similarly to the NEWTONIAN potential of a matter

[17]Editor's footnote: the original abstract was in Polish; translated by A. Krasiński.

lump, outside the lump). In this way we arrive at the notion of the *gravitational skeleton* of a charge-free material system. In a gravitational skeleton, a single pole characterizes the mass, the dipole and the quadrupole – the rotation moment. This assignment is mutually unique. From the equations obeyed by the potentials in the four-dimensional world we obtain the equations of mechanics in the form of equations with ordinary derivatives that determine the motion of the center of mass of the system and the changes of our multipoles with time. *From these equations it follows that the motion of the center of mass and the rotation are connected to each other.* The new terms in the equations of mechanics that cause this connection vanish if it is assumed that the velocity of light is infinitely large. The energy equation for the new equations of mechanics contains a new term: *the acceleration energy.*

Taking into account the multipoles, up to the quadrupole, is necessary and sufficient to obtain the laws of motion that go over into the laws of classical mechanics of a system of points and of a rigid body, when it is assumed that the velocity of light is infinitely large. The interpretation of the multipoles via classical quantities is approximate, but the equations of motion obtained for them are a mathematically exact condition of existence of multipole singularities in the gravitational solutions outside matter. Therefore, one can consider the multipoles to be an independent dynamical characteristic of the system.

The rotation axis is present here as a secondary construction (and this is a fundamental issue), while the quantity that primarily characterizes rotation is an antisymmetric tensor. We arrive at the connection between rotation and the motion of the center of mass according to a law which differs from FOKKER'S assumption about the motion of the axis of a symmetric top.

Literature

Papers by the author:

ZS. f. Phys. **67**, 270 (1931) (method of the variational equation);

ZS. f. Phys. **69**, 389 (1931) (radiating electron in exterior field);

Math. Ann. **107**, 400 (1933) and *Prace Matematyczno-Fizyczne*, **41**, 177 (1934) (in German) – referring to §2 (to the statements about solutions of the field equations with curved background.)

Chapter 8
Joshua N. Goldberg and Rainer K. Sachs 1962, connection between the Petrov types and the existence of geodesic-shearfree null congruences

Original paper: Joshua N. Goldberg and Rainer K. Sachs, A theorem on Petrov types. *Acta Phys. Polon.* **22** (supplement), 13 (1962).

Reprinted with the kind permissions of the editors of *Acta Physica Polonica*, of both authors and of Maciej Przanowski.

Previously published in *Gen. Relativ. Gravit.* **41** (2009), pages 421 – 432 (the editorial note + Goldberg's biography), 433 – 444 (the paper) and in *Gen. Relativ. Gravit.* **39**, 1929 – 1961 (2007) (Sachs' biography was contained in this in pages 1941 – 1942).

For the reprinting in this book, the second part of the editorial note was updated by Maciej Przanowski, and the biography of Rainer Sachs was included.

8.1 Editorial note to:

J. N. Goldberg and R. K. Sachs
A theorem on Petrov types

Andrzej Krasiński · Maciej Przanowski

Part 1: Explanation of some details of derivation

By Andrzej Krasiński

Assumptions about properties of congruences of curves in a spacetime have powerful implications for the Weyl tensor. A well-known example is the conclusion that follows from the propagation equations of kinematical tensors in relativistic hydrodynamics [1]: If, in a perfect fluid spacetime, there exists a family of timelike curves with zero shear, rotation and acceleration, then the spacetime must be conformally flat. The theorem proven in the paper reprinted here is another example, where limitations imposed on the Weyl tensor by properties of congruences of null curves are discussed. Namely, a geodesic and shearfree null congruence exists in a vacuum spacetime if and only if its Weyl tensor is agebraically special. The congruence is then the degenerate principal null congruence of the Weyl tensor. (The theorem is further extended to include electromagnetic field with geodesic rays and a null field.)

Although rather abstract and formal, this theorem played a powerful role in deriving the Kerr metric (see Ref. [2]), and was useful in deriving several other exact solutions of Einstein's equations [3].

The original publication is rather concise, so its calculations may be difficult to verify. Therefore, in this note we provide some details omitted by the authors. A still more detailed derivation, by the same method but in a different notation, is presented in Ref. [4].

A few rewriting errors of the original paper were corrected by the editor. All the corrections are marked by editorial footnotes, and some of them are explained in the note below. The authors' footnotes have asterisks as footnote marks; the editorial footnotes are numbered. Other features of the original article are faithfully reproduced in this reprint.

Andrzej Krasiński
N. Copernicus Astronomical Center, Polish Academy of Sciences, Bartycka 18, 00 716 Warsaw, Poland
e-mail: akr@camk.edu.pl

Maciej Przanowski
Institute of Physics, Technical University of Łódź, Wólczańska 219, 93-005 Łódź, Poland
email: maciej.przanowski@fis.cinvestav.mx

1. Sec. 1, eq. (1.8). The explanation provided in the paragraph around eq. (1.8) may be too brief for first-time readers. Here is a more detailed derivation.

A definition of a projection on a subspace locally orthogonal to a null vector k has to take into account the fact that this hypersurface contains k. Therefore, given k^a, we first define a second null vector ℓ^a that is tangent to the same light cone and obeys:

$$\ell^a k_a = 1, \qquad \ell^a \ell_a = 0. \tag{1}$$

Note that ℓ^a is not defined uniquely. If m^a is an arbitrary spacelike vector of unit length ($m^a m_a = 1$) orthogonal both to k^a and to ℓ^a, then $\ell'^a = \ell^a - \frac{1}{2}b^2 k^a + b m^a$ also obeys (1), where b is an arbitrary parameter.

Then we define the projection tensor on the (2-dimensional) surface that is orthogonal to both ℓ^a and k^a:

$$p_{ab} = g_{ab} - \ell_a k_b - k_a \ell_b \Longrightarrow p_{ab} k^b = p_{ab} \ell^b = 0. \tag{2}$$

This surface does not include ℓ^a or k^a – because a vector that is orthogonal to two linearly independent null vectors must be spacelike.

Now assume that k^a is a null geodesic vector *field*, and it is affinely parametrised so that $k^r k_{a;r} = 0$. We then define

$$A_{ab} \stackrel{\text{def}}{=} k_{r;s} p^r{}_a p^s{}_b. \tag{3}$$

Then the following holds:

$$k_{a;b} = A_{ab} + a_a k_b + k_a b_b, \tag{4}$$

where

$$\begin{aligned} a_a &\stackrel{\text{def}}{=} \ell^s k_{a;s} - \frac{1}{2} k_{r;s} \ell^r \ell^s k_a, \\ b_a &\stackrel{\text{def}}{=} \ell^s k_{s;a} - \frac{1}{2} k_{r;s} \ell^r \ell^s k_a. \end{aligned} \tag{5}$$

It follows that $a_a k^a = b_a k^a = 0$. Now we decompose A_{ab} into the trace θ (expansion), the trace-free symmetric part σ_{ab} (shear) and the antisymmetric part ω_{ab} (rotation):

$$A_{ab} = \omega_{ab} + \sigma_{ab} + p_{ab}\theta \tag{6}$$

(for some reason, tradition requires one to write the last term without the coefficient $1/2$ that would be natural here). The geometric interpretation of rotation, expansion and shear is similar to that in hydrodynamics and agrees with the everyday meaning of these words; the respective changes apply to images of an object projected by the family of light rays on 2-surfaces orthogonal to the family. As in hydrodynamics, rotation and shear do or do not vanish simultaneously with the scalars defined below. Also, the following equations are useful:

$$p^a{}_r p^r{}_b = p^a{}_b, \qquad g^{ab} p_{ab} = p^{ab} p_{ab} = 2,$$
$$p^{ab}\sigma_{ab} = g^{ab}\sigma_{ab} = 0,$$
$$k^b\omega_{ab} = k^b\sigma_{ab} = \ell^b\omega_{ab} = \ell^b\sigma_{ab} = 0. \tag{7}$$

The scalars of rotation, expansion and shear are then

$$\omega^2 \stackrel{\text{def}}{=} \frac{1}{2}\omega_{ab}\omega^{ab} = \frac{1}{2}k_{[a;b]}k^{a;b}, \tag{8}$$
$$\theta = \frac{1}{2}k^m{}_{;m}, \tag{9}$$
$$\sigma^2 \stackrel{\text{def}}{=} \frac{1}{2}\sigma_{ab}\sigma^{ab} = \frac{1}{2}k_{(a;b)}k^{a;b} - \theta^2. \tag{10}$$

As can be seen, these quantities depend only on k^a, not on the auxiliary field ℓ^a.

Now we set up a field of null vector bases over the spacetime that will include the k^a and ℓ^a. The other vectors in the basis, m^a and $\overline{m}^a$, will be complex conjugate to each other, orthogonal both to k^a and to ℓ^a, and will obey relations similar to (1):

$$g_{ab}m^a m^b = g_{ab}\overline{m}^a\overline{m}^b = 0, \qquad g_{\alpha\beta}m^a\overline{m}^b = 1,$$
$$g_{ab}m^a k^b = g_{ab}m^a\ell^b = 0, \qquad g_{ab}\overline{m}^a k^b = g_{ab}\overline{m}^a\ell^b = 0. \tag{11}$$

(As explained in the paper, these basis vectors are not uniquely defined – see eq. (1.9) of the paper. The freedom of transformations includes the nonuniqueness of ℓ^a mentioned after eq. (1) above.) We label the basis vectors as follows (hats denote tetrad indices; in what follows the hats over tetrad indices will be omitted when there is no risk of confusion):

$$e_{\hat{3}}{}^a = k^a, \qquad e_{\hat{4}}{}^a = \ell^a, \qquad e_{\hat{1}}{}^a = m^a, \qquad e_{\hat{2}}{}^a = \overline{m}^a. \tag{12}$$

The $e_\alpha{}^a$ coincide with the $k_\alpha{}^a$ of the paper, but we will keep the $e_\alpha{}^a$ notation to avoid confusion with the vector k^a introduced before. The matrix of scalar products among the basis vectors (the tetrad metric) is then given by (1.7) in the paper. Because of the form of this matrix, the upper and lower tetrad indices are related to each other as follows:

$$v_{\hat{3}} = v^{\hat{4}}, \qquad v_{\hat{4}} = v^{\hat{3}}, \qquad v_{\hat{2}} = v^{\hat{1}} = \overline{v_{\hat{1}}}, \qquad v_{\hat{1}} = v^{\hat{2}} = \overline{v_{\hat{2}}}. \tag{13}$$

Some of the Ricci rotation coefficients[1] are related in simple ways to physical quantities, and this is one of the useful features of this tetrad. For the considerations in the paper, the following formulae are useful.

The tetrad components of the "acceleration" of k^a are

$$e_\alpha{}^a k_{a;r}k^r = \Gamma_{\hat{3}\alpha\hat{3}}. \tag{14}$$

[1] Their definition in the paper is not a universally accepted one: some authors define $\Gamma_{\alpha\beta\gamma}$ with the opposite sign.

Thus, if k^a is assumed geodesic and affinely parametrised, the first of (2.4) follows.

In order to calculate the expansion, we do the following operations

$$k^m{}_{;m} = g_{mn}k^{m;n} = \eta_{\alpha\beta}e^\alpha{}_m e^\beta{}_n k^{m;n},$$

where $\eta_{\alpha\beta}$ is the flat Lorentzian metric; and then we explicitly run through all the values of α and β. Most of the terms are zero or cancel out, and what remains is

$$\theta = \frac{1}{2}k^{r;s}(m_r\overline{m}_s + \overline{m}_r m_s) = \frac{1}{2}(\Gamma_{312} + \Gamma_{321}) \equiv \frac{1}{2}\left(\Gamma_{312} + \overline{\Gamma}_{312}\right). \tag{15}$$

The tetrad components of the shear tensor are found directly from (6), with use of (15) and of the orthogonality relations (11). Their only nonzero components are:

$$\begin{aligned}\sigma_{\widehat{1}\widehat{1}} &= k_{a;b}m^a m^b = \Gamma_{311} \equiv -\Gamma_{131},\\ \sigma_{\widehat{2}\widehat{2}} &= k_{a;b}\overline{m}^a\overline{m}^b = \Gamma_{322} \equiv -\Gamma_{232}.\end{aligned} \tag{16}$$

The component $\sigma_{\widehat{1}\widehat{1}}$ defined above coincides with the complex scalar σ in (1.8) in the paper.

For the tetrad components of the rotation tensor we find

$$\omega_{12} = -\omega_{21} = \frac{1}{2}(\Gamma_{312} - \Gamma_{321}) = \overline{\omega}_{21} = -\overline{\omega}_{12}, \tag{17}$$

all other components being zero. Thus, the only nonvanishing component of the rotation tensor is pure imaginary, and, moreover, as follows from (8) and (11), it is connected to the scalar of rotation ω by ${\omega_{12}}^2 = -\omega^2$; thus $\omega_{12} = \mathrm{i}\omega$. As one can see then, the expansion and the rotation are, respectively, the real and the imaginary part of the same complex quantity

$$z \overset{\text{def}}{=} \theta + \mathrm{i}\omega = \Gamma_{312} \equiv -\Gamma_{132}. \tag{18}$$

Projecting (6) on all the tetrad vectors and using all of the above, eq. (1.8) in the paper follows.

2. Sec. 2, eqs. (2.5) – (2.6). The paragraph containing eqs. (2.5) and (2.6) in the paper does not give a readable hint on how to derive these equations. Here is the derivation.

Take the Ricci identity for the field k^a:

$$k_{a;bc} - k_{a;cb} = R_{rabc}k^r. \tag{19}$$

In the further calculations, projections of this equation on different combinations of the tetrad vectors will be considered. The following equations will be useful ($k^a{}_{;b}k^b = 0$ being assumed):

$$g^{ab}k_{a;bc} = 2\theta_{,c}, \tag{20}$$

$$k^c g^{ab}k_{a;cb} = -k^{c;b}k_{b;c} = -2\left(\sigma^2 - \omega^2 + \theta^2\right), \tag{21}$$

$$k^c m^a \overline{m}^b k_{a;bc} = k^c \left(m^a \overline{m}^b k_{a;b}\right)_{;c} - k^c m^a{}_{;c} \overline{m}^b k_{a;b} - k^c m^a \overline{m}^b{}_{;c} k_{a;b}, \tag{22}$$

$$k^c m^a \overline{m}^b k_{a;cb} = \overline{m}^b \left(k^c m^a k_{a;c}\right)_{;b} - \overline{m}^b k^c{}_{;b} m^a k_{a;c}, \tag{23}$$

$$m^a{}_{;c} k_{a;b} = g^{as} m_{s;c} m_{a;b} = \eta^{\alpha\beta} e_\alpha{}^a e_\beta{}^s m_{s;c} m_{a;b}. \tag{24}$$

Equations (20) and (21) follow from (9) and (4) – (6). In simplifying the right-hand sides of (22) and (23) the definition of $\Gamma_{\alpha\beta\gamma}$ will be used, and in simplifying the right-hand side of (24) the terms corresponding to different values of α and β will be written out explicitly.

Contract (19) with $k^c g^{ab}$; the result is the equation of evolution of θ, analogous to the Raychaudhuri equation:

$$k^c \theta_{,c} + \sigma^2 - \omega^2 + \theta^2 = \frac{1}{2} R_{rc} k^r k^c. \tag{25}$$

Now contract (19) with $k^c \left(m^a \overline{m}^b - m^b \overline{m}^a\right)$ and use the fact that $R_{r[ab]s}\, k^r k^s \equiv 0$. The result is

$$k^s \Gamma_{3[12],s} + \frac{1}{2}\left[\left(\Gamma_{312}\right)^2 - \left(\Gamma_{321}\right)^2\right] = 0. \tag{26}$$

In consequence of (17), Eq. (26) can be written as

$$\mathrm{i}k^s \omega_{,s} + 2\mathrm{i}\theta\omega = 0. \tag{27}$$

Equations (25) and (27) can be written as one complex equation, eq. (2.5) in the paper (actually, in the original text, the last term $+\sigma^2$ was missing; it was restored here).

Equation (2.6) in the paper follows on contracting (19) with $k^c m^a m^b$ and using (22) – (24). The complex conjugate of (2.6) also holds, and it follows from (19) contracted with $k^c \overline{m}^a \overline{m}^b$.

3. Sec. 2, eq. (2.10). The derivation of eq. (2.10) involves a trick. From writing out $R_{31} = 0$ we have $R_{31} = R_{4313} + R_{1312} = 0$ – and then we apply the first of (2.1) to calculate R_{1343} instead of R_{4313}. (The equality $R_{\alpha\beta\gamma\delta} = R_{\gamma\delta\alpha\beta}$ is not an identity when the rotation coefficients are used as basic objects; it has to be imposed as an extra condition.) This trick enables us to efficiently use the assumptions $\Gamma_{314} = 0$ and $\Gamma_{3\alpha 3} = 0$.

4. Sec. 2, after (2.10). The deductions following (2.5') – (2.10) in the paper are described somewhat too briefly, and are thus difficult to follow. Here is a more detailed presentation. We use the tetrad in which $\Gamma_{314} = 0$ and assume that the field $k = e_{\widehat{3}}$ is geodesic, affinely parametrised and shearfree ($\Gamma_{3\alpha 3} = 0 = \Gamma_{311} = \Gamma_{322}$). Also, $\Gamma_{411} = 0$ by (2.9).

Applying the integrability condition (2.12) to the set (2.5') – (2.10) one gets

$$k^s \left(\Gamma_{413} - \Gamma_{341}\right)_{,s} = -2z\Gamma_{413} + \left(\Gamma_{123} - \overline{z}\right)\left(\Gamma_{413} - \Gamma_{341}\right). \tag{28}$$

Then, from the equation $R_{31} = R^3{}_{313} + R^2{}_{312} = 0$ (in tetrad indices), using the first of (2.1), we get

$$-k^s\Gamma_{341,s} = \Gamma_{341}\,(z+\bar{z}-\Gamma_{123}) - \bar{z}\Gamma_{413} - m^s z_{,s}\,. \tag{29}$$

We use (29) to eliminate $k^s\Gamma_{341,s}$ from (28), and then use (2.10) to eliminate $m^s z_{,s}$ from the result. This results in eq. (2.13) (corrected here; in the original paper the last term before 0 was absent).

Next, writing out $R_{11} = R^3{}_{113} + R^4{}_{114} = 2R_{4113} = 0$ with use of the first of (2.1), we obtain (2.14). The integrability condition of (2.13) – (2.14) leads to

$$\begin{aligned}\Gamma_{413}\,(-k^s\Gamma_{121,s} + m^s\Gamma_{123,s}) = \Gamma_{413}\,[&\Gamma_{341}\Gamma_{123} - 9\Gamma_{413}z \\ &+ \bar{z}(\Gamma_{121}+\Gamma_{413}) + \Gamma_{123}\,(-\Gamma_{121}+\Gamma_{413})]\end{aligned} \tag{30}$$

(in obtaining the above, one must use (2.5'), (2.10), (2.13) and (2.14) to eliminate the derivatives of z and of Γ_{413}). Writing out $R_{13} = R_{3134} + R_{2131} = 0$ with use of (2.1), we get

$$-k^s\Gamma_{121,s} + m^s\Gamma_{123,s} = \Gamma_{413}\,(z+\bar{z}+\Gamma_{123}) + \Gamma_{341}\Gamma_{123} - \Gamma_{121}\,(\Gamma_{123}-\bar{z})\,. \tag{31}$$

Substituting (31) in (30), we obtain $z\Gamma_{413}{}^2 = 0$, as stated in the paper. (Actually, the original paper said $-10z^2\Gamma_{413} = 0$, but the exponent was misplaced.)

5. Sec. 2, eq. (2.15). The details of the proof in the paragraph containing (2.15) are as follows. In the tetrad used in the paper, the conditions for the Weyl tensor to be algebraically special, with k_3 being the double Debever vector, are (in tetrad indices):

$$C_{133\alpha} = C_{233\alpha} = 0. \tag{32}$$

Using this, the tetrad components of $C^\rho{}_{\alpha\rho\beta} = 0$ imply

$$\begin{aligned}-C_{3434} &= C_{3142} + C_{3241} = -C_{1212} = 0,\\ -C_{3441} &= C_{4112}, \qquad C_{3442} = C_{4212},\\ C_{3141} &= C_{3112} = C_{3242} = C_{3212} = C_{4142} = 0.\end{aligned} \tag{33}$$

Take the Bianchi identities $R_{ab[cd;e]} = 0$, contract them with g^{bc} and use the assumption $R_{ab} = 0$. In vacuum, the result is equivalent to $C^r{}_{abc;r} = 0$. By the equation preceding eq. (2.1) in the paper, this is equivalent to:

$$e_a{}^s C^a{}_{bcd,s} - \Gamma^s{}_{rs}C^r{}_{bcd} + \Gamma^s{}_{br}C^r{}_{scd} + \Gamma^s{}_{cr}C^r{}_{bsd} + \Gamma^s{}_{dr}C^r{}_{bcs} = 0. \tag{34}$$

Take the components $(b,c,d) = (3,3,1)$ and $(3,3,2)$ of this equation and use (32) – (33). Then, join to this set the first of (33) and the identity $C_{3[\beta\gamma\delta]} = 0$. When $\Gamma_{133} \neq 0 \neq \Gamma_{233}$, a linear homogeneous set of algebraic equations results, whose solution is

$$C_{3412} = C_{3142} = C_{3241} = C_{3434} = C_{1212} = 0. \tag{35}$$

Now take the components $(3,3,4)$ and $(3,1,2)$ of (34), with (35) assumed. Taking into account the middle part of (33), with $\Gamma_{133} \neq 0 \neq \Gamma_{233}$, the solution of that set is

$$C_{3441} = C_{3442} = C_{4112} = C_{4212} = 0. \tag{36}$$

Take the components $(3,4,1)$ and $(3,4,2)$ of (34), with (35) and (36) assumed. They say that $\Gamma_{233}C_{4141} = 0 = \Gamma_{133}C_{4242}$, i.e. $C_{4141} = C_{4242} = 0$. Together with (33), (35) and (36) this means $C_{abcd} = 0$. With $R_{ab} = 0$, this is the Minkowski spacetime, in which a congruence of shearfree null geodesics does exist, so the theorem is trivially true.

Thus only $\Gamma_{133} = \Gamma_{233} = 0$ needs to be considered.

Now take (34) with the sets of indices $(1,3,1)$ and $(2,3,2)$, and use $\Gamma_{133} = \Gamma_{233} = 0$. Then join the first of (33) and again $C_{3[\beta\gamma\delta]} = 0$ to this set. If $\Gamma_{311} \neq 0 \neq \Gamma_{322}$ then (35) results once more.

The components $(4,3,1)$ and $(4,3,2)$ of (34) now say that $\Gamma_{311}C_{3442} = \Gamma_{322}C_{3441} = 0$. With the assumed $\Gamma_{311} \neq 0 \neq \Gamma_{322}$, and with (33), this implies (36) again. Then, the components $(4,3,4)$ and $(4,1,2)$ of (34) give a set of 2 equations, whose solution, with $\Gamma_{311} \neq 0 \neq \Gamma_{322}$ is $C_{4141} = C_{4242} = 0$, i.e. again $C_{abcd} = 0$. Thus the nontrivial solution of the set considered here is $\Gamma_{311} = \Gamma_{322} = 0$, which, as seen from (16), means that the shear of the k^a congruence is zero.

Part 2: Further developments

By Maciej Przanowski

The classical Goldberg – Sachs theorem found its important generalizations to the case of non-vacuum "physical" (i.e. Lorentzian) spacetimes as well as its counterparts in the cases of 4-dimensional Riemannian manifolds of non-Lorentzian signature or 4-dimensional complex analytic Riemannian manifolds called the *complex spacetimes*.

(a) "Physical" (Lorentzian) spacetime.

The most famous, perhaps, is here the Kundt–Thompson theorem [5], [6]
THEOREM (Kundt–Thompson) *For a spacetime V_4 (i.e. a 4-dimensional differential manifold endowed with the metric of signature $+2$) two of the following statements yield the third*

(A) The Weyl tensor is algebraically special in the sense of the Petrov classification
(B) There exists a shear-free null geodesic congruence
(C)

$$V^{ea}V^{bc}C^{d}{}_{abc;d} = 0 \quad \textit{for the types II or D,}$$
$$V^{bc}C^{d}{}_{abc;d} = 0 \quad \textit{for the type III,}$$
$$V^{ea}C^{d}{}_{abc;d} = 0 \quad \textit{for the type N}$$

where C_{abcd} is the Weyl tensor, V_{ab} is a complex null bivector such that $V_{ab}k^b = 0$ with k^a being the multiple principal null vector in case (A) and the vector tangent to the shear-free null geodesic congruence in case (B), and ";" stands for the covariant derivative. ■

It is evident that the "generalization" is expressed by point (C); in this point some conditions on the covariant derivative of the Weyl tensor are imposed, but it is not assumed that the respective spacetime is vacuum, as it was in the classical Goldberg–Sachs theorem.

Consequently, employing the Einstein equations and also the field equations describing the sources of gravitational field one obtains significant information about the spacetime geometry (connection form) [7], [8], [9], [10], [11] which in many cases simplifies the analysis of the equations and enables one to get explicitly the spacetime metric. In particular, there are distinguished examples of this procedure in the cases of the Einstein–Maxwell equations or the Einstein equations with pure radiation field [3].

(b) Complex spacetime.

Complex spacetime, which is the main object of *complex relativity*, is a 4-dimensional complex analytic manifold endowed with a holomorphic metric.

Complex relativity has attracted a great deal of intrest for many years. First, a complex spacetime was found by E. Newman [12], [13] as a space of "good cuts" for asymptotically flat Lorentzian spacetimes. Such a complex spacetime has been called *$\mathcal{H}$-space* or *heavenly space*. Then, complex spacetimes emerge in a natural way in the twistor program of R. Penrose [14], [15]. $\mathcal{H}$-space appears here as a nonlinear graviton. Finally, a program was proposed by J. Plebański in which the "physical" spacetime metrics could be obtained from the complex metrics by taking the real slices.

This promising program was justified by two outstanding results:

(i) the reduction of the $\mathcal{H}$-space equations ($R_{ab} = 0$, $*C_{abcd} = \pm C_{abcd}$; where $*$ denotes the Hodge duality operation) to a single second order nonlinear partial differential equation (the *first* or the *second heavenly equation* (*$\mathcal{H}$-equation*)) for one holomorphic function [16],

(ii) the reduction of the $\mathcal{H}\mathcal{H}$-space equations ($R_{ab} = 0$, $C_{abcd} + *C_{abcd}$ or $C_{abcd} - *C_{abcd}$ is algebraically special) to a single second order nonlinear partial differential equation (*$\mathcal{H}\mathcal{H}$-equation*) for one holomorphic function [17], [18].

[Remark: Recently we learned that the first heavenly equation was found in 1935 by two Japanese mathematicians from Hiroshima University, T. Sibata and K. Moringa [19]. Of course, J. Plebański was not aware of this fact].

Looking at the considerations leading to the $\mathcal{H}$ or $\mathcal{H}\mathcal{H}$-equations one quickly realizes that the crucial point lies in a complex version of the Goldberg–Sachs theorem given by J. Plebański and S. Hacyan [20]

Before we cite that version some remarks are needed.

The Weyl tensor can be split into its self-dual and anti-self-dual parts as follows

$$C_{abcd} = \frac{1}{2}(C_{abcd} + *C_{abcd}) + \frac{1}{2}(C_{abcd} - *C_{abcd}).$$

Unlike in (real) Lorentzian spacetimes, where the self-dual part $\frac{1}{2}(C_{abcd} + *C_{abcd})$ is complex conjugate to the anti-self-dual part $\frac{1}{2}(C_{abcd} - *C_{abcd})$, in the case of complex spacetimes these two objects are independent. Consequently one can consider independently the Petrov algebraic classification of the self-dual part and the anti-self-dual part of the Weyl tensor.

Now we need also the notion of a *self-dual (anti-self-dual) null string* or α (β, resp.) *surface* according to the Penrose twistorial approach.

The self-dual (anti-self-dual) null string is a totally null 2-dimensional complex surface Σ in the complex spacetime $V_4^{\mathbb{C}}$ such that the null bivector V_{ab}, i.e. the simple 2-form V_{ab}, orthogonal to Σ is self-dual (anti-self-dual, resp.). One can quickly show that Σ is also a totally geodesic complex 2-surface.

Then, the complex version of the Goldberg – Sachs theorem can be stated as follows [20], [21]

THEOREM (Plebański–Hacyan) *The self-dual (anti-self-dual) part of the Weyl tensor of an Einstein complex spacetime $V_4^{\mathbb{C}}$, $R_{ab} = \Lambda g_{ab}$, is algebraically special with k^a being the multiple principal null vector if and only if for each point $p \in V_4^{\mathbb{C}}$ there exist an open neighbourhood U of p and a congruence of self-dual (anti-self-dual, resp.) null strings on U orthogonal to the self-dual (anti-self-dual, resp.) null bivector V_{ab}, such that the null vector k^a satisfies the condition $V_{ab}k^b = 0$* ■

Thus, in the complex case a congruence of null strings plays the role of the shearfree null geodesic congruence of the Lorentzian spacetime. (Note that nowadays it is clear that the best formalism for studying non-Lorentzian versions of Goldberg – Sachs theorem is a spinorial one. We do not use it here by reason of the fact that the classical Goldberg – Sachs theorem employs tensorial formalism)

There exist some generalizations of the Plebański–Hacyan theorem to the non-Einstein complex spacetimes which are analogous to the Kundt–Thompson theorem [11]

It is not very difficult to carry over the Plebański–Hacyan theorem and its non-Einstein generalizations to the case of a real 4-dimensional Riemannian manifold endowed with a metric of signature $(++--)$. The simplest and most direct way consists in the assumption that all the geometrical objects which appear in the Plebański–Hacyan theorem are real. So the metric is real (smooth), the null strings are real 2-dimensional totally null surfaces etc.

However, there exist also some more subtle versions of the Goldberg – Sachs theorem in the case of $(++--)$-metrics. The reader is referred to the recent review on the Goldberg – Sachs theorem in 4-dimensional Riemannian manifolds given by Gover, Hill and Nurowski [22]

Finally, we should mention that the Goldberg – Sachs theorem as adapted to the case of $(++--)$-metrics has found an important application in the theory of Walker and Osserman spaces [23], [24], [25].

(c) Euclidean spacetime

It is worthwhile to note that there exists a counterpart of the classical Goldberg – Sachs theorem in the case of 4-dimensional Riemannian manifolds of Euclidean signature $(++++)$ (*Euclidean spacetimes*). Such spaces have attracted considerable interest in quantum gravity where they appear as the so called *gravitational instantons* [26].

It was shown that also here the Goldberg – Sachs theorem enables us to reduce the Einstein field equations [27], [28], [29]. This theorem reads now [27], [28], [22]: **THEOREM** *The self-dual or anti-self-dual part of the Weyl tensor of the Einstein Euclidean spacetime* V_4^E, $R_{ab} = \Lambda g_{ab}$, *is algebraically special if and only if* V_4^E *is a locally Hermitian manifold i.e., for each point* $p \in V_4^E$ *there exist an open neighbourhood* U *of* p *and two complex coordinates* (z^1, z^2) *on* U *such that the metric* ds^2 *of* V_4^E *takes the following form on* U

$$ds^2 = g_{\alpha\bar{\beta}}(dz^\alpha \otimes dz^{\bar{\beta}} + dz^{\bar{\beta}} \otimes dz^\alpha)$$

$\overline{g_{\alpha\bar{\beta}}} = g_{\beta\bar{\alpha}}$, $\alpha, \beta = 1, 2$ *and* $dz^{\bar{\beta}} := \overline{dz^\beta}$, *where the overbar stands for the complex conjugation.* ■

Last but not least, one meets an application of the Goldberg – Sachs theorem when the so-called *Cauchy–Riemann structure* on a Riemannian manifold is considered. It has been found that there exists a close relation between the Cauchy–Riemann structure on a 4-dimensional Riemannian manifold and the Goldberg – Sachs theorem [30], [31].

Concluding our note we would like to point out the universality of the Goldberg – Sachs theorem. We can see that this theorem plays an important role not only in standard relativity, but also in the case of any 4-dimensional Riemannian manifold, giving a deep insight into its geometry.

Acknowledgments

One of us (MP) is indebted to Sebastian Formański for his assistance and interest in this note.

References

1. G. F. R. Ellis, Relativistic cosmology, in: *Proceedings of the International School of Physics 'Enrico Fermi', Course 47: General Relativity and Cosmology*. Edited by R. K. Sachs. Academic Press, New York and London 1971, pp. 104 – 182. Reprinted as a Golden Oldie in *Gen. Relativ. Gravit.* **41** 575 – 660 (2009).
2. R. P. Kerr and A. Schild, A new class of vacuum solutions of the Einstein field equations, in *Atti del Convegno sulla Relatività Generale: Problemi dell'Energia e Onde Gravitazionali*. G. Barbèra, Firenze (1965). Reprinted as a Golden Oldie in *Gen. Relativ. Gravit.* **41** 2469 – 2499 (2009).
3. H. Stephani, D. Kramer, M. MacCallum, C. Hoenselaers, and E. Herlt, *Exact Solutions of Einstein's Field Equations*. Cambridge University Press 2003.

4. J. Plebański and A. Krasiński, *An Introduction to General Relativity and Cosmology*. Cambridge University Press 2006.
5. W. Kundt and A. Thompson, *C. R. Acad. Sci.* (Paris) **254**, 4257 (1962).
6. I. Robinson and A. Schild, *J. Math Phys.* **4**, 484 (1963).
7. L. Mariot, *C. R. Acad. Sci. (Paris)* **238**, 2055 (1954); **239**, 1189 (1954).
8. I. Robinson, *J. Math Phys.* **2**, 290 (1961)
9. P. Szekeres, *J. Math. Phys.* **7**, 751 (1966).
10. J.F. Plebański, *Spinors, Tetrads and Forms*. Unpublished monograph of CINVESTAV, Mexico City 1974.
11. M. Przanowski and J.F. Plebański, *Acta Phys. Pol.* **B 10**, 573 (1979).
12. E.T. Newman, in *General Relativity and Gravitation*, Edited by G. Shaviv and J. Rosen. J. Wiley and Sons, New York 1975.
13. M. Ko, M. Ludvigsen, E.T. Newman, and K.P. Tod, *Phys. Rep.* **71**, 51 (1981).
14. R. Penrose, *Gen. Relativ. Gravit.* **7**, 31 (1976).
15. R. Penrose and R.S. Ward, Twistors for flat and curved space-time, in *General Relativity and Gravitation*. Edited by A. Held. Plenum Press, New York 1980, vol 2, pp.283-328.
16. J.F. Plebański, *J. Math. Phys.* **16**, 2395 (1975).
17. J.F. Plebański and I. Robinson, *Phys. Rev. Lett.* **37**, 493 (1976).
18. C.P. Boyer, J.D. Finley, III, and J.F. Plebański, Complex general relativity $\mathscr{H}$ and $\mathscr{H}\mathscr{H}$-spaces- a survey of one approach, in *General Relativity and Gravitation*. Edited by A. Held. Plenum Press, New York 1980, vol 2, pp.241-281.
19. T. Sibata and K. Moringa, *J. Sci. Hiroshima Univ.* **5**, 173 (1935).
20. J.F. Plebański and S. Hacyan, *J. Math. Phys.* **16**, 2403 (1975).
21. J.F. Plebański, K. Rózga, *J. Math. Phys.* **25**, 1930 (1984).
22. A.R. Gover, C.D. Hill, P. Nurowski, *Ann. Mat. Pura Appl.* **190**, 295 (2011); arXiv: 0911.3364
23. P.R. Law, *J. Geom. Phys.* **59**, 1087 (2009); arXiv: 0802.1761
24. E. García-Rio, D.N. Kupeli, R.Vázquez-Lorenzo, *Osserman Manifolds in Semi-Riemannian Geometry* (Lecture Notes in Mathematics) Springer, Berlin (2002)
25. A. Chudecki, M. Przanowski, *Class. Quant. Grav.* **25**, 145010 (2008); *Class. Quant. Grav.* **25**, 235019 (2008).
26. *Euclidean Quantum Gravity.* Edited by G.W. Gibbons and S. W. Hawking. World Scientific, Singapore 1993.
27. S. Hacyan, *Phys. Lett.* **A 75**, 23 (1979).
28. M. Przanowski and B. Broda, *Acta Phys. Pol.* **B 14**, 637 (1983).
29. M. Przanowski and B. Baka, *Gen. Relativ. Gravit.* **16**, 797 (1984).
30. A. Trautman, On complex structures in physics, in *On Einstein's Path, Essays in Honor of Engelbert Schucking*. Edited by A. Harvey. Springer, New York 1999, pp.487-501.
31. P. Nurowski and A. Trautman, *Diff. Geom. Appl.* **17**, 175 (2002).

8.2 Joshua N. Goldberg – a brief biography

By Josh Goldberg

I was born on May 30, 1925 in Rochester, NY and had my early education there. In 1942 I graduated from Benjamin Franklin High School and entered the University of Rochester to study physics. That study was interrupted in July 1944 when I was drafted into the US Navy for service in WWII. Along with many other draftees, I was part of an electronics training program to repair radio, radar, sonar, and other electronic equipment on board naval vessels. After discharge in the summer of 1946, I returned to the University of Rochester and received a BS in 1947.

In the fall, I entered Syracuse University at the same time that Peter Bergmann arrived as an Assistant Professor. Peter had his office in the same Quonset hut that housed 10 or 12 other graduate students. As a result, during the three days a week he was in Syracuse, Peter had an intimate relationship with the graduate students. While others got involved with relativity right away, I wanted to apply quantum mechanics that I had just learned. So for the MS thesis (1950), I worked on a problem in molecular beams jointly with John Trischka and Peter Bergmann. Then I joined the group (Ralph Schiller, Jim Anderson, Robb Thomson, Robert Penfield, and Henry Zatskis) working with Peter on problems in general relativity. We were concerned with constructing a Hamiltonian for the Einstein equations in preparation of quantization of the gravitational field. As part of this program, we studied the general transformations that left the field equations unchanged and led to conservation laws (Noether's theorems). Out of this came an understanding of the connection between the Einstein equations and the motion of its sources. This formed the basis of my PhD dissertation (1952). Others in the group worked on construction of the Hamiltonian, the demonstration that the Einstein psuedotensor did not give consistent results, the questions of constraints, primary and secondary, Lagrangian quantization, and the definition of angular momentum.

Following this, I worked for four years at a non-profit research institute in Chicago, Armour Research Foundation, on various problems of interest to the Defense Department. While there, I wrote a paper generalizing a paper by Einstein, Infeld, and Hoffmann. The important result was that gravitational radiation shows up in the 11th order in v/c (3 1/2 PN).

My next position (1956-63) was as Research Scientist with the Aerospace Research Laboratory at Wright-Patterson Airforce Base. ARL had small groups in nuclear and solid state physics as well as mathematics and I was engaged to study general relativity. There I was able to have a small group doing research on general relativity as well as a fund for support of research at various universities in the US and abroad. By the time I left, there were four of us in the group: Roy Kerr, Stuart Fickler, Joseph Schell, and I. ARL also established a postdoctoral research program that was supervised by the NAS. In January, 1957, I was able to support the Conference on the Role of Gravitation in Physics at the University of North Carolina, Chapel Hill. This was the first international conference that brought together the young scientists to discuss gravitational radiation and quantization of gravity. This conference initiated what became the triennial series of conferences on General Relativity and Gravitation, GRG, and later, the Society on General Relativity and Gravitation. During this time, I also established a relationship with Peter Havas and we published a long paper on a fast motion approximation to the Einstein equations. We found too many singularities when we tried to go to the second approximation and both of us drifted off in other directions. In the 1980s, this formed the basis of a successful effort by Nathalie Deruelle and Tibault Damour to push the calculation to higher orders. This and newer efforts now are yielding the templates for the search for gravitational waves by LIGO.

In 1960–1961, I was an NSF Senior Postdoctoral Fellow with Hermann Bondi at King's College. Although my proposal to NSF was to construct a two body solution of the Einstein equations by combining two Schwarzschild solutions, by the time I arrived, Bondi et al had constructed an asymptotic solution, with null cone coordinates, which exhibited gravitational radiation. Therefore, all effort at King's College was focused on developing that area. Ray Sachs was there at the time and we collaborated on the paper establishing the necessary and sufficient conditions that a solution of the Einstein equations be algebraically special in the Petrov classification. Andrzej Trautman was in London at the time and he arranged to have Professor Infeld invite me to visit his group in Warsaw, Poland to talk about equations of motion and gravitational radiation. NSF allowed me to break my Fellowship in the summer and return to King's College in the fall. For the summer, the Air Force allowed me to visit various institutions in Hamburg, Copenhagen, Stockholm, Paris, and a summer school on Lake Como. On my return to King's College, I developed an approach to conservation laws using the conformal tensor. At the time, I did not recognize its similarity to the approach by Roger Penrose using a spinor decomposition of the conformal tensor.

When I returned to ARL in January 1962, Roy and I worked out the explicit example of a system of localized electric charges interacting in Minkowski space. The idea was to show the utility and meaning of the Petrov classification with an elementary example. In the fall of '62, Roy went on leave from ARL for a year with Alfred Schild in Austin, TX. In the fall of 1963, I left the Airforce and became Professor of Physics at Syracuse University while Roy stayed in Austin as an Assistant Professor.

At Syracuse I replaced Artie Komar who left for a position at Yeshiva University. When I arrived I learned that Peter Bergmann had also gone to Yeshiva as Chair of the Physics Department. During his absences that year, I took over supervision of his students - Jeff Winicour, Tom Morgan, and Manny Leiser. Peter returned after one year at Yeshiva, but continued to live in New York and to spend three days a week in Syracuse during the term. So, I continued to look after his students in his absence. David Robinson and Ed Glass were my first students. Altogether I supervised eight dissertations - Fred Klotz, Robert Feinman, Stephanie Novak-Stekas, Elaine Schnierman-Lesser, Pantur Siliban, and Raj Nagarajan in addition to David and Ed. Five of the eight had academic positions. The remaining three had/have good positions using their physics degrees. Stephanie has had a particularly distinguished career with Bell Labs and General Dynamics. During this time, I tried several new ways to study the motion of particles in general relativity. I had hoped to find a simpler approach that would not require going to such a high order in perturbation to see gravitational radiation. At the same time, I kept active studying properties of gravitation on null surfaces. And I continued to look for global conserved quantities and formulated a definition of a local energy using the geometric quantities on general null surfaces. After Abhay Ashtekar joined the department in 1980, I became interested in his new variables and constructed a Lagrangian that led to his form of the Einstein equations. As Chair of the Physics Department, 1975–82, I oversaw the remodeling of the Physics Building and the hiring of three young faculty members – Ed Lipson in biophysics, Carl Rosenzweig in field theory, and Eric Schiff in solid state physics. Ed and Eric were seeds in their fields, both of which grew. Carl became part of an established group. During this time Peter and I shared research grants with the NSF and enjoyed many postdocs and visiting professors. In 1991, I spent six months with David Robinson. We spent the time constructing the Hamiltonian for general relativity using the Ashtekar variables. It was interesting, but did not lead to a simpler or useful formulation.

I retired in 1995, but the Physics Department has given me an office with many of the perquisites of a member of the faculty. In return, since 2006 I have been the editor of an annual newsletter for the Department, PHYSICS MATTERS. For ten years I wrote grant proposals for the Syracuse Friends of Chamber Music and for eleven years I served as a member of the board of the local chapter of the New York Civil Liberties Union.

8.3 Rainer K. Sachs – a brief biography

Compiled from the webpage of Rainer K. Sachs

Rainer K. (Ray) Sachs, Professor Emeritus of Math and of Physics
Evans Hall, MC 3840, University of California, Berkeley, CA 94720.
Email: sachs@math.berkeley.edu

Born 06/13/1932 in Germany; came to U.S. in 1937.
Bachelor's in math, Massachusetts Institute of Technology, 1953.
Ph.D. in theoretical physics, Syracuse, 1959.[2]
Flick Postdoctoral Fellow, Hamburg, Germany, 1960.
Postdoctoral Fellow, King's College, London, 1961.
Assistant Professor, Dept. Physics, Stevens Institute of Technology, Hoboken NJ., 1962-63.
Associate Professor and Professor, Dept. Physics, U. Texas at Austin, 1963-68.
Professor of Math and Physics University of California, Berkeley, 1969-present. 1993 Professor Emeritus. 1994 Research Professor of Mathematics.

Prof. Sachs writes, "I worked on general relativistic cosmology and astrophysics until about 1985; then I switched to mathematical and computational biology, especially radiation biology."

"My principal scientific accomplishment: introducing polymer models to describe the large-scale geometry of chromosomes in the cell nucleus during cell-cycle interphase."

[2] Dates in his webpage cv seem incorrect: see http://math.berkeley.edu/~sachs/cv.html

Editors' comment (by George Ellis and Andrzej Krasiński)

Professor Sachs' self-evaluation is a monumental under-appreciation, and is not borne out by the citations to his papers. His achievements in relativity include the organisation of meetings whose volumes of proceedings became important sources for decades later (like, for example, the 47th course of the Enrico Fermi School on relativistic cosmology, published in 1971 – it defined the framework of the field). Many of Professor Sachs' papers were milestones in the development of relativity, in particular in the study of gravitational radiation, the peeling-off theorem and use of optical scalars. Notions and results introduced in some of those papers are today widely known under names that include "Sachs" as their component. These include the following, in approximate chronological order:

1. The Bondi–Metzner–Sachs group of asymptotic transformations.
2. The Goldberg–Sachs theorem, proved in the paper reprinted below.
3. The Kantowski–Sachs symmetry group and spacetimes.
4. The Sachs–Wolfe effect in cosmology.[3]
5. The Ehlers–Geren–Sachs Theorem.

The Sachs–Wolfe paper is one of the most important and innovative papers in cosmology, defining the basics of study of the Cosmic Microwave Background Radiation anisotropies and providing the foundations for all later studies of this key feature of present day cosmology. The Ehlers–Geren–Sachs Theorem is an under-appreciated key result underlying our use of perturbed Robertson Walker models as the basic models of standard cosmology. The paper on observational cosmology by Kristian and Sachs[4] is no less innovative, combining Sachs' deep insights into null geodesic congruences on the one hand, resulting from his work on gravitational radiation and electromagnetic theory, and on the dynamics of cosmology on the other. This understanding was developed firstly in conjunction with Jürgen Ehlers and the other members of the Hamburg 'exact solutions' group in the early 1960's (comprising Pascual Jordan, Otto Heckmann, Engelbert Schücking, Jürgen Ehlers, Wolfgang Kundt, Rainer Sachs, and Manfred Trümper); their work is summarised in the Mainz series of papers which are due to be reproduced in the Golden Oldie series. It was then further developed through the influential Texas University group set up by Alfred Schild in the late 1960's, which again included Schücking, Sachs, and Ehlers. A key feature was that they interacted with members of the Astronomy Department at the University of Texas, particularly Gerard de Vaucouleurs, thereby learning many practicalities of observational cosmology. These insights were influential in Sachs' and Ehlers' writings on cosmology, and led to Sachs' supervision of the PhD thesis of Beatrice Tinsley, an important work that challenged the common contention at the time that galaxy evolution was unimportant in cosmological observations.

[3] The paper in which the effect was described was reprinted as a Golden Oldie in *Gen. Relativ. Gravit.* **39**, 1929 (2007).

[4] Reprinted as another "Oldie" in *Gen. Relativ. Gravit.* **43**, 331–58 (2011).

Gen Relativ Gravit (2009) 41:433–444
DOI 10.1007/s10714-008-0722-5

GOLDEN OLDIE

Republication of: A theorem on Petrov types

J. N. Goldberg · R. K. Sachs

Published online: 13 January 2009

An editorial note to this paper and a biography can be found via doi:10.1007/s10714-008-0721-6.

Original paper: J. N. Goldberg and R. K. Sachs, Acta Phys. Pol. **22** (supplement), 13 (1962). Reprinted with the kind permission of the Editorial Board of Acta Physica Polonica and of both authors.

Editorial responsibility: A. Krasiński, e-mail: akr@camk.edu.pl.

J. N. Goldberg (✉)
Department of Physics, Syracuse University, Syracuse, NY 13244, USA
e-mail: goldberg@physics.syr.edu

R. K. Sachs
Department of Mathematics, University of California, 970 Evans Hall, # 3840,
Berkeley, CA 94720, USA
e-mail: sachs@math.berkeley.edu

A THEOREM ON PETROV TYPES

By J. N. Goldberg* and R. K. Sachs**

Department of Mathematics, King's College, University of London

(Received February 23, 1962)

It is shown that a vacuum metric is algebraically special in the sense of the Petrov classification if and only if it contains a shear-free null geodesic congruence. By noting which field equations are used in the proof of the theorem, the theorem is extended to include an electromagnetic field with geodesic rays as well as a null field. From a theorem of Robinson's on the existence of a null electromagnetic "test field", one observes that such a field can be constructed in vacuum if and only if the Riemannian space is algebraically special. From the original theorem, one can easily show that the existence of two independent shear-free null congruences guarantees that the space is Petrov Type I-degenerate.

1. *Introduction*

In studying the propagation properties of the gravitational field, it has been useful to examine the geometry of a congruence of null geodesics [Sachs 1961]. Apart from the obvious relationship between propagation and null geodesics, this study is important because there is a close connection between the existence of certain preferred null directions and the Petrov classification of the Riemann tensor [Petrov 1954, Pirani 1957, Debever 1958, Penrose I960]. The purpose, of this paper, in fact, is to prove a rather technical theorem[1] identifying certain geometrical properties of a null congruence with the existence of an algebraically special Riemann tensor, in the sense of the Petrov classification.

However, before stating the theorem, it is necessary to define the phrase *algebraically special* as applied to the Riemann tensor and to define certain geometrical properties of null congruences. The remainder of the introduction will be devoted to this task. In the following section the principal theorem will be stated and proved. The final section will be devoted to a number of corrollaries related to the theorem proved.

Let R_{abcd} be the Riemann tensor, $R_{bc} = R^a_{bca}$ be the Ricci tensor, $R = R^a_a$ be the curvature scalar, $S_{ab} = R_{ab} - (1/4)g_{ab}R$, and C_{abcd} be the conformal tensor [Eisenhart 1949]:

$$C^{ab}_{cd} = R^{ab}_{cd} + 2\delta^{[a}_{[c}S^{b]}_{d]} + (1/6)\delta^{[a}_{[c}\delta^{b]}_{d]}R \Rightarrow C^{ab}_{ca} = 0. \tag{1.1}$$

*National Science Foundation Fellow, on leave from the Aeronautical Research Laboratory, Wright-Patterson Air Force Base, Ohio.

**Recipient of the stipendium of the A. E. Norman Fund, King's College, University of London.

[1]Newman-Penrose have given an independent proof of the theorem. J. math. Phys., 3, 566 (1962).

Here and throughout square brackets around indices denote antisymmetrization, for example

$$T_{[abc]} = (1/6)(T_{abc} + T_{bca} + T_{cab} - T_{bac} - T_{acb} - T_{cba});$$

round brackets will be similarly used to denote symmetrization. Debever [1958] and Penrose [1960] have shown that in general there exist four independent vectors $k_a \neq 0$ which satisfy the algebraic equations

$$k_{[a}C_{b]ij[c}k_{d]}k^i k^j = 0, \quad k^a k_a = 0. \tag{1.2}$$

This work leads to the following definition [Sachs 1961]: *A gravitational field is called "algebraically special (in the sense of the Petrov classification)" if there exists a vector $k'^a \neq 0$ which obeys*

$$k'^a k'^c C_{abc[d}k'_{e]} = 0, \quad k'^a k'_a = 0. \tag{1.3}$$

There are no solutions k'^a of Eq. (1.3) if and only if the conformal tensor belongs to the algebraically most general class (Petrov nondegenerate Type I). Unless $C_{abcd} = 0$, there are at most two directions k'^a which obey this equation.

It will be convenient to sketch the treatment of null geodesic congruences given by Sachs [1961]. Consider a space-time filling set of null geodesics defined by*

$$x^a = x^a(v, C^A),$$

where the C^A are three parameters which label a particular null geodesic and v is an affine parameter along the geodesic:

$$k^a k_a = 0, \quad k^a = \partial x^a/\partial v \quad \text{and} \quad k^a_{;b}k^b = 0. \tag{1.4}$$

Let δx^a be the connection vector between the geodesic defined by C^A and that defined by $C^A + \delta C^A$. It is then easy to show that on propagation along the null rays, δx^a undergoes a linear transformation defined by $k^a_{;b}$:

$$\frac{D(\delta x^a)}{dv} = k^a_{;b}\delta x^b. \tag{1.5}$$

From this equation it is clear that δx^a will remain in the elementary null hypersurface $k_a dx^a = 0$ if it starts out in that hypersurface. The *rate of strain* of the congruence is determined by the projection of the transformation matrix $k_{a;b}$ into a spatial two-space in the hypersurface. The terminology can be justified by imagining that the congruence is realized by light beams. Then the projection of $k_{a;b}$ determines the rate of expansion, of rotation, and of distortion (by shearing) of shadows cast by the light.

To make this picture more precise, it is convenient to work with a quasi-orthonormal tetrad system [Wong 1945, Sachs 1961, Goldberg and Kerr 1961, and Newman 1961]. Let k^a_3 be the null vector defined by Eq. (1.4); introduce

*$a, b, \ldots = 1, \ldots, 4$; $A, B, \ldots = 1, \ldots, 3$; the signature of the metric is taken to be $+2$.

any three further vectors k_1^a, k_2^a, k_4^a which obey the equations ("–" means complex conjugate)

$$k_2^a = \overline{k}_1^a, \quad k_4^a = \overline{k}_4^a$$

$$k_\alpha^a k_\beta^b g^{\alpha\beta} = g^{ab}, \quad k_\alpha^a k_\beta^b g_{ab} = g_{\alpha\beta}, \tag{1.6}$$

where $\alpha, \beta, \ldots = 1, \ldots, 4$ label the tetrad vectors and

$$g^{\alpha\beta} = g_{\alpha\beta} = \begin{bmatrix} 0 & 1 & 0 & 0 \\ 1 & 0 & 0 & 0 \\ 0 & 0 & 0 & 1 \\ 0 & 0 & 1 & 0 \end{bmatrix}. \tag{1.7}$$

The second equation in (1.6) shows that k_4^a is a (real) light-like vector and that the real and imaginary parts of k_1^a are space-like vectors orthogonal to each other, to k_3^a, and to k_4^a.[2]

The rate of strain of the null congruence is determined by the projection of $k_{3a;b}$ into the (k_1^a, k_2^a) plane. The projection operator is simply $2k_1{}^{(a}k_2^{b)}$. Thus, denoting the rate of strain matrix by A^{ab}, one has

$$A^{ab} = z k_2^a k_1^b + \overline{z} k_1^a k_2^b + \sigma k_2^a k_2^b + \overline{\sigma} k_1^a k_1^b. \tag{1.8}$$

With $z = \theta + i\omega$, θ is the rate of expansion and ω the rate of rotation of the shadow formed in the (k_1^a, k_2^a) plane by removal of certain rays from the null congruence. The rate of distortion of the shadow is determined by the complex shear σ, which is here defined so that the rate of change of its phase is equal to twice the rate of rotation of the principal axes of shear. *If* $\sigma = 0$, *the congruence will be called shear-free.*

The tetrad used above, of course, is not unique. Given a fixed direction k_3^a, which is tangent to a congruence of null geodesics, one can preserve the conditions of Eqs. (1.6) and (1.7) by means of a rotation in the (k_1^a, k_2^a) plane, a change of scale of k_3^a and k_4^a (Lorentz transformation in the (k_3^a, k_4^a) plane), and a null rotation around the k_3 direction. This group of transformations may be written concisely as

$$\begin{aligned} k'^a_3 &= A k_3^a, \quad k'^a_1 = e^{iC}(k_1^a - \overline{B} k_3^a) \\ k'^a_4 &= A^{-1}(k_4^a + B k_1^a + \overline{B} k_2^a - B\overline{B} k_3^a) \\ A &= \overline{A}, \quad C = \overline{C}, \quad A_{/3} = A_{,a} k_3^a = 0. \end{aligned} \tag{1.9}$$

These transformations will be used to simplify some of the necessary calculations. The transformation properties of z and σ are $z' = Az$, $\sigma' = Ae^{2iC}\sigma$. Although z and σ are not invariant under the transformations, $z = 0$ and $\sigma = 0$ are each invariant statements.

[2]Equation-number (1.4) in this sentence corrected to (1.6). Actually, the first part of this sentence should read: *The second equation in (1.6) shows that* k_4^a *is real, while (1.7) shows that* k_1^a *is a light-like vector and ....* The details of calculations in the next paragraph are explained in point **1** of the editorial note [editor].

2. *The Theorem*

Theorem: A vacuum metric, $R_{ab}=0$, is algebraically special if and only if it contains a shear-free null geodesic congruence; the tangent vector to the congruence obeys $k_{[a}C_{b]ijc}k^i k^j=0$.

Fragments of this theorem have appeared in many papers, among which are [Lichnerowicz 1958, Bel 1958abc and 1959, Ehlers and Sachs 1959, Sachs 1960 and 1961, JEK 1960, and JES 1961]. The present paper goes beyond these references mainly in two respects: (i) the "if" part of the theorem has not appeared in the literature; it was previously proved by Robinson and Trautman (1961) only in the special case $k_{[a;b}k_{c]}=0\neq k^a_{;a}$; (ii) by analyzing which of the field equations $R_{ab}=0$ are needed for the various portions of the theorem in some detail, it will be possible to derive several corrollaries about solutions of the Einstein-Maxwell equations.

It is convenient to work with the usual tetrad formalism [Eisenhart 1949, Schouten 1954, and Ehlers 1957]. If $T_{a...b}{}^{c...d}$ is any real tensor and $S_{\alpha...\beta}$ any real or complex scalar (which may, but need not, have tetrad indices) the following notations will be used:[3]

$$\begin{aligned} T_{\alpha..\beta\gamma..\delta} &\equiv T_{a..b}{}^{c..d}k^a_\alpha k^b_\beta k_{\gamma c}k_{\delta d} \\ &\Rightarrow T_{\alpha..\beta\gamma..\delta}g^{\beta\gamma} = T_{a..b}{}^{b..d}k^a_\alpha \cdot\cdot k_{\delta d} \\ S_{\alpha..\beta\gamma}g^{\gamma\beta} &= S_{\alpha..\beta}{}^{\beta} \\ S_{\alpha..\beta|\gamma} &= \frac{\partial S_{\alpha..\beta}}{\partial x^a}k^a_\gamma \\ T_{\alpha..\beta\gamma..\delta||\varepsilon} &= T_{a..b}{}^{c..d}{}_{;e}k^a_\alpha\cdot\cdot k^b_\beta k_{\gamma c}\cdot\cdot k_{\delta d}k^e_\varepsilon \\ T_{1..2}{}^{2..1} &= \overline{T}_{2..1}{}^{1..2} \end{aligned}$$

Moreover, the rotation coefficients will be designated by $\Gamma_{\alpha\beta\gamma}$:

$$k_{ab;c}k^b_\beta k^c_\gamma = \Gamma_{\alpha\beta\gamma} \Rightarrow \{\Gamma_{(\alpha\beta)\gamma}=0 \Leftrightarrow \Gamma_{\alpha\beta\gamma}=\Gamma_{[\alpha\beta]\gamma}\}.$$

These relations imply

$$T_{\alpha..\beta||\varepsilon} = T_{\alpha..\beta|\varepsilon} - T_{\gamma..\beta}\Gamma^{\gamma}_{\alpha.\varepsilon} - \ldots - T_{\alpha..\gamma}\Gamma^{\gamma}_{\beta.\varepsilon}.$$

One finds for the Riemann and Ricci tensors

$$\begin{aligned} &\text{(a)} & (1/2)R^\sigma_{\delta\mu\nu} &= \Gamma^\sigma_{\delta.[\mu|\nu]} - \Gamma_{\delta\omega[\mu}\Gamma^{\sigma\omega}_{..\nu]} - \Gamma^\sigma_{\delta.\omega}\Gamma^\omega_{[\mu.\nu]} \\ &\text{(a)} & R_{\alpha\beta} &= R^\sigma_{\alpha\beta\sigma}. \end{aligned} \tag{2.1}$$

That the right hand side of (2.1a) has all the usual symmetries can be inferred from the sixteen equations

$$k_{\delta[a:bc]}\equiv 0 \Leftrightarrow \Gamma_{\delta[\sigma\mu|\nu]} - \Gamma_{\delta\omega[\mu}\Gamma^\omega_{\sigma.\nu]} - \Gamma_{\omega[\mu\nu}\Gamma^\omega_{\sigma]\delta]}. \equiv 0. \tag{2.2}$$

[3] k^d_δ corrected to $k_{\delta d}$ in the second equation and $S_{\alpha...\beta}{}^\delta$ corrected to $S_{\alpha...\beta}{}^\beta$ in the third equation below [editor].

From Eq. (1.1) one sees that the conformal tensor vanishes if and only if the real and imaginary parts of the following five combinations of tetrad components of the Riemann tensor vanish (here and throughout, numerical indices are tetrad indices):[4]

$$\begin{aligned}
C^{(5)} &= (1/4)R_{3131} \\
C^{(4)} &= (-1/8)(R_{3134} + R_{3112}) \\
C^{(3)} &= (1/24)(R_{3434} - 2R_{3412} + R_{1212} + R_{3241}) \\
C^{(2)} &= (-1/8)(R_{4234} + R_{4212}) \\
C^{(1)} &= (1/4)R_{4242}
\end{aligned} \tag{2.3}$$

The numerical factors have been inserted to conform to a previous notation [Sachs 1961]. Taking tetrad components of Eq. (1.3)[5] one finds that *a field is algebraically special if and only if* $C^{(5)} = C^{(4)} = 0$ *for some* k_3^a.

The "if" half of the theorem will now be proved. Thus one supposes that

$$\Gamma_{3\alpha 3} = 0, \quad \sigma = \Gamma_{311} = 0 \tag{2.4}$$

and must show $C^{(5)} = C^{(4)} = 0$. The proof will be split into various steps to exhibit the field equations needed in order to obtain particular conclusions.[6]

A. Taking the absolute derivative of Eq. (1.5), using the definitions given in (1.6), and equating this result to that obtained from applying the Ricci identity gives the propagation equations for z and σ:

$$z_{|3} + z^2 + \sigma^2 = (1/2)R_{33} \tag{2.5}$$

and

$$\sigma_{|3} + (z + \bar{z})\sigma - 2\sigma\Gamma_{123} = 4C^{(5)}. \tag{2.6}$$

Thus, the following assertion is proven: *If* k_3^a *is tangent to a shear-free null geodesic congruence, i.e., Eqs. (2.4) are satisfied, then* $C^{(5)} = 0$. Note that this means that k_3^a satisfies Eq. (1.2). Hence, it is one of the null directions singled out by the Riemann tensor.

B. In the remaining argument it is necessary to treat separately the cases $z = 0$ and $z \neq 0$. Here we shall consider $z = 0$.

If Eqs. (2.4) *are satisfied and* $z = 0$, *then* $R_{33} = 0$; *if in addition* $R_{31} = 0$, *then* $C^{(4)} = 0$. The first statement follows trivially from Eq. (2.5). From Eqs. (2.3) and (2.1) one finds

$$C^{(4)} = (-1/8)(2R_{3112} + R_{31}).$$

[4] Two corrections were done in the formula for $C^{(3)}$: the sign of the second term was reversed and the indices at the last term were changed from the incorrect 3142. This equation was not used in the proof, so the errors had no consequences [editor].

[5] Equation-number (1.5) corrected to (1.3) [editor].

[6] The hint at the derivation of (2.5) and (2.6) in the next paragraph does not seem to be right. See the derivation in the first part of the editorial note, point **2.** Equation (2.5) was corrected – in the paper, the term $+\sigma^2$ on the left is absent [editor].

Thus, when $R_{31} = 0$, Eqs. (2.1a) and (2.4) give

$$C^{(4)} = (-1/4)(z_{|1} + z\Gamma_{431} + (\overline{z} - z)\Gamma_{314}) = 0.$$

Hence, when $z = 0$, $C^{(4)} = 0$.

C. *If Eqs.* (2.4) *hold and* $z \neq 0$, *there is precisely one direction* k_4^a *(defined up to a multiplicative scalar factor* A^{-1}*) for which* $\Gamma_{314} = 0$. Applying the transformation (1.9), one finds

$$\Gamma_{314} = e^{iC}(\Gamma_{314} + \overline{B}z).$$

Thus, when $z \neq 0$, one can adjust B in such a way that

$$\Gamma_{314} = 0. \tag{2.7}$$

This result will be assumed to hold in the following (the primes will be dropped), as the choice considerably simplifies all the calculations. In particular, from Eqs. (2.3) and (2.1b) one can also write

$$C^{(4)} = (-1/8)(2R_{3134} - R_{31});$$

when $R_{31} = 0$ and with (2.7), Eq. (2.1a) gives

$$C^{(4)} = (1/4)z\Gamma_{413}. \tag{2.8}$$

Thus, to show that $C^{(4)} = 0$ requires proving that $\Gamma_{413} = 0$.

D. *If Eqs.* (2.4) *and* (2.7) *hold, and* $z \neq 0$, *then* $R_{11} = 0$ *implies* $\Gamma_{411} = 0$. From Eq. (2.1b) for R_{11} and (2.2) with $(\delta\sigma\mu\nu) = (3411)$, one obtains

$$R_{11} = 0, \quad R_{3114} = -z\Gamma_{411} = 0;$$

Eqs. (2.1a) with (2.4) and (2.7) have also been used. Thus, when $z \neq 0$, the assertion follows:

$$\Gamma_{411} = 0. \tag{2.9}$$

Lemma 1: If a field contains a shear-free null geodesic congruence k_3^a *and if* $R_{33} = R_{31} = R_{32} = R_{11} = R_{22} = 0$, *then the field is algebraically special and* k_3^a *obeys Eq.* (1.3). From assertions A and B above, the lemma is proven for the case $z = 0$. Therefore, only $z \neq 0$ need be considered; that is, one must show that $\Gamma_{413} = 0$, Eq. (2.8). One first uses Eqs. (2.1b), together with (2.5), (2.7), and (2.9) to obtain[7]

$$(1/2)R_{33} = z_{|3} + z^2 = 0 \tag{2.5'}$$

$$R_{31} = -z_1 - z(\Gamma_{413} - \Gamma_{341}) = 0. \tag{2.10}$$

Using the cyclic relations (2.2) one finds that $R_{31} = 0$ also implies

$$\begin{aligned} &\text{(a)} \quad \Gamma_{341|3} - \overline{z}(\Gamma_{413} - \Gamma_{341}) + z\Gamma_{413} = 0,\\ &\text{(b)} \quad \Gamma_{211|3} - \overline{z}(\Gamma_{413} - \Gamma_{211}) - z\Gamma_{413} = 0. \end{aligned} \tag{2.11}$$

[7] Equation (2.10) does not follow as simply as the paper suggests – see point **3** in the editorial note [editor].

Now one examines the integrability conditions of (2.5') and (2.10). It is necessary to use the fact that "slash" derivatives do not commute, but satisfy the following relation:

$$\varphi_{|\alpha\beta} = \varphi_{|\beta\alpha} - 2\varphi_{|\gamma}\Gamma^{\gamma}_{.[\alpha\beta]}. \tag{2.12}$$

With the help of this relation, one finds that $R_{31|3} - (1/2)R_{33|1} = 0$ yields the integrability condition[8]

$$S \equiv \Gamma_{413|3} + 3z\Gamma_{413} - \Gamma_{123}\Gamma_{413} = 0. \tag{2.13}$$

The argument in D leading to Eq. (2.9) also gives

$$R_{11} = 0, \quad R_{4113} = U = 0$$

with

$$U \equiv \Gamma_{413|1} + \Gamma_{413}\Gamma_{211} - (\Gamma_{413})^2 = 0. \tag{2.14}$$

Using (2.12) and (2.11), the integrability condition for $U = S = 0$ is simply[9]

$$-10z{\Gamma_{413}}^2 = 0.$$

This relation completes the proof of the lemma.

Lemma 2: If a vacuum metric is algebraically special, then the vector $k'^a = k_3^a$ *of Eq.* (1.3) *is necessarily tangent to a congruence of null geodesics with zero shear; that is, the conditions of Eq.* (2.4) *are satisfied**.

Since a full proof of this lemma has appeared in the literature [JES 1961] only the first step of the proof will be given here to indicate how the calculations are done in the above notation.

One chooses for k_3^a the direction defined by Eq. (1.3). According to the remarks following Eq. (2.3), this choice amounts to setting $C^{(5)} = C^{(4)} = 0$. One must now prove that Ak_3^a satisfies (2.4), where A is a suitable scalar scale factor. Consider the once contracted Bianchi identities

$$R_{\alpha\beta[\gamma\delta||\varepsilon]}g^{\beta\gamma} \equiv 0. \tag{2.15}$$

Letting $\alpha = 3$, $\varepsilon = 3$, $\delta = 1$, using Eqs. (2.1) and (2.3), one finds

$$C^{(3)}\Gamma_{313} = 0.$$

Thus, if $C^{(3)} \neq 0$, then $\Gamma_{313} = 0 \Rightarrow \Gamma_{323} = \overline{\Gamma}_{313} = 0$; one can then arrange to have $\Gamma_{343} = 0$ by using the scale factor A. The remaining equation in (2.4), the shear-free condition, is proved from (2.15) by setting $\alpha = 1$, $\varepsilon = 3$, and $\delta = 1$. For the details on the case $C^{(3)} = 0$ see [JES 1961].[10]

Lemma 1 and Lemma 2 together form the proof of the theorem stated at the beginning of this section.

[8]The next equation corrected; in the paper the last term before the second equality sign was missing. The reasoning in this segment of the paper is described somewhat too briefly. See point **4** in the editorial note for more detailed hints [editor].

[9]Small correction done in the next formula, the paper had $10z^2\Gamma_{413}$ [editor].

*In the case $C_{abcd} = 0$ this lemma is not valid, but the theorem is trivially true.

[10]For the details of the case $C^{(3)} = 0$ see also the editorial note, point **5** [editor].

3. Corrollaries

Corrollary 1: *A vacuum, metric with $C_{abcd} \neq 0$ is Type D (Petrov Type I-degenerate) if and only if there exist two independent shear-free null geodesic congruences.* A Type D metric is characterized by the existence of two vectors k'^a and $k''^a, k'^a \neq \alpha k''^a$, each of which satisfies Eq. (1.3). With the choice $k'^a = k_3^a$ in the tetrad defined by (1.6), then by the theorem just proved, $C^{(5)} = C^{(4)} = 0$. Applying the transformation (1.9) does not change this situation[11]. Therefore, one may set $k''^a = k_4'^a$. With this choice of tetrad system, the theorem also gives $C^{(1)} = C^{(2)} = 0$. This concludes the "if" half of the corrollary. The "only if" half follows from Lemma 2. In addition to the two equations from the Bianchi identities considered there when $C^{(5)} = C^{(4)} = 0$, consider also $C^{(1)} = C^{(2)} = 0$ with $\alpha = 4, 1$ and $\varepsilon = 4$, $\delta = 1$. Then one obtains

$$\Gamma_{411} = \Gamma_{422} = \Gamma_{414} = \Gamma_{424} = 0$$

in addition to (2.4).

Corrollary 2: *If one has a solution of the Maxwell-Einstein equations and the Maxwell field is a null field* [*Synge* 1956], *then the gravitational field is algebraically special.* An electromagnetic null field $F^{ab} \neq 0$ is defined by the property that there is some necessarily unique direction $k^a \neq 0$ which obeys

$$(F_{ab} + i *F_{ab})k^b = 0,$$

where the star denotes the (real) dual [Synge 1956]. Robinson [1959] and Mariot [1954] have shown that k_3^a is tangent to a congruence of shear-free null geodesics because of the electromagnetic field equations. Using the expression $R_{ab} \sim F_a^c F_{cb} - (1/4) g_{ab} F_{em} F^{me}$, one finds that all the hypotheses of lemma 1 are satisfied and the corrollary follows.

Corrollary 3: *A vacuum gravitational field permits an electromagnetic null "test field" if and only if the gravitational field is algebraically special.* By a test field is meant a tensor F_{ab} that obeys the electromagnetic field equations, but makes no contribution to the Ricci tensor. Robinson [1961] has shown that a null test field can be introduced into a normal hyperbolic four dimensional Riemannian space if and only if the space contains a shear-free null geodesic congruence. Combining Robinson's results with the theorem proven here, one obtains Corrollary 3.

If one has an electromagnetic field that is not a null field, there are always precisely two distinct directions k_3^a which obey the equations

$$(F_{ab} + i *F_{ab})k^b = Kk^a, \quad k^a k_a = 0,$$

where K is a scalar. If at least one of the two directions k^a is tangent to a congruence of shear-free geodesics, one says that the electromagnetic field has shear-free geodesic rays. The field of an accelerated point particle in Lorentz covariant electrodynamics and the field of a charged point particle at rest in

[11] Equation number (2.13) in this sentence corrected to (1.9) [editor].

general relativity are examples of such electromagnetic fields [JES 1961]. For such electromagnetic fields a corrollary similar to Corrollary 2 above holds and is proved in the same fashion.

Corrollary 4: *One can always find a direction* k_4^a *which together with the direction* k_3^a *satisfying Eq.* (1.3) *is* X_2*-forming. In particular, when the metric is vacuum and algebraically special with* $z \neq 0$ *the* k_4^a *defined by Eq.* (2.7) *satisfies this condition. Furthermore, when the field is Petrov Type III or degenerate Type II (that is, if* $C^{(3)} = 0$*) these* X_2*'s have vanishing intrinsic curvature.* Two vectors u^a and v^a are X_2-forming if

$$(u^{[a}v^{b]})_{;a}w_b = 0$$

for all w_b such that $u^a w_a = v^a w_a = 0$ [Schouten 1954, p. 81]. This condition applied to k_3^a and k_4^a yields

$$\Gamma_{314} - \Gamma_{413} = 0.$$

If for a particular choice k'^a this condition is not satisfied, then from the transformation equations, Eqs. (1.9), with $A = 1$, $C = 0$, and

$$\Gamma'_{314} - \Gamma'_{413} = \Gamma_{314} - \Gamma_{413} - z\overline{B} - \overline{B}_3 = 0$$

one obtains a k_4^a which does satisfy the condition. The rest of the corrollary is easily seen from Eqs. (2.7), (2.15), and the fact that

$$C^{(3)} + \overline{C}^{(3)} = (1/24)(6R_{3434} + 2R_{21} - 4R_{43}).$$

This corrollary together with the following statements, which are easily verified, may be useful in constructing algebraically special vacuum metrics in which $z \neq \overline{z}$:

1) When $C^{(3)} = 0$, $\Gamma_{344|3} = 0$ and one may find a scale A in (1.9) such that $\Gamma_{344} = 0$ while preserving (1.4);

2) $C^{(3)} - \overline{C}^{(3)} = R_{1234} = +\Gamma_{124|3} - \Gamma_{123|4}$. When $C^{(3)} = 0$, one may find a phase C in (1.9) such that $\Gamma_{124} = \Gamma_{123} = 0$;

3) If in addition to the above, $C^{(2)} = 0$, then $\Gamma_{414|3} = 0$.

An interesting generalization of the theorem proved here was pointed out to the authors by Professor A. Schild, Dr. J. Ehlers, and Dr. W. Kundt. Suppose one goes from one space to another by a conformal transformation [Eisenhart 1949]:

$$g'_{ab} = e^{2w(x)} g_{ab}.$$

It can be shown that under such a transformation, Eq. (1.4), (1.3), and the condition $\sigma = 0$ all go over into themselves. It then follows that at each step in the above discussion one can weaken the assumptions made about the Ricci tensor. For example, Lemma 1 goes through unchanged for the field described by g'_{ab} if only the Ricci tensor R_{ab} in the field described by g_{ab} obeys the conditions stated in the lemma. The final theorem thus applies to any field that is conformal to a vacuum field.

4. Discussion

In Lorentz covariant electrodynamics one can easily construct radiation fields whose "rays" (in the intuitive sense) are shear-free. On the other hand, according to the theorem proved, the only corresponding gravitational fields are the algebraically special ones. It has recently become rather clear that algebraically special fields are too special to be more than far-field approximations to the outgoing waves from a bounded source [Misner 1957, Trautman 1958, Bondi 1960, Sachs 1961]. Thus, in any physically realistic gravitational wave, any null geodesic congruence, and, in particular, any such congruence that one could reasonably call the rays of the field, must show a certain amount of shear. This behaviour seems to be essentially due to the non-linearity of the theory. One might visualize the rays being bent by the gravitational field itself and, because the field is not spherically symmetric, picking up some shear in the process.

The idea of this proof was suggested by A. Trautman. Discussions with J. Ehlers, W. Kundt, F. A. E. Pirani, and A. Schild were very useful. Both authors are indebted to the faculty and staff of King's College, London, for their hospitality.

REFERENCES

Bel, L., *C. R. Acad. Sci. Paris*, **246**, 3015 (1958a).
Bel, L., *C. R. Acad. Sci. Paris*, **247**, 1094 (1958b).
Bel, L., *C. R. Acad. Sci. Paris*, **247**, 2096 (1958c).
Bel, L., *C. R. Acad. Sci. Paris*, **248**, 2561 (1959).
Bondi, H., *Nature*, **186**, 535 (1960).
Debever, R., *Bull. Soc. Math. Belg.*, **10**, 112 (1958).
Eisenhart, L. P., *Riemannian Geometry*, Princeton University Press, Princeton, New Jersey (1949).
Ehlers, J., *Dissertation*, University of Hamburg (1957).
Ehlers, J. and Sachs R., *Z. Phys.*, **155**, 498 (1959).
Goldberg, J. N. and Kerr, R. P., *J. math. Phys.*, **2**, 327 (1961).
Jordan, P., Ehlers, J. and Kundt, W., *Akad. Wiss. u. Lit. in Mainz, Math.-Naturwiss. Kl.*, 23–105 (1960). ([JEK 1960] in text).
Jordan, P., Ehlers, J. and Sachs, R., *Akad. Wiss. u. Lit. in Mainz, Math.-Naturwiss. Kl.* (1961), to appear ([JES 1961] in text).[12]
Kundt, W., *Z. Phys.*, **163**, 77 (1961).
Lichnerowicz, A., *C. R. Acad. Sci. Paris*, **246**, 893 (1958).
Mariot, L., *C. R. Acad. Sci. Paris*, **238**, 2055 (1954).
Misner, C, *Proceedings of the Chapel Hill Conference* (1957).
Newman, E., *J. math. Phys.*, **2**, 324 (1961).
Penrose, R., *Ann. Phys.* (New York), **10**, 171 (1960).

[12]Published: *Akad. Wiss. u. Lit. in Mainz, Math.-Naturwiss. Kl.* 3 – 62 (1961) [editor].

Petrov, A. Z., *Scientific Notices Kazan State University*, **114**, 55 (1954).[13]
Pirani, F. A. E., *Phys. Rev.*, **105**, 1089–1099 (1957).
Robinson, I., *Report to the Royaumont Conference* (1959).
Robinson, I., *J. math. Phys.*, **2**, 290 (1961).
Robinson, I. and Trautman, A., *Proc. Roy. Soc*, A **265**, 463 (1962).
Sachs, R., *Z. Phys.*, **157**, 1462 (1960).
Sachs, R., *Proc. Roy. Soc*, **264**, 309 (1961).
Schouten, J. A., *Ricci Calculus*, Springer Verlag, Berlin 1954.
Synge, J. L., *Relativity: The Special Theory*, North Holland Publishing Company, Amsterdam 1956.
Trautman, A., *Lectures on General Relativity*, King's College, London 1958.[14]
Wong, Y. Ch., *Sum. of Math.* (2) **46**, 158 (1945).

[13]An English translation of this paper was published as a Golden Oldie in *Gen. Rel. Grav.* **32**, no 8, 1665 (2000) [editor].
[14]These lectures were published as a Golden Oldie in *Gen. Rel. Grav.* **34**, no 5, 721 (2002) [editor].

Chapter 9
Jürgen Ehlers 1961, kinematics and dynamics of relativistic fluids.

Original paper: Beiträge zur relativistischen Mechanik kontinuierlicher Medien. *Akademie der Wissenschaften und Literatur (Mainz), Abhandlungen der Mathematisch-Naturwissenschaftlichen Klasse [Proceedings of the Mathematical-Natural Science Section of the Mainz Academy of Science and Literature]* Nr. 11, 792-837 (1961).

Paper translated by George F.R. Ellis, email: George.Ellis@uct.ac.za, and Peter K.S. Dunsby, Department of Mathematics and Applied Mathematics, University of Cape Town, Rondebosch 7700, Cape Town, South Africa, email: dunsby@gmail.com.

Previously published in *Gen. Relativ. Gravit.* **25**, 1225–66 (1993) (the paper) and *Gen. Relativ. Gravit.* **41**, 1899–1901 (2009) (the biography).

The editorial note was first written for this reprinting.

9.1 Editorial note to:

Jürgen Ehlers

Contributions to the relativistic mechanics of continuous media

George F. R. Ellis

This paper by Jürgen Ehlers must rank as one of the all-time classics in general relativity theory. It is one of the Mainz series of papers on Exact Solutions of General Relativity Theory published in the early 1960s by the Hamburg University group of Ehlers, Wolfgang Kundt, Rainer Sachs, and Manfred Trümper, under the auspices of Pascual Jordan and in close collaboration with Otto Heckmann and Engelbert Schücking (based at the Hamburg Observatory). This fourth paper in the series is on the kinematics and dynamics of continuous media, and – in accord with the philosophy of the series – focuses on exact properties of the full non-linear Einstein Field Equations. It is marked by an elegance and clarity of thought and presentation which is a model of how to write a scientific paper – it is concise and to the point, saying what needs to be said with great economy.

The paper is based on the 1+3 covariant formalism, whereby all quantities are projected perpendicular and parallel to a 4-velocity vector u^a, and then split into scalar, symmetric trace free, and skew parts. After developing the basic kinematic quantities (the acceleration, expansion, shear, and vorticity) the kinematics of special congruences are developed, and some basics of redshift developed and used to derive a generalised Hubble law. Then integrability conditions for general congruences are developed, leading to the energy conservation equations, a beautiful covariant derivation of curvature tensor components, and the vorticity conservation equations. Rigid and isometric flows are characterised, as well as the symmetries underlying the standard spatially homogeneous and isotropic cosmological models.

The dynamics section starts with an examination of the justification of the field equations. Then the splitting of the energy momentum tensor into its various parts is examined, leading to a study of their physical meaning. This leads to the most important equation in the paper, namely (81), which is the basic equation of gravitational attraction. This was found by Raychaudhuri for the case of dust [1] (hence geodesic motion). Here it is generalised by Ehlers to arbitrary matter and possibly non-geodesic motion. It is derived in a beautiful covariant manner (in contrast to the coordinate dependent previous derivation by Raychaudhuri). In isometric cases

George F. R. Ellis
Department of Mathematics and Applied Mathematics, University of Cape Town, Rondebosch 7700, Cape Town, South Africa
e-mail: George.Ellis@uct.ac.za

(that is, with zero shear and zero expansion), this gives the Poisson equation (82), originally found by Whittaker for the subcase $\omega = 0$ [2]. These equations make clear that the active gravitational mass density is $(\mu + 3p/c^2)$, in contrast to the inertial mass density $(\mu + p/c^2)$ (see (76)). These are crucial differences from Newtonian theory, where both are just the mass density ρ. The following examination of entropy production and linear equations of state for imperfect fluids is very useful; however it is not emphasized that there are causality problems with these equations, fixed by rather more complicated equations of state proposed much later by Stewart and Israel [3]. The nature of equations of state for perfect fluids is carefully examined. This is followed by an extremely elegant brief summary of general relativistic kinetic theory, including a derivation of the Boltzmann H-theorem and characterisation of equilibrium states – which are highly restricted.

A closing section applies the above relations to the case of pressure free matter ('dust' or 'baryonic matter') with a very elegant summary of its dynamic properties, including a statement of the basic singularity theorem for pressure free matter. The shear-free and vorticity free cases are examined, in each case leading to a generalised Friedmann equation. It was later shown [4] that in fact the shear-free case (126) cannot have rotation if the universe expands, and so reduces to the standard Friedmann equation; this was not known at the time this paper was written. The rotation free equations are examined and a basic theorem characterising Robertson–Walker spaces is proved. Various other special cases are characterised, including the Gödel stationary rotating universe [5], and the constant curvature case (128), which includes the well known Bianchi I anisotropic universes – the simplest spatially homogeneous universe models that generalise the Friedmann models with more degrees of freedom.

A later paper in this series by Wolfgang Kundt and Manfred Trümper [6] extended the formalism to include the dynamics of the the Weyl tensor, whose electric and magnetic parts[1] obey Maxwell-like equations that follow from the full Bianchi identities [7].[2] These were then used by Hawking to study perturbations of cosmological models [11], and Kristian and Sachs in their study of cosmological observations [12]. They were presented systematically in my own Varenna lectures [13], and used extensively in the 1+3 covariant approach to perturbations in cosmology,[3] which has *inter alia* been used productively in examining brane cosmology models [15].

Overall this paper is a classic: it can be recommended to every student of gravitation or cosmology as a masterly demonstration of covariant exact methods in general relativity theory.

---

1 Identified by Pirani [9, 10] in the course of his studies of the physical meaning of the Weyl tensor.

2 For a recent exposition, see [8].

3 Summarised for example in [14].

References

1. A. Raychaudhuri, "Relativistic Cosmology". *Phys. Rev.* **98**, 1123 (1955). Reprinted as Golden Oldie: *Gen. Relativ. Gravit.* **32**, no 4, 743 – 756 (2000).
2. E. T. Whittaker, *Proc. Roy. Soc. Lond.* **149**, 384 (1935).
3. W. Israel and J.M. Stewart, "Transient relativistic thermodynamics and kinetic theory". *Ann. Phys.* **118**, 341–372 (1979).
4. G. F. R. Ellis, "The dynamics of pressure-free matter in general relativity". *J. Math. Phys.* **8**, 1171–1194 (1967).
5. K. Gödel, "An Example of a New Type of Cosmological Solutions of Einstein's Field Equations of Gravitation". *Rev. Mod. Phys.* **21**, 447 (1949). Reprinted as Golden Oldie: *Gen. Rel. Grav.* **32**, no 7, 1399 (2000).
6. W. Kundt and M. Trümper, "Contributions to the theory of the gravitational radiation fields". *Abh. Akad. Wiss. Lit. Mainz* Nr. 12, 965 – 1000 (1962).
7. M. Trümper, in "Contributions to Actual Problems of General Relativity", Research Report, Air Research and Development Command, US Air Force Contract AF 61 (052)-567 (1964), with W. Beiglböck, K. Bichteler, W. Budich, P. Jordan, W. Kundt. Circulated as preprint.
8. R. Maartens and B. A. Bassett, "Gravito-electromagnetism". *Class.Quant.Grav.* **15**, 705 (1998) [arXiv:gr-qc/9704059].
9. F. A. E. Pirani, in *Recent Developments in General Relativity*. New York: Pergamon Press 1962.
10. F. A. E. Pirani, in *Gravitation: An Introduction to current research*. Ed. L Witten, New York: John Wiley and Sons 1962.
11. S. W. Hawking, "Perturbations of an Expanding Universe". *Astrophys. J.* **145**, 544–553 (1966).
12. J. Kristian and R. K. Sachs, "Observations in Cosmology". *Astrophys. J.* **143**, 379–398 (1966). Reprinted as Golden Oldie: *Gen. Relativ. Gravit.* **43**, no 1, 331–358 (2011).
13. G F R Ellis, "Relativistic Cosmology". In *General Relativity and Cosmology, Proc Int School of Physics "Enrico Fermi" (Varenna), Course XLVII*. Ed. R K Sachs, Academic Press 1971, pp. 104–179. Reprinted as Golden Oldie: *Gen. Rel. Grav*. **41**, no 3, 581 (2009).
14. G. F. R. Ellis and H. van Elst, "Cosmological Models" (Cargese Lectures 1998). In *Theoretical and Observational Cosmology*. Ed. M. Lachieze-Ray. Kluwer, Nato Series C: Mathematical and Physical Sciences, Vol. 541, 1999, pp. 1–116. [arXiv:gr-qc/9812046].
15. R. Maartens and K. Koyama "Brane-World Gravity". *Living Reviews in Relativity* **13**, 5 (2010) [arXiv:1004.3962].

9.2 Jürgen Ehlers – a brief biography

From the Biennial Report 2006/2007 of the Albert Einstein Institute

Jürgen Ehlers, one of the most distinguished and influential German scientists of his generation, died very suddenly and unexpectedly on 20 May 2008. A specialist in the study of general relativity, and at the same time a generalist and humanist with a passion for the fundamentals of the natural world, Ehlers played a leading role in the revival of research in general relativity in modern Germany from its small beginnings in the 1950s through to the flowering of the subject during the past decade. The institute he founded in Potsdam in 1995, the Max Planck Institute for Gravitational Physics (more commonly known by its subsidiary name, the Albert Einstein Institute – AEI), has become the largest institute of its kind in the world. It is a fitting memorial to his scientific life.

Born in 1929, Ehlers studied both mathematics and physics in Hamburg in the 1950s, and he finally chose physics because he could study general relativity with Pascual Jordan, one of the pioneers of quantum physics. At this time, interest in general relativity among theoretical physicists was beginning to revive after decades of neglect. Jordan was one of a handful of senior figures around the world who felt it was time to begin to understand general relativity, in order ultimately to generalize it into a full quantum theory of gravity. Key goals were to understand gravitational waves and what we now call black holes, and Jordan and his school, including Ehlers, were among the pioneers in this revival.

After visiting positions at several universities in Germany and the United States, culminating in a professorship in 1967 at the University of Texas at Austin, Ehlers moved to Munich to become a Member of the Max Planck Institute for Physics and Astrophysics in 1971. Institute director Ludwig Biermann asked Ehlers to join the Astrophysics part of the institute, because the institute was just beginning the gravitational wave activities that would eventually lead to Germany's GEO600 detector. In 1978 Ehlers organized the Ninth Texas Symposium on Relativistic Astrophysics, the principal international meeting where relativists and astrophysicists meet and update one another on their recent research. When the Astrophysics part of the Max Planck Institute moved into a new building in Garching outside Munich in 1979, Ehlers and his group, as well as the gravitational wave experimenters, went with it. Ehlers' clear commitment to astrophysics reflected his clear belief that the most important questions in general relativity were those that would be tested by astronomical observations.

Ehlers nevertheless remained a deeply mathematical physicist, and he always insisted that the great physical and astrophysical questions about relativity theory should be answered with as much rigour and care as possible. But always the important questions for him were those that the Universe itself posed. The discovery in 1974 of the first pulsar in a binary system, by Russell Hulse and Joseph Taylor, was a watershed for relativity, because it was immediately clear that the system would provide the first clean observational test of gravitational wave theory: the two stars would gradually spiral closer to one another as gravitational waves carried energy away. Ehlers quickly grasped how important this result would be, and just as quickly pointed out that the state of the theory of gravitational radiation itself was by no means satisfactory; relativity could not properly be tested against the observations until relativists sorted out the theory.

For the next ten years, Ehlers pushed his own research associates and scientists around the world to do this, with considerable success. The award of the 1993 Nobel Prize to Hulse and Taylor, the building of giant gravitational wave detectors around the world since the 1990s, and the use of modern supercomputers to predict gravitational wave emission from neutron stars and black holes all rest on secure theoretical foundations, thanks in part to Ehlers' insistence that even the complex mathematics of general relativity should be done carefully and rigorously.

Always looking ahead for the big challenges, Ehlers in the late 1980s took up research into another of Einstein's predictions, the bending of light by gravity. Again he was motivated by recent astronomical discoveries of gravitational lensing, where telescopes see multiple images of the same object, created as light takes multiple paths to the Earth through the gravitational field of an intervening galaxy. But again the theory needed work, and Ehlers stimulated young scientists in Garching to do it better. Today gravitational lensing is a central tool in astronomy, used among other things to prove that the Universe contains far more dark matter than it does stars and visible galaxies. The nature of this dark matter is not known, but it certainly is not composed of the protons, electrons, and neutrons that dominate the world we experience. Ehlers' young associates have gone on to make important and leading contributions to this branch of astronomy.

Ehlers' work at the junction between mathematics and physics had strong influences on the development of mathematics as well. He initiated a number of new research directions in analysis and differential geometry. Notable among these was his theory of reference systems, called 'frame-theory'. This provides a crucial mathematical link between concepts from classical physics and the geometric language of general relativity, and it has been successfully used to understand the precise relations between the different ways that Newton's and Einstein's theories of gravity would describe the same physical system. This is a key question, because many of the tests of the correctness of general relativity rely on measuring some of the small ways in which motions in the Solar System differ from those expected on the basis of Newton's theory of gravity. Ehlers had a rare ability to formulate fundamental physical questions in a precise mathematical language. His influence on research on the mathematics of Einstein's equations will be felt for many years to come.

In 1990 Ehlers had what he later described as the one good political idea in his life: he proposed that the Max Planck Society should create a research institute dedicated to research on gravitation. The reunification of Germany had created a need to expand the Max Planck system into former East Germany, and Ehlers felt that an institute in Potsdam, near Berlin, would not only make sense scientifically but would also finally allow Germany to make a visible and practical repudiation of the Nazis' personal vilification of Einstein, which had driven Einstein away from Berlin and Germany and had completely stopped relativity research in Germany. He used his scientific prestige to open political doors, and the result was the opening in 1995 of the AEI in Potsdam.

His vision for the research scope of the AEI showed again his perceptiveness and interest in relativity as a whole, even in work that was far from his own research. He wanted all of relativity under one roof: astrophysical research into black holes and gravitational waves, mathematical research to keep providing rigorous answers to questions raised by astronomical discoveries, and research designed to lead finally to a quantum theory of gravity, the goal that had motivated the revival of relativity in the 1950s and which even today is still not met. Today the AEI actually has two roofs: its theory branch in Potsdam/Golm, and its experimental branch in Hannover, which operates the GEO600 gravitational wave detector and plays a key role in the development of future detectors on the ground and in space. Employing two hundred staff, hosting a further two hundred scientific visitors each year, housing some of the world's fastest supercomputers, operating the GEO600 detector, hosting numerous conferences and workshops, publishing its own scientific journal and editing others, the AEI amply justifies Ehlers' initial vision that relativity best makes progress by keeping all its sub-fields connected and in communication.

In recent years Ehlers spent more time pursuing his life-long interests in the history of science and the meaning and importance of science to society. He engaged in public debates and wrote numerous articles. He strongly believed that rational thought and the scientific process were key ingredients of a civilized society, but he wanted society to understand the process as a human one, as an ongoing search for an ever deeper reality rather than as a way of manufacturing 'laws' written in stone.

Naturally, Jürgen Ehlers won many honours in his lifetime: the Max Planck Medal from the German Physical Society in 2002, the Volta Gold Medal of Pavia University in the 'Einstein Year' 2005, and recently the Chancellor's Medal of Charles University in Prague (2007). He was a member of Berlin-Brandenburg Academy of Sciences, the Mainz Academy of Sciences and Literature, the Leopoldina, and the Bavarian Academy of Sciences. In 1995 his scientific colleagues elected him President of the International Society for General Relativity and Gravitation for three years. But despite these honours and his considerable influence, Ehlers will be remembered by those who knew him as a modest man and a gentleman, a mentor who led by his example and by his deep scientific insight, a leader who always showed respect for his colleagues and co-workers.

Note from the Editors of this book: In the years 2006 – 2008 Jürgen Ehlers was a member of the Advisory Board of the Golden Oldies. He was active both in selecting the papers to reprint, and in writing the editorial notes, and thereby helped to maintain a high standard for the GOs.

General Relativity and Gravitation, Vol. 25, No. 12, 1993

Contributions to the Relativistic Mechanics of Continuous Media

Jürgen Ehlers[1]

Received July 12, 1993

This is a translation from German of an article originally published in *Proceedings of the Mathematical-Natural Science Section* of the Mainz Academy of Science and Literature, Nr. 11, 1961 (pp. 792-837) (printed by Franz Steiner and Co, Wiesbaden), which is Paper IV in the series "Exact Solutions of the Field Equations of General Relativity Theory" by Pascual Jordan, Jürgen Ehlers, Wolfgang Kundt and Rainer K. Sachs. The translation has been carried out by G. F. R. Ellis (Department of Applied Mathematics, University of Cape Town), assisted by P. K. S. Dunsby, so that this outstanding review paper can be readily accessible to workers in the field today. As far as possible, the translation[2] has preserved both the spirit and the form of the original paper.[3] Despite its age, it remains one of the best reviews available in this area.

1. INTRODUCTION

The description of matter in the General Theory of Relativity can be carried out with the help either of a model of point masses, or a model of a continuously spread out medium. Given the present state of knowledge,

1 Present address: Max Planck Institute for Astrophysics, 85740 Garching-bei-München, Germany

2. We thank Dr. Ehlers for checking and correcting the English version.

3 In accord with present day usage, the covariant derivative is represented by a semi-colon and partial derivative by a comma, instead of the double vertical line and single vertical line used for these quantities in the original paper. Thus $T_{ab;c} \equiv T_{ab||c}$, $T_{ab,c} \equiv T_{ab|c}$.

0001-7701/93/1200-1225$07.00/0 © 1993 Plenum Publishing Corporation

only the second method of description can be carried out in a objection-free mathematical way, since no one has so far succeeded even in defining the concept of a "singularity of a metric field that describes a particle", much less in deducing the consequences of such a definition.

From the physical viewpoint there is no decisive objection that gives fully convincing reasons to consider the first model as the more suitable. Indeed in our opinion this is not the case; for the considerations that led Einstein to setting up his theory related to macroscopic processes, and whether the theory can contribute something to the understanding of microphysical elementary processes is at the present time undecided. (Pauli remarked in the introduction to the new edition of his *Relativity Theory* appearing in 1958, "These differences of opinion are merging into the great open problem of the relation of relativity theory to quantum theory, which will presumably occupy physicists for a long time to come. In particular a clear connection between the general theory of relativity and quantum mechanics is not yet in sight.") In a macroscopic theory however there can be no doubt that the hydrodynamic (or elasto-mechanical) description should be regarded as primary and the point-mechanical as an approximating simplification. Hydrodynamics must therefore be considered as an essential part of gravitational theory.

While the formal carrying over of the special-relativistic expressions for the energy tensor corresponding to hydrodynamics and the conservation equations (replacing the ordinary by covariant divergence) was already carried out by Einstein in 1916, the attempt to develop a systematic general relativistic hydrodynamics was first undertaken in a beautiful work by J. L. Synge in 1937, and these investigations were carried forward with other methods by A. Lichnerowicz; the results of the last named author are presented in his volume on relativity theory that appeared in 1955. The works of these authors stimulated the present investigations and laid their essential foundations. We will not refer to the contributions of other authors here, but rather at the appropriate place in this work.

The aim of this work is to communicate a series of new results in the framework of a comprehensive presentation that combines in a unified viewpoint the contributions of various authors. This seems all the more appropriate because the standard presentations of relativity theory—with the exception of the above-named book of Lichnerowicz—at best skim over this important and beautiful part of Einstein's theory.[4] Furthermore

[4] Some important remarks are contained in the recent book by J. L. Synge [35], whose geometric presentation of relativity theory serves as a model for the present work, and in the book of V. A. Fock [29]. However these remarks do not add up to a

no presentation exists that deals with the kinematic, mechanical, thermodynamic, and kinetic gas theory aspects, linked together and with reference to special solutions of the field equations, as is done here, naturally with restriction to suitably selected topics. (The treatment of specific solutions in particular is only sketched for the case of incoherent matter, but the foundations are developed for a corresponding description of solutions with ideal fluids that will follow.)

In Section 2 neither (dynamic) conservation theorems nor field equations are supposed; rather the general geometric properties of a congruence of timelike curves in a normal-hyperbolic Riemannian space are investigated.

After an illustrative derivation of the kinematic basic variables of the theory, which completes the corresponding investigations by Synge and Lichnerowicz, in the second subsection special flows are characterised.

The identities developed in the third subsection are used in the fourth paragraph for derivation of vorticity theorems. Furthermore stationary and static space-times are characterised by the test-particle relative motions that are possible in them, and a new criterion for their Petrov type is proved.

The kinematic characterisation of cosmological models in the fifth subsection shows the usefulness of the hydrodynamic concepts and theorems for the derivation or illustrative interpretation of particular line-elements. The derivation given there of the homogeneous-isotropic models is simpler than that of Robertson and Walker, usually referred to in the literature, despite the weaker presuppositions made here.

Section 3 begins with a new version of the Weyl–Pauli way of motivating the Einstein Field Equations. The two following subsections contain a derivation for the simplest cases of the thermo-hydrodynamic foundations. The general relativistic 'Navier–Stokes' equation (76) and the 'Poisson' equation (82) have not been given up to this point.

The theory of an ideal gas given in the fourth paragraph is different from that recently published by Synge in 1958, in that here the Boltzmann equation is placed at the beginning, then an H-theorem proved and from this the equilibrium distribution won 'dynamically', while Synge (like earlier authors) uses the Boltzmann counting method (with an additional ad hoc assumption). Furthermore the determination of reversible flows in a gravitational field (Theorem 3.4.3), which is of cosmological interest, is new, as is the related gas-kinetic derivation of the Tolman law concerning its relation with the gravitational (and so also centrifugal) potential. Re-

systematic presentation.

marks by A. H. Taub [37] and M. Sasaki [38] have stimulated this part of the work.

The last subsection contains general statements about the motion of pressure-free matter and a hydrodynamic description of the known exact solutions. This subsection also contains some new remarks: it seems to us more significant that the geometric presentation, avoiding a purely formal starting point, is significantly simpler than the usual ones and at the same time leads to a better understanding of the matter concerned.

To conclude these introductory remarks I must stress that—as is shown also in this work—a truly physical understanding of Einstein's theory that is so convincing in its foundations is only now gradually emerging. The wish to contribute to this clarification has led to this work.

Conventions

Equality by definition: $\equiv$. Proportionality: $\sim$.
Tensor indices: $a, b, c, ... = 1, 2, 3, 4$; $\lambda, \mu, \nu ... = 1, 2, 3$.
Symmetrisation: ( ), e.g. $F_{(ab)} \equiv \frac{1}{2}(F_{ab} + F_{ba})$.
Antisymmetrisation: [ ], e.g. $F_{[ab]} \equiv \frac{1}{2}(F_{ab} - F_{ba})$.
Metric tensor: g_{ab}, Signature $+++-$.
Riemann tensor: R_{abcd}.
Ricci tensor: $R_{ab} \equiv R^c{}_{acb}$, trace: $R \equiv R^a{}_a$.
Einstein tensor: $G_{ab} \equiv R_{ab} - \frac{1}{2} R g_{ab}$, $G \equiv G^a{}_a$.
Weyl tensor(= Conformal tensor): C_{abcd}.
Partial differentiation w.r.t. x^c: for example $F_{ab,c}$.
Covariant differentiation w.r.t. x^c: for example $F_{ab;c}$; $F_{ab;cd} = F_{ab;c;d}$.
Covariant differentiation along $x^a(\lambda)$: $\overset{\lambda}{\nabla} k_a = k_{a;b}(dx^b/d\lambda)$.
Conventions regarding units: Speed of light $c \equiv 1$, Newtonian Gravitational constant $\gamma \equiv 1/8\pi$, thus Einstein gravitational constant $\kappa \equiv 1$.

2. KINEMATICS

2.1. Fundamental concepts. Decomposition of the velocity gradient

The history of a continuously spread out body is described in relativity theory by a three-dimensional family of timelike curves, the *worldlines* of the matter elements. With respect to an arbitrary local coordinate system,

$$x^a = x^a(y^\alpha, s) \tag{1}$$

is a parametric representation of the family: the y^α designate the matter elements, s the *proper time* along their worldlines.[5] Then

$$u^a \equiv \frac{\partial x^a}{\partial s} \equiv \dot{x}^a \qquad (u^a u_a = -1) \tag{2}$$

is the normalised tangent vector to the curve, usually referred to as the *four-velocity* of the substance element.

If

$$\delta \equiv \delta y^\nu \frac{\partial}{\partial y^\nu} \tag{3}$$

means the variation across the world lines and $(\,)^{\cdot}$ the covariant derivative with respect to s along the world lines, then

$$(\delta x^a)^{\cdot} = u^a{}_{;b}\delta x^b\,; \tag{4}$$

for both sides are vectors, and when we use locally flat coordinates (4) expresses the equality of certain second partial derivatives of the functions (1).

The tensor

$$h^a_b \equiv \delta^a_b + u^a u_b \tag{5}$$

projects the tangent vector-space at each point perpendicularly onto the three-dimensional subspace orthogonal to u^a. The vector

$$\delta_\perp x^a = h^a_b \delta x^b \tag{6}$$

is the *position vector* of the particle $(y^\alpha + \delta y^\alpha)$ with respect to the particle (y^α), and its *Fermi derivative*

$$v^a \equiv h^a_b (\delta_\perp x^b)^{\cdot} \tag{7}$$

is (by definition) the *velocity* of the particle $(y^\alpha + \delta y^\alpha)$ relative to the particle (y^α).

The (absolute) *acceleration* of the substratum elements is given by the spacelike vector field

$$\dot{u}^a \equiv u^a{}_{;b} u^b \quad (u_a \dot{u}^a = 0) \tag{8}$$

(In differential geometric terms, $\dot{u}^a$ is the first curvature vector of the corresponding curve.)

---

[5] The transformation (1) corresponds to the transformation in classical mechanics from Lagrangian to Eulerian coordinates.

From eqs. (4)–(8) one can derive the velocity field in an infinitesimal neighbourhood of a material element:

$$v^a = u^a{}_{;b}\delta_\perp x^b \tag{9}$$

According to the pattern of classical hydrodynamics we decompose the tensor $u^a{}_{;b}h^b_c$, that brings about the transformation $\delta_\perp x^a \to v^a$, into irreducible parts (with respect to the rotation group):

$$u_{a;c}h^c_b = \omega_{ab} + \sigma_{ab} + \tfrac{1}{3}\Theta h_{ab} \tag{10}$$

where

$$\omega_{(ab)} = \sigma_{[ab]} = 0, \quad \sigma^a_a = 0, \quad \omega_{ab}u^b = \sigma_{ab}u^b = 0. \tag{11}$$

The infinitesimal transformation that the vector-space $\{\delta_\perp x^b\}$ experiences in the time interval ds decomposes according to the preceding expressions into a *rotation* $\omega^a{}_b\delta_\perp x^b$, a rotation-free, volume-preserving *shearing* $\sigma^a{}_b\delta_\perp x^b$, and a rotation-free *similarity transformation* $\frac{1}{3}\Theta\delta_\perp x^a$. The distances $\delta l \equiv (g_{ab}\delta_\perp x^a\delta_\perp x^b)^{1/2} = (h_{ab}\delta x^a\delta x^b)^{1/2}$ between neighbouring particles alter during this time according to the equation

$$\frac{(\delta l)^{\cdot}}{\delta l} = \frac{1}{3}\Theta + \sigma_{ab}e^a e^b, \qquad \left(e^a \equiv \frac{\delta_\perp x^a}{\delta l}, e_a e^a = 1\right) \tag{12}$$

that follows from (7),(9)–(11), and the directions to the neighbouring particles according to

$$h^a_b \dot{e}^b = (\omega^a{}_b - \sigma^a{}_b - \sigma_{cd}e^c e^d \delta^a{}_b)e^b. \tag{13}$$

According to (12) and $\sigma^a{}_a = 0$, if $\overline{\delta l}$ means a direction average and δV the volume of an infinitesimal set of substratum particles,

$$\Theta = 3\frac{\overline{(\delta l)^{\cdot}}}{\overline{(\delta l)}} = \frac{(\delta V)^{\cdot}}{\delta V}. \tag{14}$$

As (13) shows, $h^a_b\dot{e}^b = \omega^a{}_b e^b$ holds exactly when e^a lies in a principal shearing direction. Therefore that material orthogonal triad of vectors which at any time s coincides with the main shearing axes experiences in the time interval $(s, s+ds)$ the infinitesimal rotation determined by $\omega^a{}_b$.[6]

[6] Through this, in general the triad rotates itself out of the main shear directions. An exception is given in Corollary 2 to Theorem 2.4.2 in subsection 2.4.

From the decomposition formulae (10) and the conditions (11) one can derive the following expressions for the *vortex velocity* ω_{ab}, the *shearing velocity* σ_{ab}, and the *expansion velocity* Θ:

$$\omega_{ab} = u_{[a;b]} - \dot{u}_{[a}u_{b]}, \tag{15}$$

$$\sigma_{ab} = u_{(a;b)} - \dot{u}_{(a}u_{b)} - \tfrac{1}{3}\Theta h_{ab}, \tag{16}$$

$$\Theta = u^{a}{}_{;a}. \tag{17}$$

Instead of the vorticity tensor ω_{ab} one can introduce the *vorticity vector*

$$\omega^{a} \equiv \tfrac{1}{2}\eta^{abcd}u_{b}\omega_{cd} = \tfrac{1}{2}\eta^{abcd}u_{b}u_{c,d} \quad (u_{a}\omega^{a} = 0); \tag{18}$$

ω^{a} is dual to ω_{ab} in the rest-space determined by u^{a}:

$$\omega_{ab} = \eta_{abcd}\omega^{c}u^{d}. \tag{19}$$

The consequent equation

$$\omega_{ab}\omega^{b} = 0 \tag{20}$$

implies, according to (9),(10), that ω^{a} determines the rotation axis. According to (18) the vector ω^{a} is spacelike, and by (20) lies orthogonal to the plane determined by the simple spacelike bivector ω_{ab}.

A family of timelike curves in general has exactly *nine invariants* of the first order (independent of each other): the six independent components of the vectors $\dot{u}^{a}$ and ω^{a} relative to the eigenvector basis of the tensor $\sigma^{a}{}_{b}$, two independent eigenvalues[7] of $\sigma^{a}{}_{b}$, and Θ. These magnitudes determine (for a given metric g_{ab}) the family of curves in the infinitesimal first order neighbourhood of a point uniquely up to homogeneous Lorentz transformations, as follows from the relation

$$u_{a;b} = \omega_{ab} + \sigma_{ab} + \tfrac{1}{3}\Theta h_{ab} - \dot{u}_{a}u_{b} \tag{21}$$

which is equivalent to (10).

In addition to Θ we wish to explicitly introduce for later application the scalars

$$\dot{u} \equiv (\dot{u}_{a}\dot{u}^{a})^{1/2}, \quad \omega \equiv (\omega_{a}\omega^{a})^{1/2} = (\tfrac{1}{2}\omega_{ab}\omega^{ab})^{1/2}, \quad \sigma \equiv (\tfrac{1}{2}\sigma_{ab}\sigma^{ab})^{1/2}. \tag{22}$$

These are non-negative and vanish only at the same time as their corresponding tensors.

---

[7] Or, algebraically simpler, $\sigma^{a}{}_{b}\sigma^{b}{}_{a}$ and $\sigma^{a}{}_{b}\sigma^{b}{}_{c}\sigma^{c}{}_{a}$.

While the physical meaning of the deformation of a substratum element described by the formulae (12),(14) is immediately determined by the axiomatically laid down interpretation of the metric fundamental form $g_{ab}dx^a dx^b$, the definition of a rotation given above contains an implied *convention*: this rotation relates to a Fermi-transported triad of vectors that can be thought of as carried along in the rest-frame of the central particle ($v^a = 0$ in (7) means that $\delta_\perp x^a$ is Fermi-transported). How such a rotation can make itself dynamically felt, can naturally not be decided in a purely kinematical manner.

2.2. Special Flows. Redshift

With the help of the concepts introduced in the previous subsection, and with an eye on suitable applications, it is possible to introduce particular designations for some types of flows.

We speak of an *inertial flow* if the substratum elements follow the "inertial guiding field", i.e. when the matter world lines are geodesic ($\dot{u} = 0$). A *vortex-free*, or irrotational flow is characterised by $\omega = 0$, a *volume-preserving*, or isochoric one by $\Theta = 0$. If infinitesimal substratum elements remain similar to themselves under the action of the flow, for which according to (12) $\sigma = 0$ is characteristic, a *shear-free* flow (isotropic expansion or compression) is taking place. If a flow is isochoric and shear-free, thus when all the distances of neighbouring particles are constant in time, the flow is *rigid* ($\Theta = \sigma = 0$).

Symmetry properties of a flow determined by a curve family K in the space-time W (or in a region of W) are described by groups[8] of isometric (or conformal) mappings of W into itself, which transform K into itself. If ξ^a is the generating vector of a one-dimensional Lie Group G_1 and u^a the tangent vector of a family of curves K, the invariance of K under G_1 is expressed by

$$u_{a;b}\xi^b - u_b\xi^b{}_{;a} = 0. \tag{23}$$

If K is invariant under an isometry group G_1 and the trajectories of G_1 in W are timelike, the flow is called *stationary*, and in the special case of hypersurface-orthogonal trajectories, *static*. If K is stationary and $\xi^a \sim u^a$, the flow will be named *isometric*, and in the more general case when $\xi^a \sim u^a$ and G_1 is conformal, it will be called *conformal*.

The following statements are direct consequences of the definitions.

i. Inertial flows are characterised by the equation

$$\omega_{ab} = u_{[a;b]}. \tag{24}$$

[8] Among these groups we also include *local* Lie groups, which actually are most important for us.

(We display this simple consequence because, while (24) is the *formal* analogue of the equation $\vec{\omega} = \frac{1}{2} rot \vec{v}$ of classical mechanics, that analogue only holds in restricted cases. According to (18), $(\omega^{\nu}) = \frac{1}{2} rot\, \vec{v}$ always holds in the instantaneous rest-system of a substratum element, and this corresponds *in content* to the usual formula).

ii. A flow is vortex-free exactly when there are (necessarily spacelike) orthogonal hypersurfaces to the matter world-lines [3], i.e. when (locally) there exists a non-constant scalar t such that[9]

$$\dot{t} u_a = -t_{,a} \quad (\Leftrightarrow h^b_a t_{,b} = 0). \tag{25}$$

Equation (25) is at the same time characteristic of the fact that the time functions defined along the matter world lines are synchronous in Einstein's sense.[10] Such a global time coordinate t agrees with the proper-time s everywhere only in the case of inertial flows; for such vortex-free flows $u_a = -t_{,a}$ and t is a solution of the Hamilton–Jacobi equation

$$t_{,a} t^{,a} + 1 = 0 \tag{26}$$

for free particles in a gravitational field.

iii. Vortex-free flows are volume-preserving precisely when the orthogonal hypersurfaces to the streamlines are minimal [2] ($u^a{}_{;a} = 0$ is in fact the condition for the vanishing of the first variation of the volume of the hypersurfaces orthogonal to u^a).

A similar statement, useful for the geometry of static matter fields, is the following.

iv. A vortex-free flow is rigid precisely when its orthogonal hypersurfaces are totally geodesic. (Namely let $\omega = 0$ and $x^a(\lambda)$ a geodesic, $dx^a/d\lambda = t^a$. Then

$$\frac{d}{d\lambda}(u_a t^a) = u_{a;b} t^a t^b = (\sigma_{ab} + \frac{1}{3}\Theta h_{ab} - \dot{u}_a u_b) t^a t^b,$$

so the property $u_a t^a = 0$ propagates itself along exactly when $\sigma = \Theta = 0$.)

v. Conformal flows are characterized by

$$\sigma = 0, \qquad (\dot{u}_{[a} - \tfrac{1}{3}\Theta u_{[a})_{;b]} = 0. \tag{27}$$

[9] We use the fact that $u_a dx^a$ has an integrating factor precisely when $u_{[a} u_{b,c]} = 0$.

[10] This synchronisation condition is usually formulated non-covariantly and without recognition of its world-geometric meaning, see e.g. Ref. 1, pp. 259,260.

For them there exists a scalar U such that

$$\dot{u}_a - \tfrac{1}{3}\Theta u_a = U_{,a} \qquad (\dot{U} = \tfrac{1}{3}\Theta). \tag{28}$$

(The conditions (27) show that a scalar U exists such that $\xi^a \equiv e^U u^a$ obeys the conformal Killing equation $\xi_{(a;b)} = (e^U)\dot{}\, g_{ab}$.)

This implies the special case

vi. Isometric flows are characterised by

$$\sigma = \Theta = 0, \qquad \dot{u}_{[a;b]} = 0. \tag{29}$$

For them there thus exists an "*acceleration potential*"

$$\dot{u}_a = U_{,a}. \tag{30}$$

[(29) means that a Killing vector is collinear with u^a, namely $\xi^a \equiv e^U u^a$ with U according to (30)].

The relative motion of two particles and the "absolute acceleration" of each of them described by $\dot{u}_a$ show up e.g. in the *wavelength change* that occurs when one of the particles emits monochromatic light and the other receives it. We will suppose (in view of application to cosmological models) that the matter elements of the substratum we are considering send out and receive light, and wish to express $d\lambda/\lambda$ for the light that one particle receives from its neighbour, in terms of the kinematic quantities of the flow.

A light ray is described in geometric optics by a geodesic null line[11] $x^a(v)$: $k^a \equiv dx^a/dv$, $k_a k^a = 0$, $\overset{k}{\nabla} k^a = 0$. For monochromatic light k^a can and will be chosen so that, for an observer with 4-velocity u^a, $-k_a u^a = 2\pi/\lambda$ (see, for example, Ref. 5, p.16). Therefore according to (21), along a light ray in our substratum we have:

$$\overset{k}{\nabla}(-u_a k^a) = -u_{a;b} k^a k^b = -(\sigma_{ab} - \tfrac{1}{3}\Theta h_{ab} - \dot{u}_a u_b) k^a k^b.$$

Furthermore (see Figure 1)

$$\delta l^2 = h_{ab} dx^a dx^b = (u_a k^a) dv^2 = \left(\frac{2\pi}{\lambda}\right) dv^2.$$

This leads to

$$\frac{d\lambda}{\lambda} = \frac{-\lambda}{2\pi} d\left(\frac{2\pi}{\lambda}\right) = \left(\sigma_{ab} e^a e^b - \frac{1}{3}\Theta\right)\delta l - \dot{u}_a \delta_\perp x^a,$$

[11] The substratum is supposed to have the refraction index $n = 1$.

and so from (12) finally

$$\frac{d\lambda}{\lambda} = (\delta l)^{\cdot} - \dot{u}_a \delta_\perp x^a. \tag{31}$$

From this we draw two conclusions. Firstly we conclude through (12):

vii. For a substratum particle receiving light, $d\lambda/\lambda$ is independent of the direction of the infinitesimally neighbouring radiating particles, precisely when its world line L is geodesic and the shear velocity vanishes along L; then

$$\frac{\delta\lambda}{\lambda} = \frac{1}{3}\Theta\delta l. \tag{32}$$

Secondly we combine (27) with remark v., and obtain

viii. On a conformal flow λe^{-U} is constant along a lightray.[12]

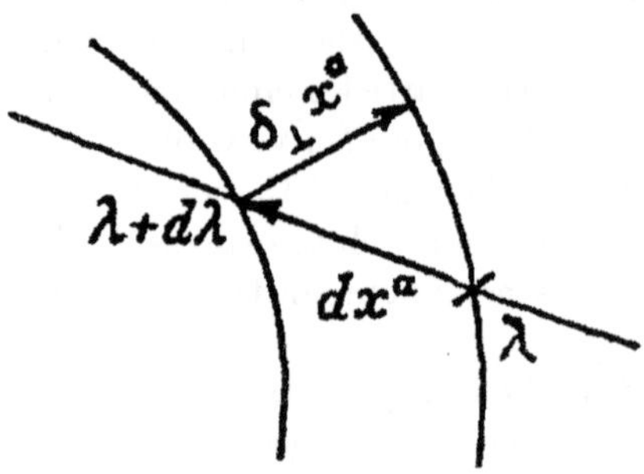

Figure 1.

The last conclusion contains the theory of redshift in stationary and static fields and in the Friedmann cosmological models (see subsection 2.5). In the first case $\Theta = 0$, $\dot{u} \neq 0$ and U (by definition) is the scalar gravitational potential,[13] and in the last case $\Theta \neq 0$, $\dot{u} = 0$ and e^U according to (28) is proportional to the radius of curvature of the space (see subsection 2.5), if in both cases the timelike orbits of the isometry (resp. conformal) group are taken as the world-lines of the substrate.

---

[12] This follows directly from the fact that the scalar product of the generating vector of an infinitesimal conformal transformation and the (parallely propagated) tangent vector of a null geodesic is constant along the geodesic.

[13] This definition is validated through (30) and proves itself both suitable and useful. In this connection see (82), Theorem 3.3.2, Theorem 3.4.4, and Ref. 32, Ch.2.

2.3. Differential Identities for the Kinematic Quantities

The derivatives of the kinematic quantities introduced in subsection 2.1 are not independent of each other. Rather, they fulfil certain identities into which inter alia the curvature tensor of space-time enters. These equations are useful for the derivation of propagation and conservation theorems, and for the investigation of the relations between field equations and families of curves.

First we determine the *relative acceleration* of neighbouring particles. We define it analogously to the relative velocity v^a [see (7)] through

$$b^a \equiv h^a_b \dot{v}^b \tag{33}$$

and obtain for it from (9), using the Ricci identity, the expression

$$b^a = (R^a{}_{bcd}u^b u^d + h^a_b \dot{u}^b{}_{;c} + \dot{u}^a \dot{u}_c)\delta_\perp x^c \tag{34}$$

which generalises the well known formula for geodesic deviation [6].

The first term on the right hand side of (34) brings to light the influence of the 'world' curvature (in Einstein's theory: the gravitational field gradient) on the relative acceleration. The tensor $R_{abcd}u^d$ contained in it can be expressed in terms of the first derivatives of the kinematic quantities ω_{ab}, σ_{ab}, Θ

$$\begin{aligned}\tfrac{1}{2}R_{abcd}u^d = {} & \omega_{c[a;b]} + \sigma_{c[a;b]} + \tfrac{1}{3}h_{c[a}\Theta_{,b]} - \dot{u}_{c;[b}u_{a]} \\ & + \tfrac{1}{3}\Theta(u_c\omega_{ab} - u_c\dot{u}_{[a}u_{b]} - u_{[a}\omega_{b]c} \\ & + \sigma_{c[b}u_{a]} + \tfrac{1}{3}\Theta g_{c[b}u_{a]}) - \dot{u}_c(\omega_{ab} - \dot{u}_{[a}u_{b]}).\end{aligned} \tag{35}$$

By contraction one obtains from this $R_{ab}u^b$ and its components $R_{ab}u^a u^b$, $h^{ab}R_{bc}u^c$. After simplification by partial differentiation, where the formulae (11),(21),(22) are used, the following formulae arise:

$$R_{ab}u^a u^b = \dot{\Theta} + \tfrac{1}{3}\Theta^2 - \dot{u}^a{}_{;a} + 2(\sigma^2 - \omega^2) \tag{36}$$

and

$$h^{ab}R_{bc}u^c = h^a_b(\omega^{bc}{}_{;c} - \sigma^{bc}{}_{;c} + \tfrac{2}{3}\Theta^{,b}) + (\omega^a{}_b + \sigma^a{}_b)\dot{u}^b. \tag{37}$$

Finally in order to express the tensor $R_{abcd}u^b u^d$ that occurs in (35) in terms of the kinematic quantities, we multiply (35) by u^b and simplify the still very complicated expression by symmetrising with respect to indices

a, c and multiplying by $h^a_e h^c_f$, which does not alter the left hand side. We obtain

$$R_{abcd}u^b u^d = \omega_a \omega_c - \omega^2 h_{ac} + \sigma_{ab}\sigma^b{}_c + \frac{\ddot{l}}{l} h_{ac}$$
$$- \dot{u}_a \dot{u}_c + h_{ab}h_{cd}(l^{-2}(l^2\sigma^{bd})^{\cdot} - \dot{u}^{(b;d)}). \tag{38}$$

Here we have also introduced a useful length scale l by

$$\frac{\dot{l}}{l} \equiv \frac{1}{3}\Theta \Rightarrow \frac{1}{3}\left(\dot{\Theta} + \frac{1}{3}\Theta^2\right) = \frac{\ddot{l}}{l}. \tag{39}$$

(l is only determined up to a scalar factor that is constant along the flow lines.[14])

While the equations (36) and (38) contain the derivatives of the quantities Θ and σ_{ab} respectively, thus far $\dot{\omega}^a$ does not occur.[15] One can however derive a propagation equation for the vorticity vector from the identity $R_{a[bc]d}u^a u^d = 0$ by some transformations, or by direct calculation from the definition (18), which is comparatively simple:

$$h^a_b(l^2\omega^b)^{\cdot} = \sigma^a{}_b l^2 \omega^b + \frac{l^2}{2}\eta^{abcd}u_b \dot{u}_{c;d} \tag{40}$$

or equivalent to this

$$h^a_b(l^3\omega^b)^{\cdot} = u^a{}_{;b} l^3 \omega^b + \frac{l^3}{2}\eta^{abcd}u_b \dot{u}_{c;d}\,. \tag{41}$$

It implies for the vorticity scalar

$$l^{-4}(l^4\omega^2)^{\cdot} = 2\sigma_{ab}\omega^a\omega^b + \omega_{ab}\dot{u}^{a;b}\,. \tag{42}$$

For completeness we give also an identity, somewhat surprising in view of classical hydrodynamics, that follows from the definition of the vorticity vector:

$$\omega^a{}_{;a} = 2\dot{u}_a \omega^a. \tag{43}$$

[14] Flow lines ≡ matter world lines.
[15] It is clear that there is no propagation equation for $\dot{u}^a$ because u^a, $\dot{u}^a$, $\ddot{u}^a$ can be given arbitrarily along a curve independently of the neighbouring curves.

2.4. Vorticity theorems and theorems for rigid motions

As a first application of the formulae collected together in the last section, we will prove some theorems that correspond to the *vorticity theorems* of classical hydrodynamics.

We designate as *vortex lines* of a flow, the (spacelike) curves that have the vorticity vector as their tangent vector; according to (8) they cut the flow-lines orthogonally.

The first result is the following lemma.

Lemma 2.4. The flow lines are geodesic with respect to the metric $\overline{g}_{ab} = w^2 g_{ab}$ $(w > 0)$ precisely when (with respect to g_{ab})

$$\dot{u}_a = -h^b_a(\log w)_{,b} \tag{44}$$

holds.

The proof follows for example from the fact that (44) represents the Euler equation for the variational problem

$$\delta \int w\, ds = 0. \tag{45}$$

According to (25) and (28), eq. (44) will hold in particular for irrotational and for conformal flows.

We recall further that two vector fields ξ^a, η^a span 2-surfaces if and only if their commutator $\xi^a{}_{,b}\eta^b - \eta^a{}_{,b}\xi^b$ is (pointwise) linearly dependent on them. From this and (41) there follows

Theorem 2.4.1. The flow- and vortex-lines span surfaces ("vortex surfaces") exactly when

$$\omega^{[b}\eta^{c]def}u_d\dot{u}_{e;f} = 0 \tag{46}$$

holds.[16]

The property stated in the theorem means that the vortex lines consist of the same substratum particles at all times ("material" conservation of vortex lines.)

Condition (46) is in particular fulfilled for inertial flows, and more generally for flows satisfying (44), since the latter equation implies that

$$\frac{1}{2}\eta^{abcd}u_b\dot{u}_{c;d} = -\frac{\dot{w}}{w}\omega^a\,. \tag{47}$$

[16] This and the following theorem agree completely with the vorticity theorems given by Synge [3], Lichnerowicz [2], Foures-Bruhat [7], Gödel [8], Schücking and Heckmann [9].

This equations permits one to simplify (40),(41) and to conclude

Theorem 2.4.2. In a flow satisfying (44) the (equivalent) propagation equations

$$h^a_b(wl^2\omega^b)^{\cdot} = \sigma^a{}_b wl^2\omega^b \quad \text{and} \quad \mathcal{L}_{w^{-1}u^b}(wl^3\omega^a) = 0 \tag{48}$$

hold, so the vector $\epsilon wl^3\omega^a$ joins neighbouring particles (for infinitesimal ϵ), and for a (thin) vortex tube the vortex strength $w\omega\delta F$ is constant in time: $(w\omega\delta F)^{\cdot} = 0$. ($\delta F$ is the surface area of a material section of the vortex tube orthogonal to ω^a).

Corollary 1. Along the world line of a fluid particle, vorticity can neither be created or destroyed.

Corollary 2. The direction of the local rotation axis is constant (in the sense of Fermi propagation) along a flow line, if and only if it lies in a principal shear direction.

Secondly, we turn to rigid flows. For them there holds

Theorem 2.4.3. In a rigid flow, u^a is constant along the vortex lines and $R_{ab}u^a u^b + \frac{1}{2}R + 3\omega^2$ is constant along the flow lines. If furthermore u^a is a Ricci eigenvector, then ω is also constant along the flow lines. If finally (44) holds and $\omega \neq 0$, then the flow is isometric [10].

Proof. Let $\Theta = \sigma = 0$. Then it follows from (21),(20) and (8) that $u^a{}_{;b}\omega^b = 0$, thus the first statement. From (37) and the definitions in subsection 2.1 there follows

$$q^a \equiv h^{ab}R_{bc}u^c = \omega^{ab}{}_{;b} + \omega^{ab}\dot{u}_b - 2\omega^2 u^a \tag{49}$$

and from this

$$q^a{}_{;a} + \dot{u}_a q^a = -2(\omega^2)^{\cdot} - \omega^{ab}\dot{u}_{a;b}. \tag{50}$$

Thus through (42) (with $l = 1$ since $\Theta = 0$)

$$q^a{}_{;a} + \dot{u}_a q^a = -3(\omega^2)^{\cdot}. \tag{51}$$

To prove the second statement, we need in addition the "conservation theorem" for the Einstein tensor $G_{ab} \equiv R_{ab} - \frac{1}{2}Rg_{ab}$. Through $\Theta = \sigma = 0$ and (21) there holds $0 = u_a G^{ab}{}_{;b} = (u_a G^{ab})_{;b} + G^{ab}\dot{u}_a u_b$. On the other hand by definition $G^{ab}u_b = q^a - u^a G_{bc}u^b u^c$, and consequently

$$q^a{}_{;a} - \dot{u}_a q^a = (G_{ab}u^a u^b)^{\cdot}. \tag{52}$$

Subtracting (51) from this, one obtains the second statement of the theorem. The third result comes from (51), since $q^a = 0$ means just that $R^a{}_b u^b$ is proportional to u^a. Finally let (44) also be true. Then (49) shows that

$$\omega^{ab}{}_{;b} - \omega^{ab}(\log\, w)_{,b} - 2\omega^2 u^a = 0 \tag{53}$$

holds. Because $\dot{\omega} = 0$, the divergence of this gives that $\omega^{ab}{}_{;a}(\log w)_{,b} = 0$. Contracting (53) with $(\log w)_{,a}$ thus gives $\omega^2 \dot{w} = 0$, so that when $\omega \neq 0$ necessarily $\dot{w} = 0$ must hold. However then according to (44), $\dot{u}_a$ is a gradient, so that according to vi. in subsection 2.2 an isometric flow occurs.

This theorem contains the noteworthy result that a rigid body in a special Einstein space ($R_{ab} = 0$) cannot change its angular velocity. This was early noticed in the case of special relativity, and was used as an argument against the validity of the concept of a rigid body in special relativity theory.

In the next theorem we will characterise the isometric flows, formally defined in vi. in subsection 2.2, by transparent kinematic properties.

Theorem 2.4.4. The isometric flows are characterised amongst the rigid flows by the fact that in them the vorticity vector is Fermi-transported along the flow lines, and the acceleration vector always points at the same neighbouring particles:

$$\Theta = \sigma = 0: \quad \dot{u}_{[a;b]} = 0 \Leftrightarrow \{h^a_b \dot{\omega}^b = 0, \quad h^a_b \ddot{u}^b = \omega^a{}_b \dot{u}^b\}. \tag{54}$$

Proof. From (40), for $\Theta = \sigma = 0$, the first of the two equivalence statements follows:

$$\begin{aligned} h^a_b \dot{\omega}^b = 0 &\Leftrightarrow u_{[a} \dot{u}_{b,c]} = 0 \\ &\Leftrightarrow \{\dot{u}_{[a,b]} = u_{[a} p_{b]} \quad \text{with } p_a = h^b_a \ddot{u}_b - \omega_{ab} \dot{u}^b.\} \end{aligned} \tag{55}$$

The second equivalence is correct, since both equations state that $\dot{u}_{[a,b]}$ is a *simple* bivector containing u_a. The value of p_a follows through contraction of the equation $\dot{u}_{[a,b]} = u_{[a} p_{b]}$ by u^b and use of the orthogonality relation $u_a p^a = 0$ and the identities $\dot{u}_a u^a = 0$ and (21). One can read off the correctness of the statement (54) from (55).

Theorem 2.4.4 contains a transparent and purely kinematic (i.e. independent of the field equations) *characterisation of stationary* and in particular static *space-times*. A universe is stationary, exactly when it is possible that (at least) one cloud of test-particles can move rigidly and with constant angular velocity under the influence of inner non-gravitational forces. For a test-particle with rest-mass m moves according to the equation $m\dot{u}^a = K^a$ $(u_a K^a = 0)$, and we can assume that in rigid motion the magnitude of the resultant force exerted on a particle by the other particles is constant and its direction with respect to the neighbouring particles does not alter. Static universes are characterised by the fact that, in them, test-body motions of the above described kind are possible without rotation.

Theorem 2.4.5. The trajectories of an isometric flow are Ricci eigenlines precisely when there exists a scalar V such that

$$\omega_a = e^{-2U} V_{,a}. \tag{56}$$

(U is defined by (30).) When that occurs, V obeys the equation

$$(e^{-4U} V^{;a})_{;a} = 0. \tag{57}$$

Proof. From the definitions of the kinematic quantities there follows the identity $h^a_b \omega^{bc}{}_{;c} + \omega^{ab}\dot{u}_b = -\eta^{abcd} u_b(\omega_{cd} - 2\omega_c \dot{u}_d)$, so we can conclude from (37) that for $\Theta = \sigma = 0$

$$u_{[a} R_{b]c} u^c = 0 \Leftrightarrow u_{[a}\omega_{b,c]} + 2u_{[a}\omega_b \dot{u}_{c]} = 0. \tag{58}$$

When we contract the equation on the right with u^a, use the supposition $\dot{u}_a = U_{,a}$ and the relation $h^a_b \dot{\omega}^b = 0$ that follows from this according to Theorem 2.4.4, and take into account $\omega_a u^a = 0$, we obtain from (58) for isometric flows the stronger equivalence

$$u_{[a} R_{b]c} u^c = 0 \Leftrightarrow \omega_{[a;b]} + 2\omega_{[a} U_{,b]} = 0 \Leftrightarrow (e^{2U} \omega_{[a})_{,b]} = 0$$

and thus the first statement of the theorem. The second statement follows directly from (56),(43) and $\dot{u}_a = U_{,a}$.

To conclude these remarks we prove a further theorem—in analogy with certain theorems from the theory of pure gravitational radiation fields[17] —which shows that curve families with specified kinematic properties in general can only exist in space-times whose conformal curvature tensor belongs to a particular *Petrov type*.

Theorem 2.4.6. In a space-time W let K be a congruence of vortex- and shear-free curves. If furthermore K is rigid or u^a is a Ricci eigenvector, then the conformal curvature tensor is of type I and has real eigenvalues, and u^a is a Weyl principal vector, or W is conformally flat.

Proof. Let $\sigma = \Theta = 0$. Then a glance at (35) shows that

$$u_{[a} R_{bc]de} u^e = 0 \tag{59}$$

and (37) reduces to

$$u_{[a} R_{b]e} u^e = 0. \tag{60}$$

[17] According to Ref. 13 and the literature cited there.

From the equation defining the conformal curvature tensor C_{abcd} and the two previous relations comes

$$u_{[a}C_{bc]de}u^e = 0 \quad \text{or} \quad -{}^*C_{abcd}u^b u^d = H_{ac} = 0. \tag{61}$$

This is however characteristic of the claimed properties. If on the other hand $\sigma = \omega = 0$ and (60) is supposed, then according to (37) $\Theta_{,a} \sim u_a$. Therefore (59) holds again, so that the proof is attained.

We remark further that under the assumptions of the last theorem, $E_{ac} \equiv C_{abcd}u^b u^d$ can be expressed in terms of R and the kinematic quantities through (36)–(38), and that the theorem generalizes a known statement about static vacuum fields [11,12].

2.5. Kinematic characterisations of cosmological models

Among the structures (W, K) consisting of a four-dimensional normal hyperbolic Riemann space W and a family K of timelike curves imbedded in it, those used in cosmological theory as models for the metric relations of the cosmos and the (average) motion of its matter, are characterised by high symmetry. We wish to derive them from the weakest and most transparent assumptions.

Definition. A flow (W, K) is called *isotropic* relative to $k \in K$ if a group G of isometric mappings of W onto itself exists with the properties

a) The mappings of G leave k pointwise fixed.

b) The mappings of G induce permutations of K.

c) G acts transitively on the directions proceeding from k, orthogonal to k.

When (W, K) is isotropic with respect to k, all tensors constructed from g_{ab} and u^a are invariant under G. Therefore along k we must have $\omega = \sigma = \dot{u}_a = 0$ and $\Theta^{,a} \sim u^a$; for otherwise preferred spatial directions would be determined by ω^a, $\sigma^a{}_b$, $\dot{u}^a$, $h^{ab}\Theta_{,b}$, which is contradictory to c).[18]

Let (W, K) be isotropic with respect to all $k \in K$. Then K is thus a vortex-free inertial flow and according to ii. in subsection 2.2 $u_a dx^a = -dt$ for a suitable scalar t. By $\Theta_{,a} \sim u_a$, $\Theta = \Theta(t)$; thus the scalar l introduced in (39) can be chosen so that $l = l(t)$. From a) and b) it follows that the (orthogonal to K) hypersurfaces $t =$ const are mapped into themselves by G. According to c) and the theorem of Schur these are hypersurfaces of constant curvature $K(t)$. Because of $\sigma = 0$ they are mapped conformally

[18] From $\sigma^a{}_a = 0$, $\sigma^a{}_b$ has at least one non-degenerate eigenvector orthogonal to u^a when $\sigma \neq 0$.

onto each other by K [see (12)]; therefore the 'expansion radius' $l(t)$, so far only determined up to a constant factor, can be chosen so that $K = \epsilon l^{-2}$ ($\epsilon = \pm 1, 0$). Since according to the definition (5) of h_{ab} the metric fundamental form G is equal to $h_{ab} - (u_a dx^a)^2$, it follows further, on use of a coordinate system (x^a) with $(x^\nu)^{\cdot} = 0$, $x^4 = t$, that

$$G = l^2(t)d\sigma^2 - dt^2, \qquad u^a = \delta_4^a \tag{62}$$

where $d\sigma^2$ is the metric (depending only on x^ν) of a 3-space of constant curvature ϵ. The (W, K) described by (62) are isotropic and furthermore homogeneous; thus there holds the

Theorem 2.5.1. (W, K) is isotropic with respect to all $k \in K$ exactly when K describes a vortex- and shear-free inertial flow in W, the orthogonal hypersurfaces R_t to K are spaces of constant curvature, and the expansion velocity is constant along R_t; a normal form for (g_{ab}, u^a) is (62). In an isotropic flow the curves of K are Ricci eigenlines.

In the literature the models (62) are usually derived, following Robertson [14], from the following postulates: i) $\dot{u} = 0$, ii) $\omega = 0$, iii) isotropy, iv) homogeneity. Occasionally it is hinted that iv) is superfluous. Our proof displayed above shows that i) and ii), and hence the existence of a cosmic time ('Weyl's Postulate', according to Refs. 14 and 15), also follow from iii), and without any assistance from group theory. The proof given here can be regarded as the completion of the sketchily presented considerations of Einstein [16].

It should still be remarked that an isotropic flow is conformal, since $u_a = -t_{,a}$, $\sigma = 0$ and $l = l(t)$ (according to v. in subsection 2.2), with lu^a as generating vector, so that by viii. in subsection 2.2, λ/l is constant along a light ray, and thus by (9),(10) respectively according to (34),(38) the local relative velocity, respectively relative acceleration field are given by

$$v^a = \frac{\dot{l}}{l}\delta_\perp x^a, \qquad b^a = \frac{\ddot{l}}{l}\delta_\perp x^a. \tag{63}$$

Kinematically distinguished special cases characterised by Theorem 2.5.1 are the following models:

a) Those that are static ($\dot{l} = 0 \Rightarrow v^a = 0$), which (independently of Theorem 2.5.1) are characterised by $\dot{u} = \omega = \sigma = \Theta = 0$ ($\Leftrightarrow u_{a;b} = 0$),

b) The de Sitter ($\Leftrightarrow$ 'Steady State') model ($L \sim e^{Ht}$, $\epsilon = 0$) characterised by stationary expansion, that is $\Theta \neq 0 = \dot{\Theta} = \dot{K}$),

c) the Milne model ($\epsilon = -1, l \sim t \Rightarrow b^a = 0$), characterised by flatness of W.[19]

Now we wish to characterise the simplest *models with a rotating substratum*.

Theorem 2.5.2. i) Suppose (W, K) describes an inertial flow with the following properties: (a) In the infinitesimal spacelike neighbourhood of each substratum element, the direction of the rotation axis is constant, i.e. $(\omega^a/\omega)_{;b}\delta_\perp x^b \equiv 0$. (b) The rotation axis is a main shear direction. Then (g_{ab}, u^a) can be transformed to the normal form

$$\begin{aligned} &u^a = \delta^a_4, \\ &G = dz^2 + \gamma_{AB}(x^A, t)dx^A dx^B - (dt + u_A(x^B)dx^A)^2 \quad (A = 1, 2) \end{aligned} \tag{64}$$

and this normal form is characteristic of the stated properties.

ii) Suppose (W, K) has the properties named in i) and also $\sigma = 0$. Then $\Theta = 0$, in (64) $\gamma_{AB,t} = 0$, and ω is constant in the vortex surfaces.

iii) If additionally to ii) u^a is a Ricci eigenvector, then ω is constant (and with this $\omega_{a;b} = 0$).

Proof. i) From $\dot{u} = 0$, (b) and Theorem 2.4.2 follows $(\omega^a/\omega)^{\cdot} = 0$; from this and from (a) it follows that $(\omega^a/\omega)_{;b} = 0$. Then W decomposes directly, that is $G = dz^2 + G'(x^A, t)$ with $\omega^a = \omega\delta^a_3$ $(z \equiv x^3)$. Through $u^a\omega_a = 0$, thus $u^3 = 0$, the natural projections of the flow-lines in the 3-space (x^A, t) are geodesic with respect to G', and thus (64) can be attained. The converse is trivial.

ii) From the normal form (64) one can read off that the neighbouring particles with equal x^A are rigidly connected, so that from $\sigma = 0$ and using (12) $\Theta = 0$ and from this further follows $\gamma_{AB,t} = 0$. According to Theorem 2.4.2, ω^a is parallelly transported along the flow lines, thus $\omega_{,a}u^a = 0$. Furthermore according to (43), $0 = (\omega^a/\omega)_{;a} = \omega_{,a}\omega^a$. Consequently $\omega = \omega(x^A)$.

iii) According to Theorem 2.4.5 now $\omega_a = \omega z_a$ is a gradient, thus $\omega = \omega(z)$. On the other hand $\omega = \omega(x^A)$; the two together give $\omega = \text{const.}$

We will use this theorem later to make possible a simple derivation of the Gödel cosmological model.

In the case of rigid inertial flows through (9),(10) and (34),(38) the relative velocity and acceleration are given by the relations

$$v^a = \omega^a{}_b\delta_\perp x^b, \qquad b^a = \omega^a\omega_b\delta_\perp x^c - \omega^2\delta_\perp x^a \tag{65}$$

[19] $R_{ab} = 0$ is sufficient, since all models (62) are conformally flat because of the existence of a three-dimensional isotropy group, see e.g. Ref. 12, p.43.

of elementary kinematics. According to viii. in subsection 2.2 no change of the frequency of light signals occurs in such motions (even in the case of distant particles). The transversal Doppler effect is here compensated by the gravitational 'redshift'. (In the Minkowski universe rigid rotational flows are necessarily non-geodesic, so that such a compensation cannot take place, cf. (36).

3. DYNAMICS

3.1. The Gravitational Field equations. Similarity law.

The local equivalence of inertial and gravitational forces, empirically proved with great accuracy, but only formulisable through an ad hoc hypothesis in the framework of the conceptual system of Newtonian mechanics and gravitational theory, led Einstein to the supposition that both these "forces" are expressions of one and the same structural property of the space-time manifold, for which Weyl coined the expression *guiding field*. The local validity of the special theory of relativity makes necessary that this structure is mathematically represented by a normal hyperbolic Riemannian metric.[20]

Since on the one hand matter generates the gravitational field and the Poisson equation describes well what is experienced, and on the other hand in the new theory g_{ab} are the state variables for the guiding field, and since in special relativity theory the mechanical properties of matter are described through the *energy-momentum-pressure tensor* T_{ab}, it is a reasonable assumption that

(1) The components T_{ab} of the matter tensor are equal to quasi-linear differential expressions of the second order in the components g_{ab} of the metric tensor.

According to a theorem proved by Weyl (Ref. 18, Appendix II), it follows from this very general assumption alone that the field equations have the form[21]

$$G_{ab} + (MG - \Lambda)g_{ab} = -\kappa T_{ab} \tag{66}$$

where Λ, M and κ ($\neq 0$) are constants with dimensions[22] $-2, 0, 0$ respectively.

This equation is *non-linear* in the g_{ab}, that is the assumption (1) already contains that the metric field mediates an *interaction* between matter-filled world-tubes.

20 The presentation of the foundations of Einstein's gravitational theory given here follows—with some modification and supplementation—Refs. 18 and 19.

21 $G_{ab} \equiv$ Einstein Tensor, $G \equiv G^a{}_a$; see our conventions given above.

22 With respect to length; $c \equiv 1$, $\gamma \equiv \frac{1}{8\pi}$, according to our conventions.

The symmetry of the matter tensor does not have to be introduced here as an independent hypothesis (equivalence of momentum- and energy-density) as in special relativity, rather it follows—if one considers postulate 1) as an implicit definition of T_{ab}—from the Weyl Theorem mentioned above: namely there exists in a Riemann space no quasi-linear, invariant bivector of 2nd order in the metric g_{ab}. In General Relativity theory, *the symmetry of the matter tensor is a necessary condition, in order that it can occur as a source of the metric field in the sense of (1).*

From (66) it follows that $-G(1+4M)+4\Lambda = \kappa T$. Therefore since $T=$ const. is not always true, $1+4M \neq 0$. From this one obtains from (66) on using the Bianchi identities the divergence equation

$$T^b{}_{a;b} = eT_{,a} \qquad \left(e \equiv \frac{M}{1+4M}\right). \tag{67}$$

For the particularly simple case of matter only interacting through gravity, with its proper rest mass density of ρ and u^a its four-velocity, according to the fundamental ideas of the theory sketched above one can set (see subsections 3.4 and 3.5)

$$T_{ab} = \rho u_a u_b \qquad (\rho > 0). \tag{68}$$

Equations (67) is then equivalent to the system of equations

$$(\rho u^a)_{;a} = e\dot{\rho}, \qquad \dot{u}_a = -eh^b_a(\log\rho)_{,b}\,. \tag{69}$$

The assumptions introduced this far thus lead to a theory in which incoherent matter moves on the geodesics of the metric $\overline{g}_{ab} = \rho^{2e}g_{ab}$ (Lemma 2.4) and in which spontaneous rest-mass creation takes place with the production density $e\dot{\rho} = (e/e-1)\rho\Theta$.

If one introduces either the *additional assumption*

(2a) *The rest mass of matter that interacts only by gravity is conserved*,

or

(2b) *The flow lines of matter that interacts only through gravity are geodesics* with respect to the 'natural' metric g_{ab},

it follows that $e = M = 0$, and we obtain from (66) the *Einstein Field Equation*

$$G_{ab} - \Lambda g_{ab} = -\kappa T_{ab} \tag{70}$$

and from this the so-called *conservation theorem*

$$T^{ab}{}_{;b} = 0 \tag{71}$$

that for (68) again leads to (2a) and (2b) as consequences.

Since at the present time there exists no reason to reject the simple assumption (2a) and independently, the hypothesis (2b) fulfils the idea of the guiding field in the simplest manner [26], we take as a basis in the further development of this work the classic equation (70).

The more general theory with $e \neq 0$ can easily be developed and leads for $|e| \ll 1$ to a theory that is similar to Jordan's 'extended gravitational theory' (Ref. 25, books 2,3) and to Hoyle's cosmological theory (Refs. 15,27 and the works cited there), but is simpler and, it seems to us, better grounded from the viewpoint of local physics (see above); we introduce no new hypothesis, rather omitting only one of the axioms of the Einstein theory, and indeed the mathematically least important one. It is easy to see that this theory is incompatible with a stationary state in the sense of the PCP of Bondi and Gold [26]. We remark further that in the theory with $e \neq 0$, (71) holds for a Maxwell field, so that the theory of light propagation remains unaltered.

Through consideration of the spherically symmetric solutions[23] of (70), it turns out (with our choice of units) that

$$\kappa = 1 \tag{72}$$

must be supposed. The positivity of κ expresses the fact that *matter attracts gravitationally rather than repels.*

The *cosmological term* Λg_{ab} in (70) can be excluded ($\Lambda = 0$) through either of the following two assumptions:

(3a) Apart from the speed of light and the Newtonian gravitational constant, the field equation contains no dimensional constants.

(3b) In a vacuum, the trace of the tensor which transforms the relative position vector of neighbouring free particles into their relative acceleration has vanishing trace[24] (corresponding to the Laplace equation of Newtonian Theory).

Although we regard the specialisation $\Lambda = 0$ as appropriate on methodological grounds as long as experience does not indicate $\Lambda \neq 0$, for mathematical generality we will set $\Lambda = 0$ only when this restriction permits derivation of theorems that otherwise would not be valid.

---

[23] Or through (82) or alternatively (123), where to begin with κ is retained and then its value determined through correspondence with the corresponding equations of Newtonian Theory.

[24] See (34) for $\dot{u} = 0$. (3b) is Pirani's characterisation of special Einstein spaces, see [28].

According to the laws of classical hydrodynamics (with gravitational interaction) for every flow there exists a two-parameter family of similar flows. This is closely connected to the fact that in the basic equations there enters only one universal constant with dimensions, the gravitational constant, so that only one of the three basic dimensions of length, mass and time (let us say the mass) can be eliminated by a 'rational' choice of units. In relativity theory, already kinematics contains a constant, namely c; one can thus expect a one-parameter family of similar flows to a given flow for $\Lambda = 0$, and none for $\Lambda \neq 0$.

In fact the field equation (70) shows that for $\Lambda = 0$ (and only then) out of each solution (g_{ab}, T_{ab}) a family of similar solutions $(\overline{g}_{ab}, \overline{T}_{ab})$ can be obtained through $\overline{g}_{ab} = m^2 g_{ab}$, $\overline{T}_{ab} = m^2 T_{ab}$, $(m > 0)$. The space-time $\overline{W}$ is equivalent to W, only if W admits a homothetic mapping onto itself.

In a classification of exact solutions of (70) with $\Lambda = 0$ it is useful to arrange the solutions into families that are similar to each other, and to choose as characterising data for a particular solution a 'similarity parameter' (proportional to the above m), together with such invariants or properties as characterise the whole family.[25]

For example within the Schwarzschild family of vacuum solutions the gravitational radius is such a similarity parameter, and the family as such is characterised by the property of 'spherical symmetry'. Examples of when the above similarity transformation leads to equivalent worlds are provided by the Levi-Civita static cylindrically symmetric vacuum fields.

3.2. Conservation of rest mass, energy theorem and equation of motion for ponderable matter

To determine if and in what sense the divergence equation (71) also describes the conservation of energy and momentum in gravitational fields ($R_{abcd} \neq 0$), one can either convert it (or the corresponding equation for the tensor density $(-g)^{1/2} T^{ab}$) into an ordinary divergence equation by addition of an affine tensor and thus set up for finite volumes two local conservation balances containing as well as material also gravitational energy and momentum—which as is known, on many grounds has not yet led to clear insights—or introduce an *average four-velocity* u^a for the matter and set up substantial balances related to matter elements. We will briefly describe the second procedure, corresponding to thermo-hydrodynamics.[26]

---

[25] To the latter belong for instance the Petrov type of the conformal tensor and the structure tensor of an isometry group.

[26] We follow essentially Eckart [20], whose results we complete through the 'Navier–Stokes' equation (76).

First one can uniquely decompose a symmetric tensor field T_{ab} with respect to an arbitrary timelike unit vector u^a thus:

$$T_{ab} = \mu u_a u_b + p h_{ab} + 2u_{(a}q_{b)} + \pi_{ab} \tag{73}$$

with h_{ab} according to (5), and

$$q_a u^a = 0, \qquad \pi_{ab}u^b = 0, \qquad \pi^a{}_a = 0. \tag{74}$$

By contraction of (71) with u_a, partial differentiation and using the kinematic quantities (defined in 1.1) formed from u_a, one can derive the equation

$$\dot{\mu} + (\mu + p)\Theta + \pi_{ab}\sigma^{ab} + q^a{}_{;a} + q_a\dot{u}^a = 0. \tag{75}$$

Correspondingly the 'spacelike' part of the equation (71) gives

$$(\mu + p)\dot{u}_a + h^b_a(\dot{q}_b + p_{,b} + \pi^c_{b;c}) + (\omega_{ab} + \sigma_{ab})q^b + \tfrac{4}{3}\Theta q_a = 0. \tag{76}$$

This pair of equations is, for a given decomposition (73), equivalent to the relation (71).

The equations (70) and (74),(75) attain a concrete physical meaning only when the nature of the matter generating the field is stated and the introduced auxiliary quantities are subject to corresponding equations of state.

For bulk matter it is permissible to assume that for a uniquely determined 4-velocity[27] u^a, the main part of the energy density μ is the positive *rest-mass proper density* ρ that remains conserved,

$$(\rho u^a)_{;a} = 0, \left(\Leftrightarrow \rho\dot{v} = \Theta, \quad v \equiv \frac{1}{\rho} = \text{ specific volume}\right). \tag{77}$$

The quantity $u = \mu/\rho$ (which in general differs but little from 1) is the *specific internal energy*. Under this assumption we can designate the relation [following from (74),(76)]

$$\rho\frac{du + pdv}{ds} + q^a{}_{;a} + \pi_{ab}\sigma^{ab} + q_a\dot{u}^a = 0 \tag{78}$$

as the *substantial thermodynamic energy balance*, the equation (76) as the *momentum balance* or *equation of motion*,[28] and the quantities p, q^a and

[27] The local 'barycentric' velocity, see e.g. [21].

[28] Lichnerowicz [2] has, through use of the auxiliary metric referred to in Lemma 2.4, proposed another relativistic equation of motion which we consider as physically unfounded.

π_{ab} respectively as the *pressure, energy-flow density* relative to the matter (arising from transport processes such as diffusion and heat conduction), and *viscous pressure* tensor.

These equations differ from the corresponding non-relativistic ones only through the very small terms $q_a\dot{u}^a$ respectively $h^b_a\dot{q}_b + (\omega_{ab} + \sigma_{ab} + \frac{4}{3}\Theta h_{ab})q^b$, necessitated by the inertia of energy, and through the fact that in (76), $(\mu + p)$ occurs as the *effective density of the inertial mass* instead of ρ.

From the definition (39) of l and (77) it follows that the conservation of rest mass can also be written in the particularly transparent form

$$\rho l^3 = \text{const. (along the flowlines).} \tag{79}$$

From the definition of the Einstein tensor, (70),(73) and (74) follows

$$R = 3p - \mu - 4\Lambda, \quad R_{ab}u^a u^b = \Lambda - \tfrac{1}{2}(\mu + 3p), \quad h^{ab}R_{bc}u^c = q^a. \tag{80}$$

Thus *the matter flow-lines are Ricci eigenlines, exactly when there is no energy flow relative to the matter.*

The last remark and (80) permit a dynamic interpretation of a statement in subsections 1.4 and 1.5.

From (36),(39) and (80) results the equation

$$3\frac{\ddot{l}}{l} + 2(\sigma^2 - \omega^2) - \dot{u}^a{}_{;a} + \frac{1}{2}(\mu + 3p) - \Lambda = 0. \tag{81}$$

This shows that in rigid motions, the magnitude $-\Lambda + \frac{1}{2}(\mu + 3p) - 2\omega^2$ is the source density of the acceleration field. This statement can be interpreted as the *relativistic analogue of the Gauss theorem* $\mathrm{div}\mathbf{F} = -\frac{1}{2}\rho + 2\omega^2$ ($\mathbf{F} \equiv$ the field strength of the gravitational and centrifugal fields in a rigidly rotating system).[29] For isometric flows one obtains via (30) the 'Poisson' equation

$$U^{;a}{}_a = -\Lambda + \tfrac{1}{2}(\mu + 3p) - 2\omega^2 \qquad (\dot{U} = 0). \tag{82}$$

Reformulating this as a three-dimensional divergence equation suggests designation of $(\mu + 3p)$ as the *effective density of the active gravitational mass* [28].

29 For the special case $\omega = 0$ compare [30].

3.3. Entropy Theorem. Ideal fluids. Isentropic flows.

An exact description of matter requires in addition to the concepts and laws introduced in the previous section further thermodynamic considerations, that will not be carried out here in full generality; for the relativistic thermodynamics of irreversible processes we refer to the beautiful *Handbuch* article by Meixner and Reik [21], the literature cited there, and to Just [22].

We will only sketch one example in anticipation of the next section, and on account of the soon to be handled case of an ideal fluid: the *one component fluid* (following Ref. 20).

For such a fluid we require first the existence of a (thermal) *equation of state*

$$u = u(p, v). \tag{83}$$

Consequently there exist (scalar) functions $T(p, v)$ and $s(p, v)$ such that

$$du + pdv = Tds. \tag{84}$$

With this, on referring back to (77) we can transform the energy equation (78) into

$$\rho\dot{s} + \left(\frac{q^a}{T}\right)_{;a} = \left(\rho s u^a + \frac{q^a}{T}\right)_{;a} = \eta,$$
$$\eta \equiv -\frac{1}{T}\left(\pi_{ab}\sigma^{ab} + q_a(\dot{u}^a + (\log T)^{,a})\right). \tag{85}$$

Under the further assumption $T > 0$ it is suggested to interpret T as the *temperature*, s as the *specific entropy*, and this equation as the *entropy balance* equation, which—in agreement with the meaning of π_{ab}, σ_{ab} [see (12)], q^a and T—leads to the phenomenological equations

$$\pi_{ab} = -\lambda\sigma_{ab} \qquad (\lambda(p, v) \equiv \text{ viscosity coefficient}), \tag{86}$$
$$q_a = -\kappa h_a^b(T_{,b} + T\dot{u}_b) \quad (\kappa(p, v) \equiv \text{ heat conduction coefficient}). \tag{87}$$

The entropy theorem demands that the *entropy production density* η is non-negative, and consequently

$$\lambda \geq 0, \qquad \kappa \geq 0 \tag{88}$$

must be true. The 4-vector

$$S^a \equiv \rho s u^a + \frac{q^a}{T} \tag{89}$$

is to be designated as the *entropy flow density*. It appears in (89) separated into a convection part $S^a \equiv \rho s u^a$ and a conduction part q^a/T, and always has non-negative divergence.

In the equations (75),(76) the terms containing π_{ab} and q^a are often very small in relation to the others; consequently one can then, in particular when studying the interaction of matter with the gravitational field (taken into account implicitly through the covariant derivatives), ignore these terms. This leads to the introduction of the model of an *ideal fluid* defined in view of (86),(87) by

$$\lambda = 0, \qquad \kappa = 0. \tag{90}$$

For an ideal fluid therefore according to (73) and (5),

$$T_{ab} = \mu u_a u_b + p h_{ab} = (\mu + p) u_a u_b + p g_{ab}. \tag{91}$$

(Since $-u^a u_b$ and h^a_b are projections orthogonal to each other, the first equation (91) is the spectral decomposition of T^a_b.)

The energy and momentum laws simplify to

$$\dot{u} + p\dot{v} = 0, \tag{92}$$

$$(\mu + p) + h^b_a p_{,b} = 0, \tag{93}$$

and for a chemically homogeneous ideal fluid[30] according to (94), always

$$s = \text{ const. along the flow lines;} \tag{94}$$

all processes in them proceed (adiabatically) reversibly. The flow-lines are Ricci eigenlines, by (91), (77).

According to the rules of thermodynamics, the *specific enthalpy*

$$w = w(s, p) = u + pv = \frac{\mu + p}{\rho} \tag{95}$$

can be used as a thermodynamic potential; through elimination of s by use of the equations

$$\frac{\partial w}{\partial s} = T, \qquad \frac{\partial w}{\partial p} = v \tag{96}$$

[30] During chemical reactions (84) must be replaced by the Gibbs fundamental equation, and in general (94) is no longer valid.

one obtains the thermal equation of state $T = T(p, v)$ and further the caloric equation of state (83). According to (95) and (96) along an isentrope

$$w(s,p) = w(s,p_0)\exp\left(\int_{p_0}^{p}\frac{dp}{\mu+p}\right). \tag{97}$$

A flow is called *isentropic* when, even in the whole domain of the flow, $s =$ const. (compare Ref. 4 as well as Ref. 38). In an isentropic flow, μ, p and ρ are pairwise functionally dependent, and according to (97) and (95)

$$\frac{dw}{w} = \frac{dp}{\mu+p}, \qquad \frac{dp}{dw} = \frac{\mu+p}{w} = \rho. \tag{98}$$

The equation of motion (93) is thus equivalent to

$$\dot{u}_a = -h_a^b(\log w)_{,b}$$

[eq. (44) above]. From this we can extract

Theorem 3.3.1. In an isentropic flow of a one-component ideal fluid, the pressure is a function of the specific enthalpy. The matter tensor, rest mass density and specific energy are determined by the formulae

$$T_{ab} = \rho w\, u_a u_b + p g_{ab}, \qquad \rho = \frac{dp}{dw}, \qquad u = w - \frac{p}{\rho} \tag{99}$$

(with $w > 0$, $(dp/dw) > 0$). The divergence of T_{ab} vanishes precisely when $(\rho u^a)_{;a} = 0$ and the streamlines are geodesics[31] of the auxiliary metric $\overline{g}_{ab} = w^2 g_{ab}$. The energy theorem (92) is identically fulfilled as a consequence of the last two equations (99).

Instead of the variables (w, u^a) and the function $p(w)$ one could also take (p, u^a) with $\mu(p)$ (e.g. Refs. 3,2) or with $\rho(p)$ (e.g. Refs. 19,20); however the representation in the theorem is the simplest because of (44). We mention also the often used (e.g. Refs. 3,2) representation following from (98),(99):

$$T_{ab} = \rho\left(\overset{0}{w} + \int_{p_0}^{p}\frac{dp}{\rho}\right) u_a u_b + p g_{ab} \quad (\overset{0}{w} \equiv w(p_0),\ \rho = \rho(p)) \tag{100}$$

31 Eisenhart [24] recognised this geodesy in another way. Through the formal analogy between (45) and the Fermat principle of geometric optics, Synge [3] designated w, whose thermodynamic meaning he did not mention, as the 'index' of the fluid.

in which usually the unnecessary and often impossible specialisation $p_0 = 0$, $w = 1$ is presupposed.

According to Theorem 2.4.1 and (47) the flow- and vortex-lines span vortex surfaces [2], and from Theorem 2.4.2 there follows for isentropic flows the dynamic vortex theorem (48) (with the present meaning for w); (97) shows the influence of the pressure on the vorticity.

In an isentropic flow, $\Theta = 0$ has the consequence that w is constant along the flow lines, so that with (44) and according to Lemma 2.2.6 there holds the following theorem.

Theorem 3.3.2. A rigid isentropic flow is only possible in a stationary gravitational field [4]. Then $\xi^a = u^a/w$ is a Killing vector and $U = -\log w$ the scalar gravitational potential [defined in 2.2.6 and occurring in (82)]. The vortex vector ω^a is (Fermi-) constant along the flow lines (Theorem 2.4.1) and satisfies the equations (56),(57).

In conclusion we remark that according to Theorem 2.4.6, a vortex- and shear-free isentropic flow is only possible in a gravitational field whose conformal tensor is of Type I and has real eigenvalues. This corresponds to the conception that such fields contain no gravitational waves, which are apparently always tied in with shearing motions of the matter that interacts with the field (according to Refs. 31 and 32, sec.2–4.8).

3.4. Kinetic gas theory. H-Theorem. Equilibrium distributions in a gravitational field.

In this section we will not idealise the matter from the beginning as a continuum, but rather describe it through the model of a statistical assembly of particles of rest mass m that only interact with each other through elastic collisions. We take it that the space-time in a 'physically infinitely small region' can be regarded as flat, as is required for the definition of the moments of the distribution function and in particular for the formulation of the Boltzmann collision equation, although the 'gas' as a whole may find itself in a gravitational field that then (expressed in analogy to electron theory) can be regarded as the 'macroscopic' field.

To define the (one-particle) *distribution function* $F(x,p)$[32] we think of an infinitesimal space element characterised by the (pseudo-) vector $d\overset{*}{x}_a$ at the (arbitrary) event x, and in momentum space[33] an infinitesimal cell at the point p^a on the mass-hyperboloid

$$p_a p^a = -m^2, \tag{101}$$

[32] We leave out the indices of the arguments: $p \equiv (p^a)$ and so on.
[33] This can be identified with the tangent space at x.

to which then analogously a $d\overset{*}{p}_a$ belongs. F is defined by the statement that

$$F(x^a, p^a)|d\overset{*}{x}_a d\overset{*}{p}^a| \tag{102}$$

is the number of those particles whose worldlines intersect the space- element $d\overset{*}{x}_a$ and whose momentum lies in the cell $d\overset{*}{p}_a$. Clearly F is a scalar.

The function F obeys the relativistic *Boltzmann equation* formulated by Sasaki (Ref. 33; see also Ref. 37)

$$p^a F_{,a} = \int\int\int (F''F''' - FF')\, W(p, p'; p'', p''')\, dP' dP'' dP''' . \tag{103}$$

The partial derivative $F_{,a}$ is to be thought of as relative to x^a with parallelly propagated p^a. The first argument in F'', F''', F, F' is always x, the second respectively p'', p''', p, p'; furthermore we set

$$d\overset{*}{p}_a \equiv p_a dP \tag{104}$$

and so on. $W(p, p'; p'', p''')$ describes the *probability* for collisions $p, p' \to p'', p'''$ (with corresponding phase space ranges). We have assumed

$$W(p, p'; p'', p''') = W(p', p; p''', p'') = W(p'', p'''; p, p'). \tag{105}$$

For the *first moment*

$$g^a(x) \equiv \int g(p) p^a F(x, p) dP \tag{106}$$

of an arbitrary function $g(p)$ there follows from (103),(105),

$$g^a{}_{;a} = \frac{1}{4} \int\int\int\int (g + g' - g'' - g''') \times \times\, (F''F''' - FF') W dP dP' dP'' dP''' . \tag{107}$$

This *balance equation* holds analogously also for tensors $g^{ab\cdots}$ instead of g.

When g is an additive collision invariant, there holds therefore $g^a{}_{;a} = 0$. From this follows:

(i) The *material flow density*

$$\rho u^a \equiv \int m p^a F dP \quad (u_a u^a = -1) \tag{108}$$

$(g = m)$ obeys the continuity equation

$$(\rho u^a)_{;a} = 0. \tag{76}$$

(ρ is according to (102),(104),(108) the density of the restmass in the local rest system determined by the four-velocity u^a.)

(ii) The *matter tensor*

$$T^{ab} = \int p^a p^b F dP \tag{109}$$

$(g \to p^a)$ is divergence-free:

$$T^{ab}{}_{;b} = 0. \tag{71}$$

Next we define through

$$S^a \equiv - \int p^a F \log F dP \tag{110}$$

the *entropy flow density* of our gas. There holds

$$S^a{}_{;a} = - \int (1 + \log F) F_{,a} p^a dP,$$

from which by the same transformation that leads to (107),

$$S^a{}_{;a} = \frac{1}{4} \int\int\int\int \log \left(\frac{F'' F'''}{F F'} \right) (F'' F''' - F F') W dP dP' dP'' dP''' \tag{111}$$

arises. Thus there holds the following covariant relativistic form of the Boltzmann H-Theorem:

Theorem 3.4.1. The entropy production density

$$S^a{}_{;a} \equiv \eta$$

is non-negative and only vanishes in a world region if there at every point F is an additive collision invariant.[34]

If on the grounds of the relativistic collision laws we make the very plausible assumption that (corresponding to the Grad theorem [34] in the

[34] Therefore in special relativity the 'total entropy at a fixed time' is only a scalar when an equilibrium distribution is at hand.

non-relativistic theory) the general collision invariant has the form $c - \xi_a p^a$, it follows from Theorem 3.4.1 that the general equilibrium distribution of a relativistic gas has the form

$$\overset{0}{F}(x,p) = C(x)\exp(\xi_a(x)p^a). \tag{112}$$

From this and (108),(109) there follows

Theorem 3.4.2. (Ref. 35, sec.14, where. further references are given.) For local equilibrium, the distribution function of an ideal gas has the form

$$\overset{0}{F}(x,p) = \frac{\rho(x)\xi}{4\pi m^3 K_2(m\xi)} \exp(\xi u_a p^a). \tag{113}$$

The corresponding matter tensor is equal to that of an ideal fluid (91) with the equation of state

$$u = G\left(\frac{1}{pv}\right) - pv, \tag{114}$$

where $G(y)$ is given by

$$G(y) \equiv \frac{2}{y} - \frac{K_2'(y)}{K_2(y)}, \quad K(y) \equiv \int_0^\infty \exp(-y\cosh\, z)\cosh(2z)dz. \tag{115}$$

The partition parameter ξ is related to the density and pressure by

$$m\xi = \frac{\rho}{p}. \tag{116}$$

We now determine what conditions the Boltzmann equation (103) imposes on the state variables $\rho(x)$, $\xi(x)$ and $u^a(x)$.

Theorem 3.4.3. An equilibrium distribution (113) is possible only in a stationary gravitational field, and indeed $\xi^a = \xi u^a$ must be a Killing vector; thus the flow is isometric. Furthermore in the flowfield

$$\frac{p\xi^2 x}{K_2(m\xi)} \equiv A = \text{ const.} \tag{117}$$

must hold. These conditions are sufficient, too.

Proof. For an equilibrium distribution (112), because of four-momentum conservation at collisions, (103) reduces to the Liouville equation

$$p^a \overset{0}{F}_{,a} = 0, \tag{118}$$

which is thus written out with $\log C = c$ to

$$c_{,a}p^a + \xi_{a;b}p^a p^b = 0. \tag{119}$$

This equation must be fulfilled at every point x for all p^a that lie in the forwards lightcone, with $p_a p^a = -m^2$. From this follows (for example by choosing some special values for the p^a)

$$c_{,a} = 0, \qquad \xi_{a;b} = 0\,, \tag{120}$$

whereby with (112),(113),(116) the theorem is proved, since from (120) naturally again (118) follows.

Comparison of this theorem with the corresponding theorem of the non-relativistic theory (Ref. 39 and cf. Ref. 36) shows:
The reversible flows of a relativistic ideal gas are kinematically more strongly restricted than those of a non-relativistic one; they are namely not only shear-free, but also volume-preserving. Thus, in the isotropically expanding models of relativistic cosmology, one cannot understand the substratum—as in the corresponding Newtonian models [40,41]—as an ideal gas in equilibrium.[35]

Next we decompose the entropy flow (110) following the example of equation (89) into a convection and a conduction part:

$$S^a \equiv \rho s u^a + s^a \qquad (u_a s^a \equiv 0). \tag{121}$$

For an equilibrium distribution, s^a vanishes according to (110),(112), and one obtains an expression for s after a short calculation (in the rest frame), that is connected to u from (114) through $\xi(du + pdv) = ds$. Thereby $1/\xi \equiv T$ is identified as the *temperature* and the connection with the phenomenological theory of the previous section recovered—at least in the equilibrium case.

The relation $\xi = 1/T$, Theorem 3.4.3, and the remark 2.2.vi give

Theorem 3.4.4. For an ideal gas in thermodynamic equilibrium in a gravitational field, there exists the relation

$$Te^U = \text{const.} \tag{122}$$

between temperature T and the gravitational potential U.

[35] Thereby a question posed by Heckmann, Ref. 40 p.50, is answered at least in a simple case.

Tolman [42] derived the relation (122) for a (phenomenologically described) isotropic radiation field in a static gravitational field; for the general stationary case this is contained in Theorem 3.3.2, since for the "fluid" considered there one can take isotropic radiation, which is permitted.

It should be pointed out that in the case of a rotating gas, the centrifugal potential is also contained in U; the temperature difference for this case asserted by (122) is of the same order of magnitude as that recently measured by Pound and Rebka [43] through the transverse Doppler effect by means of the Mössbauer effect.

Through the use of distribution functions that differ from (113) only by such a small amount that we can still set $\log F \simeq \log \overset{0}{F}$, one can still through the definitions of q^a [(73),(108),(109)] and s^a [(121),(110)] determine that $s^a = (1/T)q^a$. Thereby then the phenomenological entropy theorem [(85) with $\eta \geq 0$] is recognized as a consequence of the statistical H-theorem 3.4.1. The phenomenological equations (86),(87) have been motivated statistically by Sasaki [33] (without use of the entropy concept); his equations appear to us however not to be sufficiently general.

We remark further, that the assumptions made by Synge about the direction of time (Ref. 35, p.54) in handling of shockwaves in gases, follow from the H-theorem. It should however be stressed that because of Theorem 3.4.3, the use of equations (114)–(116) in general flows is strictly speaking not permitted.

3.5. Dynamics of incoherent matter.

Incoherent matter is formally defined by the matter tensor

$$T_{ab} = \rho u_a u_b. \tag{68}$$

One can obtain this expression if in the general representation (73) one regards the pressure p, the eigenvalues of the viscosity tensor π_{ab}, and the total energy-flow density q^a as negligibly small in comparison with the energy density μ [and then sets $\mu = \rho$ through (77)]. If one takes as a foundation the statistical model of matter handled in the previous section, formulae (108),(109) or the distribution function (113) show that (68) emerges exactly when F has a sharp maximum in momentum space at the point $p_a = m u_a$ (the limiting case $T \to 0$.).

In this subsection (68) will be taken as given.

Almost all the previously formulated theorems can be applied to (68); we collect some together here.

i. The streamlines are geodesic: $\dot{u} = 0$.
ii. The (proper) mass density is conserved: $\rho l^3 = M$, $\dot{M} = 0$.

iii. In vortex-free flows, there exists a time coordinate t that is equal to proper-time along the flow lines, synchronised to be equal to each other; the hypersurfaces $t=$ const. are geodesically parallel.
iv. Conformal, non-rigid flows are vortex-free [see (28)]
v. Rigid flows are isometric: u^a then generates a translation group, and ω^a is harmonic.[36]
vi. The usual Doppler formula $d\lambda/\lambda = (\delta l)^{\cdot}$ holds for the redshift of infinitesimally neighbouring particles.
vii. Stream- and vortex-lines span (2-) surfaces [see (46)].
viii. For shear-free motion, $(l^2\omega^a)^{\cdot} = 0$, thus $l^2\omega \equiv \Omega$, $\dot{\Omega} = 0$. (With ii.: The (Newtonian) angular momentum of a sphere (consisting of specified matter particles) of radius l is $J = (8\pi/15)M\Omega$, $\dot{J} = 0$.)
ix. The flow lines are Ricci lines: [see (37)]:

$$h^a_b(\omega^{bc}{}_{;c} - \sigma^{bc}{}_{;c} + \tfrac{2}{3}\Theta^{,b}) = 0. \tag{123}$$

x. Curvature scalar, density and cosmological constant stand in the relation $R = -\rho - 4\Lambda$ [see (80)].
xi. There holds the "expansion equation"[37] [see (81)]

$$3\frac{\ddot{l}}{l} + 2(\sigma^2 - \omega^2) + \frac{1}{2}\rho - \Lambda = 0. \tag{124}$$

xii. The field equations (70) with (68) are equivalent to ix., xi. and

$$-h_{ab}R^{bc}h_{cd} = (\Lambda + \tfrac{1}{2}\rho)h_{ad}. \tag{125}$$

(xi. arises through contraction of the field equations by $u^a u^b$, ix. with $u^a h^{bc}$, xii. with $h_{ac}h_{bd}$.)

From xi. there follows the noteworthy

Theorem 3.5.1. [44,45]. If along a flow line $\frac{1}{2}\Lambda + \omega^2 \leq \frac{1}{4}\rho + \sigma^2$ (in particular thus for $\omega = 0$, $\Lambda \leq 0$) and $\Theta(s) > 0$ at a (proper-)time s, then there exists a (finite) $s_0 < s$ such that $\lim_{s\to s_0+0}(\rho s^2) = +\infty$: at the time $s = s_0$ a singular explosion has happened. For given $\Theta(s)$, $s - s_0$, the 'age' of the geodesics, is increased by increasing ω or Λ, and is lowered by increasing σ or ρ.

[These statements follow immediately from xi., if one observes that $\ddot{l}$ is the curvature of the graph of the function $l(s)$ and $\Theta = \dot{l}/l$. For shear-free flows, $\sigma = 0$, one can through ii. and viii. obtain from xi. an ordinary differential equation for $l(s)$. This equation can be integrated once.]

36 That is to say, vortex- and divergence-free; see Theorem 2.4.5.
37 This equation was derived by Raychaudhuri [44] in a special coordinate system.

xiii. For shear-free flows there holds the simplified expansion equation

$$3\dot{l}^2 + \frac{2\Omega^2}{l^2} - \frac{M}{l} - \Lambda l^2 \equiv 10E \tag{126}$$

with parameters Ω, M, and E constant (along the flow lines).

In a vortex-free flow, (125) has the meaning of a condition for the curvature of the space-sections $t =$ const. ($u_a x^a = -dt$). From the definitions of the quantities Θ, σ_{ab} it emerges that in this case ($\omega = 0$, $\dot{u} = 0$) (up to a conventional factor) $(\sigma_{ab} + \frac{1}{3}\Theta h_{ab})\delta_\perp x^a \delta_\perp x^b$ along $t =$ const. is the second fundamental form of these space sections, so that according to the formulae of differential geometry[38] the left hand side of (124) can be reduced to the Ricci tensor of the space sections. Thus there holds

Theorem 3.5.2. In a vortex-free flow,

$$K(x^a, e^a) = \tfrac{1}{3}(\Lambda + \rho + \sigma^2 - \tfrac{1}{3}\Theta^2) + l^{-3}(l^3\sigma_{ab})^{\cdot} e^a e^b \tag{127}$$

is the Gauss curvature of that surface F which is generated by the flowlines of the geodesics (drawn with respect to the induced metric) going out from x^a, orthogonal to e^a at x^a; where $e_a u^a = 0$ (see Figure 2).

This theorem is a simple example of how in Einstein's theory the curvature of the rest-space of the matter is determined by its distribution and relative motion; the second term shows that *form-altering motions* of the matter inter alia are connected with an *anisotropy of the spatial curvature* (for the exceptions see Theorem 3.5.5.).

The equation (124) holds also in the Newtonian Mechanics of pressure-free matter interacting only through gravity; the relation (127) is the first typical of the Einstein theory. The first remark leads us to interpret the equation (126) as an energy theorem. In fact in Newtonian Theory the expansion energy, rotational energy, inner potential gravitational energy of a (small) matter sphere of radius l, consisting of substratum particles, are respectively given by the first three terms in (126) multiplied by $(2\pi/15)M$. If one adds to these a Λ-energy of magnitude $-(2\pi/15)Ml^2\Lambda$, then (126) states the conservation of energy $(4\pi/3)l^3\rho E$ of this sphere. This *mechanical* energy theorem must however be distinguished from the *thermodynamic* energy theorem (78) (trivial for incoherent matter). The first concerns itself with a finitely extended system, and correspondingly contains a gravitational part, the second to an infinitesimal substratum

[38] Equations of Gauss and Codacci, see e.g. Ref. 46, sec.43 or Ref. 12, eq. (2.2.4a).

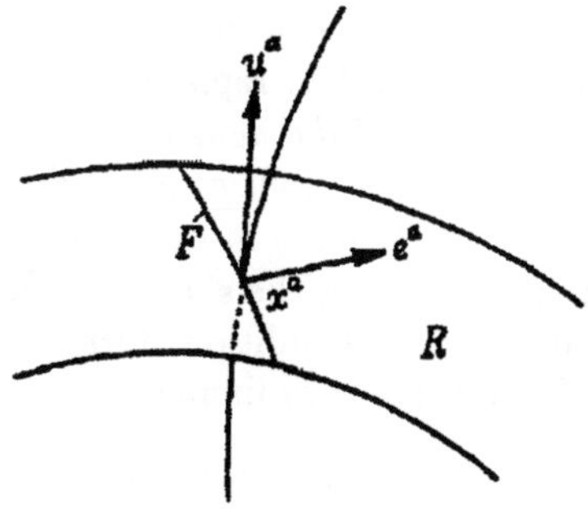

Figure 2.

particle, in which according to the equivalence principle gravitation does not make itself noticeable.

The dynamical conservation theorems viii. and xiii. holding for a matter sphere K under the supposition $\sigma = 0$, are valid although K is not a closed system. On shrinking of K through small spheres k_i the energy does not behave linearly: the E_i are related to local inertial systems moving relative to each other, and furthermore the k_i interact through gravitation.

We now seek the kinematically simplest solutions of the gravitational equations for incoherent matter.

A. ($\omega = \Theta = \sigma = 0$) The simplest flow is a vortex- free rigid motion; this is by xi. and Theorem 3.5.2 only possible if $\rho = 2\Lambda$ and the rest-space of the matter has the constant positive curvature $K = \Lambda$; and these conditions are by xii. and ix. also sufficient: Einstein-Universe.

B. ($\omega = \sigma = 0$, $\Theta \neq 0$) The next simplest possibility is that only σ and ω vanish. Then (126) and (127) hold, and give the Friedmann-Lemaître models. The rest spaces have the constant curvature $K = -(10/3)(E/l^2) \equiv (k/R^2)$ $(k = \pm 1, 0)$. This curvature is thus equal to the above defined energy of a l–sphere, divided by $(4/10)\pi\rho l^5$. [That Θ and ρ are constant in the rest spaces follows from (124) and (127)]. The theorems developed here and Theorem 2.5.1 permit, without further calculation, to draw the conclusion:

Theorem 3.5.3. (proved in Ref. 44 by other methods). The only universes with shear- and vortex-free motion of incoherent matter are those among the everywhere isotropic models characterised in Theorem 2.5.1, whose curvature radius obeys the Friedmann–Lemaître equation (126) (with $l = R$, $\Omega = 0$, $10E = -3k$).[39]

[39] For $\Lambda = 0$, k separates three classes of similar solutions (in the sense of Section 3.1)

In anticipation of the application of these models to the system of extra-galactic nebulae, it is useful to replace the suppositions $\omega = 0$ and $\sigma = 0$ respectively by ***isotropy of the density distribution*** [48] ($h_a^b \rho_{,b} = 0$) ***and redshift*** [47]; from this namely follows according to 2.2.ii. and 2.2.vii. $\omega = 0$ and $\sigma = 0$. Both properties are in principle observable.

C. ($\Theta = \sigma = 0$, $\omega \neq 0$) Among the shear-free flows with rotation, the simplest are those that obey the conditions formulated in Theorem 2.5.2. For them there holds in suitable coordinates

$$G = dz^2 + d\sigma^2 - (dt + u_A(x^B)dx^A)^2, \quad u^a = \delta_4^a, \tag{64}$$

where $d\sigma^2 = \gamma_{AB}(x^C)dx^A dx^B$ is a surface metric. According to (125) necessarily $\frac{1}{2}\rho = \Lambda + 2\omega^2$. Therefore according to Theorem 2.5.2 $\rho = \text{const.}$ and $\omega = \text{const.}$ must hold. Since G is the direct sum (see e.g. Ref. 12, Ch.1.4.) of the z-line and a V_3, from (125) necessarily $\Lambda + \frac{1}{2} = 0$ must hold. Thus there follows $\rho = 2\omega^2 = -2\Lambda$, and to show that ω_a is constant one need only further solve the equation $h_{ab}R^{bc}h_{cd} = 0$ for (64), which for example can be easily done with the help of the formulae in Ref. 12 (eq. (2.3.3), p.36), and then gives the Gödel cosmos [17,47,49–51].

Theorem 3.5.4. The Gödel cosmos is the only universe with incoherent matter in which the motion of the substratum proceeds shear-free, and the direction of the rotation axis in the infinitesimal spacelike neighbourhood of each particle is constant.

As the theorem shows, a rigid, spatially-constant rotation is only consistent with the field equations for $\Lambda < 0$; the rotation must therefore only be so strong, that the matter can be held together by the (usual) gravitational and Λ–attraction. If one however does not require constancy of ω, there are permitted for $\Lambda = 0$ flow fields with rigid rotations that with the help of remarks v.,viii.,ix.,xi.,xii. can be reduced to the static solutions of the vacuum field equations $R_{ab} = 0$ in a surprisingly simple way (Ref. 47, § 29, Theorem IV). Since this has already been presented completely in the spirit of the methods used here (Ref. 52; see also Ref. 53), we refrain from a repetition and simply remark: In these solutions, $\rho = 4\omega^2$; $\rho > 0$ therefore does not have to be assumed separately. Solutions with (spatially) constant density do not exist (Ref. 47, § 29, Theorem III)—contrary to the Newtonian theory, where from $\Theta = \sigma = 0$ it follows that $\rho = \text{const.}$ (Ref. 41, I), just as there exist no solutions with compact orientable 3-space (Ref. 47, § 29, Theorem IV). Axially symmetric solutions of this kind have been constructed—without reference to vacuum fields—by van

from each other; in each class, M is a similarity parameter.

Stockum [54]. This author also found the only singularity-free solution of this kind known up to now, a rotating cylinder with a static (!) vacuum field attached to it; solutions that contain singular points, surfaces, and so on can be obtained in great numbers (Ref. 47, § 29, Theorem IV). It is an interesting question, whether these singularities can be eliminated through introduction of physically acceptable pressures, or through joining to a vacuum field.

D. ($\omega = 0$). The next simplest flow is one that is vortex-free and volume-preserving; according to xi. this is at best possible for $\Lambda = \frac{1}{2}\rho + 2\sigma^2$ (> 0). Solutions of this type [except for $\sigma = 0$, see (A)] have not been given, to my knowledge. Under the more complicated assumptions $\omega = 0$, $\Theta \neq 0 \neq \sigma$, firstly the spherically symmetric flows can be specified, and have been exhaustively investigated by several authors [55,56]; in the framework of this presentation, they can be lucidly handled with the help of statements xi.,ix. and 3.5.2.

From Theorem 3.5.2, ix. and xi. follows the

Theorem 3.5.5. In a vortex-free flow of incoherent matter, the rest-spaces of the substratum are of constant curvature $K(t)$ if and only if the tensor $l^3\sigma_{ab}$ is constant along the flow lines. For such flows the Einstein equations are equivalent to ix., xi. (for $\omega = 0$) and

$$K(t) = \frac{k}{R^2} = \frac{1}{3}\left(\Lambda + \rho + \sigma^2 - \frac{1}{3}\Theta^2\right). \tag{128}$$

The corresponding flows are natural generalisations of the Friedmann–Lemaître models: examples of this kind have been determined (from another viewpoint and using other methods) by Schücking and Heckmann, see Ref. 9.

It is clear that in the characterisation of more general flows, the vanishing or not of kinematic quantities or their constancy is not sufficient, rather further invariants must be introduced; and that for $\omega \neq 0$ the metric determined by h_{ab} should be considered as the metric of the three-dimensional space of the flow-lines, for which circumstances theorems analogous to 3.5.2 should be formulated. We will not investigate these questions here; our aim was to show the simplicity and applicability of geometrically formulated hydrodynamics in Riemannian space for the handling of exact solutions, and presentation of the simplest examples.[40]

[40] More general solutions, that are spatially homogeneous, have been determined by Heckmann, Schücking, and Ozsvath; these works have not yet been published (private communication).

REFERENCES

1. Landau, L. D., and Lifschitz, E. M. (1962) *The Classical Theory of Fields* (Addison-Wesley, Reading, Mass.) [also reprinted (1975). (Pergamon Press, Oxford)].
2. Lichnerowicz, A. (1955). *Théories relativistes de la gravitation et de l'électromagnétisme* (Masson, Paris)..
3. Synge, J. L. (1937). *Proc. Lond. Math. Soc.* **43**.
4. Salzman, G., and Taub, A. H. (1954). *Phys. Rev.* **95**, 1659.
5. Jordan, P., Ehlers, J., Sachs, R. K. (1961). *Akad. Wiss. Mainz. Abh. math.-nat. Kl. 1961* Nr. 1.
6. Synge, J. L., and Schild, A. (1952). *Tensor Calculus* (University of Toronto Press, Toronto) [also reprinted (1978). (Dover, New York)].
7. Foures-Bruhat, Y. (1958). *Comptes Rend. Acad. Sci. (Paris) II* **246**, 3319.
8. Gödel, K. (1952). In *Proc. Int. Congress of Mathematicians (Cambridge, Mass., Aug./Sept. 1950)*, (Int. Congress Mathematicians/A.M.S., Providence, R.I.) vol. 1.
9. Schücking, E., and Heckmann, O. (1959). In *Onzième Conseil de Physique Solvay: La Structure et l'Évolution de l'Univers*, Int. Inst. de Physique Solvay (Editions Stoop, Brussels).
10. Rayner, C. B. (1959). *Comptes Rend. Acad. Sci. (Paris) II* **248**, 929, 2725; **249**, 1327.
11. Pirani, F. A. E. (1957). *Phys. Rev.* **105**, 1089.
12. Jordan, P., Ehlers, J., and Kundt, W. (1960). *Akad. Wiss. Mainz. Abh. math.-nat. Kl. 1961* Nr. 2.
13. Just, K. (1956). *Z. Physik.* **145**, 235.
14. Robertson, H. P. (1933). *Rev. Mod. Phys.* **5**, 62.
15. Bondi, H. (1960). *Cosmology* (Cambridge University Press, Cambridge).
16. Einstein, A. (1949). *The Meaning of Relativity* (3rd. ed., Princeton University Press, Princeton).
17. Gödel, K. (1949). *Rev. Mod. Phys.* **21**, 447.
18. Weyl, H. (1921/1923/1970/1988). *Raum Zeit Materie* (4th ed., 5th ed., 6th ed., 7th ed., Springer-Verlag, Berlin) [English transl. reprinted (1952). *Space-Time-Matter* (Dover, New York)].
19. Pauli, W. (1921). "Relativitätstheorie" In *Enzykl. d. Math. Wiss.* (Teubner, Leipzig) vol. 5, part 2, p. 539 [English transl. with supplements (1958). *Theory of Relativity* (Pergamon Press, London/Oxford)].
20. Eckart, C. (1940). *Phys. Rev.* **58**, 919.
21. Meixner, J., and Reik, W. (1959). In *Handbuch der Physik* **III**, 2 (Springer-Verlag, Berlin).
22. Just, K. (1958). Habilitation Thesis, FU Berlin.
23. Birkhoff, G. (1950). Hydrodynamics (Princeton University Press, Princeton).
24. Eisenhart, L. P. (1924). *Trans. Amer. Math. Soc.* **26**, 205.
25. Jordan, P. (1955). *Schwerkraft und Weltall* (2nd. ed., Vieweg & Sohn, Braunschweig).
26. Fierz, M. (1956). *Helv. Phys. Act.* **29**, 128.
27. Heckmann, O., and Schücking, E. (1959). In *Handbuch der Physik* **LIII**, 519 (??, Berlin).
28. Pirani, F. A. E. (1956). *Acta Phys. Polonica* **15**, 389.
29. Fock, V. (1960). *Theorie von Raum, Zeit und Gravitation* (Akademie-Verlag, Berlin) [English transl. by N. N. Kemmer (1964). *The Theory of Space, Time and Gravitation* (2nd. rev. ed., Macmillan, New York)].

30. Whittaker, E. T. (1935). *Proc. Roy. Soc. Lond.* **149**, 384.
31. Bondi, H., Pirani, F. A. E., and Robinson, I. (1959). *Proc. Roy. Soc. Lond.* **A251**, 519.
32. Witten L., ed. (1962). *The Theory of Gravitation: An Introduction to Current Research* (Wiley, New York).
33. Sasaki, M. (1958). In *Max Planck Festschrift* (VEB Deutscher Verlag der Wissenschaften, Berlin).
34. Grad, H. (1949). *Comm. Pure Appl. Math.* **2**, 331.
35. Synge, J. L. (1960). *Relativity: The General Theory* (North-Holland, Amsterdam).
36. Grad, H. (1959). In *Handbuch der Physik* **XII** (Springer-Verlag, Berlin).
37. Taub, A. H. (1948). *Phys. Rev.* **74**, 328.
38. Taub, A. H. (1956). *Phys. Rev.* **103**, 454.
39. Boltzmann, L. (1895). *Vorlesungen über Gastheorie* (Barth, Leipzig) [English transl. by S. G. Brush (1964). *Lectures on Gas Theory* (University of California Press, Berkeley)].
40. Heckmann, O. (1942). *Theorien der Kosmologie* (Springer-Verlag, Berlin).
41. Heckmann, O., and Schücking, E. (1955). *Z. Ap.* **38**, 95 (I); (1956). **40**, 81 (II).
42. Tolman, R. C. (1934). *Relativity, Thermodynamics and Cosmology* (Oxford University Press, Oxford) [also reprinted (1987). (Dover, New York)].
43. Pound, R. V., and Rebka, G. A. (1960). *Phys. Rev. Lett.* 4, 337.
44. Raychaudhuri, A. (1955). *Phys. Rev.* **98**, 1123.
45. Komar, A. (1956). *Phys. Rev.* **104**, 544.
46. Eisenhart, L. P. (1956). *Riemannian Geometry* Princeton University Press, Princeton).
47. Ehlers, J. (1957). Dissertation, Hamburg University.
48. Bondi, H. (1947). *Mon. Not. R. Astr. Soc.* **107**, 410.
49. Kundt, W. (1955). Diploma Thesis, Hamburg University.
50. Kundt, W. (1956). *Z. Phys.* **145**, 611.
51. Einstein, A. (ed. P. A. Schilpp) (1949). *Albert Einstein: Philosopher-Scientist* (Library of Living Philosophers, Tudor, New York).
52. Ehlers, J. (1959). In *Les Théories relativistes de la gravitation (Actes du Colloque Int., Royaumont, 21–27 Juin 1959)* (CNRS, Paris).
53. Ehlers, J. (1962). In *Recent Developments in General Relativity (Dedicated to Infeld)* (Pergamon Press, Oxford, and PWN–Polish Scientific Publishers, Warsaw).
54. van Stockum, W. J. (1937). *Proc. R. Soc. Edinburgh* **57**, 135.

Chapter 10
Matvei Bronstein 1936, quantum theory of weak gravitational fields.

Original paper: Matvei Bronstein, Quantentheorie schwacher Gravitationsfelder. *Phys. Zeitschr. der Sowjetunion* **9**, 140 – 157 (1936).

Reprinted with the kind permissions of Stanley Deser and Alexei Starobinsky. The publication rights of the original publisher expired in 2006.

Paper translated by M.A. Kurkov, translation edited by S. Deser.

Previously published in *Gen. Relativ. Gravit.* **44** (2012), pages 263 – 265 (the editorial note + biography), 267 – 283 (the paper).

Gen Relativ Gravit (2012) 44:263–265
DOI 10.1007/s10714-011-1284-5

GOLDEN OLDIE EDITORIAL

Editorial note to: Matvei P. Bronstein, Quantum theory of weak gravitational fields

S. Deser · A. Starobinsky

Published online: 8 November 2011

Keywords Quantization of gravity · Bronstein · Golden Oldie

Bronstein's is only the second work, after L. Rosenfeld's 1930 papers [1,2], on the quantization of (linearized) gravity. It was carried out in 1935, at a time when modern quantum field theory was in its infancy—well before Fierz and Pauli's treatment of spin 2, and soon after Dirac's (1931) and Fermi's (1934) formulations of QED. Its birthplace—Leningrad—was also remarkable: here a small group of young theoreticians were forming the nucleus of the future Landau school and many other eminent Russian institutions. Indeed, Bronstein had close ties, both personal and scientific, with Landau and his circle, and the spirit animating this work clearly reflects these affinities of style, recognizable to readers of the Landau & Lifshitz series. Had Bronstein's life not come to a tragic end only a few years later, in Stalin's purges, he would certainly have fulfilled his promise as a most valuable contributor to forefront theoretical physics.

The present work contains a, still quite readable and modern, exposition of second quantization of the massless spin-2 field in flat space-time, including even such

The republication of the original paper can be found in this issue following the editorial note and online via doi:10.1007/s10714-011-1285-4.

S. Deser (✉)
Brandeis University, Waltham, MA 02 454,
MC 452-48, Caltech, Pasadena, CA 91125, USA
e-mail: deser@brandeis.edu

A. Starobinsky
Landau Institute for Theoretical Physics,
Russian Academy of Sciences, Kosygina 2,
119334 Moscow, Russia
e-mail: alstar@landau.ac.ru

essential details as the proper operator ordering required to avoid zero-point energy. The system is quantized in Lorentz gauge, and shown to have positive energy for each of its transverse-traceless graviton excitations. A large part of the paper is devoted to the development of measurement theory for a quantized gravitational theory—the first attempt in this direction, too, obviously influenced by the then-new Bohr–Rosenfeld analysis of quantum electrodynamics. Though the author generally follows their treatment, he shows that the gravitational field is essentially different due to the appearance of another restriction on the accuracy of measurement of gravitational fields—in the case of large matter densities. [Here he concentrates on the Christoffel symbols, as the first derivative equivalents of the EM fields.] From these delicate and carefully treated physical effects, the author also concludes—most wisely—that it would be hardly possible to quantize full non-linear gravity without deep revision of classical space-time notions—prophetic conclusion that remains very much actual!

Going beyond the free-field analysis, Bronstein further treated gravity-matter interaction, concretely through Fock's novel covariantization of the Dirac field. Here the author first considers energy transfer by quantized gravitational waves and shows by lengthy calculation that, in the correspondence limit of large occupation number, the classical Einstein expression follows (after correcting a coefficient given by Einstein!). This shows that his quantization is not only qualitatively, but also quantitatively correct. Finally, following the, also contemporaneous, Fock—Podolsky derivation of Coulomb's law from QED, he obtains Newton's law from his linearized quantum gravity, carefully analyzing how the correct sign of the gravitational potential—i.e., opposite to that of Coulomb's—appears in this formalism; while this sign flip is already built-in at the classical level, it is a gratifying check of a new quantum model's correctness.

Finally, we note again the elegant and thorough style and notation—not only for its time, but even now; as noted at the beginning, it will appeal, not coincidentally, to anyone who has studied Landau-Lifshitz. In summary, this paper certainly deserves revival, not just historically, but for its intrinsic (and pedagogical) value. When it was published, general relativity had fallen into a dormant phase, from which it only awoke two decades later. At last, three quarters of a century since 1935, we begin the observational search for quantum-gravitational effects in primordial gravitational wave background generated during inflation. Published in German in a journal inaccessible today, this accessible Golden Oldie translation is a welcome (re-)addition to our subject!

Acknowledgment SD acknowledges support from NSF PHY-1064302 and DOE DE-FG02-164 92ER 40701 grants. We thank G. Gorelik and J. Stachel for very useful correspondence.

Translator's remark: A more detailed summary of this work was published at the same time in "Journal of Experimental and Theoretical Physics". [Bronstein M. Quantization of Gravitational Waves. *Zhurn. Eksper. Teor. Fiz.* **6**, 195–236 (1936)]. Fragments of this large article are included in the symposium "Albert Einstein and Theory of Gravity", Mir, Moscow 1979, pp. 433–445.

Matvei P. Bronstein: a brief biography

By S. Deser

Matvei Petrovich Bronstein was born December 2, 1906, and shot in a Leningrad prison in Stalin's purges on February 18, 1938 (rehabilitated posthumously in 1957). A precocious scientist, he was one of the legendary group that included Landau and Gamow at Leningrad University in 1926. His first publications date from 1925; in his one allotted decade, he covered a wide range of frontier problems in cosmology, astrophysics and application of the then new quantum mechanics. For his PhD thesis, a degree newly re-instituted in the USSR, he chose the work presented here, which he defended in front of I. Tamm and V. Fock in November 1935. In addition to his great scientific output and participation in the surge of Soviet science in those early years, he translated Dirac's "Quantum mechanics", works by Bohr and wrote several children's science books. On August 6, 1937, he was arrested in the Great Purge; he was never seen again, despite his widow's being informed that he was sentenced to ten years' prison. For a detailed biography, see Refs. [3] and [4].

References

1. Rosenfeld, L.: Ann. Physik **5**, 113 (1930)
2. Rosenfeld, L.: Z. Physik **65**, 587 (1930)
3. Gorelik, G., Frenkel, V.: Matvei Petrovich Bronstein and Soviet Theoretical Physics in the Thirties. Birkhäuser, Boston (1994)
4. Gorelik, G.: Matvei Bronstein and quantum gravity: 70th anniversary of the unsolved problem. Physics-Uspekhi **48**(10), 1039–1053 (2005)

Gen Relativ Gravit (2012) 44:267–283
DOI 10.1007/s10714-011-1285-4

GOLDEN OLDIE

Republication of: Quantum theory of weak gravitational fields

Matvei Bronstein

Published online: 10 November 2011

An editorial note to this paper and a biography can be found in this issue preceding this Golden Oldie and online via doi:10.1007/s10714-011-1284-5.

Original paper: Matvei Bronstein, Quantentheorie schwacher Gravitationsfelder, *Physikalische Zeitschrift der Sowjetunion*, Band **9**, Heft 2–3, pp. 140–157 (1936). Translated from German by M.A. Kurkov, edited by S. Deser.

Editorial responsiblility: A. Krasiński, e-mail: akr@camk.edu.pl.

Matvei Bronstein murdered in February 1938 in Stalin's purges.

M. Bronstein
Leningrad, USSR
e-mail for correspondence: akr@camk.edu.pl

QUANTUM THEORY OF WEAK GRAVITATIONAL FIELDS[1]

By M. Bronstein.

(Received on 2. January 1936)

§1. General remarks. §2. Hamiltonian form and plane waves. §3. Commutation relations and eigenvalues of the energy. §4. Let us undertake a little gedanken experiment! §5. Interaction with matter. §6. Energy transfer by gravitational waves. §7. Deduction of Newton's law of gravitation.

§ 1. General remarks

It is known that the deviations of a space-time continuum from "Euclideanness" can be characterized by the components of the Riemann - Christoffel tensor. When these deviations are small, this fourth-rank tensor field can be derived from a symmetric second-rank tensor field as follows:

$$(\mu\rho\nu\sigma) = \frac{1}{2}\left(\frac{\partial^2 h_{\mu\nu}}{\partial x_\rho \partial x_\sigma} + \frac{\partial^2 h_{\rho\sigma}}{\partial x_\mu \partial x_\nu} - \frac{\partial^2 h_{\mu\sigma}}{\partial x_\rho \partial x_\nu} - \frac{\partial^2 h_{\rho\nu}}{\partial x_\mu \partial x_\sigma}\right), \tag{1}$$

where $h_{\mu\nu}$ is the small deviation of the fundamental metric tensor from its Minkowskian value $\Delta_{\mu\nu}$ ($\Delta_{00} = 1$, $\Delta_{11} = \Delta_{22} = \Delta_{33} = -1$; $\Delta_{\mu\nu} = 0$, if $\mu \neq \nu$). In these circumstances, we consider the world as "Euclidean" with the metric tensor $\Delta_{\mu\nu}$, and $(\mu\rho\nu\sigma)$ as the components of a fourth-rank tensor field embedded in this flat world. Thereby $h_{\mu\nu}$ play the role of the "potentials", whose values can be fixed by four additional "gauge conditions"

$$[\alpha\alpha, \beta] = 0 \qquad (\beta = 0,\ 1,\ 2,\ 3). \tag{2}$$

(Here and in what follows the summation convention applies only to Greek indices, also in the sense that e.g. $A_{\alpha\alpha}$ denotes $A_{00} - A_{11} - A_{22} - A_{33}$; it allows us not to worry about the difference between co- and contravariant components; $[\alpha\beta, \gamma]$ is the usual notation for the three-index Christoffel symbol).

In empty space, the equations of gravitation read

$$(\mu\rho\nu\rho) = 0. \tag{3}$$

[1]A more detailed summary of this work appears simultaneously in "Journal of Experimental and Theoretical Physics" (russian).

Under the "gauge conditions" (2), they are fully equivalent to the usual wave equations for potentials

$$\Box h_{\mu\nu} = 0. \tag{4}$$

In what follows, we consider a quantum-mechanical continuous system, for which the classical equations of motion can be written in the form (4); along with the additional conditions (2), which, as we shall show, are compatible with the Schrödinger equation for the quantum-mechanical system under consideration, this system is identical to the gravitational field in empty space. This treatment is, to some extent, analogous to Fermi 's quantization of electrodynamics: Fermi 's depends on a non-gauge-invariant Lagrangian; our quantization of the gravitational field also relies on quantities that are not (even approximately) relativistic invariants.

§ 2. Hamiltonian form and plane waves

Let us consider a dynamical continuum with ten fields $h_{\mu\nu}$ $(\mu \leq \nu)$, which play the role of mechanical coordinates with the Lagrangian density

$$[\alpha\alpha, \beta]\,[\beta\gamma, \gamma] - [\alpha\beta, \gamma]\,[\alpha\gamma, \beta] + \frac{1}{2}\,[\alpha\alpha, \beta]\,[\gamma\gamma, \beta]\,.$$

Rather complicated calculations, which we omit here for brevity, show that the corresponding Hamiltonian density is

$$2\left(p_{00} + \sum_l \frac{\partial h_{0l}}{\partial x_l}\right)^2 - \frac{1}{4}\left(3p_{00} - \sum_l p_{ll} + 2\sum_l \frac{\partial h_{0l}}{\partial x_l}\right)^2 -$$

$$-\frac{1}{2}\sum_l \left(-p_{0l} + \frac{\partial h_{00}}{\partial x_l} + \sum_m \frac{\partial h_{ml}}{\partial x_m}\right)^2 +$$

$$+\frac{1}{4}\sum_n \left(2p_{nn} + p_{00} - \sum_l p_{ll} + 2\frac{\partial h_{0n}}{\partial x_n}\right)^2 +$$

$$+\frac{1}{2}\sum_{l<m} \left(p_{lm} + \frac{\partial h_{0m}}{\partial x_l} + \frac{\partial h_{0l}}{\partial x_m}\right)^2 + \frac{1}{8}\sum_l \left(\frac{\partial h_{00}}{\partial x_l}\right)^2 -$$

$$-\frac{1}{4}\sum_{lm} \left(\frac{\partial h_{0l}}{\partial x_m} - \frac{\partial h_{0m}}{\partial x_l}\right)^2 + \frac{1}{4}\sum_{lm} \frac{\partial h_{00}}{\partial x_l}\frac{\partial h_{mm}}{\partial x_l} - \frac{1}{2}\left(\sum_l \frac{\partial h_{0l}}{\partial x_l}\right)^2 -$$

$$-\frac{1}{8}\sum_m\left(\frac{\partial}{\partial x_m}\sum_l h_{ll}\right)^2+\frac{1}{4}\sum_{lmn}\left(\frac{\partial h_{ln}}{\partial x_m}\right)^2+$$
$$+\frac{1}{2}\sum_{lmn}\left(\frac{\partial h_{lm}}{\partial x_m}\frac{\partial h_{ln}}{\partial x_n}-\frac{\partial h_{lm}}{\partial x_n}\frac{\partial h_{ln}}{\partial x_m}\right),$$

where $p_{\alpha\beta}$ are the momenta conjugate to the coordinates $h_{\alpha\beta}$ and that the corresponding equations of motion are (4). Here and in what follows, Latin indices take the values $(1, 2, 3)$ only. For simplicity we have also set the speed of light to 1 and the Newtonian gravitational constant to $1/16\pi$. (The reader can easily verify the value of gravitational constant if he compares our formulas e.g. with the formulas 58.1 and 59.4 of Eddington's book "Relativitätstheorie in mathematischer Behandlung", Berlin, Springer 1925.)

Now let us introduce the Fourier expansion,

$$h_{\alpha\beta}=\frac{1}{(2\pi)^{3/2}}\int d\mathfrak{k}\left[h_{\alpha\beta,\mathfrak{k}}e^{-i(\omega t-\mathfrak{k}\mathfrak{r})}+h^+_{\alpha\beta,\mathfrak{k}}e^{i(\omega t-\mathfrak{k}\mathfrak{r})}\right]$$

(where $\omega=|\mathfrak{k}|$). The Hamiltonian can the be calculated to be

$$H=\int d\mathfrak{k}\omega^2\left\{\frac{1}{2}\left(h^+_{00,\mathfrak{k}}+\sum_l h^+_{ll,\mathfrak{k}}\right)\left(h_{00,\mathfrak{k}}+\sum_l h_{ll,\mathfrak{k}}\right)+\right.$$
$$\left.+\sum_{l\neq m}\left(h_{lm,\mathfrak{k}}h^+_{lm,\mathfrak{k}}-h^+_{ll,\mathfrak{k}}h_{mm,\mathfrak{k}}\right)-2\sum_l h_{0l,\mathfrak{k}}h^+_{0l,\mathfrak{k}}\right\}. \tag{5}$$

(At first sight, the factor-ordering seems to be arbitrary, but it will be clear from what follows that this is the only one that will avoid quantum"zero-point energy".)

After Fourier expansion, the conditions (2) take the following form

$$\left.\begin{aligned}\tfrac{1}{2}\omega\left(h_{00,\mathfrak{k}}+\textstyle\sum_l h_{ll,\mathfrak{k}}\right)+\textstyle\sum_l \mathfrak{k}_l h_{0l,\mathfrak{k}}=0,\\ \omega h_{0l,\mathfrak{k}}+\textstyle\sum_m \mathfrak{k}_m h_{ml,\mathfrak{k}}+\tfrac{1}{2}\mathfrak{k}_l\left(h_{00,\mathfrak{k}}-\textstyle\sum_m h_{mm,\mathfrak{k}}\right)=0\\ (l=1,\ 2,\ 3).\end{aligned}\right\} \tag{6}$$

Hence the number of independent $h_{\mu\nu,\mathfrak{k}}$ (for a given $\mathfrak{k}$) is equal to $10-4=6$. However one may easily show that for many problems where e.g., energy transfer by gravitational waves is discussed, the number of independent polarizations is even fewer. Let us consider e.g., the case $\mathfrak{k}\|z$. Then the conditions (6) become

$$h_{11,\mathfrak{k}} + h_{22,\mathfrak{k}} = h_{00,\mathfrak{k}} + 2h_{03,\mathfrak{k}} + h_{33,\mathfrak{k}} = h_{01,\mathfrak{k}} + h_{31,\mathfrak{k}} =$$

$$= h_{02,\mathfrak{k}} + h_{32,\mathfrak{k}} = 0. \tag{6'}$$

Taking into account the relation (6′) one may write the energy content of the gravitational wave in the form

$$2d\mathfrak{k}\omega^2 \left(h^+_{12,\mathfrak{k}} h_{12,\mathfrak{k}} + h^+_{11,\mathfrak{k}} h_{11,\mathfrak{k}}\right).$$

(Using the conditions (6′) the integrand in (5) can be easily brought to this form, if all $h_{\mu\nu,\mathfrak{k}}$ and $h^+_{\mu\nu,\mathfrak{k}}$ commute; later we shall see, that even when the h and h^+ are quantized and do not commute, a similar result holds.) Thus, without changing the energy content of the gravitational wave, we can subject the 10 amplitudes $h_{\mu\nu,\mathfrak{k}}$ to four additional conditions besides (6). For example, we can choose (also when $\mathfrak{k}$ is not parallel to z) the following (albeit relativistically non-invariant) additional conditions:

$$h_{00,\mathfrak{k}} = h_{01,\mathfrak{k}} = h_{02,\mathfrak{k}} = h_{03,\mathfrak{k}} = 0.$$

Thus we see among other things that for the energy transfer only the transverse gravitational waves are significant, namely those with two independent polarizations. However, the situation is quite different if one considers not energy transfer but e.g., gravitational interaction between two massive bodies: we shall see below that for this interaction only the longitudinal $h_{\mu\nu}$ waves dominate.

§3. Commutation relations and eigenvalues of the energy.

The Heisenberg – Pauli commutation relations read[2]:

$$\left.\begin{aligned} &[h_{\alpha\beta}(\mathfrak{r}), h_{\alpha'\beta'}(\mathfrak{r}')] = 0, \\ &[p_{\alpha\beta}(\mathfrak{r}), p_{\alpha'\beta'}(\mathfrak{r}')] = 0, \\ &[p_{\alpha\beta}(\mathfrak{r}), h_{\alpha'\beta'}(\mathfrak{r}')] = \tfrac{h}{i}\delta_{\alpha\alpha'}\delta_{\beta\beta'}\delta(\mathfrak{r}-\mathfrak{r}') \qquad (\alpha \leqslant \beta,\ \ \alpha' \leqslant \beta'). \end{aligned}\right\} \tag{7}$$

[2]Here h means $h/2\pi$.

After Fourier expansion, we obtain

$$\left.\begin{aligned}
&[h_{\alpha\beta,\mathfrak{k}}, h_{\alpha'\beta',\mathfrak{k}'}] = 0, \qquad [h^+_{\alpha\beta,\mathfrak{k}}, h^+_{\alpha'\beta',\mathfrak{k}'}] = 0,\\
&[h^+_{00,\mathfrak{k}}, h_{00,\mathfrak{k}'}] = [h^+_{00,\mathfrak{k}}, h_{ll,\mathfrak{k}'}] = [h^+_{ll,\mathfrak{k}}, h_{ll,\mathfrak{k}'}] = -\frac{h}{2\omega}\delta(\mathfrak{k}-\mathfrak{k}'),\\
&[h^+_{00,\mathfrak{k}}, h_{0l,\mathfrak{k}'}] = [h^+_{ll,\mathfrak{k}}, h_{0m,\mathfrak{k}'}] = [h^+_{0l,\mathfrak{k}}, h_{mn,\mathfrak{k}'}] = 0,\\
&[h^+_{00,\mathfrak{k}}, h_{lm,\mathfrak{k}'}] = [h^+_{nn,\mathfrak{k}}, h_{lm,\mathfrak{k}'}] = 0 \qquad (l \neq m),\\
&[h^+_{ll,\mathfrak{k}}, h_{mm,\mathfrak{k}'}] = \frac{h}{2\omega}\delta(\mathfrak{k}-\mathfrak{k}') \qquad (l \neq m),\\
&[h^+_{0l,\mathfrak{k}}, h_{0m,\mathfrak{k}'}] = \frac{h}{2\omega}\delta_{lm}\delta(\mathfrak{k}-\mathfrak{k}'),\\
&[h^+_{lm,\mathfrak{k}}, h_{pq,\mathfrak{k}'}] = -\frac{h}{2\omega}\delta_{lp}\delta_{mq}\delta(\mathfrak{k}-\mathfrak{k}') \quad (l < m, \quad p < q).
\end{aligned}\right\} \tag{8}$$

We introduce the operators

$$A = \frac{1}{2}\omega\left(h_{00,\mathfrak{k}} + \sum_l h_{ll,\mathfrak{k}}\right) + \sum_l \mathfrak{k}_l h_{0l,\mathfrak{k}},$$

$$B_l = \omega h_{0l,\mathfrak{k}} + \sum_m \mathfrak{k}_m h_{ml,\mathfrak{k}} + \frac{1}{2}\mathfrak{k}_l\left(h_{00,\mathfrak{k}} - \sum_m h_{mm,\mathfrak{k}}\right).$$

Calculation shows that due to (8) all eight operators A, B_l, A^+, B_l^+ $(l = 1, 2, 3)$ commute with each other. But they do not commute with the Hamiltonian (5), i.e. they are not integrals of motion. However in the particular case, when

$$A = B_l = A^+ = B_l^+ = 0,$$

one can easily show, that the Poisson brackets of any of these operators and the Hamiltonian, i.e. the rates of change of these operators, are all zero. It means that the conditions (6) (and their conjugates) are compatible with each other and with the quantum-mechanical equations of motion.

The commutation relations (8), the Hamiltonian operator (5) and the additional conditions (6) (together with the ansatz for the interaction between gravitational field and matter, which will be introduced later) form the foundation of the quantum theory of gravity proposed here. Note that the quantum-mechanical Hamiltonian operator can never be uniquely specified by the correspondence principle: it is always possible to change the Hamiltonian by introducing additional terms that go to zero when $h \to 0$

(e.g., "spin terms" in the theory of the electron); in general even relativistic requirements are not sufficient to fix these "spin terms" uniquely. Nevertheless, we believe that adding such terms is not necessary here.

Now to calculate the eigenvalues of the energy! As in the quantum electrodynamics this is accomplished by introducing the new variables ξ that satisfy the commutation relations

$$[\xi, \xi^+] = 1.$$

The eigenvalues of $\xi\xi^\dagger$ are well-known to equal $n+1$, and those of $\xi^+\xi$ equal n, where n is a positive integer or zero.

Here it is not possible to introduce these ξ-variables in a symmetric manner. One possible solution of the problem looks like this,

$$\frac{1}{2}\left(h_{00,\mathfrak{k}} + \sum_l h_{ll,\mathfrak{k}}\right) = \sqrt{\frac{h}{2\omega d\mathfrak{k}}}\xi_{00,\mathfrak{k}} e^{i\omega t},$$

$$\begin{aligned}
h_{11,\mathfrak{k}} &= \sqrt{\frac{h}{2\omega d\mathfrak{k}}}\left(\frac{\xi^+_{11,\mathfrak{k}}}{\sqrt{3}}e^{-i\omega t} + \frac{\xi_{22,\mathfrak{k}}}{\sqrt{3}}e^{i\omega t} + \xi_{33,\mathfrak{k}}e^{i\omega t}\right),\\
h_{22,\mathfrak{k}} &= \sqrt{\frac{h}{2\omega d\mathfrak{k}}}\left(\frac{\xi^+_{11,\mathfrak{k}}}{\sqrt{3}}e^{-i\omega t} + \frac{\xi_{22,\mathfrak{k}}}{\sqrt{3}}e^{i\omega t} - \xi_{33,\mathfrak{k}}e^{i\omega t}\right),\\
h_{33,\mathfrak{k}} &= \sqrt{\frac{h}{2\omega d\mathfrak{k}}}\left(\frac{\xi^+_{11,\mathfrak{k}}}{\sqrt{3}}e^{-i\omega t} - \frac{2}{\sqrt{3}}\xi_{22,\mathfrak{k}}e^{i\omega t}\right),\\
h_{lm,\mathfrak{k}} &= \sqrt{\frac{h}{2\omega d\mathfrak{k}}}\xi_{lm,\mathfrak{k}}e^{i\omega t} \qquad (l \neq m),\\
h_{0l,\mathfrak{k}} &= \sqrt{\frac{h}{2\omega d\mathfrak{k}}}\xi^+_{0l,\mathfrak{k}}e^{-i\omega t}.
\end{aligned}$$

The Hamiltonian in the new variables transforms into

$$\left.\begin{aligned}
H = \sum_{\mathfrak{k}} h\omega\,\big(&\xi^+_{00,\mathfrak{k}}\xi_{00,\mathfrak{k}} + \xi_{12,\mathfrak{k}}\xi^+_{12,\mathfrak{k}} + \xi_{23,\mathfrak{k}}\xi^+_{23,\mathfrak{k}} + \xi_{13,\mathfrak{k}}\xi^+_{13,\mathfrak{k}} -\\
&-\xi^+_{01,\mathfrak{k}}\xi_{01,\mathfrak{k}} - \xi^+_{02,\mathfrak{k}}\xi_{02,\mathfrak{k}} - \xi^+_{03,\mathfrak{k}}\xi_{03,\mathfrak{k}} - \xi_{11,\mathfrak{k}}\xi^+_{11,\mathfrak{k}} +\\
&+\xi^+_{22,\mathfrak{k}}\xi_{22,\mathfrak{k}} + \xi^+_{33,\mathfrak{k}}\xi_{33,\mathfrak{k}}\big)\,.
\end{aligned}\right\} \qquad (5')$$

Therefore the eigenvalues of the energy (for each value of $\mathfrak{k}$) are

$$h\omega\,(n_{00} + n_{12} + n_{23} + n_{31} - n_{01} - n_{02} - n_{03} - n_{11} + n_{22} + n_{33} + 2)\,,$$

where n_{00}, $n_{12}, \ldots$ are ten quantum numbers $(n = 0,\ 1,\ 2, \ldots)$.

The conditions (6) make this expression positive definite. To see this, we again consider the case $\mathfrak{k}||z$. From (6) we then obtain the following conditions:

$$\xi_{00,\mathfrak{k}} e^{i\omega t} + \xi^{\dagger}_{03,\mathfrak{k}} e^{-i\omega t} = 0, \qquad \xi^{\dagger}_{01,\mathfrak{k}} e^{-i\omega t} + \xi_{13,\mathfrak{k}} e^{i\omega t} = 0,$$

$$\xi^{\dagger}_{02,\mathfrak{k}} e^{-i\omega t} + \xi_{23,\mathfrak{k}} e^{i\omega t} = 0, \qquad \xi^{\dagger}_{11,\mathfrak{k}} e^{-i\omega t} + \xi_{22,\mathfrak{k}} e^{i\omega t} = 0.$$

From this it follows that

$$n_{01} = n_{31} + 1, \qquad n_{02} = n_{23} + 1, \qquad n_{22} = n_{11} + 1, \qquad n_{03} = n_{00} + 1.$$

The eigenvalues of the energy for this $\mathfrak{k}$ become

$$h\omega \left(\xi^{\dagger}_{12,\mathfrak{k}} \xi_{12,\mathfrak{k}} + \xi^{\dagger}_{33,\mathfrak{k}} \xi_{33,\mathfrak{k}} \right) = h\omega \left(n_{12} + n_{33} \right).$$

We see, consequently, that the energy of the gravitational field consists of positive gravitational quanta, of two polarizations for each wave vector $\mathfrak{k}$. In analogy to the classical case, also here only the transverse gravitational excitations matter: e.g., for $\mathfrak{k}||z$, these are h_{12} and $1/2\,(h_{11} - h_{22})$.

No "zero-point energy terms" arise in the process, due to the suitably chosen factor-ordering in the expression (5).

§ 4. Let us undertake a little gedanken experiment!

In order to understand somewhat better the physical content of the quantum theory of the gravitational field, let us consider the measurement of one of the field quantities appearing here, for example, of the three-index Christoffel symbol $[00, 1]$. The classical Einstein equations of motion read in our case (all $h_{\mu\nu} \ll 1$):

$$\frac{d^2x}{dt^2} = \frac{\partial h_{01}}{\partial t} - \frac{1}{2}\frac{\partial h_{00}}{\partial x} = [00, 1]. \tag{9}$$

Following Bohr and Rosenfeld [3] let us consider the measurement of a space-time average of $[00, 1]$ in volume V and time interval T. Take a test body of volume V. Let its mass be ρV. The above equation of motion, which is valid only when the speed of the test body is small compared to the speed of light, makes the following measurement possible: let us measure

[3]Bohr N. and Rosenfeld L., Dansk. Vidensk. Selskab., Math.-fys. Meddel. **12**, 8. 1933.

the momentum of the test body at the beginning and at the end of a time interval T; then by definition the required average is

$$\frac{(p_x)_{t+T} - (p_x)_t}{\rho V T}.$$

Therefore the measurement of $[00, 1]$ is associated with an uncertainty of order

$$\Delta[00, 1] \approx \Delta p_x / \rho V T, \tag{10}$$

where Δp_x is the uncertainty in momentum. Let the duration of the momentum measurement be Δt (of course, $\Delta t \ll T$); Δx be the uncertainty in the coordinate associated with measuring the momentum. The uncertainty Δp_x consists of two terms: the usual quantum-mechanical $h/\Delta x$ and one associated with the gravitational field produced by the test body itself because of its recoil due to the measurement. Because of Einstein's equation of gravitation $\Box h_{01} = \rho v_x$ the uncertainty in h_{01}, which appears as a consequence of the undetermined recoil speed $\Delta x/\Delta t$, must be of the order $\rho \frac{\Delta x}{\Delta t} \cdot \Delta t^2$. One can see from (9), that the corresponding uncertainty in $[00, 1]$ is of the order $\rho \Delta x$, and therefore during every measurement of momentum there appears an additional uncertainty in momentum connected with the gravitational field, which is of order $\rho \Delta x \cdot \rho V \Delta t$. In order to simplify the comparison with the usual units of measurement we suspend our convention $c = 1, G = 1/16\pi$ (till the end of this section). For the momentum we obtain

$$\Delta p_x \approx \frac{h}{\Delta x} + G\rho^2 V \Delta x \Delta t.$$

It can be shown (similarly to the arguments of Bohr and Rosenfeld) that the second term can be made arbitrarily small compared to the first. But in order to make the best measurement of $[00, 1]$ it seems more appropriate to bring Δp_x to its minimum, i.e. to make both terms of the same order. Thus one should choose Δx of the order

$$\Delta x \approx \frac{1}{\rho} \left(\frac{h}{G V \Delta t} \right)^{1/2}.$$

For $\Delta[00, 1]$ we obtain

$$\Delta[00, 1] \gtrsim \frac{h^{1/2} G^{1/2} \Delta t^{1/2}}{V^{1/2} T}. \tag{11}$$

Thus an absolutely precise measurement of the gravitational field would be possible only if an arbitrarily rapid measurement of momentum were possible. But two circumstances make the latter impossible: firstly, according to the definition of the measurement, $\Delta x \ll V^{1/3}$ should hold, and it leads to

$$\Delta t \gg \frac{h}{\rho^2 G V^{5/3}}.$$

Secondly, according to the theory of relativity, Δx can never be greater than $c\Delta t$, and it leads to

$$\Delta t \gtrsim \frac{h^{1/3}}{c^{2/3}\rho^{2/3}V^{1/3}G^{1/3}}.$$

It follows from (11) that $\Delta[00,1]$ can never be made smaller than

$$\frac{h}{\rho T V^{4/3}} \quad \text{or} \quad \frac{h^{2/3}G^{1/3}}{c^{1/3}\rho^{1/3}V^{2/3}T}.$$

Of these two bounds the first is the only significant one for the case of a light test body ($\rho V \lesssim h^{1/2}c^{1/2}G^{1/2}$, i.e. smaller than approximately 0.01 mg). For a heavier test body the second one is the most significant. It is clear that for the most precise measurement possible of $[00,1]$ heavy test bodies should be used, so only the second bound is of theoretical importance. We finally have

$$\Delta[00,1] \gtrsim \frac{h^{2/3}G^{1/3}}{c^{1/3}\rho^{1/3}V^{2/3}T}. \tag{12}$$

Thus it is clear that in the region where all $h_{\mu\nu}$ are small compared to 1 (this is just the meaning of the word "weak" in the title of this paper), accuracy of gravity measurements can be increased arbitrarily: as in this domain of phenomena the approximate linearized equations (1) hold, consequently the superposition principle is also valid, and it is therefore always possible to create a test body of arbitrary large ρ. From this we conclude that it is possible, just as, e.g., this paper attempts to do, to construct a completely self-consistent quantum theory of gravity within the framework of special relativity (i.e. when the space-time continuum is "Euclidean"). However, within the domain of General Relativity theory, where deviations from "Euclideanness" can be arbitrary large, the situation is quite different. Indeed the gravitational radius of the test body used for the measurement ($G\rho V/c^2$) cannot be larger than its linear dimensions $\left(V^{1/3}\right)$; this means that the upper bound on its density is ($\rho \lesssim c^2/GV^{2/3}$). Consequently, in this domain the possibilities of measurement are even more restricted than those due to the quantum-mechanical commutation relations. Without a deep revision of classical notions it seems hardly possible to extend the quantum theory of gravity also to this domain.

§5. Interaction with matter

The Ansatz for the energy of interaction between gravitational field and matter that is consistent with the correspondence principle can be obtained from the general relativistic form of the Dirac wave equation established by V. Fock.[4] When all $h_{\mu\nu}$ are small compared to 1, this equation can be written, for vanishing electromagnetic field, in the following form:

$$\frac{h}{i}\sum_{k=0}^{3} e_k \alpha_k \left(\frac{\partial}{\partial x_k} - \frac{1}{2}\sum_{l=0}^{3} e_l h_{kl} \frac{\partial}{\partial x_l} \right) \psi +$$
$$+ \left(\frac{1}{8}\frac{h}{i}\sum_{k=0}^{3} \frac{\partial h_{00}}{\partial x_k}\alpha_k - m\beta \right) \psi = 0,$$

where

$$e_0 = 1, \qquad e_1 = e_2 = e_3 = -1$$

and

$$\alpha_0 = 1, \qquad \alpha_1 = \rho_1\sigma_1, \qquad \alpha_2 = \rho_1\sigma_2, \qquad \alpha_3 = \rho_1\sigma_3, \qquad \beta = \rho_3$$

(the Dirac matrices). If we introduce two two-component functions χ and φ instead of the four-component ψ-function:

$$\psi_1 = \chi_1 e^{-imt/h}, \qquad \psi_2 = \chi_2 e^{-imt/h}, \qquad \psi_3 = \varphi_1 e^{-imt/h}, \qquad \psi_4 = \varphi_2 e^{-imt/h},$$

then for decreasing particle velocity, χ goes to zero, and φ tends to its non-relativistic wave function. From the Schrödinger equation for these φ_x one can see that the interaction energy between the particle and the gravitational field takes the form

$$V = \frac{m}{2}h_{00} + \frac{h}{i}\sum_k h_{0k}\frac{\partial}{\partial x_k} - \frac{h^2}{2m}\sum_{kl} h_{kl}\frac{\partial^2}{\partial x_k \partial x_l} - \frac{h^2}{4m}\sum_{kl}\frac{\partial h_{kl}}{\partial x_l}\frac{\partial}{\partial x_k} +$$

[4] V. Fock, ZS. f. Phys., **57**, 261, 1929.

$$+\frac{h}{4i}\sum_{l}\frac{\partial h_{0l}}{\partial x_l}+\frac{h^2}{4mi}\sum_{jklm}\sigma_j e_{jlm}\frac{\partial h_{km}}{\partial x_l}\frac{\partial}{\partial x_k}+\frac{h}{4}\sum_{jlm}\sigma_j e_{jlm}\frac{\partial h_{0m}}{\partial x_l},$$

where e_{jlm} is the skew-symmetric unit tensor (i.e. $e_{123}=1$ and e_{jlm} is antisymmetric with respect to each pair of its indices). When the wavelength of the gravitational perturbations is sufficiently large, this expression simplifies to

$$V=\frac{m}{2}h_{00}+\frac{h}{i}\sum_{k}h_{0k}\frac{\partial}{\partial x_k}-\frac{h^2}{2m}\sum_{kl}h_{kl}\frac{\partial^2}{\partial x_k\partial x_l}. \tag{13}$$

We shall use the relation (13) below. Note that even simple considerations based on the correspondence principle, without going through the D i r a c - F o c k equation, also lead to the relation(13) for the interaction energy.

§ 6. Energy transfer by gravitational waves

One of the simplest applications of the quantum theory of gravity sketched above is in the calculation of the energy radiation into gravitational waves emitted by material systems. Here we make use of the conditions (6) and (6′) together with $h_{00,\mathfrak{k}}=h_{01,\mathfrak{k}}=h_{02,\mathfrak{k}}=h_{03,\mathfrak{k}}=0$. In the ξ-variables (see above, §3) for the case $\mathfrak{k}\|z$ it leads to

$$V=\frac{m}{8\pi}\sqrt{\frac{hd\mathfrak{k}}{\pi\omega}}\left\{\left[\xi\left(\dot{x}_1^2-\dot{x}_2^2\right)+2\eta\dot{x}_1\dot{x}_2\right]e^{i\mathfrak{k}\mathfrak{r}}+\right.$$
$$\left.+\left[\xi^+\left(\dot{x}_1^2-\dot{x}_2^2\right)+2\eta^+\dot{x}_1\dot{x}_2\right]e^{-i\mathfrak{k}\mathfrak{r}}\right\}, \tag{14}$$

where for brevity $\dot{x}_k$ is written instead of $\frac{h}{mi}\frac{\partial}{\partial x_k}$, and ξ and η are written instead of ξ_{33} and ξ_{12}. Let us denote the initial or the final state of the emitting particle by k or l, the initial or the final state of the gravitational excitations by k' or l'. For the transition probability per unit time quantum mechanics gives the well-known expression

$$\frac{2\pi}{h}\delta\left(E_l+E_{l'}-E_k-E_{k'}\right)\left|\left(kk'\left|V\right|ll'\right)\right|^2.$$

Using the known values of oscillator matrix elements we conclude that the probability of spontaneous emission of a gravitation quantum per unit time, with a given $\mathfrak{k}$ ($\mathfrak{k}\|z$) and with ξ - polarization is

$$\frac{m^2}{32\pi^2}\frac{d\mathfrak{k}}{\omega}\left|\left(k\left|\dot{x}_1^2-\dot{x}_2^2\right|l\right)\right|^2\delta\left(E_l-E_k+h\omega\right),$$

and with η-polarization

$$\frac{m^2}{32\pi^2}\frac{dt}{\omega}\left|\left(k\left|2\dot{x}_1\dot{x}_2\right|l\right)\right|^2\delta\left(E_l-E_k+h\omega\right).$$

The probability of the transition $k\rightarrow l$ with the simultaneous emission of a gravitation quantum into a cone $d\Omega(\|z)$ per unit time is therefore

$$\frac{d\Omega}{8\pi^2}\frac{m^2\omega}{h}\left\{\left|\left(k\left|\frac{\dot{x}_1^2-\dot{x}_2^2}{2}\right|l\right)\right|^2+\left|\left(k\left|\dot{x}_1\dot{x}_2\right|l\right)\right|^2\right\}.$$

It is not difficult to generalize this expression to arbitrary direction (not only $\|z$) and then to integrate over all directions. The calculation leads to the following result: the total probability (per unit time) of the transition $k\rightarrow l$ with simultaneous emission of a gravitation quantum of the frequency $\omega=\left(E_k-E_l\right)/h$ is

$$\frac{m^2\omega}{10\pi h}\left\{\sum_{pq}\left|\left(k\left|\dot{x}_p\dot{x}_q\right|l\right)\right|^2-\frac{1}{3}\left|\left(k\left|\sum_p\dot{x}_p^2\right|l\right)\right|^2\right\}.$$

It is not difficult to generalize this expression also to an arbitrary system of matter particles. For the energy that such a system loses to gravitational waves radiated per unit time during the transition $k\rightarrow l$, we obtain the expression

$$\frac{\omega^2}{10\pi}\left\{\sum_{pq}\left|\left(k\left|\sum m\dot{x}_p\dot{x}_q\right|l\right)\right|^2-\frac{1}{3}\left|\left(k\left|\sum_p\sum m\dot{x}_p^2\right|l\right)\right|^2\right\} \tag{15}$$

(the summation symbol $\sum$ without indices means summation over different particles). This formula is the quantum-theoretic generalization of Einstein's known result.

Indeed, Einstein's expression for the energy radiated per unit time in the form of gravitational waves (with $G=1/16\pi$) is[5]

$$\frac{1}{80\pi}\left\{\sum_{pq}\left(\frac{d^3}{dt^3}\sum mx_px_q\right)^2-\frac{1}{3}\left(\sum_p\frac{d^3}{dt^3}\sum mx_p^2\right)^2\right\}.$$

[5]In Einstein's work (Berl. Ber., 1918, p. 154) there is $1/160\pi$ instead of $1/80\pi$ due to a calculation error. More recent calculations by Eddington (see his textbook or Proc. Roy. Soc. **102**, 281, 1922) lead to the correct coefficient.

If $I_{pq} \equiv \sum m x_p x_q$ can be represented through a Fourier series of the form

$$I_{pq} = \sum_{k=-\infty}^{+\infty} I_{pq}^{(k)} e^{ik\omega_0 t},$$

then this classical expression for the emission of energy at frequency $\omega = k\omega_0$ becomes

$$\frac{\omega^6}{40\pi}\left\{\sum_{pq}\left|I_{pq}^{(k)}\right|^2 - \frac{1}{3}\left|\sum_p I_{pp}^{(k)}\right|^2\right\}. \tag{16}$$

On the other hand we have,

$$\left(k\left|\sum |m\dot{x}_p\dot{x}_q|\right|l\right) = \sum\sum_j \left(k\left|\sqrt{m}\dot{x}_p\right|j\right)\left(j\left|\sqrt{m}\dot{x}_q\right|l\right) =$$

$$= -\sum\sum_j m\left(k\left|x_p\right|j\right)\left(j\left|x_q\right|l\right)\omega_{kj}\omega_{jl}.$$

At lower frequencies and higher quantum numbers, as one can easily calculate e.g. for the case of a rotator, we can put ω^2 instead of $2\omega_{kj}\omega_{jl}$ (ω = the emitted frequency), and it gives approximately

$$\left(k\left|\sum m\dot{x}_p\dot{x}_q\right|l\right) = -\frac{\omega^2}{2}\left(k\left|I_{pq}\right|l\right).$$

Then the expression (15) becomes

$$\frac{\omega^6}{40\pi}\left\{\sum_{pq}\left|\left(k\left|I_{pq}\right|l\right)\right|^2 - \frac{1}{3}\left|\sum_p\left(k\left|I_{pp}\right|l\right)\right|^2\right\}.$$

If we replace Fourier amplitudes for matrix elements, this expression goes over into Einstein 's (16) classical one. Thus in the limit $h \to 0$ the quantum theory of gravity agrees with classical Einstein theory.

§ 7. Derivation of Newton's law of gravitation

Dirac showed[6] that various interactions between charges can be always interpreted as realized through a mediation by an intermediate agent, namely

[6]Dirac P.A.M.- Proc. Roy. Soc. **136**, 453, 1932.

by the quantized field. Here we will show that is also the case for gravitational phenomena. At first glance, this looks a little paradoxical because both expressions for the interaction between field and matter are almost exactly the same [$e\Phi$ is the main term of interaction in the electromagnetic case; if we write $m/2$ instead of the charge e, and replace the scalar potential Φ with the scalar potential h_{00}, we obtain the main term in (13)]; yet, the very same scheme should in one case explain the repulsion of particles of the same type (Coulomb force), but in the other case the attraction (Newtonian forces). The solution of the paradox lies in the fact that in quantum electrodynamics, the commutation relations for the potential Φ are

$$\left[\Phi_{\mathfrak{k}}^{+}, \Phi_{\mathfrak{k}'}\right] = \frac{h}{2\omega}\delta(\mathfrak{k}-\mathfrak{k}'),$$

while in our case another commutation relation applies, namely (cf. (8))

$$\left[h_{00,\mathfrak{k}}^{+}, h_{00,\mathfrak{k}'}\right] = -\frac{h}{2\omega}\delta(\mathfrak{k}-\mathfrak{k}').$$

Neither commutation relations are introduced *ad hoc*, but originated quite naturally from the general quantum-mechanical formalism. As we shall see this suffices to obtain the correct sign of the gravitational interactions. *Thus, the fundamental difference between Coulomb and Newtonian forces is explained from quantum mechanics.*

Following the idea of Dirac, Fock and Podolsky[7] derived Coulomb's law. Our calculation proceeds exactly parallel to theirs. We start from the equations

$$\left(\frac{1}{2m_1}\mathfrak{p}_1{}^2 + \frac{m_1}{2}h_{00}(\mathfrak{r}_1)\right)\psi + \frac{h}{i}\frac{\partial\psi}{\partial t_1} = 0,$$
$$\left(\frac{1}{2m_2}\mathfrak{p}_2{}^2 + \frac{m_2}{2}h_{00}(\mathfrak{r}_2)\right)\psi + \frac{h}{i}\frac{\partial\psi}{\partial t_2} = 0.$$

At $t_1 = t_2 = t$ we have

$$\left(\frac{1}{2m_1}\mathfrak{p}_1{}^2 + \frac{1}{2m_2}\mathfrak{p}_2{}^2 + \frac{h}{i}\frac{\partial}{\partial t}\right)\psi = -\left(\frac{m_1}{2}h_{00}(\mathfrak{r}_1) + \frac{m_2}{2}h_{00}(\mathfrak{r}_2)\right)\psi.$$

[7] V. Fock and B. Podolsky, Sow. Phys. 1, 801, 1932 (Part II).

The expansion of the solution in the powers of m_1 and m_2 is required.

Due to the above-mentioned difference between the commutation relations, instead of the formulas (39) and (40) of Fock and Podolsky we obtain the following formulas:

$$-\left(\frac{1}{2m_1}\mathfrak{p}_1{}^2+\frac{1}{2m_2}\mathfrak{p}_2{}^2+\frac{h}{i}\frac{\partial}{\partial t}\right)\varphi_2\sim$$
$$\sim\frac{m_1}{2(2\pi)^{3/2}}\int h_{00,\mathfrak{k}}\varphi_1(\mathfrak{p}_1-h\mathfrak{k},\mathfrak{p}_2)e^{-i\omega t}d\mathfrak{k}+$$
$$+\frac{m_2}{2(2\pi)^{3/2}}\int h_{00,\mathfrak{k}}\varphi_1(\mathfrak{p}_1,\mathfrak{p}_2-h\mathfrak{k})e^{-i\omega t}d\mathfrak{k};$$
$$-\left(\frac{1}{2m_1}\mathfrak{p}_1{}^2+\frac{1}{2m_2}\mathfrak{p}_2{}^2+\frac{h}{i}\frac{\partial}{\partial t}\right)\varphi_1\sim$$
$$\sim\frac{m_1}{2(2\pi)^{3/2}}\int h^+_{00,\mathfrak{k}}\varphi_0(\mathfrak{p}_1+h\mathfrak{k},\mathfrak{p}_2)e^{i\omega t}d\mathfrak{k}+$$
$$+\frac{m_2}{2(2\pi)^{3/2}}\int h^+_{00,\mathfrak{k}}\varphi_0(\mathfrak{p}_1,\mathfrak{p}_2+h\mathfrak{k})e^{i\omega t}d\mathfrak{k}.$$

Then, [8]

$$\varphi_1(\mathfrak{p}_1,\mathfrak{p}_2)=$$
$$-\frac{m_1}{2(2\pi)^{3/2}h^3}h^+_{00,\frac{\mathfrak{p}_1{}^0-\mathfrak{p}_1}{h}}\frac{\delta(\mathfrak{p}_2-\mathfrak{p}_2{}^0)\delta_{j0}}{W-W_0+|\mathfrak{p}_1{}^0-\mathfrak{p}_1|}e^{i\frac{|\mathfrak{p}_1{}^0-\mathfrak{p}_1|-W_0}{h}t}-$$
$$-\frac{m_2}{2(2\pi)^{3/2}h^3}h^+_{00,\frac{\mathfrak{p}_2{}^0-\mathfrak{p}_2}{h}}\frac{\delta(\mathfrak{p}_1-\mathfrak{p}_1{}^0)\delta_{j0}}{W-W_0+|\mathfrak{p}_2{}^0-\mathfrak{p}_2|}e^{i\frac{|\mathfrak{p}_2{}^0-\mathfrak{p}_2|-W_0}{h}t}.$$

Let us put

$$h^+_{00,\mathfrak{k}}h_{00,\mathfrak{k}'}\sim 0\qquad\text{and}\qquad h_{00,\mathfrak{k}}h^+_{00,\mathfrak{k}'}\sim\frac{h}{2\omega}\delta(\mathfrak{k}-\mathfrak{k}').$$

After removing the infinite self-interaction term we finally obtain

$$\left(\frac{1}{2m_1}\mathfrak{p}_1{}^2+\frac{1}{2m_2}\mathfrak{p}_2{}^2+\frac{h}{i}\frac{\partial}{\partial t}\right)\sim$$
$$\sim\frac{m_1m_2}{4(2\pi)^{3/2}h}\frac{\delta(\mathfrak{p}_1-\mathfrak{p}_1{}^0+\mathfrak{p}_2-\mathfrak{p}_2{}^0)\delta_{j0}}{|\mathfrak{p}_1-\mathfrak{p}_1{}^0|^2}e^{-\frac{i}{h}W_0t}.$$

[8]For the notations see V. Fock and B. Podolsky, loc. cit.

The sign of the right-hand side is different than in the Fock - Podolsky formula (42). When we go back to the configuration space we accordingly obtain the Schrödinger equation with the potential energy

$$-\frac{m_1 m_2}{16\pi|\mathfrak{r}_1 - \mathfrak{r}_2|},$$

and thus we have recovered Newtonian gravitation as a necessary consequence of the quantum theory of gravity.

The Physical-Technical Institute
and The Physical Institute of the University.
Leningrad, August 1935.

Part III
Basic exact solutions and their interpretation

Chapter 11
Hidekazu Nariai 1950 + 1951, spherically symmetric solutions in vacuum that have no Newtonian limit

Original papers:

1. Hidekazu Nariai, On some static solutions of Einstein's gravitational field equations in a spherically symmetric case. *Scientific Reports of the Tôhoku University* **34**, 160 (1950).

2. Hidekazu Nariai, On a new cosmological solution of Einstein's field equations of gravitation. *Scientific Reports of the Tôhoku University* **35**, 46 (1951).

Reprinted with the kind permissions of the Astronomical Institute of the Tôhoku University and of Kenji Tomita. The journal no longer exists.

Previously published in *Gen. Relativ. Gravit.* **31** (1999), pages 945 – 949 (the editorial note + biography), 951 – 961 (paper 1), 963 – 971 (paper 2).

General Relativity and Gravitation, Vol. 31, No. 6, 1999

Editor's Note: On Some Static Solutions of Einstein's Gravitational Field Equations in a Spherically Symmetric Case. On a New Cosmological Solution of Einstein's Field Equations of Gravitation.

by Hidekazu Nariai
Scientific Reports of the Tôhoku University **34, 160 (1950).**
Scientific Reports of the Tôhoku University **35, 46 (1951)**

The solution presented in Nariai's papers is an illustration to a certain textbook problem that has been persistently gotten wrong in most relativity textbooks, even the recent ones (the *Exact solutions...*, Ref. 1, p. 155–158, being a notable exception). From the assumption of spherical symmetry, the following metric form follows:

$$ds^2 = \alpha(t,r)dt^2 + 2\beta(t,r)dtdr + \gamma(t,r)dr^2 + \delta(t,r)(d\vartheta^2 + \sin^2\vartheta d\varphi^2), \quad (1)$$

where α, β, γ and δ are arbitrary functions, ϑ and φ are coordinates on a sphere, r is a parameter labelling the spheres and t is a time coordinate. The functions α, β, γ and δ are determined up to the transformations:

$$t = f(t', r'), \qquad r = g(t', r'), \quad (2)$$

where f and g are arbitrary functions subject only to $\partial(f,g)/\partial(t',r') \neq 0$. Then, the argument goes, we can use the transformations (2) to simplify two of the functions α, β, γ and δ as we wish, so we choose f and g so that $\beta = 0$ and $\delta = -r^2$ after the transformation. This is where the error is. Such a transformation exists only if the gradient of δ is a spacelike vector,

0001-7701/99/0600-0945$16.00/0

which is a coordinate-independent property. The cases that are then left out of sight are:

(i) The gradient of δ being a timelike vector. Then coordinates can be chosen so that $\delta = -t^2$. This case contains, among other things, the Kantowski–Sachs [2] class of metrics (and its inhomogeneous generalizations in the $\beta' = 0$ subfamily of the Szekeres [3]–Szafron [4] metrics, see Ref. 5). Also, this case contains the Schwarzschild metric extended into the black hole region $r < 2m$. Strangely, this extension *is* discussed in most textbooks, but its inconsistency with $\delta = -r^2$ remains unnoticed.

(ii) The gradient of δ being a nonzero null vector.[1] The choice $\delta = -r^2$ is then possible, but it automatically implies $\alpha = 0$ in the new coordinates, and so one cannot achieve $\beta = 0$ in addition. In vacuum with zero cosmological constant, no solution of Einstein's equations exists in this case, but this editor is not aware of any further study of this class of metrics.

(iii) The gradient of δ being zero, i.e. δ being a constant. This is a coordinate-independent property in the class (1)–(2), and so no condition can be imposed on δ by coordinate transformations in this case. Again, no vacuum solution with $\Lambda = 0$ exists, and the vacuum solution with $\Lambda \neq 0$ was found by Nariai in the papers reprinted here (see also Ref. 6 and Ref. 1, p. 155–158).

In Nariai's first paper, the solution is hidden among other results. The main purpose of the paper was to obtain a collection of static spherically symmetric solutions of Einstein's equations, and the solution in question came up as just one element of the collection. The key observation is made in the phrase containing eq. (3), and the solution itself is given by eq. (35). The second paper is all devoted to investigating its properties.

Later research has brought more information on the solution (see Ref. 6). Nariai found it having assumed staticity from the beginning. However, this assumption is not necessary. It was shown in Ref. 6 that the collection of all spherically symmetric vacuum solutions of Einstein's equations with the cosmological constant consists of two metrics; one of them is the Kottler solution (i.e. the Schwarzschild solution generalized for Λ), and the other is the following:

$$\begin{aligned} ds^2 &= \{a(t)\cos[\ln(r/l)] + b(t)\sin[\ln(r/l)]\}^2 dt^2 \\ &\quad - (l/r)^2 dr^2 - l^2(d\vartheta^2 + \sin^2\vartheta d\varphi^2), \end{aligned} \tag{3}$$

[1] Here we refer to the case when the function δ has a nonzero null gradient in an open four-dimensional region. This should not be mixed up with the situation in the Schwarzschild solution, where the said gradient is null on a 3-dimensional hypersurface, the event horizon.

where $\Lambda = l^{-2}$. The time-dependence of (3) is spurious; a and b can be made constant by a coordinate transformation (see Ref. 6), and the resulting metric is equivalent to eq. (1) in the second paper.

The manifold of the Nariai solution is a Cartesian product of two surfaces of the same positive constant curvature, one with signature (+ −), the other with signature (++) (a sphere) (see Ref. 6). The solution can be reparametrized so that the limit $\Lambda \to 0$ (i.e. $l \to \infty$) becomes meaningful; the following coordinate transformation makes it possible:

$$r = le^{r'/l}, \qquad \vartheta = \pi/2 + \vartheta'/l, \qquad \varphi = \varphi'/l. \tag{4}$$

Then, dropping primes, the result is

$$ds^2 = [a(t)\cos(r/l) + b(t)\sin(r/l)]^2 dt^2 - dr^2 - d\vartheta^2 - \cos^2(\vartheta/l)d\varphi^2, \tag{5}$$

and the limit $l \to \infty$ is the Minkowski metric.

A certain misunderstanding concerning the so-called Bertotti [7]–Robinson [8] solution can be explained by this opportunity. The two solutions are not the same, contrary to common wisdom. The Bertotti solution is a generalization of the Nariai solution for electromagnetic field in the source. It has the same geometric structure as the Nariai solution, but the two curvatures are different, the first one is not necessarily positive, and their difference is proportional to the electromagnetic field. The Robinson solution [8] is the limit $\Lambda = 0$ of the Bertotti solution (it is conformally flat), and does not contain the Nariai solution as a subcase; the two curvatures in it are of the same absolute value, but of opposite signs. Another related problem is the so-called Birkhoff theorem (see Ref. 9, p. 167, for a discussion), which is not as strong as textbooks like to imply. The following two formulations of it are met in the literature: Every spherically symmetric solution of the Einstein equations in vacuum is:
(i) Static.
(ii) Equivalent to the Schwarzschild solution under a coordinate transformation.
The first formulation is false (the Schwarzschild solution taken inside the black hole region is the counterexample; unless the value of Λ is negative with a sufficiently large absolute value so that the horizon disappears). The second formulation is correct if $\Lambda = 0$, but false if $\Lambda \neq 0$; in the second case the Nariai solution is the counterexample.

Bonnor in Ref. 9, p. 167, proposed bypassing the problem by adding "Every *physically significant*..."; the physical significance requirement was meant to exclude the Nariai solution, very nearly rediscovered in Ref. 9.

However, it is fair to say that the *physical* meaning of the Nariai solution is still unknown; the original paper and Ref. 6 only investigated its geometry.

— *Andrzej Krasiński, Associate Editor*

Acknowledgement. The editor is grateful to K. Tomita for his help in contacting the publisher and for providing copies of the papers.

REFERENCES

1. Kramer, D., Stephani, H., MacCallum, M. A. H., and Herlt, E. (1980). *Exact Solutions of Einstein's Field Equations* (Cambridge University Press, Cambridge).
2. Kantowski, R., and Sachs, R. K. (1966). *J. Math. Phys.* **7**, 443.
3. Szekeres, P. (1975). *Commun. Math. Phys.* **41**, 55.
4. Szafron, D. A. (1977). *J. Math. Phys.* **18**, 1673.
5. Krasiński, A. (1997). *Inhomogeneous Cosmological Models* (Cambridge University Press, Cambridge).
6. Krasiński, A., and Plebański, J. (1980). *Rep. Math. Phys.* **17**, 217.
7. Bertotti, B. (1959). *Phys. Rev.* **116**, 1331.
8. Robinson, I. (1959). *Bull. Acad. Polon. Sci., Ser. Mat. Fis. Astr.* **7**, 351.
9. Bonnor, W. B. (1962). In *Recent Developments in General Relativity* (Pergamon and PWN, New York/Warszawa).

Brief biography

Hidekazu Nariai was born in Taisha-machi, Hikawa-gun, Shimane prefecture (west Japan) on February 2, 1924. He graduated from the Matsue high school in Shimane prefecture and the Tohoku university in Miyagi prefecture (east Japan). He obtained his PhD degree at this university and stayed there as an assistant. In 1953 he moved to Takehara city in Hiroshima prefecture (west Japan) as a research associate in the Research Institute for Theoretical Physics, Hiroshima university and continued his academic career there as a lecturer, an associate professor, professor (1972–1986), and emeritus professor (1987–1990). He died and was buried in Takehara city on December 5, 1990. In June 1990 his Institute (RITP) and the Research Institute for Fundamental Physics, Kyoto university (RIFP) were united into a single Institute — the Yukawa Institute for Theoretical Physics, Kyoto university (YITP).

His many and valuable works are concerned with (1) exact solutions including the Nariai solutions, (2) cosmological turbulence theory, (3) cosmological perturbation theory including Nariai and Ueno's theory of the cosmological Newtonian approximation, (4) quantum field theory in the expanding universe including Nariai and Kimura's theory, (5) junction conditions and general-relativistic dynamics in collapsing stars, and (6) renormalized gravitation theory with higher-order Lagrangians as an extension of the Einstein theory. At present the Nariai solutions seem to be most famous among his many works.

— *K. Tomita*

A more extended biography of H. Nariai can be found in Ref. 1.

— *Ed.*

REFERENCE

1. Tomita, K. (1992). In *Proc. 6th Marcel Grossman Meeting*, H. Sato and R. Ruffini, eds. (World Scientific, Singapore), p. XX.

General Relativity and Gravitation, Vol. 31, No. 6, 1999

On Some Static Solutions of Einstein's Gravitational Field Equations in a Spherically Symmetric Case†

Hidekazu Nariai*

Received November 30, 1950

Communicated by Prof. Z. Hitotuyanagi

1. INTRODUCTION

Several studies have been made on the problem of solving elementarily the gravitational field equations of the general theory of relativity.[1]

Tolman and Wyman discussed the problem starting from the following so-called standard form of line element

$$ds^2 = e^{\nu} dt^2 - e^{\lambda} dr_1^2 - r_1^2 d\theta^2 - r_1^2 \sin^2 \theta d\phi^2, \tag{1}$$

with $\nu = \nu(r_1)$ and $\lambda = \lambda(r_1)$, and assuming some suitable relations between ν and λ in order to make possible to solve the field equations. It is very difficult to find ν and λ assuming any suitable physical relation between density ρ_{00} and pressure p_0, therefore they treated the field equations by an inverse procedure and checked the obtained density and pressure from a physical point of view.

---

† Originally published in *The Science Reports of the Tohoku University* Series I, vol. **XXXIV**, No. 3 (1950), p. 160-167. Reprinted with the kind permission of the Astronomical Institute of the Tohoku University in Sendai (Japan), the current copyright owner.

* Institute of Astronomy, Faculty of Science, Tohoku University

1 R. C. Tolman, *Phys. Rev.* **55**, 364 (1939); M. Wyman, *Phys. Rev.* **75**, 1920 (1949).

0001-7701/99/0600-0951$16.00/0 ©1999 Plenum Publishing Corporation

In this paper, we shall discuss the same problem starting from the following so-called isotropic form of line element[2]

$$ds^2 = e^\nu dt^2 - e^\mu (dr^2 + r^2 d\theta^2 + r^2 \sin^2\theta d\phi^2), \tag{2}$$

with $\nu = \nu(r)$ and $\mu = \mu(r)$, and applying Tolman's procedure. Although we can always transform from the standard form to the isotropic one by putting

$$\frac{dr}{r} = e^{\lambda/2}\frac{dr_1}{r_1}, \tag{3}$$

but, as it is obvious in the case $e^\mu = \frac{\text{const.}}{r^2}$, the inverse process is not always possible, hence the isotropic form seems to be more general than the standard one. Moreover we have no systematic method of solving the field equations.

From these reasons, starting from the isotropic line element, we may have a further possibility of finding some new solutions of the field equations. Actually we can find some new types of solutions other than Tolman's. Here we do not discuss any detailed physical interpretation of them, but restrict ourselves to discuss only their mathematical characters.

2. FIELD EQUATIONS

The field equations are reduced to the following system of differential equations, provided that the matter behaves as a perfect fluid and is at rest with respect to the coordinate system of (2):[3]

$$8\pi p_0 = e^{-\mu}\left[\frac{\mu'^2}{4} + \frac{\mu'\nu'}{2} + \frac{\mu'+\nu'}{r}\right] + \Lambda, \tag{4}$$

$$8\pi p_0 = e^{-\mu}\left[\frac{\mu''}{2} + \frac{\nu''}{2} + \frac{\nu'^2}{4} + \frac{\mu'+\nu'}{2r}\right] + \Lambda, \tag{5}$$

$$8\pi \rho_{00} = -e^{-\mu}\left[\mu'' + \frac{\mu'^2}{4} + \frac{2\mu'}{r}\right] - \Lambda, \tag{6}$$

$$\frac{dp_0}{dr} = -\frac{(\rho_{00}+p_0)\nu'}{2}. \tag{7}$$

[2] Non-static case, starting from the same line element, has been studied by P. Kustaanheimo and B. Qvist. *Soc. Sci. Fenn. Comm. Phys.-Math.* **XIII**, 16 (1948). Moreover Wyman treated Schwarzschild interior solution in this line element. *Phys. Rev.* **70**, 74 (1946).

[3] R. C. Tolman, *Relativity, Thermodynamics, and Cosmology*, Oxford 1934, p. 244.

Since one of the above four equations is derived from the other three, we have, in all, three independent equations for the four unknown functions ν, μ, ρ_{00} and p_0 of r. Hence it is possible to choose arbitrarily one of the four unknown functions, or to assume a suitable independent relation between the four unknowns. Our practical procedure of treating the field equations consists of the two methods; in section 3 we adopt a direct method, and in section 4 a method of putting $\nu = \mu + \lambda$.

3. SOLUTIONS BY THE FIRST METHOD

Taking the difference of (4) and (5), and putting

$$\xi = \mu'/r \quad \text{and} \quad \eta = \nu'/r, \tag{8}$$

we have

$$\xi' + \eta' + \frac{\nu'}{2}\eta - \frac{\mu'}{2}\xi - \xi\eta r = 0\,. \tag{9}$$

By further arrangement, we can write (9) as follows:

$$\frac{d[\log\xi - \frac{1}{2}\mu]}{d\nu} + \frac{d[\log\eta + \frac{1}{2}\nu]}{d\mu} - 1 = 0\,. \tag{9$'$}$$

The above expression is the clue of the first method. The field equations have, of course, one degree of freedom. Hence we shall discuss several cases corresponding to the conditions adopted for ν and μ, respectively.

Case I. If we assume $d[\log\xi - \frac{\mu}{2}] = 0$, the soution is easily obtained as follows:

$$\begin{aligned}
e^{\mu} &= \frac{4}{(\frac{a}{2}r^2 + b)^2}\,, \\
e^{\nu} &= \frac{1}{4}\left[d - \frac{4c/a}{\frac{a}{2}r^2 + b}\right]^2, \\
8\pi p_0 &= \Lambda - \frac{1}{2}\,ab + \frac{c(\frac{a}{2}r^2 - b)}{\frac{2c}{a} - \frac{b}{2}(\frac{a}{2}r^2 + b)}\,, \\
8\pi\rho_{00} &= -\Lambda + \frac{3}{2}\,ab = \text{const.},
\end{aligned} \tag{10}$$

where a,b,c, and d are constants of integration and $a \neq 0$. This solution is identical with Tolman's solution III, i.e., the Schwarzschild interior solution, and includes, as a special case ($c = 0$), Tolman's solution I, i.e., the Einstein universe.

Case II. $\mu =$ const., or $\nu =$ const. In these cases, we must start from (9), and, moreover, we may put const. $= 0$ without any loss of generality. The case $\nu = 0$ is already found in the above Case I, and its solution is identical with Tolman's solution I. Therefore we consider only the case $\mu = 0$.

Since $\xi = \xi' = 0$ by (8), the equation (9) becomes $\eta' + \frac{\nu'}{2}\dot{\eta} = 0$. Integrating this equation, we get

$$e^{\nu} = \frac{1}{4}\left[\frac{a}{2}r^2 + b\right]^2, \tag{11}$$

where a and b are constants of integration. Substituting (11) and $\mu = 0$ into (4) and (6), we get

$$8\pi p_0 = \Lambda + \frac{2a}{\frac{a}{2}r^2 + b}, \qquad (11)'$$
$$8\pi \rho_{00} = -\Lambda.$$

This solution is identical with Tolman's solution VI putting $n^2 = 1$ in his expressions.[4]

Case III. $\nu = k\mu +$ const. and $k \neq 0$. Substituting $\eta = k\xi$ into (9)′, we get

$$\xi \equiv \frac{\mu'}{r} = 2A\exp\left[\frac{1 + 2k - k^2}{2(1+k)}\,\mu\right].$$

Here we must assume $k \neq 1$. If $k = -1$, we have from (9)′

$\mu =$ const. and $\nu =$ const., so that the line element (2) is reduced to that of the flat space-time.

From the above equation, we have

$$e^{\mu/p}d\mu = 2Ardr, \tag{12}$$

where $p = \frac{2(1+k)}{k^2-2k-1}$. If $p \neq \infty$, we have the following solution:

$$e^{\mu} = \left[\frac{Ar^2 + A_1}{p}\right]^p,$$
$$e^{\nu} = \left[\frac{Ar^2 + A_1}{p}\right]^{pk},$$
$$8\pi p_0 = \Lambda + A\left[\frac{Ar^2 + A_1}{p}\right]^q \left\{Ak^2r^2 + \frac{2(1+k)}{p}A_1\right\},$$
$$8\pi \rho_{00} = -\Lambda - \frac{A}{1+k}\left[\frac{Ar^2 + A_1}{p}\right]^q \left\{A(k^2 - k)r^2 + \frac{6(1+k)}{p}A_1\right\}, \tag{13}$$

[4] R. C. Tolman, *loc. cit.*, p. 368, formula (4,6).

where $q = (\frac{1-k}{1+k})kp$. If, particularly, $k < -1$, A and $A_1 > 0$ and $q = -2n$ ($n = 1, 2, 3, \ldots$), this solution has a possibility of physical application, because ρ_{00} and p_0, under these conditions, are positive and converge to zero as $r \to \infty$.

Therefore this solution is a new type one. If $p = \infty$, we get, similarly to the above, one solution, but the obtained ρ_{00} and p_0 diverge to infinity as $r \to \infty$. Furthermore, for the case $\frac{d[\log \xi - \frac{1}{2}\mu]}{d\nu} = c$ and $c(c-1) \neq 0$, there exists only one elementary solution which reduces to the Case III for particular integration constant.

4. SOLUTIONS BY THE SECOND METHOD

Taking the difference of (4) and (5), and putting

$$\nu = \mu + \lambda, \tag{14}$$

we get the following differential equation

$$2\left(\mu'' - \frac{\mu'^2}{2} - \frac{\mu'}{r}\right) + \left(\lambda'' + \frac{\lambda'^2}{2} - \frac{\lambda'}{r}\right) = 0, \tag{15}$$

which has a separated form for μ and λ.

Since there is one degree of freedom, we can consider, instead of (15), the following system of two differential equations:

$$\begin{aligned} \mu'' - \frac{\mu'^2}{2} - \frac{\mu'}{r} &= -a(r), \\ \lambda'' + \frac{\lambda'^2}{2} - \frac{\lambda'}{r} &= 2a(r), \end{aligned} \tag{16}$$

where $a(r)$ is a yet undetermined function. Moreover, putting (14) into (4) and (6), we get

$$8\pi p_0 = \Lambda + e^{-\mu}\left[\frac{3}{4}\mu'^2 + \frac{2\mu'}{r} + \frac{\mu'\lambda'}{2} + \frac{\lambda'}{r}\right], \tag{4$'$}$$

$$8\pi\rho_{00} = -\Lambda - e^{-\mu}\left[\mu'' + \frac{1}{4}\mu'^2 + \frac{2\mu'}{r}\right]. \tag{6$'$}$$

Thus, by regarding $a(r)$ as a parametric function, we have exactly four independent equations for four unknowns. Since both of the equations (16) are of the so-called Riccati type, we can not, in general, solve them with elementary functions. But we can find it profitable to use the following substitutions:

$$\frac{\lambda'}{r} = L \quad \text{and} \quad \frac{\mu'}{r} = M, \tag{17}$$

or

$$\lambda' = 2\frac{d\log\xi}{dr} \quad \text{and} \quad \mu' = -2\frac{d\log\eta}{dr}. \tag{18}$$

Substituting (17) or (18) into (16), we get the following two alternatives of the system of differential equations:

$$\begin{aligned} L' &= \frac{2a(r)}{r} - \frac{r}{2}L^2, \\ M' &= -\frac{a(r)}{r} + \frac{r}{2}M^2, \end{aligned} \tag{19}$$

or

$$\begin{aligned} \xi'' - \frac{\xi'}{r} - a(r)\xi = 0, \\ \eta'' - \frac{\eta'}{r} - \frac{1}{2}a(r)\eta = 0. \end{aligned} \tag{20}$$

Thus each of (19) has an ordinary form of the Riccati type, and each of (20) is a linear homogeneous differential equation of the second order. We shall discuss the solutions starting at first from (19), and later from (20).

If we assume

$$\alpha(r) = \alpha_0 r^2, \tag{21}$$

where α_0 is an arbitrary constant, we can easily perform the separation of variables and we get

$$\begin{aligned} L' &= r[2\alpha_0 - \tfrac{1}{2}L^2], \\ M' &= r[-\alpha_0 + \tfrac{1}{2}M^2]. \end{aligned} \tag{19$'$}$$

We have here three cases according to whether α_0 is positive, negative or zero. When $\alpha_0 > 0$, the obtained ρ_{00} and p_0, however, diverge to infinity as $r \to \infty$. Hence we consider only the cases $\alpha_0 \leq 0$.

Case IV. When $\alpha_0 < 0$, the equations (20) become

$$\begin{aligned} L \equiv \frac{\lambda'}{r} &= L_0 \tan\left(a - \frac{L_0}{4}r^2\right), \\ M \equiv \frac{\mu'}{r} &= M_0 \tan\left(b + \frac{M_0}{4}r^2\right), \end{aligned} \tag{22}$$

where $\pm L_0 = 2\sqrt{-\alpha_0}$, $\pm M_0 = \sqrt{-2\alpha_0}$, and a and b are constants of integration.

Integrating (22), we get

$$
\begin{aligned}
e^{\mu} &= e^{A}\left[1/\cos\left(b+\frac{M_0}{4}r^2\right)\right]^2,\\
e^{\nu} &= e^{A}\left[\cos\left(a-\frac{L_0}{4}r^2\right)\Big/\cos\left(b+\frac{M_0}{4}r^2\right)\right]^2,\\
8\pi p_0 &= \Lambda + e^{-A}\left[\frac{3}{8}M_0^2r^2\left[1-\cos 2\left(b+\frac{M_0}{4}r^2\right)\right] + M_0\sin 2\left(b+\frac{M_0}{4}r^2\right)\right.\\
&\quad + L_0\tan\left(a-\frac{L_0}{4}r^2\right)\left\{\frac{1}{2}\left[1+\cos 2\left(b+\frac{M_0}{4}r^2\right)\right]\right.\\
&\quad\left.\left. + \frac{1}{4}M_0r^2\sin 2\left(b+\frac{M_0}{4}r^2\right)\right\}\right],\\
8\pi\rho_{00} &= -\Lambda + e^{-A}\left[\frac{1}{8}M_0^2r^2\left\{-5+\cos 2\left(b+\frac{M_0}{4}r^2\right)\right\}\right.\\
&\quad\left. - \frac{3}{2}M_0\sin 2\left(b+\frac{M_0}{4}r^2\right)\right],
\end{aligned}
\tag{23}
$$

where A is an integration constant. This solution seems to be entirely a new type, but because of the appearance of trigonometric functions in ρ_{00} and p_0, the ranges where $\rho_{00} < 0$ and $p_0 < 0$ have no physical meaning.

Case V. Subsituting $\alpha_0 = 0$ into (19), and integrating them, we get

$$
\begin{aligned}
e^{\mu} &= e^{A}\left[1\Big/\left(\frac{r^2}{2}+b\right)\right]^2,\\
e^{\nu} &= e^{A}\left[\left(\frac{r^2}{2}+a\right)\Big/\left(\frac{r^2}{2}+b\right)\right]^2.
\end{aligned}
\tag{24}
$$

This solution is identical with that of the Case I. We can easily show it by putting $a = 1$, $4 = e^{A}$, $d = 4$ and $b - c \to a$ in the equations (10).

It seems to be impossible to find the other forms of $\alpha(r)$, making possible to solve (19) with elementary functions. Similarly to the above, it is, in general, difficult to solve (20) with elementary functions.

But, if we assume a particular form for $\alpha(r)$, i.e.

$$
\alpha(r) = \frac{\alpha_0}{r^2}\,, \tag{25}
$$

where α_0 is an arbitrary constant, we can easily find its solutions. From (20) and (25), we get

$$
\begin{aligned}
\xi'' - \frac{\xi'}{r} - \frac{\alpha_0}{r^2}\,\xi &= 0,\\
\eta'' - \frac{\eta'}{r} - \frac{1}{2}\,\frac{\alpha_0}{r^2}\,\eta &= 0\,.
\end{aligned}
\tag{20$'$}
$$

The general solutions of (20)′ are easily obtained, but their forms depend on the values of α_0. From the considerations of indicial equations, we must consider separately the following five cases; (VI) $\alpha_0 > -1$, (VII) $-1 > \alpha_0 > -2$, (VIII) $\alpha_0 < -2$, (IX) $\alpha_0 = -1$ and (X) $\alpha_0 = -2$.

Case VI. The resulting solution is as follows:

$$\begin{aligned}
e^{\mu} &= e^{A}\left[\frac{1}{B_1 r^{1+q} + B_2 r^{1-q}}\right]^2, \\
e^{\nu} &= e^{A}\left[\frac{A_1 r^{p} + A_2 r^{-p}}{B_1 r^{q} + B_2 r^{-q}}\right]^2, \\
8\pi p_0 &= \Lambda + e^{-A}\Bigg[(3q^2-1)\,[B_1^2 r^{2q} + B_2^2 r^{-2q}] - 2(3q^2+1)B_1B_2 \\
&\qquad - 2pq[B_1^2 r^{-2q} - B_2^2 r^{-2q}] \times \left[\frac{A_1 r^{p} - A_2 r^{-p}}{A_1 r^{p} + A_2 r^{-p}}\right]\Bigg], \\
8\pi\rho_{00} &= -\Lambda - e^{-A}[\,(q^2-1)\,[B_1^2 r^{2q} + B_2^2 r^{-2q}] - 2(5q^2+1)B_1B_2],
\end{aligned} \tag{26}$$

where $p = \sqrt{1+\alpha_0}$, $q = \sqrt{1+\frac{\alpha_0}{2}}$, and A, A_1, A_2, B_1 and B_2 are constants of integration. (In the following, A, A_1, A_2, B_1 and B_2 have the same meaning as the present ones.) This solution is a new type one. If we put $\alpha_0 = 0$, this solution is reduced to (24).

Case VII. In this case, the functional form of e^{μ} is the same as that of (26), but e^{ν} is of a different form. Therefore the solution is as follows:

$$\begin{aligned}
e^{\mu} &= e^{A}\left[\frac{1}{B_1^2 r^{1+q} + B_2 r^{1-q}}\right]^2, \\
e^{\nu} &= e^{A}\left[\frac{A_1\cos(p\log r) + A_2\sin(p\log r)}{B_1 r^{q} + B_2 r^{-q}}\right]^2, \\
8\pi p_0 &= \Lambda + e^{-A}\Bigg[(3q^2-1)\,[B_1^2 r^{2q} + B_2^2 r^{-2q}] - 2(3q^2+1)B_1B_2 \\
&\qquad - 2pq[B_1^2 r^{-2q} - B_2^2 r^{-2q}]\left[\frac{A_2\cos(p\log r) - A_1\sin(p\log r)}{A_1\cos(p\log r) + A_2\sin(p\log r)}\right]\Bigg], \\
8\pi\rho_{00} &= -\Lambda - e^{-A}[\,(q^2-1)\,[B_1^2 r^{2q} + B_2^2 r^{-2q}] - 2(5q^2+1)B_1B_2],
\end{aligned} \tag{27}$$

where $p = \sqrt{-(1+\alpha_0)}$, $q = \sqrt{1+\frac{\alpha_0}{2}}$. This solution is also a new type one.

Putting $\alpha_0 = -\frac{3}{2}$, $\Lambda = B_1 = 0$ and $B_2 = 1$, we get, as a special case,

$$\begin{aligned}
e^{\mu} &= \frac{e^A}{r},\\
e^{\nu} &= e^A \left[r^{1/2} \left\{ A_1 \cos\left(\frac{\log r}{\sqrt{2}}\right) + A_2 \sin\left(\frac{\log r}{\sqrt{2}}\right)\right\}\right]^2,\\
8\pi p_0 &= \frac{1}{4} \cdot \frac{e^{-A}}{r} \left[2\sqrt{2} \left\{ \frac{A_2 \cos(\frac{\log r}{\sqrt{2}}) - A_1 \sin(\frac{\log r}{\sqrt{2}})}{A_1 \cos(\frac{\log r}{\sqrt{2}}) + A_2 \sin(\frac{\log r}{\sqrt{2}})} \right\} - 1 \right],\\
8\pi \rho_{00} &= \frac{3}{4} \cdot \frac{e^{-A}}{r}.
\end{aligned} \tag{28}$$

If we transform from the above isotropic form to the standard one by putting $r_1^2 = e^A r$, we get the following line element:

$$ds^2 = [r_1(a \cos\sqrt{2}\, r_1 + b \sin\sqrt{2}\, r_1)]^2 dt^2 - 4dr_1^2 - r_1^2 d\theta^2 - r_1^2 \sin^2\theta d\phi^2, \tag{29}$$

where $a = A_1 \cos\frac{A}{\sqrt{2}} - A_2 \sin\frac{A}{\sqrt{2}}$ and $b = A_2 \cos\frac{A}{\sqrt{2}} + A_1 \sin\frac{A}{\sqrt{2}}$, In this line element, the coefficient of dr_1^2 is constant, similar to Tolman's solution VI, but the coefficient of dt^2 is entirely different from that of Tolman's.

Case VIII. The solution is:

$$\begin{aligned}
e^{\mu} &= e^A \left[\frac{1}{r\{B_1 \cos(q \log r) + B_2 \sin(q \log r)\}} \right]^2,\\
e^{\nu} &= e^A \left[\frac{A_1 \cos(p \log r) + A_2 \sin(p \log r)}{B_1 \cos(q \log r) + B_2 \sin(q \log r)} \right]^2,\\
8\pi p_0 &= \Lambda + e^{-A} \left[\frac{(3q^2 - 1)(B_4^2 + B_2^2)}{2} \right.\\
&\quad + (3q^2+1) \left\{ \frac{B_2^2 - B_1^2}{2} \cos 2(q \log r) - B_1 B_2 \sin 2(q \log r) \right\}\\
&\quad - 3q \left\{ \frac{B_2^2 - B_1^2}{2} \sin 2(q \log r) + B_1 B_2 \cos 2(q \log r) \right\} \times\\
&\quad \times \left. \left\{ 1 + \frac{2}{3} p \left(\frac{A_2 \cos(p \log r) - A_1 \sin(p \log r)}{A_1 \cos(p \log r) + A_2 \sin(p \log r)} \right) \right\} \right],\\
8\pi \rho_{00} &= -\Lambda + e^{-A} \left[- \frac{(5q^2 - 1)(B_1^2 + B_2^2)}{2} \right.\\
&\quad - (q^2+1) \left\{ \frac{B_2^2 - B_1^2}{2} \cos 2(q \log r) - B_1 B_2 \sin 2(q \log r) \right\}\\
&\quad \left. + q \left\{ \frac{B_2^2 - B_1^2}{2} \sin 2(q \log r) + B_1 B_2 \cos 2(q \log r) \right\} \right],
\end{aligned} \tag{30}$$

where $p = \sqrt{-(1+\alpha_0)}$ and $q = \sqrt{-(1+\frac{\alpha_0}{2})}$. In this case, the variations of ρ_{00} and p_0 depend only on the trigonometric functions.

Case IX. In this case, the indicial equation concerning the first differential equation of (20)′ has a double root. Considering this fact, we get the following solution:

$$\begin{aligned}
e^{\mu} &= e^{A}\left[\frac{1}{r(B_1 r^{\sqrt{2}/2} + B_2 r^{-\sqrt{2}/2})}\right]^2, \\
e^{\nu} &= e^{A}\left[\frac{A_1 + A_2 \log r}{B_1 r^{\sqrt{2}/2} + B_2 r^{-\sqrt{2}/2})}\right]^2, \\
8\pi p_0 &= \Lambda + e^{-A}\left[\frac{1}{2}(B_1 r^{\sqrt{2}/2} + B_2 r^{-\sqrt{2}/2})^2 - 6B_1B_2 \right. \\
&\qquad \left. - \sqrt{2}\,(B_1^2 r^{\sqrt{2}} - B_2^2 r^{-\sqrt{2}}) \times \left(\frac{A_2}{A_1 + A_2 \log r}\right)\right], \\
8\pi \rho_{00} &= -\Lambda + e^{-A}\left[\frac{1}{2}(B_1 r^{\sqrt{2}/2} + B_2 r^{-\sqrt{2}/2})^2 + 6B_1B_2\right]. \qquad (31)
\end{aligned}$$

Case X. In this case, the indicial equation concerning the second of (20)′ has a double root. Therefore we get

$$\begin{aligned}
e^{\mu} &= e^{A}\left[\frac{1}{r(B_1 + B_2 \log r)}\right]^2, \\
e^{\nu} &= e^{A}\left[\frac{A_1 \cos\log r + A_2 \sin\log r}{B_1 + B_2 \log r}\right]^2, \\
8\pi p_0 &= \Lambda - e^{-A}\Bigg[(B_1 + B_2 \log r)^2 - 2B_2(B_1 + B_2 \log r)^2 - 3B_2^2 \\
&\qquad + 2B_2\left\{1 + \left(\frac{A_2 \cos\log r - A_1 \sin\log r}{A_1 \cos\log r + A_2 \sin\log r}\right)\right\}\Bigg], \\
8\pi \rho_{00} &= -\Lambda + e^{-A}[(B_1 + B_2 \log r)^2 - 3B_2^2]. \qquad (32)
\end{aligned}$$

If we consider, in this new solution, the case $B_2 = 0$ as a special one, we get

$$\begin{aligned}
e^{\mu} &= \frac{e^{A}}{B_1^2 r^2}, \\
e^{\nu} &= \frac{e^{A}}{B_1^2}(A_1 \cos\log r + A_2 \sin\log r)^2,
\end{aligned}$$

$$8\pi p_0 = \Lambda - e^{-A} B_1^2 = \text{const.},$$

$$8\pi \rho_{00} = -\Lambda + e^{-A} B_1^2 = \text{const.} \tag{33}$$

From the last two equations, we have $\rho_{00} + p_0 = 0$. Since ρ_{00} and p_0 are intrinsically non-negative quantities, we get

$$\rho_{00} = p_0 = 0 \quad \text{and} \quad \Lambda = e^{-A} B_1^2 > 0\,. \tag{34}$$

And the form of line element becomes as follows:

$$ds^2 = \frac{1}{\Lambda}\left[(A_1 \cos\log r + A_2 \sin\log r)^2 dt^2 - \frac{1}{r^2}\left(dr^2 + r^2 d\theta^2 + r^2 \sin^2\theta d\phi^2\right)\right]. \tag{35}$$

This is a new cosmological solution which has a spherical symmetric and static character and a homogeneous (empty) matter-distribution. The geometrical structure of this line element is identical with that of Einstein's space. Moreover this form can not be transformed into the standard one.

5. CONCLUSION

By suitable arrangements of the field equations, we obtained some entirely new types of solutions together with the well-known types. The most remarkable feature is the appearance of trigonometrical and logarithmic functions. Moreover we get a new example of cosmological solutions which has a spherical symmetric and static character and contains a homogeneous (empty) matter-distribution.

The author is particularly indebted to Prof. Z. Hitotuyanagi for many valuable suggestions during the course of this work.

Nov. 24th 1950.

Astronomical Institute
Tohoku University

General Relativity and Gravitation, Vol. 31, No. 6, 1999

On a New Cosmological Solution of Einstein's Field Equations of Gravitation[†]

Hidekazu Nariai*

Received April 10, 1951

Communicated by Prof. Z. Hitotuyanagi

1. INTRODUCTION

In the previous paper,[1] the author obtained the following solution of Einstein's field equations for the homogeneous static universe exhibitin spherical symmetry:

$$ds^2 = \frac{1}{\Lambda}\left[(A\cos\log r + B\sin\log r)^2 dt^2 - \frac{1}{r^2}\left(dr^2 + r^2 d\theta^2 + r^2 \sin^2\theta d\phi^2\right)\right], \tag{1}$$

where Λ is the cosmological constant and A and B are arbitrary constants. We have shown that this solution corresponds to the empty universe and, moreover, this line element can not be transformed into the standard form.

The field equations in the empty universe which this solution must satisfy are, however, reduced to

$$R_{\mu\nu} = \Lambda g_{\mu\nu}\,, \tag{2}$$

† Originally published in *The Science Reports of the Tohoku University* Series I, vol. XXXV, No. 1 (1951), p. 46-57. Reprinted with the kind permission of the Astronomical Institute of the Tohoku University in Sendai (Japan), the current copyright owner.

* Institute of Astronomy, Faculty of Science, Tohoku University

1 H. Nariai, *Sci. Rep. Tohoku Univ.* Ser. 1, vol. **XXXIV**, No. 3, p. 160 (1950).

where $R_{\mu\nu}$ is the contracted Riemann–Christoffel tensor. In this paper, at first, we rederive the line element (1) as the solution of the field equations (2) starting from the isotropic static line element and then discuss in detail the geometrical and physical natures of the solution under consideration.

2. FIELD EQUATIONS FOR STATIC EMPTY UNIVERSE EXHIBITING SPHERICAL SYMMETRY

The static isotropic line element exhibiting spherical symmetry is written as follows:

$$ds^2 = e^\nu dt^2 - e^\mu (dr^2 + r^2 d\theta^2 + r^2 \sin^2 \theta d\phi^2), \tag{3}$$

with $\mu = \mu(r)$ and $\nu = \nu(r)$. And its field equations (2) for the empty matter-distribution are reduced to the following system of differential equations:

$$\mu'' + \frac{\nu''}{2} + \frac{\nu'^2}{4} - \frac{\mu'\nu'}{4} + \frac{\mu'}{r} = -\Lambda e^\mu, \tag{4}$$

$$\frac{\mu''}{2} + \frac{\mu'^2}{4} + \frac{\mu'\nu'}{4} + \frac{\nu' + 3\mu'}{2r} = -\Lambda e^\mu, \tag{5}$$

$$\frac{\nu''}{2} + \frac{\nu'^2}{4} + \frac{\mu'\nu'}{4} + \frac{\nu'}{r} = -\Lambda e^\mu, \tag{6}$$

where primes denote differentiation with respect to r. Taking the difference of (4) and (6), we get:

$$\mu'' - \frac{\mu'\nu'}{2} + \frac{\mu' - \nu'}{r} = 0. \tag{7}$$

The integral is

$$\mu' = 2(Ae^{\nu/2} - 1)/r, \tag{8}$$

where A is an integration constant. Substituting (8) into (5), we get after suitable arrangements

$$2\xi' + \frac{\xi^2 - 1}{r} + \frac{\Lambda}{r} \exp\left(2 \int \frac{\xi}{r} dr\right) = 0, \tag{9}$$

or

$$\xi'' + \frac{\xi'(1 - \xi)}{r} + \frac{\xi(1 - \xi^2)}{r^2} = 0, \tag{9$'$}$$

where $\xi = Ae^{\nu/2}$. It is usually very difficult to obtain the general solution of (9)′ on account of its non-linear character, but we can find several particular solutions. For example, the solutions $\xi_1 = 1$ $(A > 0)$, $\xi_2 = -1$ $(A < 0)$ and $\xi_3 = 0$ $(A = 0)$ are easily found from the formal nature of (9)′. Moreover, (9)′ is also satisfied by $\xi_4 = \frac{1-a/r}{1+a/r}$ $(A = 1,\ \Lambda = 0)$ and $\xi_5 = \frac{a^2-r^2}{a^2+r^2}$ $(A = 1)$.

The solutions ξ_1 and ξ_2 represent the flat space-time, and ξ_4 corresponds to the Schwarzschild exterior solution in the isotropic form of line element (3). From ξ_5 we obtain, putting $a = 2$ without any loss of generality,

$$ds^2 = \left(\frac{1 - r^2/4}{1 + r^2/4}\right)^2 dt^2 - R^2(dr^2 + r^2 d\theta^2 + r^2 \sin^2\theta d\phi^2)\,(1 + r^2/4)^{-2}, \quad (10)$$

where $R^2 = 3/\Lambda$. This is the static isotropic form of the de Sitter universe. These last two expressions correspond to the solutions in an isotropic form representing the special cases ($\Lambda = 0$ or $m = 0$) of the general standard-form solution

$$ds^2 = \gamma dt^2 - \gamma^{-1} dr^2 - r^2 d\theta^2 - r^2 \sin^2\theta d\phi^2, \quad (11)$$

where $\gamma = 1 - 2m/r - \Lambda r^2/3$.

Finally, from $\xi_3 = 0$, (8) and (5), we get

$$e^{\mu} = 1/\Lambda r^2, \qquad (\Lambda > 0). \quad (12)$$

Substituting (12) into (6), we get

$$\eta'' + \frac{\eta'}{r} + \frac{\eta}{r^2} = 0, \qquad (e^{\nu} = \eta^2). \quad (13)$$

The general solution is

$$\eta = A \cos\log r + B \sin\log r, \quad (14)$$

where A and B are arbitrary constants. Hence we get the line element

$$ds^2 + (A\cos\log r + B\sin\log r)^2 dt^2 - \frac{1}{\Lambda r^2}(dr^2 + r^2 d\theta^2 + r^2 \sin^2\theta d\phi^2), \quad (15)$$

which, by adjusting the scale of time-variable t, becomes to (1) obtained in the previous paper. We are thus led to the conclusion that the solution (1), which is also the same as the case $\xi_3 = 0$, is one of possible solutions other than the well-known one for the empty static universe.[2]

[2] We can prove more generally that the solution (15) is the only one which has spherical symmetry and satisfies the condition: $r^2 e^{\mu}$ = function of t only.

3. GEOMETRICAL PROPERTIES OF THE LINE ELEMENT

Performing the following transformation of coordinates in the line element (1):

$$t = \tau \left(\frac{\Lambda}{A^2 + B^2} \right)^{1/2} \tag{16}$$

and $r = \exp(\pm\chi + \tan^{-1}\frac{B}{A})$ or $r_1 = R\sin\chi$ ($R^2 = \frac{1}{\Lambda}$), we get

$$ds^2 = \cos^2\chi d\tau^2 - R^2(d\chi^2 + d\Sigma^2), \tag{17}$$

or

$$ds^2 = \left(1 - \frac{r_1^2}{R^2}\right) d\tau^2 - \frac{dr_1^2}{1 - r_1^2/R^2} - R^2 d\Sigma^2, \tag{18}$$

where $d\Sigma^2 = d\theta^2 + \sin^2\theta d\phi^2$. If we compare (17) or (18) with the corresponding line elements of the de Sitter universe:

$$ds^2 = \cos^2\chi d\tau^2 - R^2(d\chi^2 + \sin^2\chi d\Sigma^2), \tag{19}$$

or

$$ds^2 = \left(1 - \frac{r_1^2}{R^2}\right) d\tau^2 - \frac{dr_1^2}{1 - r_1^2/R^2} - r_1^2 d\Sigma^2, \tag{20}$$

where $R^2 = 3/\Lambda$, we can then find their similarities. Really the different points are concerned only with the coefficients of $d\Sigma^2$ and the values of $R^2\Lambda$, and the other parts are entirely the same. Therefore, as far as we consider the purely radial case $d\Sigma = 0$ such as the propagation of light in the radial direction, both solutions give the same results.

Moreover, the de Sitter universe can be represented as a hypersurface of imaginary radius of curvature iR embedded in the five-dimensional Euclidean space. We can also obtain a similar representation of our solution by the following transformation of coordinates:

$$
\begin{aligned}
Z_1 &= R\sin\theta\cos\phi, \\
Z_2 &= R\sin\theta\sin\phi, \\
Z_3 &= R\cos\theta, \\
Z_4 &= r_1\,, \\
Z_5 &= R\left(\cosh\frac{\tau}{R}\right)\sqrt{1-\frac{r_1^2}{R^2}}\,, \\
Z_6 &= R\left(\sinh\frac{\tau}{R}\right)\sqrt{1-\frac{r_1^2}{R^2}}\,.
\end{aligned} \tag{21}
$$

Substituting (21) into (18), we get, after suitable arrangements,

$$ds^2 = -dZ_1^2 - dZ_2^2 - dZ_3^2 - dZ_4^2 - dZ_5^2 + dZ_6^2\,, \tag{22}$$

where

$$Z_1^2 + Z_2^2 + Z_3^2 = R^2 \qquad \text{and} \qquad Z_4^2 + Z_5^2 - Z_6^2 = R^2. \tag{23}$$

We can, therefore, interpret the solution considered here as a four-dimensional hypersurface, which is a section of two five-dimensional hypersurfaces (23)[3] embedded in the six-dimensional pseudo-Euclidean space (22).

Finally, if we write the spatial part of the original line element (1) as $d\sigma^2$ we get

$$d\sigma^2 = R^2(dx^2 + dy^2 + dz^2)/r^2, \tag{24}$$

where $x = r\sin\theta\cos\phi$, $y = r\sin\theta\sin\phi$, $z = r\cos\theta$. Taking the following transformation of coordinates:

$$x/r^2 = x^*, \quad y/r^2 = y^*, \quad z/r^2 = z^* \quad (\text{i.e., } rr^* = 1), \tag{25}$$

we get

$$d\sigma^2 = R^2(dx^{*2} + dy^{*2} + dz^{*2})/r^{*2}. \tag{26}$$

Hence we can see that the spatial part of (1) is invariant under the inversion given by (25).

---

[3] We can also take, in place of (23), the following two expressions:

$$z_1{}^2 + z_2{}^2 + z_3{}^2 + z_4{}^2 + z_5{}^2 - z_6{}^2 = 2R^2, \qquad z_1{}^2 + z_2{}^2 + z_3{}^2 - z_4{}^2 - z_5{}^2 + z_6{}^2 = 0\,.$$

The former is a hyper-hyperboloid of one sheet and the latter a hypercone.

4. MOTION OF A FREE PARTICLE

Let us consider the geodesic equations of motion of a free particle in the space-time considered here. Starting from (22), (23) and applying the variational principle under the constraint conditions, we get

$$\left.\begin{aligned} \frac{d^2 z_i}{ds^2} &= \alpha z_i \qquad (i = 1, 2, 3), \\ \frac{d^2 z_j}{ds^2} &= \beta z_j \qquad (j = 4, 5, 6), \end{aligned}\right\} \tag{27}$$

where $\alpha+\beta = 1/R^2$ and α, β are parameters, of which general solution may be obtained without difficulty. But let us here start from the equation (17) directly. Among forty $\frac{\partial g_{\mu\nu}}{\partial x^\lambda}$'s in (17), only two are non-vanishing quantities, i.e., $\frac{\partial g_{33}}{\partial \theta}$ and $\frac{\partial g_{44}}{\partial \chi}$, hence as the non-vanishing Christoffel symbols $\Gamma^\lambda_{\mu\nu}$ remain:

$$\begin{aligned} \Gamma^1_{44} &= -\sin\chi\cos\chi/R^2, & \Gamma^2_{33} &= -\sin\theta\cos\theta, \\ \Gamma^3_{23} &= \cot\theta, & \Gamma^4_{14} &= -\tan\chi. \end{aligned} \tag{28}$$

Substituting (28) into the geodesic equations

$$\frac{d^2 x^\lambda}{ds^2} + \Gamma^\lambda_{\mu\nu}\frac{dx^\mu}{ds}\frac{dx^\nu}{ds} = 0, \tag{29}$$

we get the following system of differential equations:

$$\frac{d^2\chi}{ds^2} - \frac{\sin\chi\cos\chi}{R^2}\left(\frac{d\tau}{ds}\right)^2 = 0, \tag{30}$$

$$\frac{d^2\theta}{ds^2} - \sin\theta\cos\theta\left(\frac{d\phi}{ds}\right)^2 = 0, \tag{31}$$

$$\frac{d^2\phi}{ds^2} + 2\cot\theta\frac{d\theta}{ds}\frac{d\phi}{ds} = 0, \tag{32}$$

$$\frac{d^2\tau}{ds^2} - 2\tan\chi\frac{d\chi}{ds}\frac{d\tau}{ds} = 0\,. \tag{33}$$

If we put $\theta = \pi/2$ and $d\theta/ds = 0$ in (31), we get $d^2\theta/ds^2 = 0$, hence the plane motion is possible. In the following we restrict ourselves only to this case. The integral of (32) reduces to

$$\frac{d\phi}{ds} = h, \tag{34}$$

where h is an integration constant. Integrating (33), we get

$$d\tau/ds = c/\cos^2\chi, \tag{35}$$

where c is an integration constant and $c > 0$. We may use (17) in place of (30), and we get, from (17), (34), (35) and the condition $\theta = \pi/2$,

$$\frac{d\chi}{ds} = \pm\frac{\sqrt{c^2 - (1 + h^2R^2)\cos^2\chi}}{R\cos\chi}. \tag{36}$$

Eliminating ds from (34), (35) and (36), and using $r_1 = R\sin\chi$ in place of χ, we get

$$\frac{dr_1}{d\phi} = \pm\frac{\alpha}{h}(r_1^2 + \beta)^{1/2}, \tag{37}$$

$$\frac{dr_1}{d\tau} = \pm\frac{\alpha}{c}\left(1 - \frac{r_1^2}{R^2}\right)(r_1^2 + \beta)^{1/2}, \tag{38}$$

where $\alpha = (h^2 + \frac{1}{R^2})^{1/2} > 0$ and $\beta = \frac{c^2-(1+h^2R^2)}{\alpha^2} = R^2(\frac{c^2}{1+h^2R^2} - 1)$. The equations (37) and (38) are our fundamental equations of motion. Integrating (37), we get as an orbital expression

$$r_1 = (e^{\Phi} - \beta e^{-\Phi})/2, \tag{39}$$

where $\Phi \equiv \pm\frac{\alpha}{h}(\phi - \phi_0)$ and ϕ_0 is an integration constant. We can regard Φ in (39) as a new angle variable. As the typical special cases of (39), we have

$$\left.\begin{aligned} r_1 &= \sinh\Phi \qquad (\beta = +1),\\ r_1 &= \tfrac{1}{2}\exp\Phi \qquad (\beta = 0),\\ r_1 &= \cosh\Phi \qquad (\beta = -1).\end{aligned}\right\} \tag{40}$$

The spatial part of (18) is entirely different from the euclidean character, hence it seems to be impossible to treat (r_1, Φ) as ordinary polar coordinates. But regarding (r_1, Φ) as an orthogonal net of coordinates, we can see the features of the orbits. (See Fig. 1.)

When $\beta > 0$, the range $0 \le r_1 < \sqrt{-\beta}$ must be excluded from the domain of particle-motion, but when $\beta \ge 0$, there exist no restriction. And then, from (38), we get

$$T \equiv \pm\frac{\alpha}{c}(\tau - \tau_0) = \int\frac{dr_1}{(1 - \frac{r_1^2}{R^2})\sqrt{r_1^2 + \beta}}, \tag{41}$$

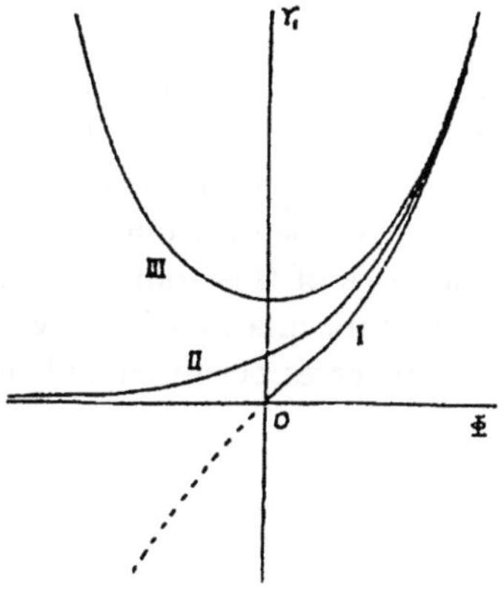

Figure 1. I: $r_1 = \sinh \Phi$, II: $r_1 = \frac{1}{2} \exp \Phi$, III: $r_1 = \cosh \Phi$.[4]

where τ_0 is an integration constant, and we may regard T as a new time variable. Integrating the right-hand side of (41) and after suitable arrangements, we get

$$r_1 = \begin{cases} R\left[\dfrac{m^2-1}{m^2(\frac{e^{2mT}+1}{e^{2mT}-1})^2-1}\right]^{1/2} & (\beta > 0,\ m^2 > 1), \\ R\left[\dfrac{1}{1+R^2e^{-2T}}\right]^{1/2} & (\beta = 0,\ m^2 = 1), \\ R\left[\dfrac{1-m^2}{1-m^2(\frac{e^{2mT}-1}{e^{2mT}+1})^2}\right]^{1/2} & (\beta < 0,\ m^2 < 1), \end{cases} \tag{42}$$

where $m^2 = 1 + \frac{\beta}{R^2} = \frac{c^2}{1+h^2R^2}$. Eliminating r_1 from (39) and (42), we get for the different cases respectively

$$\Phi = \log[R\{[\,]^{1/2} + ([\,] + m^2 - 1)^{1/2}\}], \tag{43}$$

where $[\,]$ represents the expression in the bracket on the right-hand side of (42), and $\Phi = \Phi(T)$ is equivalent to $\phi = \phi(\tau)$. Thus we can obtain the rigorous solution of the geodesic equations of motion. In a particular case of radial motion ($h = 0$), it becomes $\alpha = 1/R$, $\beta = R^2(c^2 - 1)$ and $m = c$, and the expressions (38) and (42) keep their forms. The behavior of particle-motion is the same as that in the de Sitter universe, hence the so-called "velocity-distance" relation is established.

[4] When $\beta = +1$, $\phi < 0$ seems to be impossible, but this does not mean any loss of generality, as we can see from its definition.

5. CONCLUSION

The solution considered here is a new cosmological static solution of Einstein's field equations of gravitation, which has an empty matter-distribution. Its line element can not be transformed into the standard one, and it is shown that this is the only one having such a character.

The geometrical properties of this solution are very similar to those of de Sitter universe and in particular, as far as we consider situations in the radial direction, both universes permit the identical results.

Feb. 13th 1951

Astronomical Institute
Tohoku University

Chapter 12
György Szekeres 1960, the maximal analytic extension of the Schwarzschild solution

Original paper: György Szekeres, On the singularities of a Riemannian manifold. *Publicationes Mathematicae Debrecen* **7**, 285 (1960).

Reprinted with the kind permissions of the editors of *Publicationes Mathematicae Debrecen* and of Peter Szekeres.

Previously published in *Gen. Relativ. Gravit.* **34** (2002), pages 1995 – 1999 (the editorial note + biography), 2001 – 2016 (the paper).

The editorial note was re-edited and the biography of György Szekeres was updated by Peter Szekeres for this reprinting.

12.1 Editorial note to:

György Szekeres
On the singularities of a Riemannian manifold

Peter Szekeres

This paper, written by my father George Szekeres, presents the now well known extension of the Schwarzschild solution discovered almost simultaneously by Kruskal [1]. The reason for the relative neglect of the paper, as compared to Kruskal's celebrated article, is not hard to see. While Kruskal published his work in the widely read *Physical Review*, Szekeres' paper was essentially buried in an obscure and mostly inaccessible mathematical journal – *Publicationes Mathematicae, Debrecen*. He explained his choice of journal to me in the following words:

> "At the time when this paper was written I was deeply interested in the geometric structure of General Relativity. I noticed with some surprise that General Relativity had no working definition for a singularity of a (pseudo)-Riemannian manifold, in spite of talking about the Schwarzschild "singularity". As it happened, I visited Hungary in 1958 (my first such visit after the war) and a Debrecen colleague asked me if I would not like to contribute to the volume of the Publicationes Debrecen in honour of the 50-th birthday of the Hungarian differential geometer Otto Varga. I gladly consented as I saw an opportunity to expound my definition of a singularity – one that incidentally works in any manifold with an affine connection. The Schwarzschild vacuum solution was used merely as an example to illustrate how the definition works. I knew already for some time that the Schwarzschild singularity was not really a singularity, from a remark by Lemaître quoted in a footnote by Eddington in one of his semi-popular books (I do not remember now which one it was). On the other hand I do remember how perplexed I was when my coordinate system (incidentally the same as the one discovered by Martin Kruskal around the same time) revealed a symmetry of the Schwarzschild space-time with its strange doubling up which (to me) seemed to indicate that the empty-space Schwarzschild solution is a physical absurdity, in spite of being compatible with Einstein's equations."

These remarks suggest that Szekeres' paper is interesting for much more than just the discovery of the coordinate system mentioned above. In fact it presents a bold general analysis of singularities of manifolds at a time when the problem of singularities in general relativity had barely yet raised its head.

The paper begins with a brief discussion of the Schwarzschild and de Sitter space-times. Szekeres analyzes the Schwarzschild solution near $r = 2m$ by use of a power series expansion and comes up with the transformation which is identical with Kruskal's coordinates. As quoted above, he is clearly troubled by the "doubling

Peter Szekeres
Department of Physics and Mathematical Physics, University of Adelaide, Adelaide, South Australia 5005, Australia
e-mail: szekeres@senet.com.au

up" of the Schwarzschild region exterior to $r = 2m$ and recommends an "elliptical" interpretation in which points of the mirror exterior worlds are identified. It is interesting to speculate why the relativity community has not for the most part adopted this viewpoint.

Indeed the author points out that it was known even to Lemaître [2] (reference 2 in the paper) and subsequently J.L. Synge [3] (fourth reference in the paper) that the Schwarzschild "singularity" $r = 2m$ could be transformed away by a suitable coordinate transformation and was therefore not a true singularity. While Lemaître's transformation does indeed cover the region $r = 2m$, it is achieved by means of comoving timelike coordinates and it is not a maximal extension of Schwarzschild's metric. Synge's paper is fairly heavy going and does grapple with double-null coordinate coordinate systems, but nowhere does it achieve the elegant simplicity of the Kruskal–Szekeres transformation. However, Szekeres clearly thinks very little of his "discovery" of the famous double-null coordinate system and only uses it as an *example* of the kind of problem to be encountered in any theory of singularities on manifolds. He also describes the standard removal of the singularity in de Sitter's static space-time by converting it into an exponentially expanding universe (first discovered by Eddington in his book *The Mathematical Theory of Relativity*).

Section 2 describes a method of defining singularities in a coordinate-free way. The idea is a precursor of other methods which depend upon the existence of incomplete geodesics, such as those of Geroch [4], Schmidt [5] and Scott and myself [6]. Szekeres' technique is to distinguish between geodesic arcs which terminate in points interior to a coordinate domain, and arcs which do not (called *boundary arcs*). Boundary arcs define *boundary points* of a coordinate domain D, and such a boundary point is called a *singularity* of a manifold R if it is a boundary point in every extension of the manifold R. This could well be the first rigorous definition ever given of a singularity suitable for pseudo-Riemannian manifolds. It anticipates in a remarkable way the general philosophy of the later 1960's that singularities are to be identified with the occurrence of incomplete geodesics [7]. The rest of this section attempts the fairly difficult task of defining equivalence of boundary arcs, so that separate singularities are not associated with each and every boundary arc. To achieve this, normal coordinates are associated with each boundary arc and the progress of neighbouring arcs is followed in these coordinates. The definition seems to have many similarities with Geroch's definition [4] although it is a more coordinate-oriented approach.

In this paper the word "complete" is used for a manifold to mean that it is inextendible (i.e. not a proper subset of any manifold of the same dimension), instead of today's more common usage of *geodesically* complete (all geodesics can be extended indefinitely). Szekeres makes the claim that a manifold R is complete in this sense if and only if all its boundary points are singularities. This is however at variance with a famous example of Misner [8] concerning an inextendible manifold which has incomplete geodesics. The rest of Section 2 is involved with discussing the difference between what the author terms *ordinary* and *non-ordinary* singularities such as occur at the apex of a cone. This distinction corresponds to what Clarke and Schmidt [9] call *curvature* and *quasiregular* singularities. Interestingly, the origin $u = 0, v = 0$ of Kruskal-Szekeres coordinates becomes an ordinary singularity when the "elliptic" identification is made, which may well account for its lack of popularity.

Section 3 is devoted to a detailed examination of two examples, the Schwarzschild solution and a particular Friedmann cosmological model. In the case of Schwarzschild the discussion is particularly extensive. All geodesics are calculated – spacelike, timelike and null – even those which are non-radial. The conclusion is not totally conclusive, but Szekeres is led to believe that the final manifold is complete (which, remember, means it cannot be imbedded in a larger manifold of the same dimension). Few would demur from this conclusion as the only incomplete geodesics terminate on the Schwarzschild region $r = 0$, which is a curvature singularity. However, he also suggests that all boundary arcs are associated if and only if they originate at the same point φ_0, θ_0 on u_0, v_0 with ${v_0}^2 - {u_0}^2 = 4/e$ (i.e. the region usually denoted $r = 0$ in Schwarzschild's coordinates). This is more controversial as it amounts to saying that the "natural" structure of the singular boundary is a 2-sphere. Considerable effort has gone into attempting a rigorous proof of this or any other possible structure for the Schwarzschild singularity, but with little success. A typical problem is the possibility of bizarre non-Hausdorff structures arising, such as discovered by Johnson [10] in the case of Schmidt's b-boundary construction.

The paper ends with an analysis of a particular Friedmann model, having $k = 0$ and $R(t) = t$,

$$ds^2 = -dt^2 + t^2(dx^2 + dy^2 + dz^2).$$

This model would correspond to the case $p = -\frac{1}{3}\rho$ if Einstein's equations with no cosmological constant were applied, but the author has other motivations for considering this metric which arises in earlier work of his own. Two incomplete geodesics are found which are not associated, and the reader is led to the conclusion that the structure of the singularity at $t = 0$ is not "pointlike". This is interesting, because Friedmann models are a prime test of any theory of singularities, and to this date no convincing argument exists which identifies the singularity as a point, plane, sphere or any other surface.

In view of this insightful paper the reader may well wonder whether my father went on to write more in the field of relativity. During his long and productive life he always maintained a great interest in general relativity although most of his work was in pure mathematics. He did however write a further paper with Laurence Mysak [11] on some aspects of the Schwarzschild solution. The conclusions of this paper are quite controversial. In yet other areas connected with general relativity he gave an axiomatization of Minkowski geometry [12, 13] and proposed an interesting spinorial unified field theory [14, 15, 16] to which he devoted considerably greater effort than his single published paper would indicate. The last mentioned paper was published posthumously and he was still working on it in his last days.

References

1. M.D. Kruskal, *Phys. Rev.*, **119** 1743, 1960.
2. A.G. Lemaître *Ann. Soc. Scient. Bruxelles* Ser A, **53** 51 (1933), reprinted as a "Golden Oldie" in *Gen. Rel. and Grav.*, **29** 637, (1997).
3. J.L. Synge, *Proc. Roy. Irish. Acad.* Sec A, **53** 84. 1950.
4. R.P. Geroch, *J. Math. Phys.*, **9** 450, 1968.
5. B.G. Schmidt, *Gen. Rel. and Grav.*, **1** 269, 1971.
6. S.M. Scott and P. Szekeres, *Journal of Geometry and Physics*, **13** 223, 1994.
7. S.W. Hawking and G.F.R. Ellis, *The Large Scale Structure of Space-Time*. Cambridge University Press, Cambridge, 1973.
8. C.W. Misner, Taub-NUT spaces as a counterexample to almost anything *Relativity Theory and Astrophysics I: Relativity and Cosmology*, ed J.Ehlers, Lectures in Applied Mathematics, Vol 8 (AMS) 160, 1967.
9. C.J.S. Clarke and B.G.Schmidt, Singularities: the state of the art, *Gen. Rel and Grav.* **8** 129, 1972.
10. R.A. Johnson, *J. Math. Phys.*, **18** 898, 1977.
11. L. Mysak and G. Szekeres, *Can. J. Phys.*, **44** 617, 1966.
12. G. Szekeres, *J. Aust. Math. Soc., Part 2*, **8** 134, 1968
13. J. Schutz *Foundations of special relativity: kinematic axioms for Minkowski space-time* Lecture Notes in Mathematics **361** (Berlin-Heidelberg-New York), 1973
14. G. Szekeres, *J. Math. and Mechanics*, **6** 471, 1957
15. J. Lynch *Classical Quantum Grav.*, **8** 1907, 1991
16. G. Szekeres and L. Peters *ANZIAM J.*, **50** 143, 2008

12.2 György Szekeres – a brief biography

By György Szekeres and Peter Szekeres

Born 29 May 1911 in Budapest, Hungary. Died 28 August 2005 in Adelaide, Australia. Graduated in 1932 with a degree of Chemical Engineering at the Technological University of Budapest (not a conventional training for a mathematician, but was largely self-taught in that discipline with the help of some illustrious colleagues such as Paul Erdös and Paul Turán). Industrial employment in Hungary until 1939, then as a 'political' refugee in Shanghai until 1948. Academic history: lecturer, Senior Lecturer, then Reader at the University of Adelaide 1948 – 63, Professor of Pure Mathematics at the University of New South Wales (63 – 76), Emeritus Professor from 1977 onwards. Honorary D.Sc. from UNSW (1976), Fellow of the Australian Academy of Science since 1963 (Lyle medal 1968). President of the Australian Mathematical Society 1972/74, external member of the Hungarian Academy of Sciences since 1986.

Scientific work: Nearly 100 papers in Combinatorial Geometry, Group Theory, Analytic Number Theory, Partitions, Iteration Theory, General Relativity.

General Relativity and Gravitation, Vol. 34, No. 11, November 2002 (© 2002)

On the Singularities of a Riemannian Manifold[1]

Dedicated to Professor Ottó Varga, at the occasion of his 50th birthday.
Gy. Szekeres (Adelaide, S. Australia).
[*Publicationes Mathematicae Debrecen* **7**, 285 (1960)]

1. INTRODUCTION

Riemannian manifolds, such as the ones which find application in the theory of general relativity, often have regions of singularity where the determinant of the metric tensor becomes zero or the components (or derivatives of components) of the metric tensor become infinite. Well known examples are the singular hypersurfaces which appear in Schwarzschild's[2] and de Sitter's centrosymmetrical solutions of the gravitational field equations. The Schwarzschild field is a Riemannian manifold with the line element[3]

$$\pm ds^2 = \left(1 - \frac{2m}{r}\right)^{-1} dr^2 + r^2 d\omega^2 - \left(1 - \frac{2m}{r}\right) dt^2, \tag{1}$$

$$d\omega^2 = d\theta^2 + \cos^2\theta d\varphi^2; \tag{2}$$

it consists of two disjoint regions, $0 < r < 2m$ and $r > 2m$, separated by the singular hypercylinder $r = 2m$. In de Sitter's Universe

$$\pm ds^2 = (1 - r^2/R^2)^{-1} dr^2 + r^2 d\omega^2 - (1 - r^2/R^2)\, dt^2, \tag{3}$$

there is a singular hypersurface at $r = R$, the "mass horizon" of Eddington.

Neither the Schwarzschild nor the de Sitter hypercylinders are true singularities. In the Schwarzschild case this was first noticed by LEMAITRE [2] and

[1] Reprinted with the kind permission of the author and of the Editor of the *Publicationes Mathematicae Debrecen*.

[2] In the original paper the line element has a mistake in the factor before dr^2 which is raised to power -2 instead of -1. 2. Original paper received May 26 1959.

[3] In the original paper, Schwarzschild's name was spelled as "Schwartzschild." The spelling was corrected throughout. A few minor obvious typos have been corrected, too (Editor).

2001

0001-7701/02/1100-2001/0 © 2002 Plenum Publishing Corporation

worked out in greater detail by SYNGE [4]; in the case of the de Sitter Universe by EDDINGTON ([1], Chapter V). In each case the argument was based on the observation that the singularities can be transformed away in a suitable coordinate system; but an exact definition of what should be regarded as a true singularity of a Riemannian manifold has, to my knowledge, never been proposed.[4]

Superficially, we may conceive of a singularity as a place in coordinate space where something goes wrong with the metric tensor. But surely this is not a sufficient criterion, since apparent singularities can easily be produced by the simple device of introducing "bad" coordinates in otherwise perfectly well-behaved manifolds. Perhaps the most trivial example of an apparent singularity is the one at $r = 0$ of the line element of Euclidean space in polar coordinates,

$$ds^2 = dr^2 + r^2 d\omega^2. \tag{4}$$

Clearly the singularity is due to the coordinate system, and not to any irregularity of the manifold itself. If we make the substitution $r = \bar{r} - 2m$ in (4), where m is a positive number, we get

$$ds^2 = d\bar{r}^2 + (\bar{r} - 2m)^2 d\omega^2. \tag{5}$$

Here we have an apparent singularity on the sphere $\bar{r} = 2m$, due to a spreading out of the origin over a sphere of radius $2m$. Since the exterior region $\bar{r} > 2m$ represents the whole of Euclidean space (except the origin), the interior $\bar{r} < 2m$ is entirely disconnected from it and represents a distinct manifold.

In some respects the line elements (4) and (5) are not unlike the Schwarzschild metric (1). This becomes more apparent if we make the substitution

$$\bar{r} = \frac{1}{2m} \int_{2m}^{r} \left(1 - \frac{2m}{r}\right)^{-\frac{1}{2}} dr, \tag{6}$$

valid for $r > 2m$; it transforms (1) into

$$\pm ds^2 = 4m^2 d\bar{r}^2 + r^2 d\omega^2 - \left(1 - \frac{2m}{r}\right) dt^2 \tag{7}$$

where $r = f(\bar{r})$ is given by (6). In the neighbourhood of $\bar{r} = 0$,

$$r = f(\bar{r}) = 2m\left(1 + \frac{1}{4}\bar{r}^2 - \frac{1}{48}\bar{r}^4 + \cdots\right),$$

$$1 - \frac{2m}{r} = \frac{1}{4}\bar{r}^2 - \frac{1}{12}\bar{r}^4 + \cdots$$

[4] SYNGE ([4], p. 100) gives a definition which however depends on the coordinate system, and refers therefore to singularities of coordinate systems rather than of the manifolds themselves.

and the coefficient of dt^2 becomes zero, of the same order as the coefficient of $d\omega^2$ in (4) or (5); in fact the singularity disappears if we make the substitution

$$u = h \cosh \frac{t}{4m}, \qquad v = h \sinh \frac{t}{4m} \tag{8}$$

where $h = h(\bar{r})$ is a solution of

$$d \log h / d\bar{r} = \frac{1}{2}\left(1 - \frac{2m}{r}\right)^{-\frac{1}{2}}, \qquad h'(0) = 1.$$

In terms of r,

$$h^2 = 4pe^p, \qquad p = (r - 2m)/2m. \tag{9}$$

In the new coordinate system

$$\pm ds^2 = 4m^2\{(1+p)^{-1}e^{-p}(du^2 - dv^2) + (1+p)^2 d\omega^2\} \tag{10}$$

where p is determined from

$$4pe^p = u^2 - v^2. \tag{11}$$

Although the transformations (8) and (9) which carry (1) into (10) are only valid for $u^2 > v^2$, $u > 0$, the line element (10) is regular in a whole neighbourhood of the line $u = r, \theta = 0, \varphi = 0$, indeed for $u^2 - v^2 > -4/e$ and $-\frac{1}{2}\pi < \theta < \frac{1}{2}\pi$, $-\pi < \varphi < \pi$. This shows that the Schwarzschild "singularity" is just as apparent, brought about by an improper choice of coordinates, as the origin of polar coordinates.[5]

Incidentally, the transformation

$$u = h \sinh \frac{t}{4m}, \qquad r = h \cosh \frac{t}{4m}, \tag{8'}$$

$$h^2 = -4pe^p, \qquad p = \frac{r}{2m} - 1 \tag{9'}$$

carries the region $0 < r < 2m$ of (1) into the region $0 < v^2 - u^2 < 4/e$ of (10), so that the interior of the Schwarzschild hypercylinder is a natural continuation of the exterior.

The manifold (10) differs from (1) in one important respect: each point of (1) is represented exactly twice in (10). Hence in order to obtain a manifold which represents physical reality, it seems to be necessary to identify all pairs of opposite points (u, v) and $(-u, -v)$; this is permissible since the metric at (u, v) is identical whith the metric at $(-u, -v)$. The situation is analogous to the construction of elliptic space from hyperspherical space,[6] but there is a significant

[5]This of course is not an entirely valid conclusion as long as we do not have a precise definition of a singularity; it has the same heuristic character as the similar conclusions of LEMAITRE and SYNGE. The transformations (8) and (8′) are essentially due to SYNGE.

[6][3], p. 7.

difference. Whereas elliptic space has no singularities, the Schwarzschild identification introduces an artifical singularity at $u = 0$, $v = 0$, essentially of the same kind as the singularity at the vertex of a cone obtained by identifying the points (x, y) and $(-x, -y)$ of Euclidean plane. Since (10) is a perfectly well-behaved Riemannian manifold which satisfies the gravitational field equations in the whole region $u^2 - v^2 > -4/e$, including the origin (0, 0), it seems difficult to find any physical justification (apart from a purely utilitarian one) for this identification process.

The problem of continuation of Riemannian manifolds is fundamental to relativistic cosmology; for on it depends the answer to the question whether a given cosmological frame covers the whole of the Universe, or only part of it. In the case of the de Sitter Universe (3) the answer is well known: if one replaces r, t by new coordinates $\chi, \bar{t}$, given by

$$r = R\cos\chi\cosh\bar{t}, \qquad \tanh(t/R) = \tanh\bar{t}/\sin\chi,$$

the new system extends beyond the mass horizon[7] and the line element becomes

$$\pm ds^2 = R^2\cosh^2\bar{t}\{d\chi^2 + \cos^2\chi(d\theta^2 + \cos^2\theta d\varphi^2)\} - R^2 d\bar{t}^2. \tag{11}$$

This is the metric on a 4-hyperboloid

$$x_1^2 + x_2^2 + x_3^2 + x_4^2 - x_5^2 = R^2$$

in pseudo- Euclidean 5-space with signature $(+ + + + -)$ and there is no way to extend it any further.[8]

The substitution

$$x_1 = \bar{r}\cos\theta\cos\varphi, \qquad x_2 = \bar{r}\cos\theta\sin\varphi, \qquad x_3 = \bar{r}\sin\theta,$$
$$\bar{r} = r(1 - r^2/R^2)^{-\frac{1}{2}}e^{-t/R}, \qquad \bar{t} = t + \frac{1}{2}R\log(1 - r^2/R^2)$$

transforms (3) into the Robertson frame with the line element

$$\pm ds^2 = e^{2\bar{t}/R}\left(dx_1^2 + dx_2^2 + dx_3^2\right) - d\bar{t}^2. \tag{12}$$

This frame covers the same portion of the Universe as (3) but the metric has no singularities in the finite part of the frame. Since the manifold (12) obviously admits extension, it is clear that for a consistent theory of continuation it will be necessary to consider also singularities which lie at infinity in coordinate space.

[7] For a full discussion see SCHRÖDINGER [3], Chapter I.

[8] Apart from a trivial extension to the 2-hyperboloid $x_1 = 0, x_2 = 0, x_3^2 + x_4^2 - x_5^2 = R^2$ which is not covered by the system (11).

2. DEFINITION OF A SINGULARITY

In analogy with the theory of analytic functions, we may define a singularity of a Riemannian manifold as a boundary point into which the manifold cannot be continued in a reasonable manner. The definition can be made precise if we know what a boundary point is and what we should mean by a reasonable continuation. In the case of the Riemann surface of an analytic function the problem is fairly trivial as the surface is superimposed upon the number sphere and a "reasonable" continuation can be accomplished by means of power series. In Riemannian manifolds the analytic structure is carried by the Riemannian metric, and it is in terms of this metric that we have to formulate the process of continuation.

Let us briefly review the conventions that we are going to use. An n-dimensional manifold is a connected Hausdorff space R with the property that each point p of R has a neighbourhood U which is homeomorphic to an open set V in real Euclidean n-space E_n. Let φ be a homeomorphism of U onto V, $\varphi(p) = \mathbf{x} = \{x_\mu; \mu = 1, \ldots, n\}$. U and φ determine a local coordinate system or coordinate frame X and we say that p is covered by X. In a Riemannian manifold there is also given a symmetrical metric tensor $g_{\mu\nu}(\mathbf{x})$ which obeys the usual transformation laws. In the non-definite case we agree to choose the sign of the line element

$$ds^2 = \pm \sum_{\mu,\nu} g_{\mu\nu} dx_\mu dx_\nu \tag{13}$$

so that always $ds \geq 0$. We also assume that R is of class ∞, that is, it can be covered by a set of "admissible" frames so that in these frames the $g_{\mu\nu}(\mathbf{x})$ have continuous partial derivatives of any order. Our standpoint is that a point which cannot be covered by such a frame is a singularity.

A further restriction on admissible frames is that they cover a connected domain in which the determinant of the metric tensor does not vanish. Thus in the example of (1), the exterior $r > 2m$ and interior $r < 2m$ are covered by distinct frames, even if it is the same analytical formula which expresses the metric tensor in both parts. If $\bar{R}$ is a submanifold of R in the relative topology of R, of the same dimension as R and relatively to the same set of admissible frames, then R will be called an *extension* of $\bar{R}$. In particular, we shall call R *complete* if it cannot be immersed in any proper extension.

An open set D of R will be called a *domain* if and only if it can be covered by a single admissible frame. Our first purpose is to define boundary points of a domain, and this will be done by means of geodesic arcs emanating from that boundary point.

Let D be a domain, φ the coordinate mapping of a frame X which covers D. By a *geodesic arc in D* we shall always mean a semi-open geodesic arc[9]

$$L : p(s) = \varphi^{-1}(\mathbf{x}(s)), \qquad p(s) \in D, \qquad 0 < s \leq b \tag{14}$$

where $0 < b < \infty$ and s is a distinguished parameter in which the geodesic equations have the form

$$\frac{d^2 x_\mu}{ds^2} + \left\{ \begin{matrix} \mu \\ \varrho\sigma \end{matrix} \right\} \frac{dx_\varrho}{ds} \frac{dx_\sigma}{ds} = 0, \qquad \mu = 1, \ldots, n. \tag{15}$$

Here $\mathbf{x}(s) = \{x_\mu(s)\}$ and $\left\{ \begin{smallmatrix} \mu \\ \varrho\sigma \end{smallmatrix} \right\}$ is the Christoffel affinity derived from $g_{\mu\nu}$. The parameter s is uniquely determined apart from a constant factor; if L is not a null geodesic, we make the parameter unique by requiring that $b - s$ be equal to the length of arc measured from the endpoint $p(b)$ of L.

A geodesic arc will be called an *interior arc* of D if $\mathbf{y} = \lim_{s\to 0} \mathbf{x}(s)$ exists and $\mathbf{y} \in \varphi(D)$. The point $q = \varphi^{-1}(\mathbf{y})$ is called the *origin* of L. If $\lim_{s\to 0} \mathbf{x}(s)$ does not exist, or it does exist but is not in $\varphi(D)$, then L will be called a *boundary arc*. Note that a boundary arc has always a finite length.

Clearly the definition of an interior arc or boundary arc is independent of the frame which covers D. If L is a boundary arc, we say that it determines a *boundary point*[10] of D; if also $\mathbf{y} = \lim_{s\to 0} \mathbf{x}(s)$ exists in a certain coordinate frame, we say that the boundary point (or the origin of L) lies at $\mathbf{y}$ (possibly at infinity). We agree that every *subarc* of L, i.e. every boundary arc of the form

$$L' : p(s) = \varphi^{-1}(x(s)), \qquad 0 < s \leq b_1 < b$$

determines the same boundary point as L. A more general concept of equivalence of boundary arcs will be needed and established later on.

It may happen that L is a boundary arc in a certain domain D but an interior arc in a domain D^*. We say that L is a boundary arc of R if the following is true: (i) L is a boundary arc in some domain D of R; (ii) no subarc of L is an interior arc of any domain D^* of R. The boundary point defined by L is then a boundary point of R; and it is called a *singularity* if it remains a boundary point in every extension of R. This at the moment is a rather loose definition; the final form must take account of the equivalence of boundary arcs. Note that if a proper extension of R exists at all then R must necessarily have non-singular boundary points; and R is complete if and only if all its

[9] By definition, L is a bicontinuous map of the interval $0 < s \leq b$.

[10] We disregard possible boundary points which are not accessible by a geodesic arc. They would not contribute anything new to the extension problem.

boundary points are singularities. Thus every R which has no boundary arcs is complete.

In order to arrive at a useful theory of equivalence of boundary arcs, we introduce the concept of *normal coordinates* relatively to a geodesic arc L. These are combinations of Fermi and Riemann coordinates, and are defined as follows. Suppose first that L is a non-null geodesic. Then normal coordinates relatiiely to L are coordinates y_μ with the following properties:

(i) The metric tensor takes Galilean values $g_{\mu\nu} = \pm\delta_{\mu\nu}$ at all points of L.
(ii) $\partial g_{\mu\nu}/\partial y_\varrho = 0$ for every μ, ν, ϱ at all points of L.
(iii) The equation of L is $y_\mu = \delta_{\mu n} s$ $(0 < s \leq b)$.
(iv) The equation of any geodesic G which radiates from a point $p(\sigma)$, $0 < \sigma \leq b$ of L, perpendicularly to L, is

$$y_\mu = \gamma_\mu s \quad \text{for} \quad 1 \leq \mu < n, \qquad y_\mu = \sigma,$$

where the γ_μ are constants.

To construct coordinates with these properties, select $n-1$ mutually perpendicular non-null unit vectors $\mathbf{a}^1, \ldots, \mathbf{a}^{n-1}$, each perpendicular to L, at the endpoint $p(b)$. Through Fermi propagation of these vectors (i.e. congruent displacement along L) we obtain a set of coordinate directions $\mathbf{a}^\nu(\sigma)$ at each $p(\sigma)$. For the y_n-coordinate along L choose $y_n = s = b - \lambda$, where b is the total length of L (hence finite) and λ is the length of arc measured from $p(b)$; for the coordinate plane $y_n = \sigma$ take the hyperplane $P(\sigma)$ generated by all the geodesics running from $p(\sigma)$ in the directions spanned by the vectors $\mathbf{a}^\nu(\sigma)$. Finally, in $P(\sigma)$ itself we choose Riemannian coordinates corresponding to the $n-1$ predetermined coordinate directions $\mathbf{a}^\nu(\sigma)$. Clearly these coordinates will have the required properties (i)–(iv), the only freedom in their choice being a rotation of the axes of y_μ, $\mu < n$, about the axis of y_n.

If L is a null arc, some modification is necessary. For if $\mathbf{a}^2, \ldots, \mathbf{a}^{n-1}$ are mutually perpendicular non-null vectors at $p(b)$, each perpendicular to L, then $\mathbf{a}^1$ cannot have this same property. However, we can choose $\mathbf{a}^1$ to be a null-vector perpendicular to $\mathbf{a}^2, \ldots, \mathbf{a}^{n-1}$ and not in the direction of L. This condition specifies the direction of $\mathbf{a}^1$ uniquely and yields an admissible set of coordinate directions. The rest of the construction is same as above, except that s is not the length of arc of L but a distinguished parameter. By adjusting the magnitude of $\mathbf{a}^1$ if necessary, we obtain normal coordinates with the following properties:

(i) $g_{1\mu} = 1, g_{11} = g_{\mu n} = 0, g_{\mu\nu} = \pm\delta_{\mu\nu}$ $(2 \leq \mu \leq \nu \leq n-1)$ at all points of L;
(ii) $\partial g_{\mu\nu}/\partial y_\varrho = 0$ for every μ, ν, ϱ at all points of L;
(iii) the equation of L is $y_\mu = s\delta_{\mu\nu}$ $(0 < s \leq b)$;

(iv) the equation of any geodesic G which radiates from $p(\sigma)$ on L in any direction in the coordinate hyperplane of $y_1, \ldots, y_{n-1}$ is

$$y_\mu = \gamma_\mu s \quad \text{for} \quad 1 \leq \mu < n, \qquad y_n = \sigma,$$

where the γ_μ are constants.

Again these coordinates are uniquely determined, apart from a constant rotation of the coordinate axes of y_μ, $1 < \mu < n$, in the plane of $y_2, \ldots, y_n$.

Among all coordinate systems which cover L, normal coordinates are undoubtedly the best behaved in the neighbourhood of L. Two things in particular should be noted about normal coordinates. First, their construction is entirely independent of the original frame which covered L. From any other frame we would have arrived to exactly the same family of normal systems. Secondly, the construction does not require any knowledge of the behaviour of the manifold at the origin of L itself. This property is extremely useful from the point of view of detecting singularities. Suppose that L is an interior arc; then there exists a geodesic arc L^* with the property that (i) L is contained in L^*, (ii) $q = \lim_{s\to 0} p(s)$ is an interior point of L^*. Let Y be a normal frame relatively to L. We can introduce normal coordinates y^*_μ relatively to L with the property that $y^*_\mu = y_\mu$ ($\mu = 1, \ldots, n-1$), $y_n = y^*_n + c$ for a positive constant c in a neighbourhood of L. In other words, we can continue the normal frame of L so that it should also cover the origin of L. It follows therefore that if $g^{(k)}_{\mu\nu}(\mathbf{y}(\sigma))$ denotes any partial derivative of $g_{\mu\nu}$ of order $k \geq 1$ in the system Y at the point $p(\sigma) = \varphi^{-1}(y(\sigma))$ of L and $g^{(k)}_{\mu\nu}(\mathbf{y}^*(\sigma + c))$ the corresponding derivative in the system Y^*, we have $g^{(k)}_{\mu\nu}(\mathbf{y}(\sigma)) = g^{(k)}_{\mu\nu}(\mathbf{y}^*(\sigma + c))$ and in particular

$$g^{(k)}_{\mu\nu}(y^*(c)) = \lim_{\sigma\to 0} g^{(k)}_{\mu\nu}(\mathbf{y}(\sigma)). \tag{16}$$

Hence if for some derivative $g^{(k)}_{\mu\nu}$,

$$\lim_{\sigma\to 0} g^{(k)}_{\mu\nu}(\mathbf{y}(\sigma)) \tag{17}$$

does not exist (or is infinite) then L is certainly a boundary arc of R and defines a singularity. Thus normal coordinates provide us with a convenient method to detect singularities.

We shall call a singularity determined by L *ordinary* if there exists a derivative $g^{(k)}_{\mu\nu}$ for which the limit (17) does not exist (or is infinite). There exist non-ordinary singularities for which all limits (17) exist and are finite for every $k > 0$. Such is the case for instance with the origin $\mathbf{0}: u = 0, v = 0$ of the Schwarzschild manifold, obtained from (10) by identifying opposite points. Any geodesic arc whose origin is at $\mathbf{0}$ determines a non-ordinary singularity; this is obvious from the fact that in the complete manifold (10), $\mathbf{0}$ is an interior point.

In this last example it is clearly undesirable to have a separate singularity for each boundary arc which originates at $\mathbf{0}$. This leads us immediately to the problem of equivalence of boundary arcs. What we need is an equivalence relation of geodesic arcs, sufficiently strong to identify all interior arcs with the same origin, but sufficiently weak to keep interior and boundary arcs apart.

Let Y be a normal coordinate system relatively to the geodesic arc L, and D the domain covered by Y. Let L^* be a second geodesic arc in D, not necessarily distinct from L, Y^* a normal system relatively to L^* and suppose that Y^* covers a sub-arc L' of L. We say that L^* is *associated directly* with L if the following is true:

(i) If $y_\mu(\sigma^*)$ $(\mu = 1, \ldots, n, 0 < \sigma^* \leq b^*)$ are the coordinates in Y of the point $p^*(\sigma^*)$ of L^* where σ^* is a distinguished parameter, then

$$\lim_{\sigma^* \to 0} y_\mu(\sigma^*) = 0 \quad \text{for} \quad \mu = 1, \ldots, n;$$

(ii) If $y_\mu(\sigma^*)$ $(\mu = 1, \ldots, n, 0 < \sigma^* \leq b)$ are the coordinates in Y^* of the point $p(\sigma)$ of L' where σ is a distinguished parameter, then

$$\lim_{\sigma^* \to 0} y^*_\mu(\sigma) = 0 \quad \text{for} \quad \mu = 1, \ldots, n.$$

It follows from the definition that if L^* is associated directly with L then L is associated directly with L^*. Condition (ii) is essential for symmetry and cannot be omitted; in the last section we shall give a counterexample which will show that (ii) is in fact independent of (i).

Generally, given two geodesic arcs L and L^* of R, we say that they are *associated* if and only if there exists a finite chain of geodesic arcs $L = L_0, L_1, \ldots, L_m = L^*$ with the property that L_ν and $L_{\nu+1}$ $(\nu = 0, 1, \ldots, m-1)$ are directly associated. It is clear from the definition that being associated is an equivalence relation and that the equivalence class of an interior arc L consists of exactly those interior arcs which have the same origin as L. As a corollary we find that the equivalence class of a boundary arc contains only boundary arcs; hence it is permissible to identify boundary points which are defined by associated boundary arcs. Thus by definition, a boundary point of R is a class of associated boundary arcs. The definition is relative to R and distinct boundary points may become fused in a suitable extension of R.

If the boundary point defined by the arc L is not a singularity of R then none of the associated arcs will define a singularity. Hence we can safely speak of a boundary point being a singularity, and also of being an ordinary singularity, namely if at least one of the associated arcs defines an ordinary singularity.

Non-ordinary singularities cannot be detected by an examination of the limits (17) alone, and study of these singularities should present many interesting problems. More favourable is the situation with analytic manifolds which have the property that the $g_{\mu\nu}$ are determined in a whole neighbourhood of a non-singular boundary point at which the partial derivatives are known. This remark may be utilised to construct a unique greatest extension or universal covering manifold of an analytical Riemannian manifold. However, we shall not go further into this question but conclude with the discussion of some examples.

3. EXAMPLES

To illustrate the concepts developed in the previous section, we shall examine once more the singularities of the de Sitter and Schwarzschild manifolds. The de Sitter Universe offers no particular problems: all the boundary points of (3) can be made interior points in a suitable frame, and the complete manifold (11) is free of singularities.[11]

The Schwarzschild manifold (1) is more interesting and we shall examine its boundary points in greater detail. The geodesic equations of (1) are well known; they are (in the hyperplane $\theta = 0$, which is obviously no restriction in generality),

$$\frac{d\varphi}{ds} = \frac{h}{r^2} \tag{18}$$

$$\frac{dt}{ds} = \frac{ar}{r - 2m} \tag{19}$$

$$\frac{dr}{ds} = \left(a^2 - \left(\varepsilon + \frac{h^2}{r^2}\right)\left(1 - \frac{2m}{r}\right)\right)^{\frac{1}{2}} \tag{20}$$

where a, h are constants and ε has the value $+1$, -1 or 0, depending on whether the geodesic is time-like, space-like or null.

There are several types of boundary arcs, in both the exterior and interior regions. In the exterior region $r > 2m$ we have:

a) Space-like arcs

$$t = t_0, \quad \frac{d\varphi}{dr} = h(r^2 - h^2)^{-\frac{1}{2}}(r^2 - 2mr)^{-\frac{1}{2}}, \quad h < 2m, \quad 2m < r \leq 2m + d,$$

originating at $t = t_0, \varphi = \varphi_0, r = 2m$.

b) Arcs of the form

$$\frac{d\varphi}{dr} = hr^{-2}\left(a^2 - \left(\varepsilon + \frac{h^2}{r^2}\right)\left(1 - \frac{2m}{r}\right)\right)^{\frac{1}{2}}$$

[11] It is not entirely free of boundary points, as there are exceptional points not covered by (11), but these can easily be transformed into interior points by a suitable rotation of the frame.

$$\frac{dt}{dr} = a\left(1 - \frac{2m}{r}\right)^{-1}\left(a^2 - \left(\varepsilon + \frac{h^2}{r^2}\right)\left(1 - \frac{2m}{r}\right)\right)^{-\frac{1}{2}},$$
$$2m < r \leq 2m + d$$

where $a \neq 0$, $\varepsilon = \pm 1$ or 0, and d is a suitable positive number. It originates at $r = 2m$, $\varphi = \varphi_0$, $t = \pm\infty$.

In the interior region $0 < r < 2m$ we have:

c) Time-like arcs

$$t = t_0, \quad \frac{d\varphi}{dr} = h(r^2 + h^2)^{-\frac{1}{2}}(2mr - r^2)^{-\frac{1}{2}}, \quad 0 < d \leq r < 2m.$$

d) Space-like arcs

$$t = t_0, \quad \frac{d\varphi}{dr} = h(h^2 - r^2)^{-\frac{1}{2}}(2mr - r^2)^{-\frac{1}{2}}, \quad h > 2m, \quad 0 < d \leq r < 2m.$$

e) Null arcs

$$t = t_0, \quad \frac{d\varphi}{dr} = \pm(2mr - r^2)^{-\frac{1}{2}}, \quad 0 < d \leq r < 2m.$$

f) Arcs of the form

$$\frac{d\varphi}{dr} = hr^{-2}\left(a^2 + \left(\varepsilon + \frac{h^2}{r^2}\right)\left(\frac{2m}{r} - 1\right)\right)^{-\frac{1}{2}},$$

$$\frac{dt}{dr} = ar(2m - r)^{-1}\left(a^2 + \left(\varepsilon + \frac{h^2}{r^2}\right)\left(\frac{2m}{r} - 1\right)\right)^{-\frac{1}{2}},$$

$$a \neq 0, \qquad 0 < d \leq r < 2m, \qquad \varepsilon = \pm 1 \quad \text{or} \quad 0.$$

g) Time-like arcs

$$\varphi = \varphi_0, \quad \frac{dt}{dr} = ar(2m - r)^{-1}\left(\frac{2m}{r} - 1 + a^2\right)^{-\frac{1}{2}},$$

a arbitrary, $0 < r \leq d$.

h) Null arcs

$$\varphi = \varphi_0, \quad \frac{dt}{dr} = \pm r(2m - r)^{-1}, \qquad 0 < r \leq d.$$

i) Arcs of the form

$$\frac{d\varphi}{dr} = hr^{-2}\left(a^2 + \left(\frac{h^2}{r^2} + \varepsilon\right)\left(\frac{2m}{r} - 1\right)\right)^{-\frac{1}{2}},$$

$$\frac{dt}{dr} = ar(2m-r)^{-1}\left(a^2 + \left(\frac{h^2}{r^2}+\varepsilon\right)\left(\frac{2m}{r}-1\right)\right)^{-\frac{1}{2}},$$

a arbitrary, $h \neq 0, 0 < r \leq d$.

The first three of these originate at $t = t_0, \varphi = \varphi_0, r = 2m$, f) originates at $r = 2m, \varphi = \varphi_0, t = \pm\infty$, g), h) and i) at $r = 0, \varphi = \varphi_0, t = t_0 \neq \pm\infty$.

It is easy to verify that all boundary points defined by these arcs, with the exception of the last three, are interior points in (10) and therefore do not represent singularities. In fact they all lie on the lines $u^2 = v^2$; a), c), d) and e) at $u = v = 0$, b) and f) at $u^2 = v^2 \neq 0$. Of course a), c), d) and e) become non-ordinary singularities if we identify (u, v) with $(-u, -v)$.

The arcs g), h), i) originate at $r = 0$, i.e. on the hyperbola $v^2 - u^2 = 4/e$; it is reasonable to expect that they represent true singularities. Consider first a time-like arc of the type g); for convenience we take $a = 1, \varphi = 0$, and the origin at $t = 0$. If s is a distinguished parameter, we have from (19) and (20)

$$dt/ds = r/(2m-r), \qquad dr/ds = (2m/r)^{\frac{1}{2}}.$$

They give on integration

$$L: r = 2m\sigma^2, \quad t = 2m\log\frac{1+\sigma}{1-\sigma} - 4m\sigma - \frac{4}{3}m\sigma^3, \quad 0 < \sigma \leq d, \tag{21}$$

where

$$\sigma = (3s/4m)^{\frac{1}{3}}. \tag{22}$$

Let us first study a two-dimensional reduced model

$$\pm ds^2 = \left(\frac{2m}{r}-1\right)dt^2 - \frac{r}{2m-r}dr^2, \qquad 0 < r < 2m. \tag{23}$$

Clearly (21) is also a boundary arc of (23). The equation of a geodesic which radiates from the point $p(\sigma)$ of L, perpendicularly to L, is

$$\frac{dt}{ds} = \frac{1}{\sigma}\frac{r}{2m-r}, \qquad \frac{dr}{ds} = \left(\frac{1}{\sigma^2}+1-\frac{2m}{r}\right)^{\frac{1}{2}} \tag{24}$$

with

$$r = 2m\sigma^2, \qquad t = 2m\log\frac{1+\sigma}{1-\sigma} - 4m\sigma - \frac{4}{3}m\sigma^3 \quad \text{at} \quad s = 0. \tag{25}$$

Normal coordinates are obtained by setting

$$y_1 = s, \quad y_2 = \frac{4}{3}m\sigma^3, \tag{26}$$

by (22). Equations (24) give

$$\frac{\partial t}{\partial y_1} = \frac{1}{\sigma} \frac{r}{2m - r}, \qquad \frac{\partial r}{\partial y_1} = \left(\frac{1}{\sigma^2} + 1 - \frac{2m}{r} \right)^{\frac{1}{2}} \tag{27}$$

where

$$\sigma = (3y_2/4m)^{\frac{1}{3}}, \tag{28}$$

and (25), (26) give

$$\frac{\partial r}{\partial y_2} = \frac{1}{\sigma}, \quad \frac{\partial t}{\partial y_2} = \frac{\sigma^2}{1 - \sigma^2} \quad \text{at} \quad y_1 = 0. \tag{29}$$

These formulae allow us to calculate $\partial^{n+1} r/\partial y_1^n \partial y_2$, $\partial^{n+1} t/\partial y_1^n \partial y_2$ at $y_1 = 0$ (hence on L) for every n.

Now

$$g_{\mu\nu} = \left(\frac{2m}{r} - 1 \right) \frac{\partial t}{\partial y_\mu} \frac{\partial t}{\partial y_\nu} - \frac{r}{2m - r} \frac{\partial r}{\partial y_\mu} \frac{\partial r}{\partial y_\nu}$$

in the normal system, hence $g_{11} = 1$ by (27). A simple calculation gives furthermore

$$g_{22} = -1 \quad \partial^2 g_{22}/\partial y_1^2 = \frac{8}{9} y_2^{-2} \quad \text{on} \quad L; \tag{30}$$

the last expression tends to infinity when $y_2 \to 0$, so that L defines an ordinary singularity. But the partial derivatives with respect to y_1 on L are the same in the complete as in the reduced model, so that L defines an ordinary singularity of (1), as expected.

Next we consider a null arc

$$L_0 : t = -r + 2m \log \frac{2m}{2m - r}, \qquad 0 < r \leq d \tag{31}$$

of the type h). Straightforward calculation gives in the reduced model the following normal coordinates relatively to L_0 :

$$\begin{aligned} r &= \frac{1}{2} y_1 + y_2 - m y_1/y_2, \\ t &= -y_2 + \frac{1}{2} y_1 - m y_1/y_2 + 2m \log \frac{1 + y_1/2y_2}{1 - y_2/2m}. \end{aligned} \tag{32}$$

The metric tensor in this system is

$$\begin{aligned} &g_{11} = 0, \quad g_{12} = 1, \\ &g_{22} = 8m y_2^2 (2m - y_2)^{-2} \left(\frac{1}{r} - \frac{1}{y_2} \right) - 4m y_1 y_2^{-1} (2m - y_2)^{-1} \end{aligned} \tag{33}$$

and

$$\partial^2 g_{22}/\partial y_1^2 = 4my_2^{-3} \quad \text{on} \quad L$$

which again indicates an ordinary singularity.

It can be verified easily that in the system (32), the arc (21) is represented by

$$y_1/m = -(\sqrt{2}-1)(y_2/m)^2 + \frac{4}{5}2^{-3/4}(y_2/m)^{5/2} + 0(y_2^3) \tag{34}$$

where $0(y_2^3)$ refers to $y_2 \to 0$. This suggests that (21) and (31) are associated arcs. For this to be true, it would be necessary to verify that in a normal frame relatively to (21) the arc corresponding to (31) also originates at $y_1 = 0$, $y_2 = 0$. That (34) alone is not sufficient to establish equivalence is shown by the fact that there is a boundary arc of the form

$$y_1 = -y_2 + \frac{1}{2m}y_2^2 + 0(y_2^3), \quad 0 < y_2 \leq d \tag{35}$$

in the frame (33), corresponding to the interior arc

$$t = m(1 - 2\log 2), \quad m < r \leq m + d \tag{36}$$

in (23), and this is clearly not associated with L_0 since the latter determines a singularity.

It seems likely that the manifold (10) is complete and all boundary arcs originating at the same point $\varphi_0, \theta_0, u_0, r_0$ with $r_0^2 - u_0^2 = 4/e$, are associated. We shall not attempt to prove this here, but conclude with the discussion of another example,

$$\pm ds^2 = t^2(dx_1^2 + dx_2^2 + dx_3^2) - dt^2, \quad t > 0. \tag{37}$$

This line element was suggested by the author some time ago [5] as a possible cosmological model, with a singularity at the time origin $t = 0$. An interesting feature of this line element is that the metric tensor itself has no singularities, and the only suspicious place is at $t = 0$ where the determinant vanishes.

The geodesic equations of (37) have for general solution

$$\begin{aligned} x_k &= \frac{1}{2}\lambda_k \log\frac{s+\alpha-\varepsilon\beta}{\beta+\varepsilon(s+\alpha)} + \xi_k, \qquad k = 1, 2, 3, \\ t &= [(s+\alpha-\varepsilon\beta)(\beta+\varepsilon(s+\alpha))]^{\frac{1}{2}} \end{aligned} \tag{38}$$

where $\lambda_1^2 + \lambda_2^2 + \lambda_3^2 = 1$ and ξ_1, ξ_2, ξ_3 are arbitrary constants. Again $\varepsilon = +1$ if the geodesic is time-like, $\varepsilon = -1$ if it is space-like and $\varepsilon = 0$ if it is a null-line.

There are two types of boundary arcs: time-like arcs

$$t = s, \quad x_k = \xi_k, \quad k = 1, 2, 3, \quad 0 < s \leq b, \tag{39}$$

obtained by setting $\beta = 0$ in (38), and arcs with $\beta \neq 0$, originating at $t = 0$, $r = \infty$. To examine the first type, we assume that it originates at $x_1 = x_2 = x_3 = 0$, i.e.

$$L: t=\sigma, \quad 0<\sigma\le b, \quad x_k=0, \quad k=1,2,3. \tag{40}$$

The non-null geodesics emanating from L perpendicularly to L are space-like and have the form

$$t=(\sigma^2-s^2)^{\frac{1}{2}}, \qquad x_k=\frac{1}{2}\lambda_k \log\frac{s+\sigma}{\sigma-s}. \tag{41}$$

Hence $(x_1^2+x_2^2+x_3^2)^{\frac{1}{2}}=r=\frac{1}{2}\log\frac{s+\sigma}{\sigma-s}$, $s=\sigma\tanh r$. Normal coordinates are obtained by setting

$$y_k=\lambda_k s, \qquad k=1,2,3, \qquad y_4=\sigma,$$

i.e.

$$s=\bar{r}, \quad \lambda_k=y_k/\bar{r}$$

where $\bar{r}^2=y_1^2+y_2^2+y_3^2$. The transformation equations are

$$x_k=(y_k/\bar{r}) \quad \operatorname{arctanh} \quad (\bar{r}/y_4), \quad k=1,2,3$$
$$t=\left(y_4^2-\bar{r}^2\right)^{\frac{1}{2}} \tag{42}$$

and the metric tensor in the new system is

$$\begin{aligned} g_{mn}&=(y_4^2/\bar{r}^2-1)\operatorname{arctanh}^2(\bar{r}/y_4)(\delta_{mn}-y_my_n/\bar{r}^2)\\ &\quad+y_my_n/\bar{r}^2, \qquad 1\le m<n\le 3,\\ g_{44}&=-1. \end{aligned} \tag{43}$$

Hence L determines an ordinary singularity.

Among the boundary arcs which originate at $r=\infty, t=0$, we shall only consider the null arc

$$L_0: x_1=\frac{1}{2}\log\sigma, \quad x_2=x_3=0, \quad t=\sigma^{\frac{1}{2}}, \quad 0<\sigma\le 1. \tag{44}$$

For the normal coordinate directions we choose the space-like vectors $(0,\sigma^{-\frac{1}{2}},0,0)$, $(0,0,\sigma^{-\frac{1}{2}},0)$ and the null vector $(-1,0,0,\sigma^{\frac{1}{2}})$. They are derived by Fermi propagation along L_0. The general form of space-like geodesics in the normal hyperplane spanned by these vectors is

$$\begin{aligned} x_1&=\frac{1}{2}\frac{\alpha}{\beta}\log\left[\left(1+\frac{s}{\beta+\alpha}\right)\Big/\left(1-\frac{s}{\beta-\alpha}\right)\right]+\frac{1}{2}\log\sigma,\\ x_m&=\frac{1}{2}\lambda_m\log\left[\left(1+\frac{s}{\beta+\alpha}\right)\Big/\left(1-\frac{s}{\beta-\alpha}\right)\right], \quad m=2,3,\\ t&=(\sigma-2\alpha s-s^2)^{\frac{1}{2}} \end{aligned} \tag{45}$$

where $\beta^2 - \alpha^2 = \sigma$ and $\lambda_2^2 + \lambda_3^2 = \sigma/(\alpha^2 + \sigma)$. Normal coordinates are obtained by setting

$$\sigma = y_4, \quad \alpha = \frac{1}{2} y_1 y_4/\bar{r}, \quad \lambda_m = y_m \left(\frac{1}{4} y_4 y_1^2 + \bar{r}^2\right)^{-\frac{1}{2}}, \quad m = 2, 3,$$

$$s = \bar{r}, \quad \bar{r}^2 = y_2^2 + y_3^2$$

and substituting in (45). The transformation equations are

$$x_1 = \frac{1}{4}(y_1/\eta) \log\left[\left(1 + \eta - \frac{1}{2}y_1\right) \Big/ \left(1 - \eta - \frac{1}{2}y_1\right)\right] + \frac{1}{2} \log y_4$$

$$x_m = \frac{1}{2}\left(y_m/y_4^{\frac{1}{2}}\eta\right) \log\left[\left(1 + \eta - \frac{1}{2}y_1\right) \Big/ \left(1 - \eta - \frac{1}{2}y_1\right)\right], \quad m = 2, 3, \quad (46)$$

$$t = \left(y_4 - y_1 y_4 - y_2^2 - y_3^2\right)^{\frac{1}{2}}$$

with

$$\eta = \left(\frac{1}{4}y_1^2 + (y_2^2 + y_3^2)/y_4\right)^{\frac{1}{2}}. \quad (47)$$

A lengthy computation gives, for $m = 2, 3$,

$$g_{mn} = t^2 \frac{x_2^2 + x_3^2}{y_2^2 + y_3^2} \delta_{mn} + \frac{1}{y_4 \eta^2}(y_m y_n - t^2 x_m x_n) \quad (48)$$

where η, x_2, x_3, t are given by (46), (47). Again it is seen that L_0 determines an ordinary singularity.

From (44) it can be verified easily that in the system (42), L_0 is represented by

$$y_1 = \frac{1}{2}(\sigma - 1), \quad y_2 = y_3 = 0, \quad y_4 = \frac{1}{2}(\sigma + 1),$$

$$0 < \sigma \leq 1.$$

Hence L and L_0 are not associated directly, and probably not associated at all.

BIBLIOGRAPHY

[1] A. S. Eddington, The Mathematical Theory of Relativity, *Cambridge*, 1937.
[2] G. E. Lemaitre, *Ann. Soc. Scient. Bruxelles* Ser A, **53** (1933), 51–85.
[3] E. Schrödinger, Expanding Universes, *Cambridge*, 1956.
[4] J. L. Synge, *Proc. Roy. Irish Acad.* **53** (1950), 83–114.[12]
[5] G. Szekeres, *Phys. Rev.* **97** (1955), 212–223.

(Received May 26, 1959)

[12]The name of the journal was corrected from the erroneous "*Proc. Roy. Irish. Soc.*" (Editor).

Chapter 13
Igor D. Novikov 1964, geometrical interpretation of spherically symmetric manifolds, the Schwarzschild solution among them

Original paper: Igor D. Novikov, R- i T-oblasti v prostranstve-vremeni so sphericheski symmetrichnim prostranstvom. *Soobshcheniya GAISh [Communications of the State Shternberg Astronomical Institute]* **132**, 3–42 (1964).

Reprinted with the kind permissions of the Director of the State Shternberg Astronomical Institute and of Igor D. Novikov.

Paper translated by S. B. Beisekeev and M. A. H. MacCallum,
email: m.a.h.maccallum@qmul.ac.uk

Previously published in *Gen. Relativ. Gravit.* **33** (2001), pages 2255 – 2258 (the editorial note + biography), 2259 – 2295 (the paper).

The editorial note was re-edited for this reprinting, and the biography was updated by I. D. Novikov.

13.1 Editorial note to:

Igor D. Novikov

R- and T-Regions in a spacetime with a spherically symmetric space

Andrzej Krasiński

Once there was a time when scientists were not pushed to publish instantly whatever they could in the leading journals. The ISI citation index (and the ISI itself) did not yet exist, so authors were free to choose where to submit their papers. As a result, brilliant papers were occasionally published in inconspicuous local journals where the authors assumed they naturally belonged. Still, somehow the most important results were able to find their way to public knowledge.

Novikov's paper reprinted below is an example. Few people have had the chance to see it (which is one good reason to republish it), and yet most researchers in relativity have heard about the Novikov coordinates for the Schwarzschild solution, defined and discussed in it.

Some statements and results in the paper need to be related to the remaining literature. The starting point of the paper is the observation that for a general spherically symmetric metric:

$$\mathrm{d}s^2 = a(t,r)\mathrm{d}t^2 + b(t,r)\mathrm{d}t\mathrm{d}r + c(t,r)\mathrm{d}r^2 + d(t,r)(\mathrm{d}\vartheta^2 + \sin^2\vartheta\mathrm{d}\phi^2) \tag{1}$$

the "standard" spherical coordinates, in which:

$$b = 0, \qquad d = -r^2, \tag{2}$$

exist only if the gradient of $\sqrt{-d}$ is a spacelike vector. The points and regions of spacetime in which this can be done are called R-points and R-regions. If the gradient of $\sqrt{-d}$ is timelike, then $\sqrt{-d}$ may be chosen as the time coordinate. The corresponding points and regions are called T-points and T-regions. These notions were defined in an earlier paper by the same author [1], and Novikov's papers were apparently the first effort to systematically explore the geometry of the R- and T-regions. However, related problems appeared, without being explicitly named, in at least three earlier papers. Datt [2] first found a solution of Einstein's equations with a dust source in which the underlying spacetime is globally a T-region. Datt's solution is uniquely determined by the properties that it is a spherically symmet-

Andrzej Krasiński
N. Copernicus Astronomical Center, Polish Academy of Sciences, Bartycka 18, 00 716 Warsaw, Poland
e-mail: akr@camk.edu.pl

ric dust solution in which the coordinates defined by $t = \sqrt{-d}$ are at the same time comoving. The well-known Kantowski-Sachs solution [3] is a spatially homogeneous limit of the Datt solution. Kantowski and Sachs are rightly credited for exploring the geometry of such spatially homogeneous spacetimes in which the complete symmetry group is 4-dimensional, multiply transitive and has no simply transitive 3-dimensional subgroups (see also Ref. 4). However, the solution of Einstein's equations that they found is a subcase not only of Datt's solution, but also of the Kompaneets–Chernov solution [5], in which the symmetry was the same as in the Kantowski–Sachs model, but the source was a general perfect fluid. The Datt solution was rediscovered, discussed in much detail and generalized (for the cosmological constant and for the electric charge) in a series of papers by Ruban [6–9]. The so far most elaborate generalization of the Datt solution is the Szekeres dust solution [10] that has no symmetry.

The other two papers in which the problem of the T- and R-regions implicitly appeared are those of Nariai [11]. Nariai discussed the vacuum solution with a Λ-term for the case when d is constant, which is an invariant property under all the admissible coordinate transformations in eq. (1). Hence, the Nariai solution is neither an R- nor a T-region. The existence of the Nariai solution does not contradict the remark in Novikov's paper that follows his eq. (2.7). Novikov referred there to the case when the gradient of $\sqrt{-d}$ is a nonzero null vector, and in the Nariai solution this gradient is identically zero. Also, the Nariai solution does not contradict what Novikov says in his sec. 3 because it has no limit $\Lambda \to 0$. As far as this editor is aware, the case of $(\sqrt{-d})_{,\alpha}$ being null but nonzero in an open 4-dimensional region has been investigated only in Ref. 17 [info from Malcolm A. H. MacCallum].

Apart from the Schwarzschild solution, discussed in Novikov's paper, and the Datt solution which is a T-region globally, a natural area of application of the definitions of R- and T-regions is the Lemaître–Tolman (L–T) cosmological model that was discussed in many papers (Novikov's Ref. 15, see Ref. 12 below for an overview). In a few other papers, Novikov proved several theorems that apply to the L–T model (see Ref. 12 again). In particular, his result (stated after his eq. (2.3)) that the boundary between the R- and T-regions in the L–T model is given by the equation $R = F$ fits in well with the result by Barnes [13] that the hypersurface $R = F$ is the outer boundary of the region of trapped surfaces.

The "throat" that Novikov discussed at the end of his sec. 5 was also identified and discussed by other authors a few years later, see Refs. 13 and 14 (where the "throat" was named a "neck" and discussed for the L–T model) and Ref. 15 (where it was named "bottleneck" and discussed for a spherically symmetric perfect fluid model).

As Novikov stated in his sec. 9, the coordinates that are initially used there were first introduced by Lemaître [16]. Lemaître found that the Schwarzschild solution is the subcase of the L–T solution corresponding to $F = \text{const}$ and $f = 0$ (in Novikov's notation) and that this limit taken in the L–T model defines such coordinates in which the singularity at the Schwarzschild horizon just disappears. In fact, the condition $f = 0$ is superfluous: the Schwarzschild solution results with $F = m = \text{const}$ with arbitrary form of f. Novikov extended the discussion to the cases $f \neq 0$ and provided a geometrical interpretation in each case. The Lemaître coordinates are comoving coordinates of such freely falling observers in the Schwarzschild spacetime whose velocity at infinity is zero. The other two cases correspond to freely falling observers who never reach infinity (when $f < 0$) and those whose velocity at infinity is still nonzero (when $f > 0$).

Acknowledgements. The Editors are grateful to D. Gal'tsov and I. Dymnikova for their help in reconstructing the full names of the Soviet publishers for the bibliography, and to Alexei Starobinsky for his help in getting permissions for this reprinting from Igor Novikov and from the Shternberg Astronomical Institute.

References

1. I. D. Novikov, *Vestn. Mosk. Univ.* no 6, 66 (1962).
2. B. Datt, Z. *Physik* **108**, 314 (1938); reprinted in *Gen. Relativ. Gravit.* **31**, 1619 (1999).
3. R. Kantowski, R. K. Sachs, *J. Math. Phys.* **7**, 443 (1966).
4. R. Kantowski, *Gen. Relativ. Gravit.* **30**, 1665 (1998).
5. A. S. Kompaneets, A. S. Chernov, *ZhETF* **47**, 1939 (1964); *Sov. Phys. JETP* **20**, 1303 (1965).
6. V. A. Ruban, *ZhETF, Pisma v Red.* **8**, 669 (1968); *Sov. Phys. JETP Lett.* **8**, 414 (1968); reprinted in *Gen. Relativ. Gravit.* **33**, 369 (2001).
7. V. A. Ruban, *ZhETF* **56**, 1914 (1969); *Sov. Phys. JETP* **29**, 1027 (1969); reprinted in *Gen. Relativ. Gravit.* **33**, 375 (2001).
8. V. A. Ruban, in: *Tezisy Dokladov 3y Sovetskoy Gravitatsyonnoy Konferentsii [Theses of Lectures of the 3rd Soviet Conference on Gravitation]*. Izdatelstvo Erevanskogo Universiteta, Erevan 1972, p. 348.[1]
9. V. A. Ruban, *ZhETF* **85**, 801 (1983); *Sov. Phys. JETP* **58**, 463 (1983).
10. P. Szekeres, *Commun. Math. Phys.* **41**, 55 (1975).
11. H. Nariai, *Sci. Rep. Tôhoku. Univ.* **34**, 160 (1950); **35**, 46 (1951); reprinted in *Gen. Relativ. Gravit.* **31**, 951 and 963 (1999).
12. A. Krasiński, *Inhomogeneous cosmological models*. Cambridge University Press 1997.
13. A. Barnes, *J. Phys.* **A3**, 653 (1970).
14. C. Hellaby, *Class. Quant. Grav.* **4**, 635 (1987).
15. Y. Suto, K. Sato, H. Sato, *Progr. Theor. Phys.* **72**, 1137 (1984).
16. G. Lemaître, *Ann. Soc. Sci. Bruxelles* **A53**, 51 (1933); English translation in *Gen. Relativ. Gravit.* **29**, 641 (1997).
17. J. M. Foyster, C. B. G. McIntosh; *Commun. Math. Phys.* **27**, 241 (1972).

[1] For a more accessible presentation and discussion of this Ruban solution see A. Krasiński and G. Giono, arXiv 1107.4897, *Gen. Relativ. Gravit.*, **44**, 239 (2012).

13.2 Igor D. Novikov – a brief biography

By Igor D. Novikov

I was born on November 10, 1935 in Moscow. I got my M.Sc. degree from the Moscow State University in 1959 and Ph.D. degree in 1963.

I worked at the Institute of Applied Mathematics, Moscow; Space Research Institute, Moscow; Lebedev Physical Institute, Moscow as a Research Fellow, a Senior Research Fellow and head of the Department. Simultaneously I worked as a Professor of the Moscow State University and Moscow Pedagogical University. Some of my students have gone on to become Professors in Russia and many other countries.

I visited and gave lectures in many universities around the globe.

In 1991 I left Moscow to take up a Professorship at the Copenhagen University where I continue to research and teach. In 1994 I became Director of the Theoretical Astrophysics Center, Copenhagen.

In 2006 I became the Deputy Director of Astro Space center of the Lebedev Physical Institute, Moscow.

My researches are mainly devoted to the theory of gravity, physics and astrophysics of black holes, manifestations of relativistic objects in the Universe, and different aspects of Cosmology.

A more extended scientific biography of Professor Novikov can be found in Ref. [1] below [editor].

References

1. Relativistic Astrophysics. Proceedings of the Relativistic Astrophysics Conference in honour of Prof. I.D. Novikov's 60th birthday. Edited by B.J.T. Jones and D. Markovic. Cambridge University Press, 1997.

General Relativity and Gravitation, Vol. 33, No. 12, December 2001 (© 2002)

R- and T-Regions in Space-Time with Spherically Symmetric Space[1]

I. D. Novikov

Abstract.[2] The properties of solutions of Einstein's equations for spherically symmetric gravitational fields are studied. The notions of R- and T-universe regions are introduced. T-regions are shown to be regions of essential instability in the distinguished direction of the flow of time. The geometrical properties and those of the deformation of reference frames in these regions are discussed. Birkhoff's theorem for the Schwarzschild field is stated more precisely. In a T-region of this field there exists a frame of reference with the interval in the form

$$ds^2 = \left(1 - \frac{2\gamma m}{c^3 t}\right)^{-1} c^2 dt^2 - \left(1 - \frac{2\gamma m}{c^3 t}\right) dr^2 - c^2 t^2 (d\theta^2 + \sin^2\theta d\varphi^2),$$

where t is the time coordinate, r the radial space coordinate, γ the Newtonian gravitational constant, c the fundamental velocity and m the mass producing the field. In this system, by comparison with Schwarzschild's, the space and time coordinates have changed places. The properties of the motion of light rays and test particles in R- and T- regions are discussed. The conclusions should be taken into account in considering cosmological questions and in the investigation of superdense stellar configurations.

[1] Original title: *R- i T-oblasti v prostranstve-vremeni so sphericheski symmetrichnim prostranstvom. Soobshchenija GAISh [Communications of the Shternberg State Astronomical Institute]*, **132** (3), 3-42 (1964).
Reprinted with the kind permission of the author and of the publisher of *Soobshchenija GAISh*.
Translated by S.B. Beisekeev, People's Friendship University, Moscow, and M.A.H. MacCallum, School of Mathematical Sciences, Queen Mary, Mile End Road, London E1 4NS, U.K. E-mail: M.A.H.MacCallum@qmul.ac.uk. The translators thank Dr. A.G. Polnarev for help with the final draft.
Translators' note: the footnote numbering of the original has not been maintained as it depended on the pagination. Here all footnotes have been numbered sequentially. The notation for the hyperbolic functions and their inverses in this paper, ch, sh, Arsh, ... follows a standard East European usage rather than the cosh, sinh, arcsinh, ..., common, though far from universal, in English.

[2] This is an amended version of the original English abstract which translated the Russian abstract.

0001-7701/01/1200-2259/0 © 2002 Plenum Publishing Corporation

The investigation of solutions of Einstein's equations describing spherically symmetric (S.S.) distributions of matter presents particular interest. On one hand, many astronomical objects of differing scales, starting from planetary satellites and finishing at the least at clusters of galaxies, clearly manifest spherical symmetry. On the other hand, the assumption of spherical symmetry of the gravitational field significantly simplifies Einstein's equations and allows us to investigate problems which considered in the general case run into big difficulties.

1. FUNDAMENTAL EQUATIONS

We shall make use of coordinate systems which possess spatial spherical symmetry (from now on we shall use these without comment). If for coordinates on the spheres we make use of the usual angular coordinates θ and φ, then the most general expression for ds^2 in the coordinates t, r, θ, φ, where t is the time and r is the radial spatial coordinate, is written down in the following form (see for example, [1])

$$ds^2 = a(t,r)dt^2 + b(t,r)dt\,dr + c(t,r)dr^2 + d(t,r)(d\theta^2 + \sin^2\theta d\varphi^2). \quad (1.1)$$

Let us suppose that a, b, c, d satisfy the usual conditions accepted in the theory of general relativity [2]. The relations between the coordinate systems considered are given by the following transformations (omitting inessential spatial rotations),[3]

$$\tilde{t} = \tilde{t}(t,r), \quad \tilde{r} = \tilde{r}(t,r), \quad \theta = \tilde{\theta}, \quad \varphi = \tilde{\varphi}. \quad (1.2)$$

If a motion of a reference body is given (generally speaking, a deformation) then the freedom in the choice of the spatial system of coordinates **in this given reference body** and the choice of time coordinate does not reflect any physical change [3]. We call the set of all coordinate systems that belong to the given reference body the system of reference. The relation between one coordinate system and another in the system of reference is given by the transformations

$$\tilde{t} = \tilde{t}(t,r), \quad \tilde{r} = \tilde{r}(r), \quad \theta = \tilde{\theta}, \quad \varphi = \tilde{\varphi}. \quad (1.3)$$

Let us make the convention that Latin indices take the values 1, 2, 3 and Greek the values 0, 1, 2, 3. It is easy to see that in any system of reference in which the expression for the line interval can be written in the form (1.1), the condition

$$A_{ik} \equiv 0,$$

[3] Additionally, in the case of homogeneous and isotropic space, space translations are allowed.

is satisfied, where A_{ik} is the chronometrically invariant tensor for the angular velocity of rotation (for definitions and formulae see [4]). And this means [5] that in any spherically symmetric system of reference at any world point it is possible to construct a space slice everywhere orthogonal to the time lines determining the motion of the reference system, i.e. it is possible, with the help of a transformation (1.3), to bring the expression for the line interval to the form

$$ds^2 = e^{\nu(t,r)}dt^2 - e^{\lambda(t,r)}dr^2 - e^{\mu(t,r)}(d\theta^2 + \sin^2\theta d\varphi^2). \tag{1.4}$$

Let us choose a system of units so that $c = 1$ and $k = 1$ where c is the fundamental speed and k is the Newtonian constant of gravitation. Then Einstein's equations are written in the following form (see, for example, [6])

$$-8\pi T_1^1 = \frac{1}{2}e^{-\lambda}\left(\frac{\mu'^2}{2} + \mu'\nu'\right) - e^{-\nu}\left(\ddot{\mu} - \frac{1}{2}\dot{\mu}\dot{\nu} + \frac{3}{4}\dot{\mu}^2\right) - e^{-\mu} + \Lambda, \tag{1.5}$$

$$-8\pi T_2^2 = -8\pi T_3^3 = \frac{1}{4}e^{-\lambda}\left(2\nu'' + \nu'^2 + 2\mu'' + \mu'^2 - \mu'\lambda' - \nu'\lambda' + \mu'\nu'\right) + \frac{1}{4}e^{-\nu}\left(\dot{\lambda}\dot{\nu} + \dot{\mu}\dot{\nu} - \dot{\lambda}\dot{\mu} - 2\ddot{\lambda} - \dot{\lambda}^2 - 2\ddot{\mu} - \dot{\mu}^2\right) + \Lambda, \tag{1.6}$$

$$8\pi T_0^0 = e^{-\lambda}\left(\mu'' + \frac{3}{4}\mu'^2 - \frac{\mu'\lambda'}{2}\right) - \frac{1}{2}e^{-\nu}\left(\dot{\lambda}\dot{\mu} + \frac{\dot{\mu}^2}{2}\right) - e^{-\mu} - \Lambda, \tag{1.7}$$

$$8\pi T_0^1 = \frac{1}{2}e^{-\lambda}\left(-2\dot{\mu}' - \dot{\mu}\mu' + \dot{\lambda}\mu' + \nu'\dot{\mu}\right), \tag{1.8}$$

where a dot means a partial derivative with respect to t and a prime means a partial derivative with respect to r; T_α^β are the components of the energy-momentum tensor; and Λ is the cosmological constant. The expression for the energy-momentum tensor of an electromagnetic field can be written in the form [7]:

$$T_\alpha^\beta = F_{\alpha\delta}F^{\delta\beta} + \frac{1}{4}F_{\delta\gamma}F^{\delta\gamma}g_\alpha^\beta, \tag{1.9}$$

where $g_{\alpha\beta}$ are the components of the fundamental tensor, and $F_{\alpha\beta}$ the components of the electromagnetic field tensor, satisfying the generally covariant Maxwell equations. In the case of vanishing viscosity and energy flux relative to the matter, the energy-momentum tensor of a continuous medium can be written in the form [8]

$$T^{\alpha\beta} = (\rho + p)u^\alpha u^\beta - pg^{\alpha\beta}, \tag{1.10}$$

where ρ is the density and p the pressure of matter and $u^\alpha = \dfrac{dx^\alpha}{ds}$.

Let us assume that the following conditions on the energy-momentum tensor of matter are satisfied [9]:

$$\rho \geq 3p \geq 0 \tag{1.11}$$

and the sign of $\dot{\rho}$ coincides with the sign of $\dot{p}$ and is opposite to the sign of D where $D = D^j_j$ is the trace of the chronometrically invariant tensor of the rate of deformation of comoving space (for definitions and formulae see [4]).[4]

As is well known, the components $T_{\alpha\beta}$ satisfy the conservation laws:

$$\nabla_\alpha T^\alpha_\beta = 0, \tag{1.12}$$

where ∇_α is the covariant differentiation operator. If in the region filled by the matter a reference system comoving with the matter [3], in which $u^i = 0$, is used, then the expressions (1.12) reduce to the following two equations

$$\dot{\lambda} + 2\dot{\mu} = -\frac{2\dot{\rho}}{\rho + p}, \tag{1.13}$$

$$\nu' = -\frac{2p'}{\rho + p}. \tag{1.14}$$

2. DEFINITION AND EXISTENCE CRITERIA OF R- AND T-REGIONS

When investigating spherically symmetric gravitational fields, it is usually assumed (see, for example, [1]) that it is always possible with the help of the transformation (1.2) to bring the interval (1.4), and consequently also (1.1), to the form

$$ds^2 = e^{\tilde{\nu}(\tilde{t},\tilde{r})} d\tilde{t}^2 - e^{\tilde{\lambda}(\tilde{t},\tilde{r})} d\tilde{r}^2 - \tilde{r}^2 (d\tilde{\theta}^2 + \sin^2 \tilde{\theta} d\tilde{\varphi}^2). \tag{2.1}$$

Since g_{22} and g_{33} are **invariant** under the transformation (1.2), for this it is necessary to make the transformation $\tilde{r}^2 = e^{\mu(t,r)}$, $\tilde{\theta} = \theta$, $\tilde{\varphi} = \varphi$, and to choose $\tilde{t}$ so that the spatial slices are orthogonal to the time lines (as we saw in § 1, this is always possible). However, it is clear that a coordinate $\tilde{r}$ obtained by a transformation of the type

$$\psi(\tilde{r}) = e^{\mu(t,r)}, \tag{2.2}$$

where $\psi(\tilde{r})$ is some function of $\tilde{r}$ can be one of the spatial coordinates[5] only in the case where the line $\tilde{r} = \text{const}$, $\tilde{\theta} = \text{const}$, $\tilde{\varphi} = \text{const}$ is timelike, i.e. if the line $r = r(t)$, defined from the first expression in (2.2) by the condition $\tilde{r} = \text{const}$, fulfils the inequality

[4] Recently, Zel'dovich showed [10] that the most general condition still compatible with relativity theory is $\rho \geq p \geq 0$. Instead of (1.11) it is possible to take this condition, which does not alter the conclusions of the present paper.

[5] That is, the reference system can be realized by "test particles" for which $\tilde{r} = \text{const}$, $\tilde{\theta} = \text{const}$, $\tilde{\varphi} = \text{const}$. We mention that in the general case in relativity theory such systems of coordinates are possible in which at least one of the four coordinate lines is timelike at any point.

$$e^{\nu-\lambda} > \left(\frac{dr}{dt}\right)^2. \tag{2.3}$$

This condition means that the vector (dr, dt) tangent to the curve at a given point lies inside the light cone.

The condition (2.3) can also be obtained in the following way. Let the relation between the coordinates r, t and $\tilde{r}$, $\tilde{t}$ be given by the functions $t = t(\tilde{t}, \tilde{r})$, $r = r(\tilde{t}, \tilde{r})$. Then from the transformation law of components of a tensor we get the following expression for $\tilde{g}_{00}$:

$$\tilde{g}_{00} = e^{\nu}\left(\frac{\partial t}{\partial \tilde{t}}\right)^2 - e^{\lambda}\left(\frac{\partial r}{\partial \tilde{t}}\right)^2. \tag{2.4}$$

But for a system traced out by real bodies, $\tilde{g}_{00} > 0$ (see [1]). From this inequality and (2.4) we get the condition

$$e^{\nu-\lambda} > \frac{\left(\frac{\partial r}{\partial \tilde{t}}\right)^2}{\left(\frac{\partial t}{\partial \tilde{t}}\right)^2},$$

which for $\tilde{r} = \text{const}$ coincides with (2.3).

Since the condition $\mu(r, t) = \text{const}$ is equivalent to the condition $\tilde{r} = \text{const}$, the inequality (2.3) can be written in the form

$$e^{\nu-\lambda} > \left(\frac{\dot{\mu}}{\mu'}\right)^2. \tag{2.5}$$

Thus a system of coordinates with $e^{\mu} = r^2$ [6] is only possible in a region of spacetime where for any system of coordinates with the interval in the form (1.4) the inequality (2.5) is satisfied.

If the inequality (2.5) is satisfied at a certain world point then by virtue of continuity ($e^{\nu-\lambda}$, $\dot{\mu}$, μ' cannot be discontinuous, see [2], [11]) it is satisfied in some neighbourhood of this point.[7] In this neighbourhood a system of coordinates in which $e^{\mu} = r^2$ is possible. Let us call such world-points R-points and a set of R-points an R-region.

If the inequality

$$e^{\nu-\lambda} < \left(\frac{\dot{\mu}}{\mu'}\right)^2. \tag{2.6}$$

[6] Such a form e^{μ} implies a monotone dependence of e^{μ} on r. But as will be shown in § 3 the equations of gravitation do not allow a nonmonotone relation of e^{μ} to r when $\dot{\mu} \equiv 0$.

[7] This neighbourhood obviously includes the whole sphere $r = \text{const}$, $t = \text{const}$.

is satisfied, then the coordinate considered cannot be one of the spatial coordinates, but could be the time coordinate,[8] and in the corresponding region of spacetime a system with the interval in the form (2.1) is impossible but a system with $e^{\mu} = \psi(\tilde{t})$ is possible.[9]

Let us call the points in the neighbourhood of which the system with $e^{\mu} = \psi(\tilde{t})$ is possible T-points and a set of them a T-region.

Note the covariance of the conditions (2.5) and (2.6) relative to the transformations which preserve the form (1.4) for ds^2.

The strict equality

$$e^{\nu-\lambda} = \left(\frac{\dot{\mu}}{\mu'}\right)^2 \tag{2.7}$$

is a condition additional to the system of the gravitational field equations and the equations of state, and therefore for an arbitrary (S.S.) distribution and motion of the matter and the field, and also, as we shall see, in vacuum, it may not be possible to fulfil it in any four-dimensional region. This equality determines the boundary between R- and T-regions.

3. TYPES OF WORLD POINTS

From now on we shall assume, if there is no special note, that $\Lambda = 0$. The criteria (2.5) and (2.6) are to be used only on condition that $\dot{\mu}$ and μ' do not become zero simultaneously. Now let the equalities

$$\dot{\mu} = 0, \quad \mu' = 0, \tag{2.8}$$

be true in some system of coordinates.[10] Then these equalities will be true in any system of coordinates. Indeed, since μ is invariant relative to the transformation (1.2), in the new system of coordinates $(\tilde{t}, \tilde{r}, \tilde{\theta}, \tilde{\varphi})$

$$\frac{\partial \mu}{\partial \tilde{t}} = \dot{\mu}\frac{\partial t}{\partial \tilde{t}} + \mu'\frac{\partial r}{\partial \tilde{t}} = 0,$$
$$\frac{\partial \mu}{\partial \tilde{r}} = \dot{\mu}\frac{\partial t}{\partial \tilde{r}} + \mu'\frac{\partial r}{\partial \tilde{r}} = 0.$$

[8] Of course we do not demand that the time coordinate is necessarily chosen so that the slice t = const is spacelike even in the case where the space slices are not orthogonal to the time lines. But this requirement is obligatory when the space slices are orthogonal to the time lines.

[9] In § 3 it will be proved that this function may be non-monotonic.

[10] World points where these equalities hold exist for example in a closed homogeneous isotropic cosmological model. When considering homogeneous and isotropic models, the origin of coordinates is always assumed to be fixed.

From the equation (1.5) and the expressions (1.10) and (1.11) it follows, when (2.8) is satisfied, that the inequality

$$\ddot{\mu} < 0 \tag{2.9}$$

is fulfilled. Consequently the equalities (2.8) cannot be satisfied simultaneously in some four-dimensional region. From the inequality (2.9) it follows in particular that for a static distribution of matter the comoving space cannot be closed. This inequality proves the statement in footnote 6.[11] Our problem is to clarify the types of worldpoints where the equalities (2.8) are satisfied.

Suppose the energy flux relative to the matter is negligibly small and the energy-momentum tensor is in the form (1.10). In the region occupied by matter let us make use of a system of reference comoving with the matter. Then, taking into account the equality (2.8), the equations (1.5), (1.7), (1.8) are written in the form

$$8\pi p = -e^{-\nu}\ddot{\mu} - e^{-\mu}, \tag{2.10}$$

$$8\pi \rho = -e^{-\lambda}\mu'' + e^{-\mu}, \tag{2.11}$$

$$\dot{\mu}' = 0. \tag{2.12}$$

Due to the spherical symmetry, all properties of the spacetime are independent of the coordinates θ and φ. So it is sufficient to restrict ourselves to two dimensions: t and r. Let us consider the two-dimensional manifold of (t, r). Our problem reduces to determining the character of the curve $\mu(t, r) = \text{const}$ at a point where the equalities (2.8) are satisfied and in its neighbourhood. From the general theory of implicit functions (see, for example, [12]) it follows that the condition $\mu(t, r) = \text{const}$, when the equalities (2.8) and (2.12) are satisfied, defines in the neighbourhood of the point considered:

(1) When $\mu''\ddot{\mu} > 0$, or, taking into account (2.9), when $\mu'' < 0$, an isolated point.

(2) When $\mu'' > 0$, two branches of the curve $r(t)$.

It is known that

$$\left(\frac{dr}{dt}\right)^2 = -\frac{\ddot{\mu}}{\mu''}. \tag{2.13}$$

Adding (2.10) and (2.11) we obtain

$$-e^{-\nu}\ddot{\mu} - e^{-\lambda}\mu'' = 8\pi(\rho + p) \geq 0. \tag{2.14}$$

[11] Hence it also follows that e^{μ} cannot be a constant.

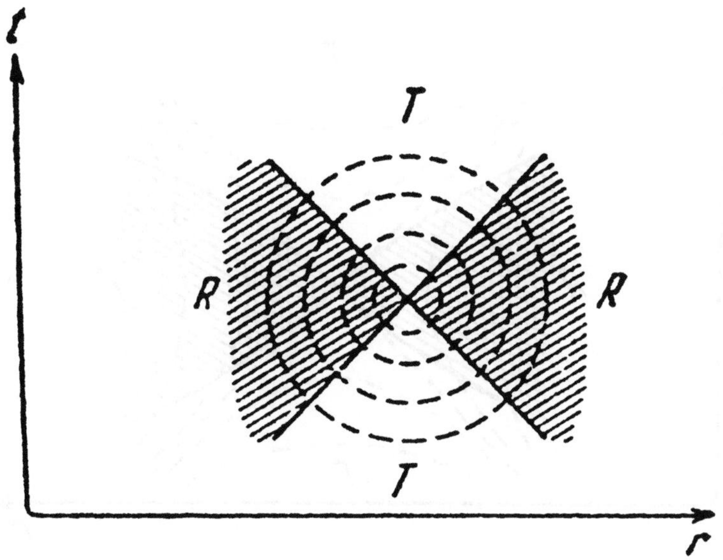

Figure 1. R- and T-regions in the neighbourhood of a special point ($\dot{\mu} = \mu' = 0$) where $\mu'' < 0$. Full lines - boundary of R- and T-regions. Dashed lines - lines of $\mu(t, r) = \text{const}$.

The equality sign in the last expression (2.14) holds only in the vacuum case when $\rho = 0$ and $p = 0$. Therefore from (2.13) and (2.14) it follows that in a given case when matter exists the vectors tangent to the branches of the curve at the self-crossing point are spacelike and in the vacuum case are null.

(3) When $\mu'' = 0$ we have either an isolated point or two branches of the curve with a common tangent, and in particular cases this reduces to one curve or a turning point. In the last two cases the vector (dt, dr) at the point considered is spacelike, as in case 2. [12]

The character of the one-parameter family of curves $r = r(t, \text{const})$ in the neighbourhood of an isolated point, a crossover point, or self-touching branches of the curve and a turning point, are shown respectively in Figs. 1, 2 and 3. In all cases in a sufficiently small neighbourhood of the point considered there are both points from the R-region and points from the T-region. Only in the case when the two branches of the curve with a common tangent reduce to one curve is our point inside the T-region. Let us show that the gravitational equations allow such a possibility.

[12] Note that an analogous classification of the points at which the conditions (2.8) are satisfied can be carried out for the case $\Lambda \neq 0$ too. In this case, the inequality (2.9) alone can turn out not to

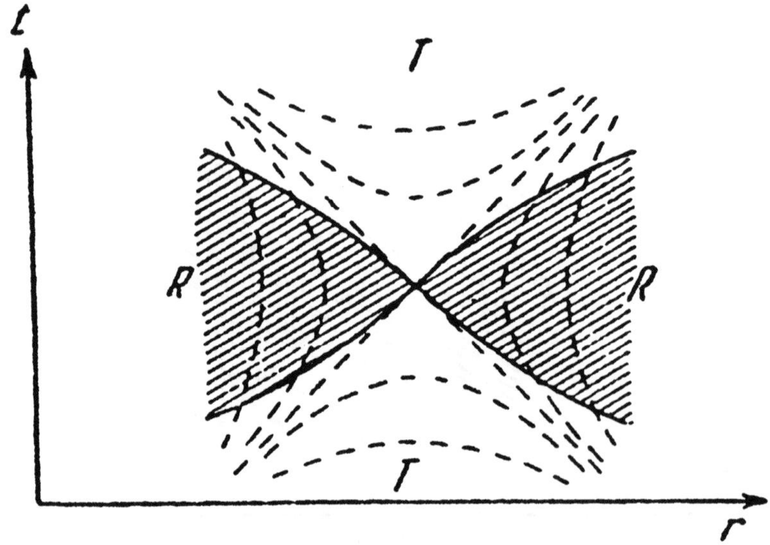

Figure 2. The same as in Fig. 1, under the condition $\mu'' > 0$.

Suppose that in some four-dimensional region in the comoving system of reference $p' \equiv 0$, $\rho' \equiv 0$, $\lambda' \equiv \mu' \equiv \nu' \equiv 0$. Then the equations (1.5-1.8) are written in the form:

$$8\pi p = -e^{-\nu}\left(\ddot{\mu} - \frac{1}{2}\dot{\mu}\dot{\nu} + \frac{3}{4}\dot{\mu}^2\right) - e^{-\mu}, \tag{2.15}$$

$$8\pi p = \frac{1}{4}e^{-\nu}\left(\dot{\lambda}\dot{\nu} + \dot{\mu}\dot{\nu} - \dot{\lambda}\dot{\mu} - 2\ddot{\lambda} - \dot{\lambda}^2 - 2\ddot{\mu} - \dot{\mu}^2\right), \tag{2.16}$$

$$8\pi \rho = \frac{1}{2}e^{-\nu}\left(\dot{\lambda}\dot{\mu} + \frac{\dot{\mu}^2}{2}\right) + e^{-\mu}. \tag{2.17}$$

With the help of a transformation of the form $\tilde{t} = \tilde{t}(t)$ it is possible in a given case to make ν zero. Let the equation of state be given in the form $p = p(\rho)$. Then the equations (2.15)-(2.17) can be brought into the form

$$\ddot{\mu} = F_1(\mu, \dot{\lambda}, \dot{\mu}), \tag{2.18}$$
$$\ddot{\lambda} = F_2(\mu, \dot{\lambda}, \dot{\mu}), \tag{2.19}$$
$$\rho = F_3(\mu, \dot{\lambda}, \dot{\mu}). \tag{2.20}$$

be satisfied (in the case $\Lambda > 0$) and therefore instead of the sign of μ'' it is necessary to take into account the sign of the product $\mu''\ddot{\mu}$.

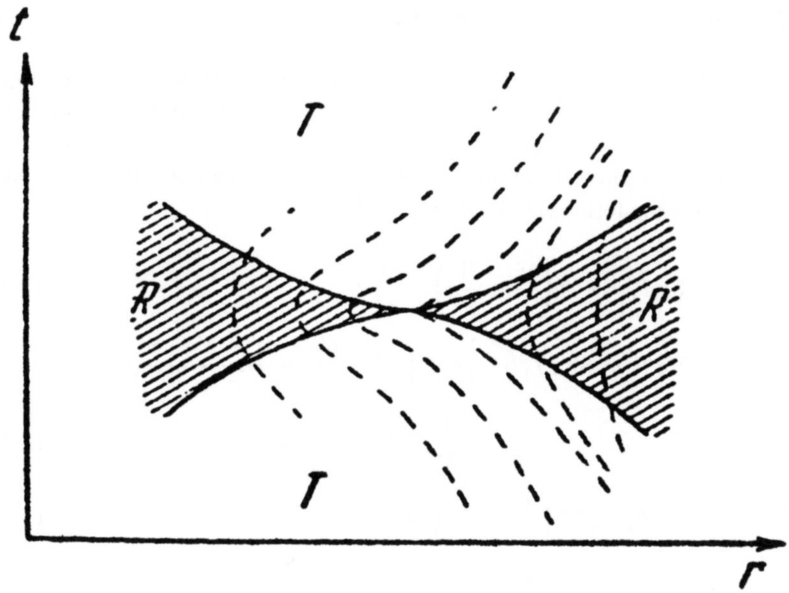

Figure 3. The same as in Fig. 1, under the condition $\mu'' = 0$.

Let us give initial conditions: when $t = t_0$, $\lambda = a$, $\dot{\lambda} = b$, $\mu = c$, $\dot{\mu} = 0$, where a, b, and c are arbitrary. From Cauchy's theorem in some neighbourhood of t_0, there is a solution of the system (2.18)-(2.19) satisfying the given initial conditions. Substituting the expressions obtained for λ and μ into (2.20), we obtain ρ. It is known that when at $t = t_0$ the relation $\rho = \frac{1}{8\pi e^{\mu}} > 0$ is fulfilled, then by continuity $\rho > 0$ also in some neighbourhood of t_0. From the equation of state we find p. We have proved the existence of a solution of the gravitational equations which in some neighbourhood of t_0 satisfies all the ordinary requirements adopted in the theory of general relativity. This solution proves the assertions above. Indeed, for $t = t_0$ and any $r = r_0$ situated in the region considered, $\mu' = \dot{\mu} = 0$, $\mu'' = 0$, and the point (t_0, r_0) lies inside the T-region, since in the neighbourhood of (t_0, r_0) μ is a function only of t.

This solution proves also the statement given in footnote 9 that $\psi(t)$ may be a non-monotonic function.

Thus if $\mu' = \dot{\mu} = 0$, then when $T_{\alpha\beta}$ has the form (1.10), the relation $\ddot{\mu} < 0$ is fulfilled and the considered point (t, r) is on the boundary, as a rule, of the R- and T-regions, as happens, for example, in closed homogeneous isotropic models of the universe, and only in exceptional cases is it in the T-region, as was shown by the theorem proved. **A point where the conditions (2.8) are fulfilled cannot lie inside an R-region.**

4. R- AND T-SYSTEMS OF REFERENCE AND THEIR PROPERTIES

In an R-region there is a system of reference which includes the coordinate system where $e^{\mu} = r^2$. Let us call this system of reference an R-system, and in the R-system we will use in particular such a system of coordinates that $e^{\mu} = r^2$. The gravitational equations in such a coordinate system simplify significantly and take the following form (see, for example, [1]):

$$-8\pi T_1^1 = e^{-\lambda}\left(\frac{\nu'}{r} + \frac{1}{r^2}\right) - \frac{1}{r^2}, \tag{2.21}$$

$$-8\pi T_2^2 = -8\pi T_3^3 = \frac{1}{2}e^{-\lambda}\left(\nu'' + \frac{\nu'^2}{2} + \frac{\nu' - \lambda'}{r} - \frac{\nu'\lambda'}{2}\right) - \frac{1}{2}e^{-\nu}\left(\ddot{\lambda} + \frac{\dot{\lambda}^2}{2} - \frac{\dot{\lambda}\dot{\nu}}{2}\right), \tag{2.22}$$

$$8\pi T_0^0 = e^{-\lambda}\left(\frac{\lambda'}{r} - \frac{1}{r^2}\right) + \frac{1}{r^2}, \tag{2.23}$$

$$8\pi T_0^1 = -e^{-\lambda}\frac{\dot{\lambda}}{r}. \tag{2.24}$$

An R-system is impossible inside a T-region (see § 2).

In a T-region a system of reference including the coordinate system where $e^{\mu} = \psi(t)$ is possible. In a region of monotone variation of $\psi(t)$ it is possible to set[13] $\psi(t) = t^2$. Let us call this system of reference a T-system and use within it a system of coordinates where $e^{\mu} = t^2$. In this system the gravitational equations take the following form:

$$-8\pi T_1^1 = e^{-\nu}\left(\frac{\dot{\nu}}{t} - \frac{1}{t^2}\right) - \frac{1}{t^2}, \tag{2.25}$$

$$-8\pi T_2^2 = -8\pi T_3^3 = \frac{1}{2}e^{-\lambda}\left(\nu'' + \frac{\nu'^2}{2} - \frac{\nu'\lambda'}{2}\right) - \frac{1}{2}e^{-\nu}\left(\ddot{\lambda} + \frac{\dot{\lambda}^2}{2} - \frac{\dot{\nu} - \dot{\lambda}}{t} - \frac{\dot{\nu}\dot{\lambda}}{2}\right), \tag{2.26}$$

$$8\pi T_0^0 = e^{-\nu}\left(\frac{\dot{\lambda}}{t} + \frac{1}{t^2}\right) + \frac{1}{t^2}, \tag{2.27}$$

$$8\pi T_0^1 = -e^{-\lambda}\frac{\nu'}{t}. \tag{2.28}$$

[13] Taking into account the possibility of changing $t = -\tilde{t}$.

A T-system is impossible in an R-region (see § 2).[14]

Note the uniqueness of R- and T-systems, i.e. if in some neighbourhood of a point there is an R- or T-system, then this system is unique. Indeed any transformation $\tilde{t} = \tilde{t}(t, r)$, $\tilde{r} = \tilde{r}(t, r)$, preserving the orthogonality of the coordinate system and describing the transition to another system of reference disturbs the form of e^{μ}. Thus the statement has been proved.

Hence it follows that the same R- (or T-) system can cover a whole connected R- (or, respectively, T-) region. In fact, let us take any connected closed region A, entirely lying in an R-region. In the neighbourhood of each point of the region A let us introduce an R-system. From the infinite covering of the closed region A by these neighbourhoods let us choose a finite number of neighbourhoods covering A (that is possible, as is well known). Each of the chosen neighbourhoods overlaps with at least one more neighbourhood and passing through the overlapping parts it is possible to get from a given neighbourhood to any other. Due to uniqueness, in an overlap the R-systems of the neighbourhoods coincide. Thus the R-system of one neighbourhood could be extended to all others, i.e. to the whole region A, which could cover all points of the connected R-region. This proves the statement given.

Obviously there exist such distributions of matter for which the R-system appears also to be a system comoving with the matter. Indeed any static S.S. distribution of matter allows such a system of reference. If the energy flux relative to the matter is absent, then from the equation (2.24) it follows that only a static distribution of matter allows a comoving system which is also simultaneously an R-system.[15]

Let us prove that there exist such distributions of matter for which (in a T-region) the T-system appears also to be a comoving system of reference. As a proof let us show two examples of such distributions of the matter. The first example can be the solution of Einstein's equations whose existence was proved in the previous section (in this solution we are limited to the region of monotone variation of $\mu(t)$). The second example is the following solution of Einstein's equations, written down in an explicit form.

By direct calculation it is possible to verify that the $g_{\mu\nu}$ from the expression

$$ds^2 = \frac{24t^{12}}{t^{12} + 24B}dt^2 - t^{10}dr^2 - t^2(d\theta^2 + \sin^2\theta\, d\varphi^2)$$

[14] Note that in an R-region it is possible to introduce a system of reference with the coefficient of the angular part equal to t^2, but necessarily with non-orthogonality of the time lines to the space slices. Obviously it will not be a T-system. In a T-region it is impossible to introduce a system of reference with $g_{22} = r^2$ even by violating the orthogonality of the space slices to the time lines (see footnote 5).

[15] Everything said here up to this point in this section is true also in the case $\Lambda \neq 0$. For this it is necessary to add Λ to the right hand sides of the equations (2.21), (2.22), (2.25) and (2.26), and to the right hand sides of equations (2.23) and (2.27) to add $(-\Lambda)$.

for the interval, where B is a constant, satisfy the system of equations (2.25)-(2.28), written down for the system of reference comoving with the matter and for an energy-momentum tensor of the form (1.10). In this case, the equation of state is time dependent and for the t satisfying the inequalities $11B > 1.04t^{12} \geq 5B$ the conditions

$$\rho > 0,\ p > 0;\ \dot{\rho} < 0,\ \dot{p} < 0,\ \rho \geq 3p,\ D > 0,$$

are fulfilled, i.e. all the requirements accepted in the theory of general relativity are fulfilled. Here we will not consider the properties of the comoving space for that distribution and motion of matter. We note only that these properties largely coincide with the properties of the space of the system of reference which are investigated in part III.

Finally let us prove in addition the following statement concerning a comoving T-system. If $T_{\alpha\beta}$ is in the form (1.10) then an expansion of the system along the radial coordinate cannot be changed to compression, i.e. if $D_{11} = 0$ then $\frac{{}^*\partial D_{11}}{\partial t} > 0$, where D_{11} is the component of the chronometrically invariant tensor D_{ik} of the deformation rate of the comoving space and $\frac{{}^*\partial}{\partial t}$ is the chronometrically invariant derivative operator (for definitions and formulae see [4]).

For a proof it is sufficient to show that if $\dot{\lambda} = 0$, then $\ddot{\lambda} > 0$. Since in the comoving system $T_0^1 = 0$, then from (2.28) it follows that $\nu' = 0$. Then from the system of equations (2.25)-(2.27), written down for the comoving system of reference, follow the inequalities

$$p = e^{-\nu}\left(\frac{\dot{\nu}}{t} - \frac{1}{t^2}\right) - \frac{1}{t^2} > 0, \tag{2.29}$$

$$\rho - 3p = e^{-\nu}\left(3\frac{\dot{\lambda}}{t} - \frac{\dot{\nu}}{t} - \dot{\nu}\dot{\lambda} + 2\ddot{\lambda} + \dot{\lambda}^2\right) > 0. \tag{2.30}$$

From (2.29) it follows that $\dot{\nu} > 0$. Then from (2.30) it follows that when $\dot{\lambda} = 0$ the inequality $\ddot{\lambda} > 0$ is fulfilled, as it was necessary to show.

Hence it also follows that if $\dot{\lambda} > 0$ then

$$D = D_j^j = \frac{1}{2}e^{-\frac{\nu}{2}}\left(\dot{\lambda} + \frac{4}{t}\right) > 0$$

and an increase of the volume element could not be changed to a decrease.

By direct calculation it is not difficult to convince oneself that in a comoving T-system the tensor of the anisotropic deformation rate [4]

$$\Pi_{ik} = D_{ik} - \frac{1}{3}Dh_{ik}; \qquad h_{ik} = -g_{ik} + \frac{g_{0i}g_{0k}}{g_{00}}$$

cannot be identically equal to zero, i.e. the deformation is essentially anisotropic.

Note in conclusion that if the energy flux relative to matter is absent then in any comoving T-system the gradient of the pressure is equal to zero. Indeed, as was pointed out, from (2.28) it follows that in a comoving system $\nu' = 0$ and then from (1.14) it follows that $p' = 0$.

5. THE PROPERTIES OF THE COEFFICIENT OF THE ANGULAR PART OF THE INTERVAL

In a T-region, in whatever system of reference, $\dot{\mu}$ cannot be identically equal to zero. Indeed, in this region there exists a system of reference in which $\mu = \mu(t)$. But in any other S.S. system of reference $(\tilde{t}, \tilde{r}, \tilde{\theta}, \tilde{\varphi})$:

$$\frac{\partial \mu}{\partial \tilde{t}} = \dot{\mu} \frac{\partial t}{\partial \tilde{t}} \tag{2.31}$$

(we recall once more that μ is invariant relative to the transformation (1.2)). But from the requirements accepted in the theory of general relativity (see, for example, [1], [8]) it follows that $\frac{\partial t}{\partial \tilde{t}} \neq 0$ and as $\dot{\mu}$ cannot be identically zero in the T-system of reference (μ cannot be a constant. See footnote 11), then from (2.31) it follows that

$$\frac{\partial \mu}{\partial \tilde{t}} \neq 0$$

in any other system of reference. Thus in a T-region with any distribution of matter any S.S. system of reference is certainly deforming in directions transversal to the radial coordinate and in this sense it is possible to call the T-regions regions of essential nonstationarity.

Let us show that in any system of reference in an R-region at a fixed instant of time $t = t_0$ the coefficient of the angular part of the interval cannot be a non-monotonic function of r.

Let $\mu' = 0$ on some sphere (t_0, r_0), and then if $\dot{\mu} \neq 0$, the inequality (2.6) is fulfilled and this sphere lies in a T-region. But if on (t_0, r_0), $\mu' = 0$ and $\dot{\mu} = 0$, then, as was shown in § 3, this sphere cannot be in the R-region. So the spheres on which $\mu' = 0$ cannot be in the R-region and this means that a change of the monotonicity of the function $e^{\mu(t_0, r)}$ cannot take place in an R-region.

If in an R-region in some system of reference one of the inequalities $\mu' \lessgtr 0$ is fulfilled on the sphere (t_0, r_0), then the sign is conserved on transforming into any other system of reference (the reference direction of the radial coordinate is presumed, of course, to be the same as in the old system of reference, i.e. $\frac{\partial r}{\partial \tilde{r}} > 0$). In fact for definiteness let $\mu' > 0$. Then in the new system of reference

$$\frac{\partial \mu}{\partial \tilde{r}} = \dot{\mu} \frac{\partial t}{\partial \tilde{r}} + \mu' \frac{\partial r}{\partial \tilde{r}}, \tag{2.32}$$

where we take into account that in the new system of reference the space slices are orthogonal to the time lines. Let us transform the right side of (2.32) thus:

$$\dot{\mu}\frac{\partial t}{\partial \tilde{r}}+\mu'\frac{\partial r}{\partial \tilde{r}}=\left(\frac{\dot{\mu}}{\mu'}\frac{\dfrac{\partial t}{\partial \tilde{r}}}{\dfrac{\partial r}{\partial \tilde{r}}}+1\right)\mu'\frac{\partial r}{\partial \tilde{r}} \tag{2.33}$$

But since the space slices $\tilde{t}=\text{const}$ are orthogonal to the time lines, they should be spacelike, i.e. should satisfy the inequality

$$\left|\frac{\dfrac{\partial t}{\partial \tilde{r}}}{\dfrac{\partial r}{\partial \tilde{r}}}\right|<e^{-\frac{\nu-\lambda}{2}}$$

But due to the condition that the sphere considered belongs to an R-region, consequently

$$\left|\frac{\dot{\mu}}{\mu'}\right|<e^{\frac{\nu-\lambda}{2}}$$

From the last two inequalities it follows that the expression in brackets on the right side of (2.33) is greater than zero and since $\mu'>0$ and $\frac{\partial r}{\partial \tilde{r}}>0$, then, consequently, $\frac{\partial \mu}{\partial \tilde{r}}>0$, as it was required to show.

Absolutely analogously it is proved that if in a T-region in some system of reference one of the inequalities $\dot{\mu}\gtrless 0$ is fulfilled on the sphere (t_0, r_0), then the sign is conserved when transforming into any other system of reference (the reference direction of time remaining the same: $\dfrac{\partial t}{\partial \tilde{t}}>0$).

If in a T-region on some sphere $\dot{\mu}>0$ then this sign is preserved in any system of reference in a certain world neighbourhood of the given sphere, such that this neighbourhood does not depend on the choice of the system of reference. In fact according to what was said in this section and in § 3, the neighbourhood considered could be extended provided the boundary of the R- and T-regions or a point where $\dot{\mu}=0$ and $\mu'=0$ is not crossed. But these objects do not depend on the choice of the system of reference.

Therefore in some world region situated in a T-region the coordinate spheres of **any** S.S. system are only expanding (or only contracting), the properties of this region are not invariant under the transformation $t=-t$, and there is an asymmetry in the direction of the flow of time. The analogous asymmetry in empty space was pointed out by Finkelstein [13] (cf. § 9).

If in a T-region in some system of reference $\dot{\mu}\neq 0$ on the sphere (t_0, r_0) and $\mu'\gtrless 0$ then it is always possible to choose the system of reference so that on that sphere: 1) μ has an extremum with respect to r, 2) the sign of μ' is changed to the opposite one, 3) if $\mu'=0$, then, by a choice of reference system, it is possible to make $\mu' \neq 0$ and with either sign (everywhere we assume $\frac{\partial r}{\partial \tilde{r}} > 0$).

Indeed let, for example, $\mu' \neq 0$. Let us prove the first statement. Transform into the T-system of reference, so $e^{\mu} = t^2$. Then let us change the time coordinate: $t = t[\cos(\tilde{r} - \tilde{r}_0)]$, where $\tilde{r}_0$ is the coordinate of the sphere considered at the given moment, and make the corresponding change of the radial coordinate in order to preserve the orthogonality of the space slices to the time lines. It is not difficult to verify that these transformations in the neighbourhood of the given sphere are allowed in the sense which was pointed out in §2. As a result of all the transformations that were made we get:

$$\left.\frac{\partial \mu}{\partial \tilde{r}}\right|_{\tilde{r}=\tilde{r}_0} = \dot{\mu} \left.\frac{\partial t}{\partial \tilde{r}}\right|_{\tilde{r}=\tilde{r}_0} = -\frac{2}{t}\tilde{t}\sin(\tilde{r} - \tilde{r}_0)\big|_{\tilde{r}=\tilde{r}_0} = 0,$$

$$\left.\frac{\partial^2 \mu}{\partial \tilde{t}^2}\right|_{\tilde{r}=\tilde{r}_0} = \ddot{\mu}\left.\left(\frac{\partial t}{\partial \tilde{r}}\right)^2\right|_{\tilde{r}=\tilde{r}_0} + \dot{\mu}\left.\frac{\partial^2 t}{\partial \tilde{r}^2}\right|_{\tilde{r}=\tilde{r}_0} = -\frac{2}{t^2}\tilde{t} < 0,$$

and the coefficient of the angular part on this sphere has an extremum with respect to r.

Obviously a reciprocal procedure is also possible: in the neighbourhood of a sphere on which at the instant $t = t_0$ the relations $\mu' = 0$ and $\dot{\mu} \neq 0$ hold, it is possible to choose a system of reference where on the given sphere $\frac{\partial \mu}{\partial \tilde{r}} \neq 0$.

Indeed, let us make a transformation of the coordinates (which is possible in the sense of §2): $t = t(\tilde{t}, \tilde{r})$, $r = r(\tilde{t}, \tilde{r})$ such that $\frac{\partial t}{\partial \tilde{r}} \neq 0$. Then

$$\frac{\partial \mu}{\partial \tilde{r}} = \frac{\partial \mu}{\partial t}\frac{\partial t}{\partial \tilde{r}} \neq 0$$

Obviously, it is possible to give any sign to $\frac{\partial \mu}{\partial \tilde{r}}$, which proves the statement (3). The proof of the statement (2) is analogous to the proof of the statement (1) and we need not dwell on it.

Thus an extremum of $e^{\mu(t_0, r)}$ as a function of r can only be inside a T-region or on the boundary between R- and T-regions. The latter happens, for example, in a closed homogeneous isotropic model of the universe in a comoving reference system at the moment of its maximum volume. In a T-region we are able to create an extremum of $e^{\mu(t_0, r)}$ or destroy it at any world point by choice of the system of reference.

We point out that an extremum of the function $e^{\mu(t, r_0)}$ as a function of t when $\mu'|_{r=r_0} \neq 0$ can happen only inside an R-region. Indeed, if on a certain sphere $\dot{\mu} = 0$ and $\mu' \neq 0$ then the inequality (2.5) holds and this sphere lies in an R-region.

Let us introduce a system of reference comoving with the matter. Suppose that in this system there exists a sphere (t, r) where the conditions $\mu' = 0$, $\mu'' > 0$ hold at the given moment. Obviously this sphere, which we call a throat, cannot lie in an R-region. The r coordinate of this sphere is changing as time passes, i.e. $r = r(t)$. Let us determine the speed of motion of this sphere relative to the matter. Since the system of reference is comoving with the matter, the coordinate speed of this motion is

$$\frac{dr}{dt} = -\frac{\dfrac{\partial \mu'}{\partial t}}{\dfrac{\partial \mu'}{\partial r}} = -\frac{\dot{\mu}'}{\mu''} \tag{2.34}$$

From the equations (1.8) and (1.14) it follows that at $\mu' = 0$

$$2\dot{\mu}' = \nu'\dot{\mu}; \qquad \nu'(\rho + p) = 2p'. \tag{2.35}$$

From (2.34) and (2.35) we get

$$\frac{dr}{dt} = -\frac{p'\dot{\mu}}{(\rho + p)\mu''} \tag{2.36}$$

Thus the matter is "leaking out" from the throat or "leaking in", depending on the sign of $p'\dot{\mu}$.

Let us determine the rate of change of the value e^{μ} at the throat, which we will call the radius of curvature of the throat. Since $\mu' = 0$,

$$\frac{d\mu}{dt} = \dot{\mu}$$

If $\dot{\mu} = 0$, $\mu' = 0$ and $p \geqslant 0$ then, as was shown in § 3, the inequality $\ddot{\mu} < 0$ holds and from (2.36) it follows that $\frac{dr}{dt} = 0$. Therefore where $\dot{\mu} = 0$ and $\mu' = 0$ the relation

$$\frac{d^2\mu}{dt^2} = \ddot{\mu} < 0,$$

holds.

Thus the throat can exist only outside an R-region, consequently only in a nonstatic matter distribution (if matter exists in the region), and, while varying, it cannot pass through a regular minimum of the throat's radius of curvature.

In part III and [14] there are solutions of Einstein's equations describing those systems of reference in vacuum and distributions and motions of matter for which in the comoving system the throat really exists.

Obviously, analogous conclusions are correct also for spheres on which $\mu' = 0$ and $\mu'' < 0$. Such spheres exist, for example, in the closed homogeneous isotropic model of the universe in the comoving system of reference.

6. SUFFICIENT CONDITIONS FOR EXISTENCE OF A T-REGION FOR DUST MATTER

Let us prove a statement of general character for the case of vanishingly small pressure of matter, important in cosmological applications.

The energy flux relative to matter we still assume to be negligibly small. Let us show that if under these conditions the density of matter at a certain moment of time does not fall very fast with distance and the coefficient of the angular part of the interval takes sufficiently large values, then starting from some distance the points of the comoving system of reference at that moment lie in a T-region.[16]

Let $p=0$ at all values of r in a certain interval of t variation. Let us introduce the notation $e^{\frac{\mu}{2}} \equiv R$. Then if at $t=t_0$ which belongs to the interval considered the condition

$$\lim_{r\to r_0} R(t_0, r) = \infty \tag{2.37}$$

holds (r_0 may be infinite) and for $r > r_1$ the condition

$$\rho \geqslant \frac{A}{8\pi R^2}, \tag{2.38}$$

where $A = \text{const} > 1$, then there exists r_2 such that the points of the system with $r > r_2$ at $t = t_0$ are in a T-region.

Proof. For $p=0$ the gravitational equations are completely integrable (Tolman [15]) and it is possible to write the solution in the form (see [1])

$$\nu = 0, \quad e^{\lambda} = \frac{(R')^2}{1+f(r)}, \tag{2.39}$$

$$\dot{R}^2 = f(r) + \frac{F(r)}{R}, \tag{2.40}$$

$$8\pi\rho = \frac{F'(r)}{R'R^2}, \tag{2.41}$$

where $f(r)$ and $F(r)$ are arbitrary functions subject only to the general requirements accepted in the theory of general relativity. From the expressions (2.39) and (2.40) we get

$$e^{\nu-\lambda} = \frac{\dot{R}^2(1+f)}{(R')^2\left(\frac{F}{R}+f\right)}. \tag{2.42}$$

When the conditions (2.38) hold, from (2.41) it follows that

$$\frac{F}{R} + \frac{\text{const}}{R} \geqslant A > 1$$

[16]If e^{μ} is not monotone in r, then the sphere with $\mu' = 0$ is already outside the R-region.

Whence when (2.37) holds it follows that starting from a certain[17] $r = r_2$ the condition

$$\frac{F}{R} > 1 \tag{2.43}$$

holds. From (2.42) and (2.43) it follows that when $r > r_2$ and $t = t_0$

$$e^{\nu-\lambda} < \left(\frac{\dot{R}}{R'}\right)^2 = \left(\frac{\dot{\mu}}{\mu'}\right)^2,$$

and this means (see § 2) that the points considered are in a T-region. The boundary between the R- and T-regions is determined by the equality $R = F$.

Let us prove also the following statement which pertains to the case $p = 0$.

As before, we use the comoving system of reference and adopt the notation $e^{\frac{\mu}{2}} \equiv R$.

Let us show that for any $r = r_2$ there exist t_1 and t_2 (dependent on r_2) such that the points (r_2, t), where t belongs to the interval (t_1, t_2), are in the T-region. In other words, each point of comoving space eventually as time passes lies in a T-region.

Proof. From Tolman's solution (2.39)-(2.41) it follows (see [1]) that for each $r = r_2$ there exists a t_1 such that

$$\lim_{t \to +t_1} R(r_2, t) = 0$$

Hence it follows that there exists a t_2 (dependent on r_2) such that for each t in the interval (t_1, t_2) the inequality

$$\frac{F(r_2)}{R(r_2, t)} > 1$$

holds. Together with (2.42) this again brings us to the inequality (2.6) which proves the statement that was made.

Note that this theorem requires the equality $p = 0$ to hold **at all moments of time** and therefore has limited application. On the other hand, in the first theorem of this section it is only required that the equality $p = 0$ holds in a certain interval of time including t_0.

[17]Translators' note: the original has r_1 here instead of r_2 but the latter appears from the context to be correct.

7. R- AND T-REGIONS IN THE SCHWARZSCHILD SPACE-TIME

Let $T_{\alpha\beta}=0$ in a certain four-dimensional finite or infinite region. Consider an R-region in vacuum. In this region it is possible to introduce an R-system of reference (see § 4). In this system the equations (2.21)-(2.24) for the vacuum are written down in the following well-known form:

$$e^{-\lambda}\left(\frac{\nu'}{r}+\frac{1}{r^2}\right)-\frac{1}{r^2}=0, \tag{3.1}$$

$$\frac{1}{2}e^{-\lambda}\left(\nu''+\frac{\nu'^2}{2}+\frac{\nu'-\lambda'}{r}-\frac{\nu'\lambda'}{2}\right)-\frac{1}{2}e^{-\nu}\left(\ddot{\lambda}+\frac{\dot{\lambda}^2}{2}-\frac{\dot{\lambda}\dot{\nu}}{2}\right)=0, \tag{3.2}$$

$$e^{-\lambda}\left(\frac{\lambda'}{r}-\frac{1}{r^2}\right)+\frac{1}{r^2}=0, \tag{3.3}$$

$$e^{-\lambda}\frac{\dot{\lambda}}{r}=0 \tag{3.4}$$

It is possible to present the solution of this system of equations, known as the exterior Schwarzschild solution [16], in the following expression for the interval

$$ds^2=\left(1-\frac{2m}{r}\right)dt^2-\left(1-\frac{2m}{r}\right)^{-1}dr^2-r^2(d\theta^2+\sin^2\theta\,d\varphi^2), \tag{3.5}$$

where m is the constant of integration, which coincides with the mass creating the field. Birkhoff [17] has proved the uniqueness of this solution for the vacuum up to an arbitrary transformation of the coordinates (see also the simple proof of this theorem by Tolman [6]).

The R-system of reference in the vacuum possesses a well-known physical singularity at $r=2m$[18] which is manifested in particular in that at this value of r the gravito-inertial force F_1 (definition and formulae in [5]) becomes infinite.[19] It is obvious that by no coordinate transformation in the given system of reference, i.e. without a transition to a system of reference moving relative to the given one, is it possible to eliminate this singularity, since the physical properties do not depend on such coordinate transformations [3].

[18] This value of r is called the gravitational radius of mass m.

[19] We underline that at the gravitational radius not only has the coordinate system of three-dimensional space a singularity, which could be eliminated by a scale transformation of the radial coordinate, but, which is very important, g_{00} becomes zero. This singularity is already a physical singularity of the given system of reference.

Below we shall see that the value $r = 2m$ determines the boundary between the R- and T-regions and therefore the considered system of reference, which is an R-system, cannot be extended beyond this boundary, since in a T-region an R-system is impossible.

Consider a T-region in the vacuum. In this region it is possible to introduce a T-system of reference (see §4). In this system the equations (2.24)-(2.28) are written down in the following form:

$$e^{-\nu}\left(\frac{\dot{\nu}}{t}-\frac{1}{t^2}\right)-\frac{1}{t^2}=0, \tag{3.6}$$

$$\frac{1}{2}e^{-\lambda}\left(\nu''+\frac{\nu'^2}{2}-\frac{\nu'\lambda'}{2}\right)-\frac{1}{2}e^{-\nu}\left(\ddot{\lambda}+\frac{\dot{\lambda}^2}{2}-\frac{\dot{\nu}-\dot{\lambda}}{t}-\frac{\dot{\nu}\dot{\lambda}}{2}\right)=0, \tag{3.7}$$

$$e^{-\nu}\left(\frac{\dot{\lambda}}{t}+\frac{1}{t^2}\right)+\frac{1}{t^2}=0, \tag{3.8}$$

$$e^{-\lambda}\frac{\nu'}{t}=0 \tag{3.9}$$

It is possible to present the solution of this system of equations by the following expression for the interval

$$ds^2=\left(\frac{2m}{t}-1\right)^{-1}dt^2-\left(\frac{2m}{t}-1\right)dr^2-t^2(d\theta^2+\sin^2\theta\, d\varphi^2) \tag{3.10}$$

The properties of this system of reference are considered in [18] and [19]. Here we only point out that this system of reference can be a system of free test particles. Birkhoff's theorem, which was considered above, is not applicable any more in the T-region since in its proof it is tacitly assumed that the coefficient of the angular part of the interval can always be made equal to r^2. However, it is possible to give a proof of the uniqueness of the solution (3.10) (up to an arbitrary transformation of the coordinates) in the vacuum in a T-region.

First let us show that in vacuum the sphere (t_0, r_0) on which $\mu'=0$ and $\dot{\mu}=0$ cannot lie inside a T-region and consequently lies on the boundary of the R- and T-regions.[20] Indeed, as was shown in §4 (see formula (2.10)), in vacuum when the conditions $\mu'=0$ and $\dot{\mu}=0$ hold, then the equality

$$-e^{-\nu}\ddot{\mu}-e^{-\lambda}\mu''=0 \tag{3.11}$$

is achieved, and since with that $\ddot{\mu}<0$ (see (2.9)), from (3.11) it follows that $\mu''>0$ and this point (t_0, r_0) of the (t, r) manifold belongs to the type

[20] In §9 it will be shown that such a sphere really can exist in the vacuum.

which is shown in Fig. 2. The curve $r = r(t)$ which is determined from the condition $\mu(t, r) = \text{const}$, here has a self-crossing point and vectors tangent to the curve at this point are null. Consequently, this point lies on the boundary of the R- and T-regions. In the world region including the sphere (t_0, r_0), in any system of reference μ essentially depends on both t and r.

From the proof it follows that in the vacuum the case of non-monotone dependence of μ on t is impossible in a T-system of reference (see § 3), and in the neighbourhood of any T-point it is possible to introduce a T-system of reference with $e^{\mu} = t^2$.

Then the gravitational equations become the system (3.6)-(3.9), the solution of which is given by the expression (3.10), which it was required to show.

In the proven theorem (as in Birkhoff's theorem) it is the local character of the space-time in the vacuum case which is spoken about, and nothing is said about the topological properties of space-time (or its separate parts).

Comparing the expression for the interval (3.10) with the expression (3.5) it is not difficult to convince oneself that these expressions will coincide if r and t change places. In order not to mix the radial and time coordinates of Schwarzschild with the analogous coordinates of the T-system, we will denote the latter by X^1 and X^0. Then (3.10) will be rewritten in the form

$$ds^2 = \left(\frac{2m}{X^0} - 1\right)^{-1} dX^{0^2} - \left(\frac{2m}{X^0} - 1\right) dX^{1^2} - X^{0^2}(d\theta^2 + \sin^2\theta\, d\varphi^2) \quad (3.10a)$$

8. R- AND T-REGIONS IN THE KOTTLER AND REISSNER-NORDSTRÖM SPACE-TIMES

Suppose now that $\Lambda \neq 0$ and $T_{\alpha\beta} = 0$. The solution of the gravitational equations, written for this case in an R-system of reference, describes the space-time of Kottler [20]. The interval in the R-system has the form (see, for example, [7])

$$ds^2 = \left(1 - \frac{2m}{r} - \frac{\Lambda}{3}r^2\right) dt^2 - \left(1 - \frac{2m}{r} - \frac{\Lambda}{3}r^2\right)^{-1} dr^2 - r^2(d\theta^2 + \sin^2\theta\, d\varphi^2) \quad (3.12)$$

When some conditions for Λ and m, mentioned in [19], hold in this space-time, then there exists a T-region (and under specified conditions (see [19]) only a T-region exists). The solution of the gravitational equations for a T-system of reference brings us to the following expression for the interval in this system:

$$ds^2 = \left(\frac{2m}{t} + \frac{\Lambda}{3}t^2 - 1\right)^{-1} dt^2 - \left(\frac{2m}{t} + \frac{\Lambda}{3}t^2 - 1\right) dr^2 - t^2\left(d\theta^2 + \sin^2\theta\, d\varphi^2\right) \tag{3.13}$$

The properties of this system are considered in [19].

Now let $\Lambda = 0$, but in the space let there be a field created by a charged particle. The solution of the gravitational equations together with the generally covariant Maxwell equations, written in the R-system for this case, describes the spacetime of Reissner-Nordström [21]. The interval in the R-system has the form (see, for example, [7])

$$ds^2 = \left(1 - \frac{2m}{r} + \frac{4\pi\varepsilon^2}{r^2}\right) dt^2 - \left(1 - \frac{2m}{r} + \frac{4\pi\varepsilon^2}{r^2}\right)^{-1} dr^2 - r^2(d\theta^2 + \sin^2\theta\, d\varphi^2), \tag{3.14}$$

where $4\pi\varepsilon$ is the electric charge. The non-zero components of T_β^α have the following values

$$T_0^0 = T_1^1 = -T_2^2 = -T_3^3 = \frac{1}{2}\frac{\varepsilon^2}{r^4}$$

When some conditions on m and ε hold (see [19]) in this space-time, there is a T-region. In the T-region the values $g_{\alpha\beta}$ from the expression

$$ds^2 = \left(\frac{2m}{t} - \frac{4\pi\varepsilon^2}{t^2} - 1\right)^{-1} dt^2 - \left(\frac{2m}{t} - \frac{4\pi\varepsilon^2}{t^2} - 1\right) dr^2 - t^2(d\theta^2 + \sin^2\theta\, d\varphi^2), \tag{3.15}$$

together with the values for T_β^α

$$T_0^0 = T_1^1 = -T_2^2 = -T_3^3 = \frac{1}{2}\frac{\varepsilon^2}{t^4}; \quad T_\beta^\alpha = 0, \quad \alpha \neq \beta$$

are the solutions of the gravitational equations and Maxwell's equations in a T-system of reference. The properties of the T-system of reference in which the interval has the form (3.15) are considered in [19].

9. FREELY FALLING SYSTEMS OF REFERENCE IN THE SCHWARZSCHILD SPACE-TIME

There are systems of reference including both an R- and a T-region (see [22], [13], [23], [24], [25]). Freely falling systems are examples of such systems of reference. The expressions for the interval for all freely falling systems in the Schwarzschild space-time can be obtained from Tolman's solution (see §6, formulae (2.39)-(2.41)).[21] For this it is necessary to put $F = \text{const}$ in (2.40) and (2.41). Then if $F = 0$, the space-time will be flat (see [26]), so this case is trivial. We are not considering it here. The case $F < 0$ corresponds in the ordinary interpretation to the gravitational field of a particle with negative mass and here we will also not consider this possibility. If $F > 0$, then without loss of generality it is possible to put $F = 1$.

The integral of the equation (2.40) will have different forms depending on which of the following relations holds:

$$f(\xi) = 0, \tag{3.16}$$

$$f(\xi) > 0, \tag{3.17}$$

$$f(\xi) < 0 \tag{3.18}$$

In this section the notation ξ is adopted for the radial space coordinate of the freely falling system, and τ for the time coordinate.

Let the condition (3.16) hold. Then from the relation (2.40) we obtain

$$R = \left\{\frac{3}{2}[\tau + \Phi(\xi)]\right\}^{\frac{2}{3}} \tag{3.19}$$

The arbitrary function $\Phi(\xi)$ cannot be non-monotonic, and also cannot be constant, since from the second relation of (2.39) it follows that if $\Phi'(\xi) = 0$ then $e^{\lambda} = 0$. But if Φ is monotonic, then we can choose the radial coordinate ξ so that $\Phi(\xi) = \xi$. Then the relation (3.19) takes the form

$$R = \left[\frac{3}{2}(\tau + \xi)\right]^{\frac{2}{3}} \tag{3.20}$$

The expression for the interval is now written in the form

$$ds^2 = d\tau^2 - \left[\frac{3}{2}(\xi + \tau)\right]^{-\frac{2}{3}} d\xi^2 - \left[\frac{3}{2}(\xi + \tau)\right]^{\frac{4}{3}} (d\theta^2 + \sin^2\theta\, d\varphi^2) \tag{3.21}$$

Such an expression for the interval was obtained for the first time, probably, by Lemaître [25], and then independently by Rylov [22] from considerations different from those stated here. We will call a coordinate system where the interval has the form (3.21) a Lemaître system.

This system includes both R- and T-regions. It can be realised by a system of free test particles falling from infinity towards the central singularity and such that its speed at spatial infinity in Schwarzschild's system

[21] The exceptions are T-systems.

of reference is equal to zero. The connection between Schwarzschild coordinates r and t and Lemaître coordinates ξ and τ is written down in the following form [22]

$$\pm\sqrt{r} = \left[\tfrac{3}{2}(\xi+\tau)\right]^{\frac{1}{3}}, \tag{3.22}$$

$$t = \tau + 2\left[\tfrac{3}{2}(\xi+\tau)\right]^{\frac{1}{3}} + \ln\left|\frac{\left[\frac{3}{2}(\xi+\tau)\right]^{\frac{1}{3}}-1}{\left[\frac{3}{2}(\xi+\tau)\right]^{\frac{1}{3}}+1}\right| \tag{3.23}$$

Changing the sign on the left hand side of (3.22) is equivalent to changing the time direction (the direction of increase of the time coordinate is the direction from past to future) and, as noted in [22], depending on the sign chosen, the point (t, r) could correspond to a Lemaître system consisting either of particles falling towards the centre or of particles moving from the centre.

Analogously to (3.22) and (3.23) it is possible to write the connection between Lemaître coordinates and the coordinates of a T-system of reference with the interval (3.10a)

$$\pm\sqrt{x^0} = \left[\tfrac{3}{2}(\xi+\tau)\right]^{\frac{1}{3}}, \tag{3.24}$$

$$x^1 = \tau + 2\left[\tfrac{3}{2}(\xi+\tau)\right]^{\frac{1}{3}} + \ln\left|\frac{\left[\frac{3}{2}(\xi+\tau)\right]^{\frac{1}{3}}-1}{\left[\frac{3}{2}(\xi+\tau)\right]^{\frac{1}{3}}+1}\right| \tag{3.25}$$

From the relation (2.7) we obtain in the Lemaître system the equation for the boundary between R- and T-regions

$$\left[\frac{3}{2}(\xi+\tau)\right]^{\frac{2}{3}} = 1 \tag{3.26}$$

From (3.22)-(3.23) and (3.24)-(3.25) we obtain that this equation corresponds to $r = 2m$ in the R-system and $x^0 = 2m$ in the T-system. Therefore, (3.26) is the equation of motion of the singular sphere of Schwarzschild in the Lemaître system. The points of this sphere, which is the boundary between the R- and T-regions, move with the fundamental speed.

For the R-region

$$\left[\tfrac{3}{2}(\xi+\tau)\right]^{\frac{2}{3}} > 1, \tag{3.27}$$

and for the T-region

$$\left[\frac{3}{2}(\xi+\tau)\right]^{\frac{2}{3}} < 1 \tag{3.28}$$

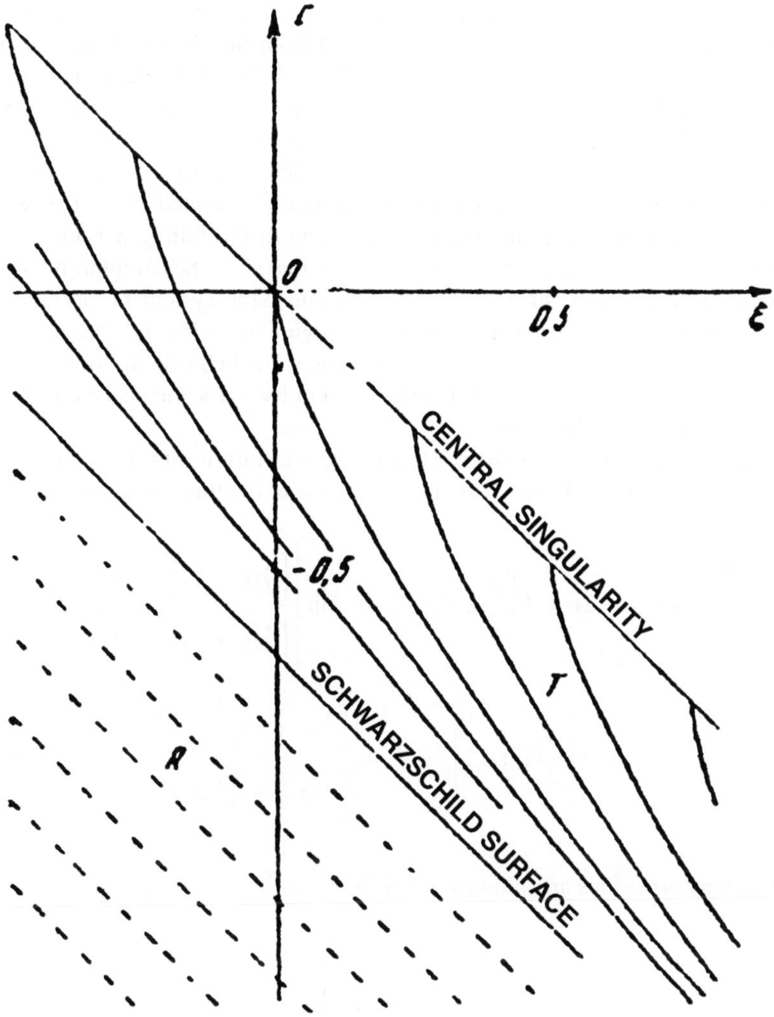

Figure 4.

In Fig. 4 the R- and T-regions of the space-time considered are represented in the Lemaître coordinates. The region of allowed values of ξ and τ lies below the line

$$\xi = -\tau, \tag{3.29}$$

which is the equation of the central singularity. This line turns out to be spacelike. Any space slice $\tau = \text{const}$ contains points of both the R- and T-regions. The trajectories of the test particles realising the R- and T-systems are also plotted in this diagram (the minus sign was chosen in (3.22) and (3.24), which corresponds to the fall of particles towards the central singularity).

In the T-region the interpretation of the double sign in (3.24) given in [22] as the existence of a system of test particles which are "reflected" from the central singularity and moving in the opposite direction to the falling particles is not correct. Let us show this. Namely, let us show that if in the T-region there is a system of reference ξ, τ, such that for the transition to this system it is necessary to choose the sign (+) in (3.24), then in the same region there does not exist a system of reference such that for the transition to it one should choose the sign (−). We shall denote the coordinates of this latter system by χ for the space coordinate and ψ for the time coordinate.

Let the system ξ, τ be given. Then in order to turn it into the system χ, ψ one should, as follows from (3.24) and (3.25), make the transformation

$$\chi = \xi + 2\tau + 4\left[\frac{3}{2}(\xi+\tau)\right]^{\frac{1}{3}} + 2\ln\left|\frac{\left[\frac{3}{2}(\xi+\tau)\right]^{\frac{1}{3}} - 1}{\left[\frac{3}{2}(\xi+\tau)\right]^{\frac{1}{3}} + 1}\right| \tag{3.30}$$

$$\psi = \tau + 4\left[\frac{3}{2}(\xi+\tau)\right]^{\frac{1}{3}} + 2\ln\left|\frac{\left[\frac{3}{2}(\xi+\tau)\right]^{\frac{1}{3}} - 1}{\left[\frac{3}{2}(\xi+\tau)\right]^{\frac{1}{3}} + 1}\right|. \tag{3.31}$$

These transformations are allowed (see § 2) if

$$e^{\nu-\lambda} > \left(\frac{\frac{\partial\chi}{\partial\tau}}{\frac{\partial\chi}{\partial\xi}}\right)^2. \tag{3.32}$$

Making the calculations, we obtain that (3.32)[22] holds in the case when

$$\left[\frac{3}{2}(\xi+\tau)\right]^{\frac{2}{3}} > 1,$$

[22] The Russian text says (3.34) here, but it follows from the context that it was a typo [Editor].

i.e. only in an R-region. This is what we were required to show.[23]

Note here that a T-region in the vacuum may arise, for example, when compressing a sphere to a size, as it is conventional to say in such cases, below its gravitational radius (about that, see [1], and also the next paper of the present communications).

Now let the condition (3.17) hold. Then from the equation (2.40) we obtain

$$\tau + \Phi(\xi) = f(\xi)^{-1}\sqrt{f(\xi)R^2 + R} - f(\xi)^{-\frac{3}{2}} \operatorname{Arsh} \sqrt{f(\xi)R}. \qquad (3.33)$$

It is true also here that R' cannot be equal to zero at any point. The boundary between R- and T-regions is determined by the condition

$$R = 1, \qquad (3.34)$$

which (when $F = 1$) follows directly from the relation (2.43).[24]

Substituting this value of R into (3.33) we obtain an equation for the boundary between the R- and T-regions

$$\tau + \Phi(\xi) = \frac{\sqrt{f(\xi) + 1}}{f(\xi)} - \frac{\operatorname{Arsh} \sqrt{f(\xi)}}{f(\xi)^{\frac{3}{2}}}. \qquad (3.35)$$

Since R is a monotonic function of ξ and τ (it follows from the equation (2.42), when the condition (3.17) holds, that $\dot{R}$ also does not become zero at any of the points), the picture of the location of the R- and T-regions will be qualitatively the same as in the case investigated above (see Fig. 4). Here too the boundary between the R- and T-regions is a sphere the points of which are moving in the same direction with the fundamental speed, and each spatial point of the system of reference is passed through once by this sphere.

The equation of the central singularity is obtained from the formula (3.33) if in the latter we put $R = 0$. This equation has the form

$$\tau = -\Phi(\xi). \qquad (3.36)$$

The line which is determined by equation (3.36), like the line described in the Lemaître system by equation (3.29), turns out to be spacelike.

Finally, let the condition (3.18) hold. Integrating (2.40) we obtain[25]

[23] About the possible motions of the particles in the T-region, see § 10. Note that the transition from ξ, τ to χ, ψ, as is easy to see, corresponds to changing the direction of time flow and the reference direction of the radial coordinate (about this see also § 5).

[24] Translators' note: the original refers to (2.47) here but there is no such equation.

[25] The form of the solution given includes only one branch with monotonic variation of R with t. The other form of the general Tolman solution is given in [11].

$$\tau + \Phi(\xi) = \frac{1}{f(\xi)}\left(\sqrt{R^2 f(\xi) + R} + (-f(\xi))^{-\frac{1}{2}} \arcsin\sqrt{-f(\xi)R}\right). \quad (3.37)$$

Here, unlike the cases which were considered above, it is possible for R' to be zero. For this it is necessary that on this sphere the condition $f(\xi) = -1$ holds. From the results obtained in § 5 it follows that in vacuum the sphere on which R' becomes zero is not moving relative to a freely falling system of reference.

The boundary between R- and T-regions determined by the condition (3.34) is described by the following function

$$\tau + \Phi(\xi) = \frac{\sqrt{f(\xi) + 1}}{f(\xi)} + \frac{\arcsin\sqrt{-f(\xi)}}{[-f(\xi)]^{\frac{3}{2}}}, \quad (3.38)$$

and the equation for the central singularity has the form (3.36).

Consider, as an example, which shows the singularities which appeared here, the solution which is defined in the following way

$$\Phi(\xi) = \frac{\pi}{2}(\xi^2 + 1)^{\frac{3}{2}}, \quad f = -\frac{1}{\xi^2 + 1}. \quad (3.39)$$

The expressions (3.37), (3.38) and (3.36) take the following forms:

$$\tau = -(\xi^2 + 1)\sqrt{-\frac{R^2}{\xi^2 + 1} + R} + (\xi^2 + 1)^{\frac{3}{2}}\left[\arcsin\sqrt{\frac{R}{\xi^2 + 1}} - \frac{\pi}{2}\right], \quad (3.40)$$

$$\tau = -(\xi^2 + 1)\sqrt{1 - \frac{1}{\xi^2 + 1}} + (\xi^2 + 1)^{\frac{3}{2}}\left[\arcsin\sqrt{\frac{1}{\xi^2 + 1}} - \frac{\pi}{2}\right], \quad (3.41)$$

$$\tau = -\frac{\pi}{2}(\xi^2 + 1)^{\frac{3}{2}}. \quad (3.42)$$

On the sphere $\xi = 0$, $R' = 0$. At $\tau = 0$ on this sphere $\dot{R} = 0$ also. As we know, this point of the manifold (ξ, τ) is the self-crossing point of the boundary of the R- and T-regions (see § 7). The location of the R- and T-regions in the system of coordinates considered is shown in Fig. 5. The T-regions as well as the R-regions are not connected.

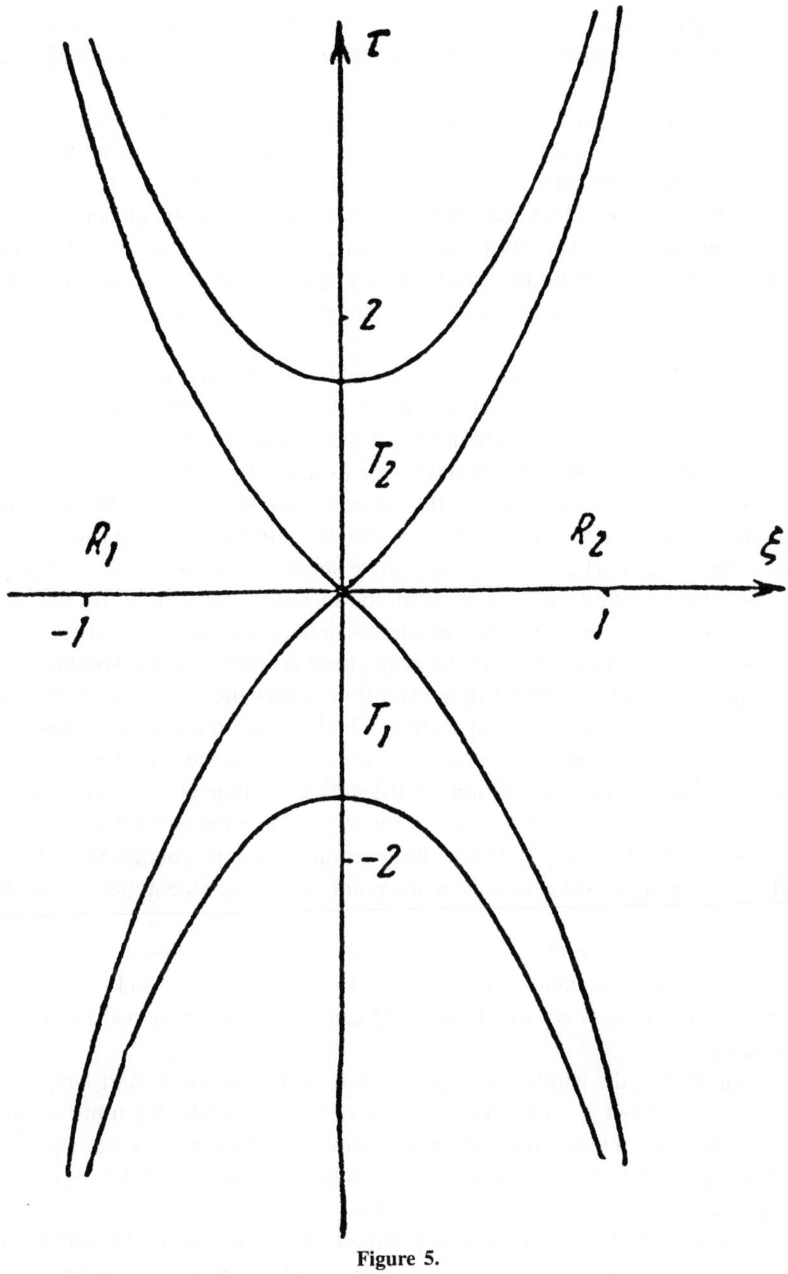

Figure 5.

The space slice $\tau = 0$ entirely lies in the R-region (except for the point $\xi = 0$ which is on the boundary of the R- and T-regions), $\dot{R}$ is everywhere equal to zero, and $R = \xi^2 + 1$.

The boundary of the R- and T-regions is two spheres, which are moving with the fundamental speed, at the start towards each other, and then, after meeting on the coordinate sphere $\xi = 0$ at the moment $\tau = 0$, away from each other. Each point of the reference space meets the boundary of the R- and T-regions twice. The existence of each point of the reference space starts and ends on the central singularity. In any space slice $\tau =$ const the coefficient of the angular part R^2 nowhere becomes zero, i.e. the space slice has no centre.

Note that when the condition (3.18) holds and for any (assumed) choice of the functions $\Phi(\xi)$ and $f(\xi)$ each point of the comoving space starts and ends its existence at the central singularity and meets the border between the R- and T-regions twice, as is well illustrated by the example considered.

In § 7 it was proven that in the vacuum case the parts of the R- and T-regions which are described by any solution are equal to each other in the sense that in the neighbourhood of any coordinate sphere consisting of R- or T-points, it is possible to introduce the interval in the form (3.5) or (3.10). But the whole space-time which is described by the solution under the conditions (3.16) or (3.17) cannot completely coincide with the space-time described by the solution under the condition (3.18). Indeed if in the solutions under the condition (3.18), $f(\xi)$ takes the value -1, then in the spacetime which is described by this solution is a sphere on which $\dot{R} = 0$ and $R' = 0$. This sphere is not in the spacetime described by any solution under the conditions (3.16) and (3.17). Indeed, if the contrary were to hold, then on this sphere in any system of reference the equalities $\dot{R} = 0$ and $R' = 0$ would hold (see § 3). But in the solutions under the conditions (3.16) and (3.17) these equalities do not hold at any point. So the statement that was made is proved.

The system of reference which was set out in the works of Kruskal [24] and Fronsdal [25] includes a sphere (t_0, r_0) on which $\dot{R} = 0$ and $R' = 0$. This system of reference is not free falling and can be realised by non-freely moving test particles.

Although in the whole, the picture of the location of the R- and T-regions in the case considered is symmetric and does not change under the transformation $t = -\tilde{t}$ (the global time symmetry is underlined by Kruskal [24], see also [13]), as was noted in § 5, in any region which is contained in a connected T-region there is an asymmetry in the direction of time flow.

It is possible to show (see [23], [23a] and § 10) that the null geodesic lines passing through a T-point cross the singular worldline of the central singularity. Consequently a particle cannot get from the region R_1 to the region R_2 (see Fig. 5). But of course both a particle and a light ray could pass through the throat ($\xi = 0$). From the region T_1 a particle could reach any part R_1, R_2 or T_2.

In the work of Ya.B. Zel'dovich [27] a system of reference with a non-monotonic dependence of R on r is considered, which is assembled as a whole from two Schwarzschild systems of reference which are "glued" together on the singular Schwarzschild surface. This study is analogous to the study made by Einstein and Rosen [27a]. But the system of reference considered has a **physical** singularity at $r = 2m$ (see § 9) and consequently cannot include the sphere with radial coordinate corresponding to the gravitational radius. Therefore from our point of view the "glueing" of two Schwarzschild systems is not possible.

The fact that in the Schwarzschild system of reference it is possible to approach arbitrarily close to the singular sphere and "only" one sphere is removed from our consideration is not essential. A similar exception could lead to whole space-time regions being dropped from consideration (in the given case, the T-regions). Let us consider an example, which particularly vividly illustrates this circumstance.

Let there be a Minkowski space-time. The interval can be written in the form

$$ds^2 = dt^2 - dr^2 - r^2(d\theta^2 + \sin^2\theta\, d\varphi^2).$$

Transform into another system of reference

$$r = r' \operatorname{ch} t', \theta = \theta'$$
$$t = r' \operatorname{sh} t', \varphi = \varphi',$$

in which the interval is written in the form

$$ds^2 = r'^2 dt'^2 - dr'^2 - r'^2(\operatorname{ch} t')^2(d\theta'^2 + \sin^2\theta'\, d\varphi'^2). \qquad (3.43)$$

Let us call the system t, r, θ, ϕ the A system, and the system t', r', θ', ϕ' the B system.

Consider together with these systems the system C, determined in the following way:

$$r = r'' \operatorname{sh} t'', \theta = \theta''$$
$$t = r'' \operatorname{ch} t'', \varphi = \varphi''.$$

The interval in this system is written

$$ds^2 = dt''^2 - t''^2 dr''^2 - t''^2(\operatorname{sh} r'')^2(d\theta''^2 + \sin^2\theta''\, d\varphi''^2). \qquad (3.44)$$

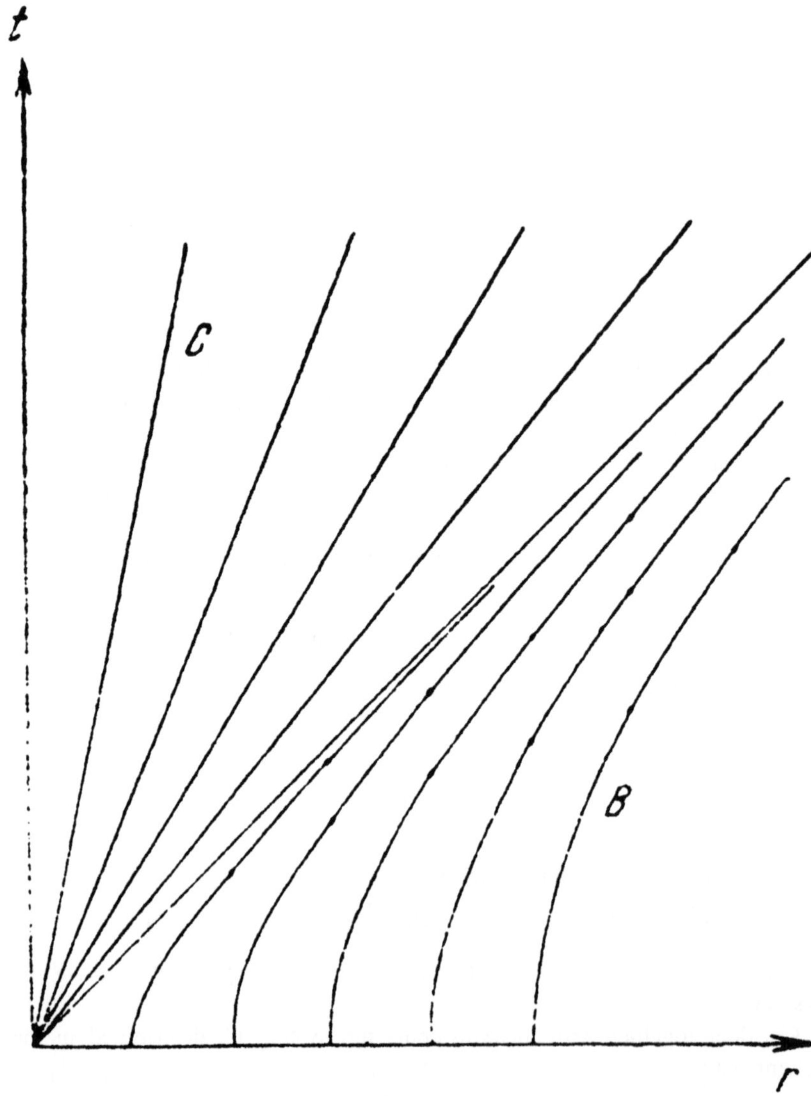

Figure 6.

The worldlines of the rest particles creating the B and C systems are represented in Fig. 6 (in the coordinates of the A system).

From the expression (3.43) it follows that the B system is usable at all values t' and r' except $r' = 0$, i.e. "only" one point – the centre of spherical symmetry – is excluded. But, as shown by Fig. 6, the whole world region which contains the C system is excluded. Note (see [28]) that the proper space volume of the C system is infinite.

Therefore we see that even in flat space-time the exclusion of one space point in some system of reference could lead to excluding an infinite space-time region.

10. NOTES ON THE RELATIVITY OF SPATIAL EXTENSION AND ON MOTION OF BODIES IN SCHWARZSCHILD SPACE-TIME

The systems of reference considered are of interest in connection with the problem of the relativity of spatial extension (infinite and finite) which was stated by A.L. Zel'manov [28].

Consider the ensemble of Lemaître systems of reference and T-systems in the Schwarzschild spacetime. As appears from the transformations (3.24), (3.25) and also from Fig. 4, at any moment of time of a Lemaître system the T-system occupies a limited volume in it, although the proper volume of a T-system at each time moment is infinite.

The proper time of a T-system's existence is finite (see [19]) but in the Lemaître system each point of the T-system of reference exists for a unlimitedly long time. As is known, in a bounded time in the Schwarzschild R-system, a light beam or particle could not reach the singular sphere or, moving in the opposite direction, escape from it (see, for example, [29]), but these both hold for a bounded proper time of a particle, for example, freely falling [29]. Let a Lemaître system of reference be possible in the space-time. Then in this system of reference, when the condition $\dot{R} < 0$ holds, in bounded time it is possible to pass from the R-region into the T-region, but not the contrary (in this system the points of the boundary sphere move away from the centre with the fundamental speed). But if in the space-time considered there is a Lemaître system of reference in which $\dot{R} > 0$, then in such a system it is possible (and if the body's speed is not tending towards the fundamental speed, it is necessary) to pass from the T-region into the R-region (in this system the points of the boundary sphere are moving towards the centre with the fundamental speed).[26] A test particle which is situated in the T-region obligatorily reaches the central singularity in bounded **proper time** if in the T-region there is a Lemaître system of reference which is realised by particles moving towards the centre. If in this region a Lemaître system of reference exists which is realised by particles moving away from the centre then a test particle whose speed does not tend towards the fundamental speed will cross Schwarzschild's singular sphere in a bounded **proper time**.[27]

[26] About this see [13] and the remarks about this work in [19].
[27] See the previous footnote.

Indeed the fact that both events can happen in bounded proper **time** $\Delta\tau$ **in a Lemaître system**, obviously follows from the solution of the equation of light propagation in the direction of the radial coordinate ξ

$$d\tau^2 = \left[\frac{3}{2}(\xi + \tau)\right]^{-\frac{2}{3}} d\xi^2, \tag{3.45}$$

if we take into account that the world line of a particle cannot exit across the boundaries of the light cone. The corresponding period of **proper time** ΔT of the particle is expressed in the following way:

$$\Delta T = \int_{\tau_1}^{\tau_2} \sqrt{1 - \left[\frac{3}{2}(\xi + \tau)\right]^{-\frac{2}{3}} v^2} \, d\tau < \Delta\tau < \infty,$$

where v is the coordinate speed, as it was required to show.

11. CONCLUSIONS

In the present paper it is shown that in spherically symmetric distributions of matter, there may be T-regions, i.e. world regions where in any S.S. system of reference the coefficient of the angular part in the expression for the interval essentially depends on time.

T-regions essentially differ from R-regions in their physical properties.

If in an R-region the coefficient of the angular part in the expression for the interval cannot be a non-monotonic function of the space's radial coordinate, in a T-region it can happen, and this property is not invariant under the transition from one system of reference to another. If in the system of reference which is comoving with matter there is an extremum with respect to the space's radial coordinate of the coefficient of the angular part in the expression for the interval, then the matter distribution is non-static and the sphere on which the extremum occurs, while changing with time, cannot pass through a regular minimum of the curvature radius (this is true also in the vacuum). T-regions are asymmetric relative to the time direction: in them the coordinate spheres of any system of reference are only expanding, or only contracting.

A sufficiently extended distribution of dust matter with density not falling very fast with increasing distance from the centre necessarily contains a T-region.

In the vacuum a space-time region which is lying, as we usually say, inside the singular sphere of Schwarzschild is a T-region. In this world region the coordinate system which is realised by real particles and clocks is obtained from the Schwarzschild system by exchanging the roles of the radial and time coordinates. A uniqueness theorem for Schwarzschild space-time in the T-region is proved.

The existence of the physical singularity at the gravitational radius of the Schwarzschild system of reference, and the similar singularities in the space-times of Kottler and Reissner-Nordström, is explained in that the rigid system of reference could not be extended outside the R-region.

An observer situated in Schwarzschild's space-time in the R-region can get information about events in the T-region only in the case that in this region the coordinate spheres are expanding. An observer situated in a T-region can get information about events going on in the R-region only if the coordinate spheres in the T-region are contracting.

The T-regions in vacuum which are realised as a result of compressing the surface of the gravitating sphere to sizes below its singular sphere and realised as a consequence of the sphere's surface having not yet expanded up to the size of the singular sphere are essentially different from each other. In the first region all the coordinate spheres are contracting and in the second are expanding.

In a Lemaître system of reference a T-system at any moment of time occupies a limited volume, but at the same time the proper volume of the T-system at any moment of time is infinite. The proper time of existence of the T-system is finite, but in the Lemaître system each point of the T-system exists for an unboundedly long time.

The conclusions given should be essential in considering systems like meta-galaxies and superdense configurations.

The author is grateful to A.L. Zel'manov for constant interest in the work, suggestions and discussions and and wishes to use this opportunity to thank Ya.B. Zel'dovich for many discussions and the kindly offered possibility of getting to know the paper [27] before its publication.

REFERENCES

[1] Landau, L.D., Lifshitz, E.M., *Teoriya polya. [Field theory]*. Fizmatgiz,[28] Moscow, 1962.

[2] Lichnerowicz, A., *Theories relativistes de la gravitation et de l'electromagnetisme*. Paris, 1955.

[3] Zel'manov, A.L., *O deformacii i krivizne soputstvuyushchego prostranstva. [On the deformation and curvature of a comoving space]*. Ph.D. thesis, Moscow, 1944.

[4] Zel'manov, A.L., *Trudy 6go soveshchaniya po voprosam kosmogonii [Proceedings of the 6th conference on the problems of cosmogony]*. Izd. Akad. Nauk. SSSR, Moscow, 1959, p. 144.

[5] Zel'manov, A.L., *Dokl. Akad. Nauk. SSSR*, **107**, 815, 1956.

[6] Tolman, R., *Relativity, thermodynamics and cosmology*. Oxford, 1934

[28] The full name of the publisher is Gosudarstvennoe Izdatelstvo Fiziko-Matematicheskoy Literatury [State Publishers of Physical and Mathematical Literature]. "Fizmatgiz" is a colloquial abbreviation commonly used in Russia (Editor).

[7] Eddington, A., *Teoriya otnositelnosti [The theory of relativity]*. ONTI,[29] M–L.,[30] 1934.
[8] Fok, V.A., *Teoriya prostranstva, vremeni i tyagoteniya [The theory of space, time and gravitation]*. Gostechizdat,[31] Moscow, 1955.
[9] Zel'manov, A.L., *Dokl. Akad. Nauk. SSSR*, **135**, 1367, 1960.
[10] Zel'dovich, Ya.B., *J.E.T.P.* **41**, 1609, 1961.
[11] Bonnor, W., *Z. Astrophys.*, **39**, 143, 1956.
[12] Fikhtengolc, G.M., *Kurs differenciyalnogo i integralnogo ischisleniya, I [A course of differential and integral calculus, I]*. Fizmatgiz, Moscow, 1947.
[13] Finkelstein, D., *Phys. Rev.*, **110**, 965, 1958.
[14] Novikov, I.D., *Vestnik Moskov. Univ. [Moscow Univ. Bulletin]* ser. **III**, 1962, No. 5, p. 90.
[15] Tolman, R., *Proc. Nat. Acad. Sci.*[32] **20**, 169, 1943.
[16] Schwarzschild, K., *Berl. Ber.*,[33] 1916, p. 189.
[17] Birkhoff, G., *Relativity and modern physics*. Cambridge, Harvard University Press, 1923, p. 253.
[18] Novikov, I.D., *Astron. Zh.* **38**, 564, 1961.
[19] Novikov, I.D., *Soobshcheniya GAISh [Communications of the Shternberg State Astron. Inst.]*, **120**, 42, 1962.
[20] Kottler, F., *Ann. Phys.* **56**, ser. 4, 1918.
[21] Reissner, H., *Ann. Phys.*, **50**, 106, 1916.
[22] Rylov, Yu.A, *J.E.T.P.* **40**, 1755, 1918.
[23] Fronsdal, G., *Phys. Rev.* **116**, 778, 1959.
[23a] Wheeler, J., *Gravitaciya, neitrino i Vselennaya [Gravitation, neutrino and the Universe]*. IL,[34] Moscow, 1962.
[24] Kruskal, M., *Phys. Rev.* **119**, 1743, 1960.
[25] Lemaitre, G., *Ann. Soc. Sci. Bruxelles*,[35] **A53**, 51, 1933.
[26] Lifshitz, E.M., Khalatnikov, I.M., *J.E.T.P.* **39**, 149, 1960.
[27] Zel'dovich, Ya.B., *J.E.T.P.* **43**, 1037, 1962.
[27a] Einstein, A., Rosen, N., *J. Franklin Inst.* **223**, 43, 1937.
[28] Zel'manov, A.L., *Dokl. Akad. Nauk. SSSR*, **124**, 1030, 1959.
[29] Bergmann, P., *Vvedenye v teoriyu otnositelnosti [Introduction to the theory of relativity]*. IL,[36] 1947.

[29] This is again a popular abbreviation. The name of the publisher as printed on the title page of the book is Gosudarstvennoe Tekhniko-Teoreticheskoe Izdatelstvo [State Technical and Theoretical Publishers] (Editor).

[30] This is an abbreviation for "Moskva–Leningrad" (Editor).

[31] The full name of the publisher is Gosudarstvennoe Izdatelstvo Tekhniko-Teoreticheskoy Literatury [State Publishers of Technical and Theoretical Literature] (Editor).

[32] This paper was recently reprinted in *Gen. Rel. Grav.* **29**, 935 (1997) (Editor.)

[33] *Berliner Berichte* is a colloquial abbreviation. The full title of the journal is *Sitzungsberichte der Königlich–Preussischen Akademie der Wissenschaften zu Berlin*. This journal had no volume–numbering and it appeared in two sections. Physicists meet only the *Physikalisch-Mathematische Klasse* (Editor).

[34] The name of the publisher is Izdatelstvo Inostrannoy Literatury [Publishers of Foreign Literature] (Editor).

[35] This paper was recently reprinted in *Gen. Rel. Grav.* **29**, 637 (1997) (Editor.)

[36] Library records show that the name of the publisher of this book was Gosudarstvennoe Izdatelstvo Inostrannoy Literatury [State Publishers of Foreign Literature]. The "Gosudarstvennoe" = "State" word seems to have disappeared from the name between 1947 and 1962 (Editor).

Chapter 14
Roy Patrick Kerr and Alfred Schild 1965, derivation of the Kerr solution from the Einstein equations

Original paper: Roy Patrick Kerr and Alfred Schild, A new class of vacuum solutions of the Einstein field equations. In: *Atti del Convegno sulla Relativita Generale: Problemi dell' Energia e Onde Gravitazionali*, G. Barbera Editore, Firenze 1965, pp. 1 – 12.

Previously published in *Gen. Relativ. Gravit.* **41** (2009), pages 2469 - 2484 (the editorial note and Kerr's biography), 2485 – 2499 (the paper) and *Gen. Relativ. Gravit.* **8**, 955 – 956 (1977) (Schild's biography).

The editorial note was updated for this reprinting and Schild's biography, published separately from this article, was included here, but without Schild's photo that accompanied the original text.

14.1 Editorial note to:

R. P. Kerr and A. Schild

A new class of vacuum solutions of the Einstein field equations

Andrzej Krasiński · Enric Verdaguer · Roy Patrick Kerr

Part 1: Explanation of some details of derivation

By Andrzej Krasiński

The Kerr solution is today textbook material and a basic element of education in relativity. However, derivations of it are not easy to find in the literature – most textbooks and monographs simply quote it as a given thing. There exist several uniqueness theorems (see part 2 of this note) that refer to physical and geometrical properties of the Kerr solution, but none of them gives a hint on how to derive it starting from the Einstein equations. Given the great physical and astrophysical significance of this solution,[1] this is a rather frustrating situation. Therefore, we (the editors) decided to go back to the historical origins. In Part 3 of this note, Roy Kerr gives a first-hand account on how he arrived at his famous result. Here, some comments on the Kerr – Schild paper.

The original publication [3] reported only the final result and some of its basic properties, but gave no hint on how to derive it. A derivation was published two years later, in a volume of conference proceedings that has never been easily accessible, and was not frequently referred to. This is the text reprinted here. The idea of the derivation is mostly self-explanatory, but not the details. The authors warn the reader at one point that the calculations are not simple, but even where they say that something follows by "simple calculation", the results are not necessarily easy to

Andrzej Krasiński
N. Copernicus Astronomical Center, Polish Academy of Sciences, Bartycka 18, 00 716 Warsaw, Poland
e-mail: akr@camk.edu.pl

Enric Verdaguer
Departament de Física Fonamental, Universitat de Barcelona, Av. Diagonal 647, 08028 Barcelona, Spain
email: verdague@ffn.ub.es

Roy Patrick Kerr
Mathematics and Statistics Department, University of Canterbury, Private Bag 4800, Christchurch 1, New Zealand
email: roy.kerr@canterbury.ac.nz

[1] This is attested by the very large number of papers discussing its various astrophysical applications – see Refs. [1] and [2].

reproduce. This note is meant to be a guide for those readers who wish to verify all the details.[2] We will use the following notation for the tetrad vectors:

$$\{e_1, e_2, e_3, e_4\} = \{m, \overline{m}, \ell, k\}. \tag{1}$$

The following auxiliary results are helpful in verifying the "simple direct calculation" at the beginning of Sec. 2:

The combinations $\left\{ {\alpha \atop \beta\gamma} \right\} l^\beta l^\gamma$, $\left\{ {\alpha \atop \beta\gamma} \right\} l_\alpha l^\gamma$ and $\left\{ {\alpha \atop \beta\alpha} \right\}$ have the same value, no matter whether $g_{\alpha\beta}$ or $\eta_{\alpha\beta}$ is used to calculate the Christoffel symbols.

Then:

$$\begin{aligned} \left\{ {\alpha \atop \beta\gamma} \right\} l^\gamma &= \left\{ {\alpha \atop \beta\gamma} \right\}(\eta) l^\gamma - \frac{1}{2} l^\rho \left(l^\alpha l_\beta \right)_{;\rho}, \\ \left\{ {\alpha \atop \beta\gamma} \right\} l_\alpha &= \left\{ {\alpha \atop \beta\gamma} \right\}(\eta) l_\alpha + \frac{1}{2} l^\rho \left(l_\beta l_\gamma \right)_{;\rho}, \end{aligned} \tag{2}$$

where $\left\{ {\alpha \atop \beta\gamma} \right\}$ are calculated from $g_{\alpha\beta}$, and $\left\{ {\alpha \atop \beta\gamma} \right\}(\eta)$ are calculated from $\eta_{\alpha\beta}$ (remember: $\eta_{\alpha\beta}$ is flat, but expressed in arbitrary coordinates, so in general $\left\{ {\alpha \atop \beta\gamma} \right\}(\eta) \neq 0$). Using these formulae in (2.1), after a lot of algebra, one is led to (2.2).

The other "simple calculation" mentioned before (2.8) is again not really simple. To carry it out one needs to note that, in consequence of (2.6), eqs. (2) above simplify to:

$$\begin{aligned} \left\{ {\alpha \atop \beta\gamma} \right\} k^\gamma &= \left\{ {\alpha \atop \beta\gamma} \right\}(\eta) k^\gamma - k^\rho H_{,\rho}\, k^\alpha k_\beta, \\ \left\{ {\alpha \atop \beta\gamma} \right\} k_\alpha &= \left\{ {\alpha \atop \beta\gamma} \right\}(\eta) k_\alpha + k^\rho H_{,\rho}\, k_\beta k_\gamma, \end{aligned} \tag{3}$$

while the general Christoffel symbols of the two metrics are related by

$$\begin{aligned} \left\{ {\alpha \atop \beta\gamma} \right\} = \left\{ {\alpha \atop \beta\gamma} \right\}(\eta) &- k^\alpha k_\beta H_{,\gamma} - H k^\alpha k_{\beta;\gamma} - H k^\alpha{}_{;\gamma} k_\beta - H_{,\beta}\, k^\alpha k_\gamma \\ &- H k^\alpha k_{\gamma;\beta} - H k^\alpha{}_{;\beta}\, k_\gamma + \eta^{\alpha\rho} H_{,\rho}\, k_\beta k_\gamma + \eta^{\alpha\rho} H k_{\beta;\rho} k_\gamma \\ &+ \eta^{\alpha\rho} H k_\beta k_{\gamma;\rho} + 2H k^\rho H_{,\rho}\, k^\alpha k_\beta k_\gamma. \end{aligned} \tag{4}$$

To derive (3.9) one observes that $R_{12} = 0$ is equivalent to $k^\rho H_{,\rho}(z + \overline{z}) + H\left(z^2 + \overline{z}^2\right) = 0$, while z obeys $k^\rho z_{,\rho} = -z^2$. This last equation is not trivial either. To obtain it, one must take the Ricci identities for the vector field k^α:

$$k_{\alpha;\beta\gamma} - k_{\alpha;\gamma\beta} = R_{\rho\alpha\beta\gamma} k^\rho \tag{5}$$

and contract them first with $k^\gamma g^{\alpha\beta}$, obtaining an equation analogous to the Raychaudhuri equation:

---

[2] According to the information from R. P. Kerr (see part 3 of this note), the Kerr – Schild paper is a text specially adapted to deriving the Kerr metric. The derivation by the original method was published still later, see Ref. [4].

$$k^\gamma \theta_{,\gamma} + \sigma^2 - \omega^2 + \theta^2 = -\frac{1}{2} R_{\rho\gamma} k^\rho k^\gamma, \tag{6}$$

where

$$\omega^2 \stackrel{\text{def}}{=} \frac{1}{2} k_{[\alpha;\beta]} k^{\alpha;\beta} \tag{7}$$

$$\theta \stackrel{\text{def}}{=} k^\alpha{}_{;\alpha} \tag{8}$$

$$\sigma^2 \stackrel{\text{def}}{=} \frac{1}{2} k_{(\alpha;\beta)} k^{\alpha;\beta} - \theta^2 \tag{9}$$

are the rotation, expansion and shear of the null geodesic k-congruence. Then one contracts eq. (5) with $k^\gamma \left(m^\alpha \overline{m}^\beta - m^\beta \overline{m}^\alpha\right)$ and uses the fact that $R_{\rho[\alpha\beta]\gamma} k^\rho k^\gamma \equiv 0$. The result is:

$$k^\gamma \left(\Gamma_{4[12]}\right)_{,\gamma} + \frac{1}{2} \left[(\Gamma_{412})^2 - (\Gamma_{421})^2\right] = 0. \tag{10}$$

But θ and ω are the real and the imaginary part of the same Ricci rotation coefficient Γ_{412}, which is another exercise for the reader (eqs. (2.17) in the paper seem to have misplaced indices). Knowing this, one can rewrite (10) as

$$k^\gamma \omega_{,\gamma} + 2\theta\omega = 0, \tag{11}$$

and then observe that (6) and (11) can be written as one complex equation:

$$k^\gamma z_{,\gamma} + z^2 + \sigma^2 = -\frac{1}{2} R_{\rho\gamma} k^\rho k^\gamma \equiv -\frac{1}{2} R_{44}. \tag{12}$$

In the shearfree vacuum case this reduces to $k^\rho z_{,\rho} = -z^2$, as stated above. Thus, $k^\rho H_{,\rho} (z + \overline{z}) + H\left(z^2 + \overline{z}^2\right) = 0$ is equivalent to $k^\rho [H/(z+\overline{z})]_{,\rho} = 0$, whose solution is (3.9).

Deriving (3.11) is again tricky. The set (3.10) implies the integrability condition $P_{|12} - P_{|21} = P_{|m} (\Gamma^m{}_{12} - \Gamma^m{}_{21})$ (from (2.21)), which reads explicitly

$$\begin{aligned} S \stackrel{\text{def}}{=} & \frac{z}{\overline{z}^2} Y_{,u} m^\rho \overline{z}_{,\rho} - \frac{z}{\overline{z}} m^\rho Y_{,u\rho} - \frac{\overline{z}}{z^2} \overline{Y}_{,u} \overline{m}^\rho z_{,\rho} + \frac{\overline{z}}{z} \overline{m}^\rho \overline{Y}_{,u\rho} \\ + & \left(\frac{z}{\overline{z}} - \frac{\overline{z}}{z}\right) Y_{,u} \overline{Y}_{,u} - \ell^\rho P_{,\rho} (\overline{z} - z) = 0. \end{aligned} \tag{13}$$

Other useful formulae are:

$$m^\rho z_{,\rho} = (\overline{z} - z) \overline{Y}_{,u}, \tag{14}$$

$$\ell^\rho z_{,\rho} = Y_{,u} \overline{Y}_{,u} - H z^2 + \overline{m}^\rho \overline{Y}_{,u\rho}, \tag{15}$$

and their complex conjugates. Also, it is important to note that $R_{33} = R^1{}_{313} + R^2{}_{323} \equiv 2R_{1323} = 2R^0{}_{323}$. Although R_{33} must be real, the equation $R_{1323} = R_{2313}$

does not hold identically.[3] Consequently, $2R_{2313}$ is not identically equal to $R_{1323} + R_{2313}$. In the equation $R^4{}_{132} = 0$ one must use the analogue of (2.23) for the full Riemann tensor and all the other equations derived so far. The result is:

$$\begin{aligned}&\overline{z}\ell^\rho \left(e^{3P}\right)_{,\rho}(z+\overline{z}) + 3H\frac{z+\overline{z}}{z} Y_{,u}\overline{Y}_{,u} + 3ze^{3P}m^\rho Y_{,u\rho} \\ &-3\frac{\overline{z}^2}{z}e^{3P}\overline{m}^\rho\overline{Y}_{,u\rho} + 3\frac{\overline{z}^2}{z^2}e^{3P}\overline{Y}_{,u}\overline{m}^\rho z_{,\rho} - 3\frac{z}{\overline{z}}e^{3P}Y_{,u}m^\rho\overline{z}_{,\rho} = 0. \qquad (16)\end{aligned}$$

Taking the imaginary part of the above we obtain $(z+\overline{z})S = 0$, where S is the expression defined in (13). Thus, the imaginary part of (16) vanishes by virtue of the integrability condition of (3.10). Using now (13) in (16) to eliminate the $m^\rho Y_{,u\rho}$ terms and recalling that $Y_{,u} = \ell^\rho Y_{,\rho}$, we obtain (3.11).

From here on, until the end of Sec. 4, the derivation is general and covariant, and relatively easy to follow. The non-covariant arbitrary assumption is made at the beginning of Sec. 5. What the authors say here is not strictly correct: eq. (5.1) does not result for a general quadratic polynomial $\Phi(Y)$. One must assume $\Phi = \alpha Y^2 + \beta Y - \overline{\alpha}$, where α and β are arbitrary complex constants; only then can (5.1) be achieved by a translation of ξ and u or of ξ and v.

Also (5.5) requires a comment: the equation $F = 0$ has two solutions for Y, the other one follows from (5.5) by replacing $\rho \to -\rho$. However, what then results in (5.8) is a continuation of the Kerr metric to negative ρ, and is thus not a new solution.

Another instructive derivation of the Kerr metric can be found in the paper by Carter [5]. The Carter paper, and that of Boyer and Lindquist [6], are also good sources of information about geometrical properties of the Kerr metric.

The Kerr–Schild paper reprinted here is at the same time an important early source of knowledge on the, so-called today, Kerr–Schild metrics. That metric ansatz was quite popular in its time. A good overview of the exact solutions obtained on the basis of that ansatz is Chapter 32 in Ref. [7].[4]

Part 2: Later research stimulated by the Kerr solution

By Enric Verdaguer

The relevance of the Kerr solution in the development of general relativity during the 1970s cannot be overemphasized. On the one hand it stimulated the search for

[3] The identities obeyed by the Riemann and Weyl tensors are deduced under the assumption that these tensors arise from commutators of second derivatives of tensors. However, if the basic objects in the theory are the Ricci rotation coefficients, as in the Newman–Penrose formalism, then the curvature tensors are present in first-order equations, and not all the 'identities' will automatically be fulfilled (in fact, only $R_{ijkl} = -R_{ijlk} = -R_{jikl}$ and $C^i{}_{jil} = 0$). Thus, some of those other 'identities' must be imposed as equations to fulfil. The equation $R_{1323} = R_{2313}$ holds by virtue of (13).

[4] One remark should be added to the account in Ref. [7]: All the vacuum solutions obtained via the Kerr–Schild ansatz have nontrivial isometry groups that are subgroups of the Poincaré group (J. Ehlers, private communication to AK).

solutions describing spinning isolated bodies, both in the exterior as well as in the interior, and also the development of new solution-generating techniques. On the other hand the development of black hole physics, and the so called uniqueness or "no hair" theorems, places the Kerr solution in a unique position as a key piece in our understanding of gravitational theory. Let us briefly comment on those points.

Solution-generating techniques

The Kerr–Schild ansatz and the Kerr solution played an inspiring role in the development of the powerful solution-generating techniques that were introduced into general relativity in the late 1970s in order to solve Einstein equations in the stationary and axisymmetric context. Thus Maison [8] proved that Einstein equations in vacuum for stationary and axisymmetric spacetimes are an integrable system. At about the same time Belinski and Zakharov [9, 10] extended the Inverse Scattering Method (ISM), which had been developed to solve some nonlinear wave equations in fluid dynamics, to general relativity under certain symmetry assumptions. They showed that the ISM could be applied to solve Einstein equations in spacetimes with two commuting Killing vector fields that admit the existence of 2-surfaces orthogonal to the group orbits, i.e. spacetimes that admit an orthogonally transitive two-parameter group of isometries. This includes the stationary axisymmetric spacetimes when one of the Killing fields is timelike (stationary) and the second Killing field generates closed curves and vanishes on the symmetry axis (axial symmetry).

In the stationary and axisymmetric context the ISM is closely related to other solution-generating techniques such as different Bäcklund transformations, like Harrison's [11, 12] and Neugebauer's Bäcklund transformations [13, 14], Kinnersley-Chitre transformations [15, 16, 17], or the Hauser-Ernst formalism [18, 19]. All these techniques provide algorithms to obtain new solutions starting from some known solutions. Repeated applications of these transformations to a given solution leads to a large class of new solutions with an increasing number of parameters. Generally these solution-generating techniques can also be extended to the Einstein-Maxwell equations.

As an example, and to be specific, let us summarize briefly the soliton transformation of the ISM. In the stationary axisymmetric context it is convenient to write the metric in coordinates adapted to the Killing vector fields, ∂_t and ∂_φ, in the form:

$$ds^2 = f(\rho,z)(d\rho^2 + dz^2) + g_{ab}(\rho,z)dx^a dx^b, \tag{17}$$

where $a,b = 0,1$ with $x^0 = t, x^1 = \varphi$. We can impose on the 2×2 matrix g the condition that

$$\det g = -\rho^2, \tag{18}$$

so that ρ and z are Weyl coordinates. The Einstein equations in vacuum for this metric separate into two sets of equations. The first set determines the matrix g,

$$\left(\rho g_{,\rho} g^{-1}\right)_{,\rho} + \left(\rho g_{,z} g^{-1}\right)_{,z} = 0, \tag{19}$$

and the second set determines f, once g is known. The ISM focuses on the nonlinear equation (19): the integration of the second set to determine f is rather simple once g has been obtained. The idea is to associate to the nonlinear system (19) some "spectral equations" (a set of two linear equations involving g and its derivatives) for a "generating matrix" $\psi(\lambda,\rho,z)$ in such a way that the Einstein equations (19) are the integrability conditions of the spectral equations. The matrix g is then given in terms of the generating matrix when the "spectral" parameter λ is zero:

$$g(\rho,z) = \psi(0,\rho,z). \tag{20}$$

The procedure for the integration of the spectral equations is to assume a particular solution g_0 and then to find, by integration, the corresponding solution $\psi_0(\lambda,\rho,z)$. This is the only non algebraic step in the procedure, which is made easier by assuming a simple "background" solution g_0. A particularly simple and useful choice is the Minkowski metric, for which $f_0 = 1$ and $g_0 = \mathrm{diag}\,(-1,\rho^2)$, in the Weyl coordinates. The generating matrix is in this case

$$\psi_0 = \mathrm{diag}\,(-1,\rho^2 - 2z\lambda - \lambda^2). \tag{21}$$

Next one looks for solutions of ψ of the form $\psi = \chi\psi_0$, and the spectral equations become equations for the "dressing matrix" $\chi(\lambda,\rho,z)$. The fact that g must be symmetric and real imposes some restrictions on χ.

The "n-soliton transformation" of the ISM corresponds to the presence of n pole singularities of the dressing matrix in the complex plane of the spectral parameter λ. That is, when χ takes the form:

$$\chi = I + \sum_{k=1}^{n} \frac{R_k}{\lambda - \mu_k}, \tag{22}$$

where I is the unit matrix and R_k are 2×2 matrices independent of λ. The spectral equation for χ and the restriction that it must satisfy completely determine the so called "pole trajectories" $\mu_k(\rho,z)$ and the matrices $R_k(\rho,z)$. The pole trajectories are

$$\mu_k = w_k - z \pm \sqrt{(w_k - z)^2 + \rho^2}, \tag{23}$$

where w_k are arbitrary, generally complex, constants. The matrices R_k have the form

$$(R_k)_{ab} = n_a^{(k)} m_b^{(k)}, \tag{24}$$

where the two-component vectors $m_a^{(k)}$ are given by

$$m_a^{(k)} = m_{0b}^{(k)} \left[\psi_0^{-1}[\mu_k,\rho,z)\right]_{ba} \tag{25}$$

with arbitrary constants $m_{0b}^{(k)}$, and the vectors $n_a^{(k)}$ are solutions of the algebraic equations,

$$\sum_{l=1}^{n} \Gamma_{kl} n_a^{(l)} = \mu_k^{-1} m_c^{(k)} (g_0)_{ca}, \tag{26}$$

where the symmetric matrix Γ_{kl} is defined by

$$\Gamma_{kl} = m_c^{(k)} (g_0)_{cb} m_b^{(l)} (\rho^2 + \mu_k \mu_l)^{-1}. \tag{27}$$

All this leads to the following "n-soliton solution" of the nonlinear equations (19):

$$g = \psi(0) = \chi(0)\psi_0(0) = \left(I - \sum_{k=1}^{n} R_k \mu_k^{-1} \right) g_0. \tag{28}$$

This matrix g satisfies equations (19) but not the condition (18) and therefore is not a solution of the Einstein equations. However, a simple normalization procedure, which involves the use of the differential equation satisfied by $\det g$, shows that the matrix

$$g^{(ph)} = \rho^{-n} \left(\prod_{k=1}^{n} \mu_k \right) g, \tag{29}$$

where g is given by (28) satisfies equations (18) and (19). It corresponds to the n-soliton solution of Einstein equations. An important property of the soliton transformation is that each pole trajectory changes the sign of $\det g$ so that if we start from a physical background the number of pole trajectories must be even. Furthermore if a pole trajectory is complex its complex conjugate must also be included to ensure that the final solution is real.

So far we have ignored the metric coefficient $f(\rho, z)$ but the differential equation for f is easily integrated when the n-soliton solutions for g are known. The final coefficient, see [9, 10] for the details, is given by the following algebraic expression:

$$f^{(ph)} = f_0 \rho^{-n^2/2} \left(\prod_{k=1}^{n} \mu_k \right)^{n+1} \left[\prod_{k>l=1}^{n} (\mu_k - \mu_l)^{-2} \right] \det \Gamma_{kl}, \tag{30}$$

where the second product above is equal to 1 for $n = 1$.

The Kerr metric and some generalizations

That the Kerr solution could describe the exterior gravitational field of a spinning mass was already noticed in Kerr's original paper [3], where it was shown that the two parameters of the solution represented the mass and the angular momentum with no higher order multipole moments, and that it reduced to the Schwarzschild solution when the angular momentum vanished. This by itself was an important achievement since we had for the first time a solution describing the exterior field of stationary rotating isolated sources. The search for a matching interior solution soon began but, so far, this search has been unsuccessful.

The Kerr solution has some important generalizations. One of them is the Kerr-NUT solution and its extensions [7]. The Kerr-NUT solution is not asymptotically flat but it reduces to the Kerr solution when the NUT parameter vanishes.

We may now use the soliton transformation of the ISM sketched above to derive explicitly the Kerr-NUT solution as a 2-soliton transformation on the Minkowski background. Since the pole trajectories come in pairs, this is the simplest solution we can obtain from the Minkowski background. Thus, let us take μ_1 and μ_2 with the w_k parameters in equation (23) written as

$$w_1 = z_1 + \sigma, \quad w_2 = z_2 - \sigma, \tag{31}$$

where we assume that the new parameters z_1 and σ are both real, which means that the pole trajectories are real. Instead of Weyl's coordinates ρ and z it is convenient to introduce Boyer-Lindquist coordinates r and θ as

$$\rho = \sqrt{(r-m)^2 - \sigma^2}\sin\theta, \quad z - z_1 = (r-m)\cos\theta, \tag{32}$$

where m is a new parameter whose value will be specified below. The pole trajectories take now the form

$$\mu_1 = 2(r-m+\sigma)\sin^2(\theta/2), \quad \mu_2 = 2(r-m-\sigma)\sin^2(\theta/2), \tag{33}$$

where we have chosen the plus signs in front of the square roots of (23) for the two pole trajectories.

Since the generating matrix ψ_0 for the Minkowski background is given by (21) we get from (25) the following components for the vectors $m_a^{(k)}$:

$$m_0^{(k)} = C_0^{(k)}, \quad m_1^{(k)} = C_1^{(k)}\mu_k^{-1}, \tag{34}$$

where now $k = 1, 2$ and $C_0^{(k)}$ and $C_1^{(k)}$ are arbitrary constants. Without loss of generality we can impose the following two conditions on these constants

$$C_1^{(1)}C_0^{(2)} - C_0^{(1)}C_1^{(2)} = \sigma, \quad C_1^{(1)}C_0^{(2)} + C_0^{(1)}C_1^{(2)} = -m. \tag{35}$$

The first equation is possible because there is a normalization freedom on these constants: $C_a^{(k)} \to \zeta^{(k)}C_a^{(k)}$. The second condition is the definition of m. We can also introduce two new arbitrary constants a and b, defined by

$$C_1^{(1)}C_1^{(2)} - C_0^{(1)}C_0^{(2)} = -b, \quad C_1^{(1)}C_1^{(2)} + C_0^{(1)}C_0^{(2)} = a. \tag{36}$$

It follows from (35) and (36) that

$$\sigma^2 = m^2 - a^2 + b^2. \tag{37}$$

Finally, we can substitute the above expressions for the vectors $m_a^{(k)}$ and the pole trajectories μ_k into (27), solve (26) for $n_a^{(k)}$ and find the metric coefficients $g_{ab}^{(ph)}$ from (28) and (29). The coefficient f of the metric (17) is easily obtained from equation (30). The resulting expressions for the metric contain only those combinations of the constants $C_a^{(k)}$ which are expressible through the three independent arbitrary parameters m, a and b according to (35)-(37). One also needs to write the line element $d\rho^2 + dz^2$ in the new r and θ variables and the result is,

$$\begin{aligned} ds^2 &= \omega(\Delta^{-1}dr^2 + d\theta^2) - \omega^{-1}(\Delta - a^2\sin^2\theta)(dt + 2ad\varphi)^2 \\ &+\omega^{-1}[4\Delta b\cos\theta - 4a\sin^2\theta(mr + b^2)](dt + 2ad\varphi)d\varphi \\ &-\omega^{-1}[\Delta(a\sin^2\theta + 2b\cos\theta)^2 - \sin^2\theta(r^2 + b^2 + a^2)^2]d\varphi^2, \end{aligned} \tag{38}$$

where ω and Δ are defined as

$$\omega = r^2 + (b - a\cos\theta)^2, \quad \Delta = r^2 - 2mr + a^2 - b^2. \tag{39}$$

These formulas are the standard expression for the Kerr-NUT solution in the Boyer-Lindquist coordinates if we take $t + 2a\varphi$ as the new time variable.

Since σ is real $m^2 + b^2 > a^2$ which corresponds to the usual Kerr-NUT solution with an event horizon. Note that if one uses complex conjugate poles, $\mu_1 = \mu_2^*$ this is equivalent to assume that σ in equation (31) is pure imaginary and one obtains $m^2 + b^2 < a^2$, which corresponds to the Kerr-NUT solution with naked singularities. Note also that if opposite signs in front of the square roots in (23) are taken for the two pole trajectories, the final solution is equivalent to that obtained here provided appropriate relations are written between the constants $C_a^{(k)}$ and the parameters σ, m, a and b. When $b = 0$ the metric (38) is the Kerr metric. The metric (38) is not asymptotically flat in the presence of the NUT parameter b, when $b \neq 0$. When $a = 0$ the metric (38) is the Taub-NUT metric. Of course, $b = a = 0$ corresponds to the Schwarzschild metric, and $b = 0$ and $a = m$ corresponds to the "extreme" Kerr solution.

Another important generalization is the Tomimatsu-Sato solution [20] and some extensions. The Tomimatsu-Sato family of solutions are asymptotically flat and include an arbitrary real "deformation parameter", δ, that may be related to the quadrupole moment of a massive isolated source. The solutions for integer deformation parameters may be obtained as the n-soliton transformation on the Minkowski background and a limiting procedure of pole fusion. When $n = 4$ this procedure leads to the extended Tomimatsu-Sato metric with five arbitrary parameters that was obtained by Kinnersley and Chitre [15, 16]. It includes the original two parameter Tomimatsu-Sato solution with deformation parameter $\delta = 2$, after two of the five parameters are taken to be zero and two other parameters are combined to impose asymptotic flatness of the metric. Therefore the Tomimatsu-Sato solution corresponds to an overlap of two identical Kerr solutions centred at the same point. The same limiting procedure can be made to obtain an overlap of $n/2$ identical Kerr solutions. The distorsion parameter is then $\delta = n/2$ and the solution can be made asymptotically flat, with a suitable choice of parameters.

Many of the generalizations of the Kerr metric obtained by the solution-generating techniques derived from a search for stationary asymmetric solutions with higher multipole moments, which could describe the exterior gravitational field of realistic rotating bodies. Unlike the case of spherical symmetry, where Birkhoff's theorem guarantees that the exterior solution is uniquely described by the Schwarzschild metric, one expects that a rotating axisymmetric body should contain an arbitrary number of multipole moments.

The Kerr solution and black holes

The Kerr solution, however, acquired a new and unique significance with black hole physics, in particular, from a new understanding of black holes as describing the spacetime geometry resulting from the gravitational collapse of isolated bodies. The great importance of the Kerr metric is not only that it describes the entire spacetime geometry of a spinning black hole but that it describes, in some sense, the most general final state of an uncharged black hole. This is the result of work by several authors such as Price, Israel, Carter, Robinson and Hawking among others, which lead to the formulation of the uniqueness theorems. In fact, one would expect that a black hole formed by gravitational collapse would settle down to a stationary final state. The uniqueness theorems then conclude that a stationary vacuum black hole is uniquely described by the Kerr metric.

The chain of theorems leading to this result may be summarized as follows. It begins with Hawking's demonstration that the event horizon of a stationary black hole has S^2 topology, or more precisely, that the intersection of the horizon with a Cauchy surface has the topology of a two-sphere. This is followed by the result also proved by Hawking [21] that the event horizon of a stationary asymptotically flat spacetime is a Killing horizon, both in vacuum and in Einstein-Maxwell; see the Hawking and Ellis book [22] for a complete exposition of these results. The corresponding Killing vector field may be normal to the horizon or not. When it is normal the spacetime is static and thus spherically symmetric and, consequently, it is described by the Schwarzschild solution. When it is not normal the spacetime is axisymmetric.

The chain is finally closed with theorems due to Carter [23] and Robinson [24]; see Chandrasekhar's book [25] for a detailed demonstration of Robinson's version. This theorem establishes that a stationary and axisymmetric vacuum solution of Einstein equations with a regular convex event horizon which is asymptotically flat and not singular outside the horizon is uniquely specified by two parameters which correspond to the mass and the angular momentum. Since the Kerr metric satisfies the conditions of this theorem it provides a unique specification of a stationary black hole.

It is interesting that, in recent years, research in string theory has stimulated the search for black hole solutions in higher dimensions, and that solution-generating techniques such as the ISM are playing an important role in this search. Thus, in five dimensions the analog of the Kerr solution was found by Myers and Perry [26]. It is a vacuum stationary axisymmetric and asymptotically flat metric representing a rotating five dimensional black hole with an event horizon of topology S^3. Remarkably the uniqueness of the four dimensional Kerr black hole is somewhat relaxed in higher dimensions, thus Emparan and Reall [27] discovered a black ring solution: a vacuum stationary axisymmetric and asymptotically flat solution in five dimensions that has an event horizon of topology $S^1 \times S^2$. The black ring rotates along the direction of the S^1. Both the rotating five dimensional black hole and the rotating black ring solutions have been recently generated and generalized by the ISM [28, 29].

Part 3: A brief history of the discovery of the Kerr solution and the Kerr-Schild metrics

By Roy Patrick Kerr

The motivation for my study of algebraically special (AS) metrics was very simple. The only known *physically significant* solution of the Einstein equations was Schwarzschild. The Weyl solutions generalised this but were still static and added little to our understanding of stellar structures. Relativists had been searching for a stationary rotating solution for decades but had found nothing. The Schwarzschild metric is AS but *static*, so it was possible that there were interesting *stationary* AS metrics which were asymptotically flat (like Schwarzschild) but necessarily rotating.

When I realised that one previous attempt to find all rotating AS spaces had foundered and another seemed to have stopped at the static ones, I rushed headlong into the search for solutions that fit the conditions of the previous paragraph, found the now standard coordinates for algebraically special metrics in spring 1963, solved the easy field equations and then calculated the remaining hard ones. Since these were impossibly difficult, I then "solved Killings equation", looking for all possible symmetries. There are two distinct types of Killing vectors with very simple canonical forms, $\mathrm{d}/\mathrm{d}\phi$ and $P\mathrm{d}/\mathrm{d}u$. I was fairly sure that if there were two commuting symmetries one of which was asymptotically timelike then canonical coordinates could be chosen for which each of the above vectors is a symmetry. Instead of wasting time proving this conjecture, I assumed it to be true and forged ahead. The complete solution with these conditions contains four parameters. These are m, a, the NUT parameter and a further garbage parameter. Since the metric was not asymptotically flat when either of the last two parameters is nonzero, I discarded them, kept m and a and the rest is history.

This metric was originally found in a rather nasty coordinate system, but it did have a visible Kerr–Schild structure. I was able to transform the flat space part to $dx^2 + dy^2 + dz^2 - dt^2$, giving a final metric that was asymptotically flat. At this point I told Alfred that I was about to calculate its angular momentum. He came to my office and sat smoking his pipe while I compared its asymptotic expansion with the well-known linear approximation for any slow moving body (time independent in our case). When I found that the angular momentum is nonzero he was even more excited than myself.

When this solution was first found it was clear to everyone that it must have an event horizon for small a since it was a generalisation of Schwarzschild. For this the light cones at $r = m$ are tilted inwards. A small perturbation of the metric, $a = 10^{-3}m$ say, could not change this. Physicists knew that a spherically symmetric body that collapsed had to fall inside the event horizon and quickly become singular. However, it had not been known what would happen if the body had spin. Some thought that this would stop the formation of a horizon and the collapse to a singularity. After 1963 we knew that this did not happen for small angular momentum at least.

The first Texas Symposium was held in Dallas on December 18, 1963, a few months after the Kerr metric was found. Before I went I calculated the geodesics up and down the axis of rotation and found that there were two distinct horizons. I also attempted to calculate these off the axis, but I made a mistake and got the wrong surfaces, perhaps because I was in a rush and K–S coordinates are not the ones to use. This attempt was published in Kerr [30]. I stated there that the two horizons are the roots of the equation $r^4 + a^2z^2 = 2mr^3$, instead of the correct equation $r^4 + a^2r^2 = 2mr^3$.

Although I did not discuss the topology in Kerr [3], I did point out in Kerr [30] that the points on the axial ring $x^2 + y^2 = a^2$, $z = a$ are the branch points of the metric, and that the disc bounded by this ring is the corresponding branch cut. The two branch sheets are each asymptotically flat. The mass is positive on one branch and negative on the other, and the two event horizons are both in the first sheet.

One morning in autumn, 1963, I looked to see if there were more general situations where an AS metric splits in the same way as the Kerr metric does into two parts, $ds_o^2 + m_o k^2$, where ds_o^2 is itself an Einstein space (but not necessarily Lorentzian) and m_o is an arbitrary constant, i.e. where a first-order approximation is actually exact. The final result appeared to be that ds_o^2 is necessarily Lorentzian and the second term contains an arbitrary function of a complex variable. I did not check this result, just put the calculations aside.

Sometime after Christmas, 1963, Jerzy Plebański visited Austin. Alfred held a party for him during which I heard him talking to Jerzy about metrics of the K–S type. I said "This may be rubbish because it was a fairly rough calculation, but I think I know of a generalisation of spinning Schwarzschild with an extra function of a complex variable".

Alfred was quite excited so he and I went into his office and calculated the connexion and the easy curvature components for a metric of K-S type. The simplest equation, $R_{ab}k^a k^b = 0$, reduced to "**k** is geodesic"[5]. We then calculated those components of the Riemann tensor which determine whether the metrics were AS. Bingo! They were! This meant that even if I had made a mistake, we could soon find all (diverging) K–S metrics. We redid my previous calculations next day and found that they were correct.

Since the work on algebraically special spaces had not been published, we then worked everything out from first principles. This was a much better treatment anyway, one that could be extended to Einstein-Maxwell fields. The original derivation was published much later in Debney et al. [4].

Sometime in early 1964 I added an electromagnetic field to the Kerr–Schild ansatz. The analysis for this is given in Debney et al. [4]. The first stumbling block was that I could not prove that the special null vector was geodesic (and still cannot). I therefore took this as an additional assumption. The "easy" equations then showed that the Einstein-Maxwell field depends on three functions, M,A and γ, all constant along the rays and restricted by the "hard" field equations. I could not solve the latter unless $\gamma = 0$ so I temporarily took this as an additional assumption and continued.

[5] This does not work for charged K–S.

This led to the complete charged Kerr–Schild metrics including, of course, charged Kerr. The null congruence for the latter is the same as the Kerr congruence but the EM field depends on an arbitrary function of a complex variable. If this is a constant, $e+ib$, then e is the charge and b the magnetic charge. If it is a more complicated function then the EM field will probably be singular[6].

At that point I turned the problem over to G. C. Debney, who had just completed his prelims for a PhD, to see if he could solve the problem when $\gamma \neq 0$. Naturally, it took some time for him to get up to speed, but in the end it became clear that the three of us could not solve the general problem. George then shifted to "Symmetries in Algebraically Special Spaces", extending the analysis that was used to find the Kerr metric to cover all possible symmetry groups [33]. This became the basis for his PhD thesis.

Each year the American Mathematics Society holds a symposium on applied mathematics. The XVII one, held in New York on April 20–23 1964, included two talks on relativity, one by A. Lichnerowicz and one by myself, but written by Alfred, on algebraically degenerate solutions of the Einstein equations [32]. This was to an audience of applied mathematicians and emphasised the K–S metrics. The manuscript had to be written before the conference and so anything in it was discovered by early April at the latest. It included the following sentence:

Together with their graduate student, Mr. George Debney, the authors have examined solutions of the nonvacuum Einstein-Maxwell equations where the metric has the form (2.1).[7] *Most of the results above apply to this more general case. This work is continuing.*

This tells me that the charged metrics were known by the end of March since they were discovered before George joined the project, and that was before the Symposium paper was written. Later that year, Newman went ahead and published his own version, in which he considered the simplified problem where the background metric is Kerr and the electric field is generated by a real constant rather than an arbitrary complex function.

Acknowledgements. A. K. is grateful to Mauro Carfora for his help in contacting the publisher, to M. MacCallum and G. Ellis for useful corrections, and to J. Ehlers for providing bibliographical data on several sources.

References

1. V. P. Frolov and I. D. Novikov, *Black Hole Physics: Basic Concepts and New Developments*. Kluwer, Dordrecht 1998, ISBN: 978-0-7923-5146-7.
2. N. Straumann, *General Relativity: With Applications to Astrophysics*. Springer, Berlin 2004, ISBN 978-3540219248.

[6] In Kerr and Wilson [34] it is proved that the only K-S metric that is nonsingular at infinity is Kerr. I suspect the same is true for charged K-S and that the proof is very simple, but I have not tried to prove it.

[7] The usual Kerr–Schild form.

3. R. P. Kerr, Gravitational field of a spinning mass as an example of algebraically special metrics, *Phys. Rev. Lett.* **11**, 237 (1963).
4. G. C. Debney, R. P. Kerr and A. Schild, Solutions of the Einstein and Einstein-Maxwell equations, *J. Math. Phys.* **10**, 1842 (1969).
5. B. Carter, Black hole equilibrium states. Part I: Analytic and geometric properties of the Kerr solutions, in: *Black Holes – les astres occlus*. Edited by C. de Witt and B. S. de Witt. Gordon and Breach, New York, London, Paris, 61 (1973). Reprinted as a Golden Oldie: *Gen. Relativ. Gravit.* **41**, no 12, 2874–938 (2009).
6. R. H. Boyer and R. W. Lindquist, Maximal analytic extension of the Kerr metric, *J. Math. Phys.* **8**, 265 (1967).
7. H. Stephani, D. Kramer, M. MacCallum, C. Hoenselaers and E. Herlt, *Exact Solutions of Einstein's Field Equations*. 2nd Edition. Cambridge University Press 2003.
8. D. Maison, *Phys. Rev. Lett.* **41**, 521 (1978).
9. V. A. Belinski and V. E. Zakharov, *Sov. Phys. JETP* **48**, 985 (1978).
10. V. A. Belinski and V. E. Zakharov, *Sov. Phys. JETP* **50**, 1 (1979).
11. B. K. Harrison, *Phys. Rev. Lett.* **41**, 1197 (1978).
12. B. K. Harrison, *Phys. Rev. D* **21**, 1695 (1980).
13. G. Neugebauer, *J. Phys. A* **12**, L67 (1979).
14. G. Neugebauer, *J. Phys. A* **13**, L19 (1980).
15. W. Kinnersley and D. M. Chitre, *J. Math. Phys.* **18**, 1538 (1977).
16. W. Kinnersley and D. M. Chitre, *J. Math. Phys.* **19**, 1926 (1978).
17. C. Hoenselaers, W. Kinnersley and B. C. Xanthopoulos, *J. Math. Phys.* **20**, 2530 (1979).
18. I. Hauser and F. J. Ernst, *Phys. Rev. D* **20**, 362 (1979).
19. I. Hauser and F. J. Ernst, *J. Math. Phys.* **21**, 1126, 1418 (1980).
20. A. Tomimatsu and H. Sato, *Phys. Rev. Lett.* **29**, 1344 (1972).
21. S. W. Hawking, *Commun. Math. Phys.* **25**, 152 (1972).
22. S. W. Hawking and G. F. R. Ellis, *The large scale structure of space-time*, Cambridge University Press, Cambridge (1973).
23. B. Carter, *Phys. Rev. Lett.* **26**, 331 (1972).
24. D. C. Robinson, *Phys. Rev. Lett.* **34**, 905 (1975).
25. S. Chandrasekhar, *The mathematical theory of black holes*, Clarendon Press, Oxford (1983).
26. R. C. Myers and M. J. Perry, *Ann. Phys. (N.Y.)* **172**, 304 (1986).
27. R. Emparam and H. S. Reall, *Phys. Rev. Lett.* **88**, 101101 (2002).
28. A. A. Pomeransky, *Phys. Rev. D* **73**, 044004 (2006).
29. H. Iguchi and T. Mishima, *Phys. Rev. D* **73**, 121501 (2006).
30. R. P. Kerr, Gravitational collapse and rotation, *Quasi-stellar Sources and Gravitational Collapse, including the Proceedings of the First Texas Symposium on Relativistic Astrophysics*, Edited by I. Robinson, A. Schild and E.L. Schucking. The University of Chicago Press, Chicago and London, 99-109 (1965).
31. R. P. Kerr and A. Schild, Atti del Convegno sulla Relativita Generale: Problemi dell'Energia e Onde Gravitazionali. G. Barbèra Editore, Firenze, 1 - 12 (1965) (this Golden Oldie).
32. R. P. Kerr and A. Schild, Some algebraically degenerate solutions of Einstein's Gravitational Field Equations, *Proc. Sym. in Applied Math.* **XVII**, Edited by R. Finn. American Math. Soc, Providence, Rhode Island 1965, 199-209 (1965).
33. R. P. Kerr and G. C. Debney, Einstein spaces with symmetry groups, *J. Math. Phys.* **11**, 2807 (1970).
34. R. P. Kerr and W. B. Wilson, Singularities in the Kerr-Schild Metrics, *Gen. Relativ. Gravit.* **10**, 273 (1979).

Golden Oldie Editor's comment: Professor Kerr gave his permission to announce the following (this is a quotation from his email to A.K.):

"Fulvio Melia and I are writing a book 'Cracking the Einstein Code' which will probably be published by Farrar Straus and Giroux. It will give the history of Black holes from early last century to the present, with several chapters on the events of the 60's. It will be a popular book (we hope) for the general public, rather than for physicists."

Comment added by AK in 2012: The book was published in the meantime, with the following bibliographical details:
Fulvio Melia, *Cracking the Einstein Code. Relativity and the Birth of Black Hole Physics*. With an afterword by Roy Kerr. The University of Chicago Press, Chicago 2009, 150 pages.

14.2 Roy Patrick Kerr – a brief biography

By Andrzej Krasiński, compiled from Refs. [1] and [2].[8]

Roy Kerr was born on May 16, 1934, in Kurow, New Zealand.

He received his M.S. in 1954 from the New Zealand University (now dissolved) and his Ph.D. in 1960 at Cambridge University.

Kerr's colleague (see Ref. [1]) describes his early years as follows: "His undergraduate career was not given wholly to mathematics and science; he admits to having played a lot of billiards, and in 1952 represented his College in boxing at the Easter Tournament, as a light-welterweight. I recall W.W. Sawyer, then a lecturer at Canterbury, expressing alarm and dismay over Roy's pugilism, on the ground that he didn't want the best brain he'd encountered in a student scrambled by a well-thrown punch; but history seems to confirm that Roy came to no lasting harm over it.

In 1955 he received a M.Sc with first class honours, and went to Cambridge with a Sir Arthur Sims Empire Scholarship. He was awarded a Ph.D. in 1960, for a thesis on the equations-of-motion problem in general relativity. This work appeared in a series of three ... papers in Nuovo Cimento, and although later overshadowed by the Kerr metric, was extensively cited. He went on to a post-doctoral post at Syracuse University, and then to work with a US Air Force relativity group at Wright-Patterson Field, in Ohio. The USAF were interested in antigravity devices; one of the tasks of the relativity group was to assess and report on such devices proposed to it by inventors."

In 1962 Kerr joined the relativity group at the University of Texas in Austin. He found the famous Kerr solution in his first year there. The solution was publicly presented in 1963, during the First Texas Symposium on Relativistic Astrophysics in Austin, but did not seem to be properly appreciated by the public. Universal recognition came later. S. Chandrasekhar, in his Ryerson Lecture of 1978, said: "In my entire scientific life, extending over forty-five years, the most shattering experience has been the realization that an exact solution of Einstein's equations of general relativity, discovered by the New Zealand mathematician Roy Kerr, provides

[8] Repeated attempts by the Editor of this series to get a first-hand autobiography of Roy Kerr were unsuccessful.

the absolutely exact representation of untold numbers of massive black holes that populate the universe." (quotation after Ref. [1]).

Kerr returned to New Zealand in 1971. In 1983 he became the head of the Department. His colleague describes his activities so: "Roy's style as HOD was at once uncompromising and dashing; in a series of moves which affronted some of our colleagues in other departments, who had grown comfortable with the traditional Canterbury view that Mathematics should be a low-cost department devoted to service teaching, he contrived to reduce student-staff ratios, encourage research, and equip the department with a computer system at the sort of cost hitherto associated with spectrographs. Morale rose markedly. In many respects Roy was an unusual figure in University administration; he had very little patience for the practice of wrapping self-interest up in politically correct pieties, and was perfectly willing to offend entrenched privilege. But he was successful, and we are the better for his efforts, and we love him for them." [1]

Roy Kerr retired from his position as Professor of Mathematics at the University of Canterbury in February 1993.

He received many awards, culminating in the Hughes Medal of the Royal Society of London in 1984.

Roy Kerr's extra-scientific activities are described in some detail in Ref. [3]. The most notable of them is his long career as "a national representative and champion bridge player".

References

1. B. Woods, http://ifs.massey.ac.nz/mathnews/centrefolds/58/Aug1993.shtml
2. The New Encyclopaedia Britannica (Micropaedia), vol. 6, 15th edn, p. 815, the entry "Kerr, Roy (Patrick)". The University of Chicago, Chicago (1990).
3. Head, M.: http://www.listener.co.nz/issue/3359/features/2629/man_of_mystery.html

14.3 Alfred Schild – a brief biography

By Lawrence Shepley
University of Texas at Austin
shepley@physics.utexas.edu

In Memoriam
Alfred Schild
1921 – 1977

Alfred Schild of The University of Texas at Austin died on May 24, 1977 at the age of 55. At the time, he and his wife Winnie were living in Illinois, where he was Argonne Universities Association Distinguished Appointee for 1976 – 1977.

Alfred was born in 1921 in Istanbul, Turkey. In World War II he was studying in England when he was interned because of his German passport and was sent to Canada, where he completed his education, receiving B.A., M.A., and Ph.D. degrees from the University of Toronto. There he worked with Leopold Infeld, with whom he maintained a lifelong relationship. In 1946 Alfred joined the faculty of the Carnegie Institute of Technology. He arrived at Texas in 1957 as a professor of mathematics and in 1963 was named Ashbel Smith Professor of Physics. He founded the Center for Relativity Theory at Texas in 1962, was cofounder (with Ivor Robinson and Engelbert Schucking) of the Texas Symposia on Relativistic Astrophysics, and from 1965 to 1974 was a member of the International Committee on General Relativity and Gravitation.

Even in his earliest papers one can see the geometrical viewpoint which characterized Alfred's work. In his discussions of the foundations of relativity, in particular in his treatment of the equivalence principle, appears the well-known "Schild's Ladder". This diagram, more concretely than any formula, portrays the role of particle and light paths in determining the curved geometry of space-time. Alfred's best-known works, on foundations of relativity, on conformal techniques, on quantization, on algebraically special solutions, all demonstrate his physical-geometrical insight. His expository papers, his lectures on general relativity, the *Tensor Calculus* book written with J. L. Synge, his more popular papers on time and the twin paradox, the paper on lattice space-time, all are deservedly popular with students. His last works, on Fokker action principles and string models of particles, also bring a powerful and provocative geometrical viewpoint to physical problems.

Alfred's goal was to explain nature. The mathematics used and the geometrical intuition developed were for that purpose, not for purposes of elegance or even consistency. In his review of J. L. Synge's *Relativity: The General Theory*, Alfred wrote, "To the physicist logical consistency, as such, is no virtue. He is groping along a dark road towards a theory which will explain all of nature ... inconsistency is merely one of many indications that he is still ... far from his destination. A theory which explains a large region of experimental and observational facts, even though it is inconsistent on the blurred boundary of the region, tells him that he is on the right road. It is this road which is the physicist's truth; to him the great sin is to be lured away from it towards bright and beautiful mirages on either side."

Alfred was outspoken in favor of peace, against injustice, in support of the young and of freedom of expression. As a member of the GRG Committee he played a sober and decent role but not a conspicuous part. He was a promoter of the GRG Organization, yet he also supported André Mercier against the criticism of Mercier's refusal to attend the 1968 meeting at Tbilisi. In all his battles Alfred was ever courteous. Most of his opponents admired him, and his friends were a beautiful mixture of people.

Alfred's health had deteriorated these past couple of years. His appointment at Argonne allowed him to rest from teaching duties, and he did look much better this past spring; though he was working harder than ever on research and was looking forward to a lecture tour in early summer. But early in the morning of Tuesday, May

24, he awoke with an apparent attack of indigestion. Shortly thereafter, he died of a massive heart attack.

The loss of a friend is a shock, particularly when it happens with so little warning. Alfred was more than a friend. He was one whose insight into physics, academic life, politics, personal matters, could always be trusted: a guide and leader. We need people like him.

Comment from the Editors: The other source for Alfred Schild's biography, cited in *Gen. Relativ. Gravit.* **8**, 955 (1977), namely

Amanda Oren, http://www.tsha.utexas.edu/handbook/online/articles/SS/fsc48.html

is now accessible under the modified address

Amanda Oren, http://www.tshaonline.org/handbook/online/articles/fsc48

Gen Relativ Gravit (2009) 41:2485–2499
DOI 10.1007/s10714-009-0857-z

GOLDEN OLDIE

Republication of: A new class of vacuum solutions of the Einstein field equations

R. P. Kerr · A. Schild

Published online: 31 July 2009

An editorial note to this paper and a biography can be found in this issue preceding this Golden Oldie and online via doi:10.1007/s10714-009-0856-0.

Original paper: R. P. Kerr and A. Schild, in: Atti del Convegno sulla Relativita Generale: Problemi dell'Energia e Onde Gravitazionali. G. Barbèra Editore, Firenze 1965, pp. 1–12. Reprinted with the kind permission of R. P. Kerr. The publishers of the original text (their present name is Giunti) no longer hold the copyright.

Editorial responsibility: A. Krasiński, e-mail: akr@camk.edu.pl.

R. P. Kerr (✉)
Mathematics and Statistics Department, University of Canterbury,
Private Bag 4800, Christchurch 1, New Zealand
e-mail: roy.kerr@canterbury.ac.nz

A. Schild (Deceased May 24, 1977)
Austin, TX, USA

COMITATO NAZIONALE PER LE MANIFESTAZIONI CELEBRATIVE
DEL IV CENTENARIO DELLA NASCITA DI GALILEO GALILEI

ATTI

DEL

CONVEGNO SULLA RELATIVITÀ GENERALE: PROBLEMI DELL'ENERGIA E ONDE GRAVITAZIONALI

R. P. KERR and A. SCHILD

A New Class of Vacuum Solutions of the Einstein Field Equations

ESTRATTO

FIRENZE
G. BARBÈRA EDITORE
1965

A New Class of Vacuum Solutions of the Einstein Field Equations.(*)

R.P. Kerr and A. Schild

University of Texas – Austin

1. - Introduction.

In this paper the general solution of Einstein's empty space field equations, $R_{\mu\nu} = 0$, is obtained for a space where the metric has the form

$$g_{\mu\nu} = \eta_{\mu\nu} + l_\mu l_\nu. \tag{1.1}$$

Here $\eta_{\mu\nu}$ is the metric of Minkowski space in co-ordinates which are Cartesian but not necessarily rectangular, *i.e.*, $\eta_{\mu\nu}$ are constants, with signature $+ + + -$, and l_μ is null:

$$g^{\mu\nu} l_\mu l_\nu = 0. \tag{1.2}$$

The reason for considering vacuum solutions of the form (1.1) is that the contravariant components of the metric are easily expressed in terms of the covariant components. In fact,

$$g^{\mu\nu} = \eta^{\mu\nu} - l^\mu l^\nu, \tag{1.3}$$

where

$$l^\mu = g^{\mu\nu} l_\nu = \eta^{\mu\nu} l_\nu, \tag{1.4}$$

and the determinant

$$(-g) = -\det(g_{\mu\nu}) = 1. \tag{1.5}$$

It follows that if l_μ is null with respect to one of the two metrics $g_{\mu\nu}$ and $\eta_{\mu\nu}$, then it is null with respect to the other, *i.e.*, eq. (1.2) implies

$$\eta^{\mu\nu} l_\mu l_\nu = 0, \tag{1.6}$$

(*) This research has been supported by the Aerospace Research Laboratory, Office of Aerospace Research, and the Office of Scientific Research, U. S. Air Force.

and vice versa.

The new vacuum solutions have the following properties:

a) They include as special cases the Schwarzschild solution and the exterior solution of a rotating body which was recently discovered by one of us [1].

b) All vacuum solutions of the form (1.1) are algebraically degenerate in the sense of the Petrov-Pirani classification, l_μ being a multiple Debever-Penrose vector and thus geodesic and shear-free.

c) All vacuum solutions of the form (1.1) admit a one parameter group of motions. The Killing vector K^μ is at the same time a Killing vector of the flat Minkowski metric $\eta_{\mu\nu}$. In fact, with respect to $\eta_{\mu\nu}$, the motion is just a translation along a direction which may be time-like, space-like, or null in the Minkowski space. There are thus three cases:

$$\eta_{\mu\nu}K^\mu K^\nu \begin{cases} < 0, & \text{(Case I)} \\ > 0, & \text{(Case II)} \\ = 0. & \text{(Case III)} \end{cases} \tag{1.7}$$

d) In each of the three cases, the general solution is determined by one arbitrary analytic function of one complex variable.

e) If $g_{\mu\nu}$ admits a Killing vector other than K^μ, it also must be a Killing vector with respect to $\eta_{\mu\nu}$.

f) There are at most two essentially different ways of representing a vacuum metric in the form (1.1). For case I, apart from the special case mentioned in *a*), the representation (1.1) is, in fact, unique, so that the Riemannian space $g_{\mu\nu}$ determines uniquely the null vector field l_μ and the Minkowski space $\eta_{\mu\nu}$. Similar statements hold for cases II and III.

Together with their graduate student, Mr. George Debney, the authors have examined solutions of the (non-vacuum) Einstein-Maxwell equations where the metric has the form (1.1). Most of the results mentioned above apply to this more general case. This work is continuing.[1]

In the following sections, the derivation is outlined of the most important of the above properties, *i.e.*, properties *a*) to *d*). The full details of all the proofs will be published elsewhere.[2]

[1]The results of that work were published in the paper: G. C. Debney, R. P. Kerr and A. Schild, Solutions of the Einstein and Einstein–Maxwell equations, *J. Math. Phys.* **10**, 1842 (1969) [editor].

[2]According to R. Kerr (private communication), the publication announced here is the Debney – Kerr – Schild paper mentioned in footnote 1 [editor].

2. - Outline of derivation.

A simple direct calculation of the Christoffel symbols $\left\{ {\rho \atop \mu\nu} \right\}$ for the metric (1.1), (1.3) and of $\left\{ {\nu \atop \mu\nu} \right\}, \left\{ {\rho \atop \mu\nu} \right\} l^\nu, \left\{ {\rho \atop \mu\nu} \right\} l^\nu l^\mu{}_{,\rho}$(*), and substitution into the vacuum field equation

$$R_{\mu\nu} l^\mu l^\nu = 0, \tag{2.1}$$

yields

$$l_{\mu,\nu} l^\nu l^\mu{}_{,\rho} l^\rho = 0. \tag{2.2}$$

By differentiating eq. (1.6), we also have

$$l_{\mu,\nu} l^\nu l^\mu = 0. \tag{2.3}$$

Thus, in the Minkowski metric $\eta_{\mu\nu}$, the vector $l^\mu{}_{,\nu} l^\nu$ is null and orthogonal to the null vector l^μ. In a four dimensional space with the signature $3+1$ of space-time, a vector ν^μ orthogonal to a null vector l^μ is either space-like or else a multiple of l^μ. Therefore

$$l^\mu{}_{,\nu} l^\nu = \mu l^\mu. \tag{2.4}$$

Also, it is easily shown that

$$l^\mu{}_{;\nu} l^\nu = l^\mu{}_{,\nu} l^\nu = \mu l^\mu. \tag{2.5}$$

Thus the field eq. (2.1) implies that the null field l^μ is geodesic. We can therefore define a new null vector field $k^\mu = l^\mu/(2H)^{1/2}$, so that

$$k_\mu k^\mu = 0, \qquad k^\mu{}_{;\nu} k^\nu = 0, \tag{2.6}$$

and

$$\begin{cases} g_{\mu\nu} = \eta_{\mu\nu} + 2H k_\mu k_\nu, \\ g^{\mu\nu} = \eta^{\mu\nu} - 2H k^\mu k^\nu, \\ k^\mu = g^{\mu\nu} k_\nu = \eta^{\mu\nu} k_\nu. \end{cases} \tag{2.7}$$

Another simple calculation gives[3]

$$R^\sigma{}_{\rho\mu\nu} k^\rho k^\nu = -\ddot{H} k^\sigma k_\mu, \tag{2.8}$$

(*) A comma denotes partial differentiation, *e.g.*, $l^\mu{}_{,\rho} = \partial l^\mu / \partial x^\rho$, a semi-colon, *e.g.*, $l^\mu{}_{;\nu}$ denotes covariant differentiation with respect to the metric $g_{\mu\nu}$.

[3]The index at the first k on the left corrected from σ to ρ [editor].

where the dot denotes differentiation in the null direction k^μ:

$$\dot{H} = H_\mu k^\mu. \tag{2.9}$$

A Riemannian space-time is caled algebraically degenerate (or algebraically special), in the sense of the Petrov-Pirani classification [2], if and only if there exists a null vector field k^μ which obeys the equations

$$k_{[\alpha} C_{\sigma]\rho\mu\nu} k^\rho k^\nu = 0, \tag{2.10}$$

where $C_{\sigma\rho\mu\nu}$ is Weyl's conformal curvature tensor. In vacuum, where the field equations $R_{\mu\nu} = 0$ hold, the Weyl tensor and the Riemann tensor coincide, *i.e.*, $C_{\sigma\rho\mu\nu} = R_{\sigma\rho\mu\nu}$. Thus the condition (2.10) reduces to

$$k_{[\alpha} R_{\sigma]\rho\mu\nu} k^\rho k^\nu = 0. \tag{2.11}$$

By (2.8), this is clearly satisfied. Thus all vacuum solutions of the form (1.1), or equivalently (2.7), are algebraically degenerate, k_μ being a multiple Debever-Penrose vector.

The Goldberg-Sachs [3] theorem immediately implies that the null field k^μ is not only geodesic but also shear-free.

We now introduce at each point of space-time a quasiorthogonal tetrad of null vectors $e_a{}^\mu$, where Latin or tetrad suffixes range and sum over 1, 2, 3, 4, and serve to label the different vectors of the tetrad, while Greek suffixes are, as before tensor suffixes. The tetrad vectors $e_1{}^\mu$ and $e_2{}^\mu$ are complex conjugates,

$$e_2{}^\mu = \overline{e}_1{}^\mu, \tag{2.12}$$

$e_3{}^\mu$ and

$$e_4{}^\mu = k^\mu \tag{2.13}$$

are real, and they satisfy the quasiorthogonality relations

$$e_a{}^\mu e_{b\mu} = g_{ab} = \begin{pmatrix} 0 & 1 & 0 & 0 \\ 1 & 0 & 0 & 0 \\ 0 & 0 & 0 & 1 \\ 0 & 0 & 1 & 0 \end{pmatrix} = g^{ab}. \tag{2.14}$$

We define the Ricci rotation coefficients by

$$\Gamma_{abc} = -e_{a\mu;\nu} e_b{}^\mu e_c{}^\nu = -\Gamma_{bac}. \tag{2.15}$$

The geodesic and shear-free character of $e_4{}^\mu = k^\mu$ is expressed analytically by

$$\Gamma_{m44} = \Gamma_{411} = \Gamma_{422} = 0. \tag{2.16}$$

The rotation coefficients

$$\Gamma_{241} = z = \theta - i\omega, \quad \Gamma_{142} = \overline{z} = \theta + i\omega \tag{2.17}$$

are Sachs' complex expansion [cf. Goldberg and Sachs [3], eq. (1.8)], $\theta = \mathrm{Re}[z]$ being the expansion rate of the congruence of null geodesics which have k^μ as tangents, and $\omega = -\mathrm{Im}[z]$ being the rotation rate. Throughout the following we shall consider only the general case

$$z \neq 0. \tag{2.18}$$

The tetrad or *slash* derivative of an invariant is defined by

$$T_{/a} = T_{;\mu} e_a{}^\mu = T_{,\mu} e_a{}^\mu. \tag{2.19}$$

By (2.9) and (2.13)

$$T_{/4} = \dot{T}. \tag{2.20}$$

For the commutator of two successive slash derivatives we obtain

$$T_{/ab} - T_{/ba} = T_{/m}(\Gamma^m_{ab} - \Gamma^m_{ba}), \tag{2.21}$$

where

$$\Gamma^m_{ab} = g^{mn}\Gamma_{nab}. \tag{2.22}$$

The tetrad components of the Ricci tensor are

$$\begin{aligned} R_{bc} &= R_{\mu\nu} e_b{}^\mu e_c{}^\nu = \\ &= \Gamma^a_{ba/c} - \Gamma^a_{bc/a} + \Gamma^m_{ba}\Gamma^a_{mc} - \Gamma^m_{bc}\Gamma^a_{ma} + \Gamma^a_{bm}(\Gamma^m_{ca} - \Gamma^m_{ac}). \end{aligned} \tag{2.23}$$

3. - The field equations.

We choose null co-ordinates in Minkowski space, complex conjugate co-ordinates $x^1 = \overline{\zeta}$, $x^2 = \zeta$, and real co-ordinates $x^3 = v$, $x^4 = u$, such that

$$\eta_{\mu\nu} = \eta^{\mu\nu} = \begin{pmatrix} 0 & 1 & 0 & 0 \\ 1 & 0 & 0 & 0 \\ 0 & 0 & 0 & 1 \\ 0 & 0 & 1 & 0 \end{pmatrix}, \tag{3.1}$$

and

$$\text{(3.2)} \qquad \mathrm{d}s^2 = g_{\mu\nu}\mathrm{d}x^\mu \mathrm{d}x^\nu = 2\mathrm{d}\zeta\mathrm{d}\overline{\zeta} + 2\mathrm{d}u\mathrm{d}v + 2H(k_\mu \mathrm{d}x^\mu)^2.$$

A general field of[4] real null vectors in Minkowski space [cf. eq. (1.6)] is given by(*)

$$\text{(3.3)} \qquad k_\mu \mathrm{d}x^\mu = e_{4\mu}\mathrm{d}x^\mu = \mathrm{d}u + Y\mathrm{d}\overline{\zeta} + \overline{Y}\mathrm{d}\zeta - Y\overline{Y}\mathrm{d}v,$$

where Y is an arbitrary complex function of position. The tetrad is now completed as follows,

$$\text{(3.4)} \qquad \begin{cases} e_{1\mu}\mathrm{d}x^\mu &= \mathrm{d}\overline{\zeta} - \overline{Y}\mathrm{d}v, \\ e_{2\mu}\mathrm{d}x^\mu &= \mathrm{d}\zeta - Y\mathrm{d}v, \\ e_{3\mu}\mathrm{d}x^\mu &= \mathrm{d}v - Hk_\mu \mathrm{d}x^\mu, \end{cases}$$

and is easily seen to satisfy the quasiorthogonality relations (2.14).

The conditions (2.16) for k_μ to be geodesic and shear-free become

$$\text{(3.5)} \qquad Y_{/2} = Y_{/4} = \overline{Y}_{/1} = \overline{Y}_{/4} = 0.$$

We then find that

$$\text{(3.6)} \qquad \Gamma_{m14} = \Gamma_{m24} = \Gamma_{m34} = 0.$$

Geometrically, this means that the tetrad vectors (3.4) are propagated parallelly along each curve of the congruence of null geodesics which have k^μ as tangents. Sachs' complex expansion, eq. (2.17), becomes

$$\text{(3.7)} \qquad z = Y_{/1}, \quad \overline{z} = \overline{Y}_{/2}.$$

From here on, numerical suffixes on the Ricci tensor will always refer to its tetrad components R_{bc} as given by eq. (2.23).

The geodesic and shear-free condition (3.5) ensures that the following five field equations are satisfied:

$$\text{(3.8)} \qquad R_{11} = R_{22} = R_{44} = R_{14} = R_{24} = 0.$$

The field equation $R_{12} = 0$ gives

$$\text{(3.9)} \qquad H = e^{3P}(z + \overline{z}), \quad P_{/4} = 0,$$

[4] "or" corrected to "of" by the editor.

(*) The special case $k_\mu \mathrm{d}x^\mu = \mathrm{d}v$ can be included by a limiting process, *e.g.*, by starting with $k_\mu/(1 + Y\overline{Y})$ and then letting $Y \to \infty$. The complex function Y is the ratio of the two components of the spinor which corresponds to the null vector k_μ.

where P, like H, is real. The field equation $R_{34} = 0$ is now automatically satisfied. The field equation $R_{31} = 0$ and its complex conjugate $R_{32} = 0$ give

$$(3.10) \qquad P_{/1} = (z/\overline{z})\overline{Y}_{/3}, \quad P_{/2} = (\overline{z}/z)Y_{/3}.$$

Finally, the field equation $R_{33} = 0$ gives

$$(3.11) \qquad P_{/3} = \left(\frac{1}{z} + \frac{1}{\overline{z}}\right) Y_{/3}\overline{Y}_{/3}.$$

It is well worth pointing out that the calculations giving these results are by no means simple. In each of the above equations, all previously mentioned field equations have been assumed to hold, and frequent use has been made of the commutator relation (2.21) for slash derivatives.

It is now easy to show that P, Y and $\overline{Y}$ satisfy the pair of partial differential equations

$$(3.12) \qquad X_{/4} = 0, \quad X_{/3} - X_{/1}Y_{/3}/z - X_{/2}\overline{Y}_{/3}/\overline{z} = 0,$$

and that Y and $\overline{Y}$ are functionally independent because $z \neq 0$ has been assumed. It follows that P is a function of Y and $\overline{Y}$:

$$(3.13) \qquad P = P(Y, \overline{Y}),$$

and that the field eqs. (3.9), (3.10), (3.11) simplify to

$$(3.14) \qquad P_Y = \overline{Y}_{/3}/\overline{z}, \quad P_{\overline{Y}} = Y_{/3}/z.$$

Differentiating the first of these equations in the direction of the tetrad vector $e_1{}^{\mu}$, we obtain, again using commutator relations for the slash derivatives, $zP_{YY} = zP_Y{}^2$, or equivalently

$$(3.15) \qquad (\exp[-P])_{YY} = 0.$$

Similarly $(\exp[-P])_{\overline{Y}\overline{Y}} = 0$, so that $\exp[-P]$ must be bilinear in Y and $\overline{Y}$:

$$(3.16) \qquad \exp[-P] = b + \overline{a}\overline{Y} + aY + cY\overline{Y},$$

b and c being arbitrary real constants, and a and $\overline{a}$ arbitrary complex conjugate constants.

It is now easy to show that the operator

$$(3.17) \qquad K = K^{\mu}\frac{\partial}{\partial x^{\mu}} = b\frac{\partial}{\partial u} + a\frac{\partial}{\partial \overline{\zeta}} + \overline{a}\frac{\partial}{\partial \zeta} - c\frac{\partial}{\partial v}$$

satisfies

$$(3.18) KY = 0, \quad K\overline{Y} = 0, \quad Kz = KY_{/1} = 0, \quad K\overline{z} = K\overline{Y}_{/2} = 0.$$

Thus, by eqs. (3.2), (3.3), (3.9) and (3.13), K^μ is a Killing vector of our curved space-time with metric $g_{\mu\nu}$. Also, since K^μ has constant components, it is at the same time a translational Killing vector of the flat Minkowski space with metric $\eta_{\mu\nu}$.

4. - The general solution.

We are still free to perform Lorentz transformations which preserve the form $\mathrm{d}s_0{}^2 = 2\mathrm{d}\zeta\mathrm{d}\overline{\zeta} + 2\mathrm{d}u\mathrm{d}v$ of the Minkowski metric. By a suitable choice of such a Lorentz transformation, we can simplify our solutions as follows:

Case I: $\eta_{\mu\nu}K^\mu K^\nu < 0, \quad a = \overline{a} = 0, \quad b = c > 0.$

Then $K^\mu = b(0, 0, -1, 1)$ points along the time axis of Minkowski space. Equations (3.5) and (3.18) give

$$(4.1) \qquad \frac{\partial Y}{\partial v} + Y\frac{\partial Y}{\partial \zeta} = 0, \quad \frac{\partial Y}{\partial \overline{\zeta}} - Y\frac{\partial Y}{\partial u} = 0, \quad \frac{\partial Y}{\partial u} - \frac{\partial Y}{\partial v} = 0.$$

The general solution is

$$(4.2) \qquad F = 0,$$

$$(4.3) \qquad F(Y, \overline{\zeta}, \zeta, u+v) \equiv \Phi(Y) + [Y^2\overline{\zeta} - \zeta + (u+v)Y],$$

where Φ is an arbitrary analytic function of the complex variable Y.

It is now easy to calculate

$$(4.4) \qquad z = Y_{/1} = -(1 + Y\overline{Y})/F_Y,$$

$$(4.5) \qquad \exp[3P] = \sqrt{2}m(1 + Y\overline{Y})^{-3},$$

where F_Y is the partial derivative with respect to Y of the function (4.3), where $m = 2^{-1/2}b^{-3}$ is an arbitrary real constant, and where Y is determined as a function of the co-ordinates by the equation $F = 0$. The solution takes the final form

$$(4.6) \qquad \mathrm{d}s^2 = 2\mathrm{d}\zeta\mathrm{d}\overline{\zeta} + 2\mathrm{d}u\mathrm{d}v - - 4\sqrt{2}m\,\mathrm{Re}\left[\frac{1}{F_Y}\right]\cdot\left[\frac{\mathrm{d}u + Y\mathrm{d}\overline{\zeta} + \overline{Y}\mathrm{d}\zeta - Y\overline{Y}\mathrm{d}v}{1 + Y\overline{Y}}\right]^2.$$

Case II: $\eta_{\mu\nu}K^\mu K^\nu > 0, \quad a = \overline{a} = 0, \quad b = -c > 0.$

Then $K_\mu = b(0, 0, 1, 1)$ points along a space axis of Minkowski space.

Arguments similar to those above, give the solution in final form

$$\begin{aligned} ds^2 &= 2d\zeta d\overline{\zeta} + 2dudv + \\ &+ 4\sqrt{2}m\,\mathrm{Re}\left[\frac{1}{F_Y}\right]\cdot\left[\frac{du + Yd\overline{\zeta} + \overline{Y}d\zeta - Y\overline{Y}dv}{1 - Y\overline{Y}}\right]^2, \end{aligned} \tag{4.7}$$

where $m = 2^{-1/2}b^{-3}$ is an arbitrary real constant, Y is a complex variable, F_Y is the partial derivative with respect to Y of the function

$$F(Y, \overline{\zeta}, \zeta, u - v) \equiv \Phi(Y) - [Y^2\overline{\zeta} + \zeta + (u - v)Y], \tag{4.8}$$

where Φ is an arbitrary analytic function of Y, and where Y is determined as a function of the co-ordinates by the equation

$$F = 0, \quad \Phi(Y) = Y^2\overline{\zeta} + \zeta + (u - v)Y. \tag{4.9}$$

Case III: $\eta_{\mu\nu}K^\mu K^\nu = 0, \quad a = \overline{a} = c = 0, \quad b > 0, \quad K^\mu = b(0, 0, 0, 1).$

With the same notation and explanations as above, we have the solution in final form

$$\begin{aligned} ds^2 &= 2d\zeta d\overline{\zeta} + 2dudv - \\ &- 4\sqrt{2}m\,\mathrm{Re}\left[\frac{1}{F_Y}\right]\cdot(du + Yd\overline{\zeta} + \overline{Y}d\zeta - Y\overline{Y}dv)^2, \end{aligned} \tag{4.10}$$

$$F(Y, \overline{\zeta}, \zeta, v) \equiv \Phi(Y) + \zeta - vY, \tag{4.11}$$

$$F = 0, \quad \Phi(Y) = vY - \zeta. \tag{4.12}$$

5. - Gravitational field of rotating body and Schwarzschild solutions.

We consider here a particular solution with a Killing vector which is time-like with respect to the Minkowski metric $\eta_{\mu\nu}$. Equation (4.2) can be solved easily and explicitly if Φ is a quadratic polynomial in Y. We have chosen the direction of the time axis of Minkowski space so as to simplify the general solution to the form (4.6). We are, however, still free to perform translations and rotations in the 3-space orthogonal to the time direction. By a suitable choice of such a translation and rotation, the function F, eq. (4.3), with an arbitrary quadratic polynomial

$\Phi(Y)$ can be reduced to the standard form(*)

$$\Phi(Y) = -\sqrt{2}iaY, \quad a \geq 0. \tag{5.1}$$

We introduce rectangular Cartesian co-ordinates x, y, z, t in Minkowski space which are related to the null co-ordinates as follows

$$\sqrt{2}\zeta = x + iy, \quad \sqrt{2}v = z - t, \quad \sqrt{2}u = z + t. \tag{5.2}$$

Then the function F of eq. (4.3) is given by

$$F = 2^{-1/2}\left[(x - iy)Y^2 + 2(z - ia)Y - (x + iy)\right]. \tag{5.3}$$

We define a real function ρ by

$$\frac{x^2 + y^2}{\rho^2 + a^2} + \frac{z^2}{\rho^2} = 1. \tag{5.4}$$

The solution of $F = 0$ is

$$Y = \frac{(\rho - z)(\rho + ia)}{\rho(x - iy)}, \tag{5.5}$$

and for this Y we have

$$F_Y = \sqrt{2}\,[(x - iy)Y + (z - ia)] = \sqrt{2}(\rho^2 - iaz)/\rho, \tag{5.6}$$

$$Y\overline{Y} = \frac{\rho - z}{\rho + z}. \tag{5.7}$$

Substituting into eq. (4.6), we obtain

$$(5.8)\quad ds^2 = \mathrm{d}x^2 + \mathrm{d}y^2 + \mathrm{d}z^2 - \mathrm{d}t^2 +$$
$$+ \frac{2m\rho^3}{\rho^4 + a^2z^2}\left[\mathrm{d}t + \frac{z}{\rho}\mathrm{d}z + \frac{\rho}{\rho^2 + a^2}(x\mathrm{d}x + y\mathrm{d}y) + \frac{a}{\rho^2 + a^2}(x\mathrm{d}y - y\mathrm{d}x)\right]^2.$$

This is the exterior gravitational field of a rotating body [1]. When we put $a = 0$, so that, by eq. (5.4), $\rho = r = (x^2 + y^2 + z^2)^{1/2}$, this reduces to the Schwarzschild metric

$$\mathrm{d}s^2 = \mathrm{d}x^2 + \mathrm{d}y^2 + \mathrm{d}z^2 - \mathrm{d}t^2 + \frac{2m}{r}(\mathrm{d}r + \mathrm{d}t)^2 \tag{5.9}$$

in a form first given by Eddington [4].

(*) In this section we have changed notation slightly: a is now a real number which lias nothing to do with the complex a of eq. (3.16); z is now one of the rectangular Cartesian co-ordinates x, y, z, t in Minkowski space, and not Sachs' complex expansion, eq. (2.17).

REFERENCES

[1] R. P. Kerr: *Phys. Rev Lett.*, **11**, 237 (1963); *Proceedings of the First Texas Conference on Gravitational Collapse* (Dallas, Dec. 1963) (University of Chicago Press, in print).[5]

[2] A. Z. Petrov: *Scientific Notices Kazan State University*, **114**, 55 (1954);[6] F. A. E. Pirani: *Phys. Rev.*, **105**, 1089 (1957): R. Debever: *Bull. Soc. Math. Belg.*, **10**, 112 (1958); R. Penrose: *Ann. Phys.*, **10**, 171 (1960); R. K. Sachs: *Proc. Roy. Soc.*, A **264**, 309 (1961).

[3] J. N. Goldberg and R. K. Sachs: *Acta Phys. Polonica Suppl.*, **22**, 13 (1962);[7] I. Robinson and A. Schild: *Proc. Int. Conf. Rel. Theor. of Gravitation* (Warsaw, July 1962) (in print);[8] *Journ. Math. Phys.*, **4**, 484 (1963).

[4] A. S. Eddington: *Nature*, 113, 192 (1924).

INTERVENTI E DISCUSSIONI

– W. D. Beiglböck:

You gave us two solutions in your class, that have a fairly simple physical interpretation – namely the Schwarzschild and the Kerr solutions. Do you know whether there are any further solutions that would allow a physical interpretation?

– A. Schild:

The answer is «no». When $\Phi(Y)$ is not a quadratic polynomial the solutions are complicated, and we have not examined them.[9]

---

[5] The full reference for the second paper is: *Quasi-stellar sources and graviational collapse.* Edited by I. Robinson, A. Schild, and E.L. Schucking, Chicago, University Press, 1965, p. 99 [editor].

[6] The correct full reference to the Petrov paper is: A. Z. Petrov, Klassifikacya prostranstv opredelyaushchikh polya tyagoteniya [The classification of spaces defining gravitational fields]. *Uchenye Zapiski Kazanskogo Gosudarstvennogo Universiteta im. V. I. Ulyanovicha-Lenina* **114**(8), 55 (1954). English translation (in the Golden Oldies series): *Gen. Rel. Grav.* **32**, 1665 (2000) [editor].

[7] The Goldberg – Sachs paper has been reprinted as a Golden Oldie, *Gen. Rel. Grav.* **41**, 433 (2009) [editor].

[8] The full reference to the first Robinson–Schild paper is: I. Robinson and A. Schild, Degeneracy and shear. In: *Proceedings on theory of gravitation. Conference in Warszawa and Jabłonna, 25–31 July 1962.* Edited by L. Infeld. Gauthier–Villars, Paris and Państwowe Wydawnictwo Naukowe, Warszawa 1964, pp. 340–341 [editor].

[9] More solutions are known today. Information on them can be found, for example, in the following sources:

– A. Komar:

Ehlers, in his Royaumont paper[10], presented a general procedure for obtaining a Ricci flat stationary space from a given Ricci flat static space. What is the relationship (if any) between his procedure and yours?

– J. Ehlers:

The Schild-Kerr procedure is a definite generalization. The earlier procedure gives only one parametric sub-family of the Kerr-metrios.

– A. Schild:

Answer by Ehlers.

– J. Ehlers:

If my method of constructing stationary vacuum fields from static ones (described in my Royaumont lecture) is applied to the Schwarzschild solution a one-parameter (sub-) family of the Kerr-solution is obtained.

2) In connection with the problem of filling matter into the Kerr-solution I should like to point out: If the matter is to be a perfect fluid,

$$T_{\mu\nu} = (\rho + p)u_\mu u_\nu - pg_{\mu\nu}, \quad (u_\lambda u^\lambda = 1),$$

and if the interior metric is to be stationary (like the exterior one) with Killing vector

$$\xi^\lambda = e^U u^\lambda$$

and if, moreover, g_{ab} is supposed to satisfy the Lichnerowicz junction conditions along the (hyper-) surface S of the body with ξ^λ of class

a) J. Bičak, in *Einstein's field equations and their physical implications. Selected essays in honour of Jürgen Ehlers.* Lecture Notes in Physics, vol. 540. Edited by B. Schmidt. Springer, Berlin 2000, p. 1.

b) N. Straumann, *General relativity.* Springer, Berlin 2004; section 7.2 (complete derivation via Ernst's equation).

c) H. A. Buchdahl, *17 Simple Lectures on General Relativity Theory*, Wiley, New York (N. Y.) 1981 (Lecture 13, p. 114).

d) H. Stephani, D. Kramer, M. MacCallum, C. Hoenselaers and E. Herlt, *Exact Solutions of Einstein's Field Equations.* 2nd Edition. Cambridge University Press 2003. (Chapter 32, pp. 485 – 505: discussion of the Kerr–Schild metrics, Chapters 19 to 21, pp. 292 – 340: description of the stationary-axisymmetric solutions).

e) S. W. Hawking and G. F. R. Ellis, *The Large-scale Structure of Spaceetime.* Cambridge University Press 1973 [editor].

[10]The reference to this paper is: J. Ehlers, Transformations of static solutions of Einstein's gravitational field equations into different solutions by means of conformal mappings. In: *Les theories relativistes de la gravitation, Royaumont 1959*, edition CNRS, Paris 1965, pp. 275-283 [editor].

C^1, then S is given by $U = \text{cost}$. In fact, it follows from the above assumption that

$$U_{\lambda;\mu}u^\mu = -U_{,\lambda}, \quad p_{,\lambda}u^\lambda = 0,$$

consequently because of «Euler's» equation

$$(\rho + p)u_{\lambda;\mu}u^\mu = (\delta^\mu_\lambda - u^\mu u_\lambda)p_{,\mu}$$

we obtain $(\rho + p)\mathrm{d}u = -\mathrm{d}p$ which implies, since $p = 0$ along S, that $u = \text{cost}$ along S also. In consequence of this fact it is possible to study the «shape» of the field-producing body by means of data of the exterior field alone, at least under the hydrostatic circumstances stated above. In this way one may be able to get further support of the assumption that this exterior solution actually corresponds to a body in a state of stationary rotation.[11]

– A. SCHILD:

This is a very interesting remark. We shall look at the surfaces on which the stationary Killing vector has constant magnitude. Since the Kerr metric is known explicitly, this should be a straight forward problem.

[Alla discussione parteciparono anche G. MC VITTE, E. T. NEWMAN, C. MØLLER, H. J. TREDER, E. DEBEVER.]

– A. SCHILD (*additional remark*):

Professor Misner in a letter, and Professors Debever and Lichnerowicz at this meeting, kindly drew my attention to papers by Vaidya (*Nature* (1953))[12] and J. Hély (*Compt. Rend.* (1960))[13] which examine some space-times with electromagnetic radiation where the metric has the form of our eq. (1.1).

[11]No perfect fluid source for the Kerr solution has been found until today [editor].

[12]The full reference consists of three papers by P. C. Vaidya: The external field of a radiating star in general relativity, *Current Science* **12**, 183 (1943); The gravitational field of a radiating star, *Proceedings of the Indian Academy of Sciences* **A33**, 264 (1951); and "Newtonian" time in general relativity, *Nature* **171**, no 4345, 260 (1953). All three papers were reprinted in the Golden Oldies series in *Gen. Rel. Grav.* **31**, no 1, 137 (1999) [editor].

[13]The full reference is: J. Hély, *C. R Acad. Sci. Paris* **251**, pages 1981 and 2300 (1960) [editor].

Part IV
Appendix

Chapter 15
The list of Golden Oldies published so far

The rules for accepting papers to the Golden Oldies series, and the rules of printing them, can be found in:

A. Krasiński, Editorial: Golden Oldies – a reactivation. *Gen. Relativ. Gravit.* **39**, 1043 – 1045 (2007) and online at http://dx.doi.org/10.1007/s10714-006-0308-z

with the change that Jürgen Ehlers was replaced in the Advisory Board by Hubert Goenner, goenner@theorie.physik.uni-goettingen.de

The rules have been further modified in:

A. Ashtekar and R. Maartens: Editorial. *Gen. Relativ. Gravit.* **44**, 2419 (2012) and online at http://dx.doi.org/10.1007/s10714-012-1444-2.

The changes are:

1. One Editor in Chief is now automatically a member of the Advisory Board (at the time of printing this book this is Roy Maartens).
2. The final approval of each Golden Oldie for publication is done by an Editor in Chief.

The list copied below was created and is maintained on the web by Charles Hellaby, Department of Mathematics and Applied Mathematics, University of Cape Town, Rondebosch 7700, Cape Town, South Africa,
email: Charles.Hellaby@uct.ac.za.

−1. Three papers by Hans Thirring and Josef Lense, 1918 – 1921[1]

−1 (a) On the effect of rotating distant masses in Einstein's theory of gravitation

By Hans Thirring

[1] This item carries the number −1 because the translations of these papers were printed long before the idea of Golden Oldies was announced. We recognized it *ex post* as a precursor of our series, which was true to its purpose and rules.

Originally published: Über die Wirkung Rotierender Ferner Massen in der Einsteinschen Gravitationstheorie. *Physikalische Zeitschrift* **19**, 33 (1918).

−1 (b) Correction to my paper: "On the effect of rotating distant masses in Einstein's theory of gravitation"

By Hans Thirring

Originally published: Berichtigung zu meiner Arbeit: "Über die Wirkung Rotierender Massen in der Einsteinschen Gravitationstheorie". *Physikalische Zeitschrift* **22**, 29 (1921).

−1 (c) On the influence of the proper rotation of central bodies on the motions of planets and moons according to Einstein's theory of gravitation

By Josef Lense and Hans Thirring.

Originally published: Über den Einfluß der Eigenrotation der Zentralkörper auf die Bewegung der Planeten und Monde nach der Einsteinschen Gravitationstheorie. *Physikalische Zeitschrift* **19**, 156 (1918).

All three papers translated by: Bahram Mashhoon, Friedrich W. Hehl and Dietmar S. Theiß.

Reprinted: B. Mashhoon, F.W. Hehl and D.S. Theiß, On the gravitational effects of rotating masses: the Thirring – Lense papers. *Gen. Relativ. Gravit.* **16**, no 8, 711–50 (1984). (The Thirring – Lense papers were combined with editorial remarks and partial biographies of the authors.)

---

0. Contributions to the relativistic mechanics of continuous media[2]

By Jürgen Ehlers

Originally published: Beiträge zur Relativistischen Mechanik Kontinuierlicher Medien. *Akademie der Wissenschaften und Literatur (Mainz), Abhandlungen der Mathematisch-Naturwissenschaftlichen Klasse* Nr. 11, 792–837 (1961).

Translated by: George F.R. Ellis and Peter K.S. Dunsby.

Reprinted in: *Gen. Relativ. Gravit.* **25**, no 12, 1225–66 (1993).

Editorial note: A. Held, *Gen. Relativ. Gravit.* **25**, 1225 (1993) (positioned as the abstract to the Ehlers paper).

---

1. On a stationary cosmology in the sense of Einstein's theory of gravitation

By Kornel Lanczos

Originally published: Über eine Stationäre Kosmologie im Sinne der Einsteinschen Gravitationstheorie. *Zeitschrift für Physik* **21**, 73 (1924).

Translated by: Cornelius Hoenselaers.

Reprinted in: *Gen. Relativ. Gravit.* **29**, no 3, 363 (1997).

Editorial note: Andrzej Krasiński, *Gen. Relativ. Gravit.* **29**, 359 (1997).

Biography: Andrzej Krasiński, *Gen. Relativ. Gravit.* **29**, 360 (1997).

---

2. The expanding Universe

By the Abbé Georges Lemaître

[2] This item carries the number 0 because it was printed before the series began to be published systematically.

Originally published: L'Univers en Expansion. *Annales de la Société Scientifique de Bruxelles* **A53**, 51 (1933).
Translated by: Malcolm A. H. MacCallum.
Reprinted in: *Gen. Relativ. Gravit.* **29**, no 5, 641 (1997).
Editorial note: Andrzej Krasiński, *Gen. Relativ. Gravit.* **29**, 637 (1997).
Biography: Andrzej Krasiński, *Gen. Relativ. Gravit.* **29**, 639 (1997).

---

3. Effect of Inhomogeneity on Cosmological Models
By Richard C. Tolman
Originally published: *Proceedings of the National Academy of Sciences of the USA* **20**, 169 (1934).
Reprinted in: *Gen. Relativ. Gravit.* **29**, no 7, 935 (1997)
Editorial note: Andrzej Krasiński, *Gen. Relativ. Gravit.* **29**, 931 (1997).
Biography: Andrzej Krasiński, *Gen. Relativ. Gravit.* **29**, 932 (1997).

---

4. On the stability of cosmological models
By N. R. Sen
Originally published: *Zeitschrift für Astrophysik* **9**, 215 (1934).
Reprinted in: *Gen. Relativ. Gravit.* **29**, no 11, 1477 (1997).
Editorial note: Andrzej Krasiński, *Gen. Relativ. Gravit.* **29**, 1473 (1997).
Biography: Amal Kumar Raychaudhuri, *Gen. Relativ. Gravit.* **29**, 1474 (1997).

---

5. Observable relations in relativistic cosmology. II
By William Hunter McCrea
Originally published: *Zeitschrift für Astrophysik* **18**, 98 (1939).
Reprinted in: *Gen. Relativ. Gravit.* **30**, no 2, 315 (1998).
Editorial note: Andrzej Krasiński, *Gen. Relativ. Gravit.* **30**, 311 (1998).
Biography: William Hunter McCrea, *Gen. Relativ. Gravit.* **30**, 312 (1998); see additional note in *Gen. Relativ. Gravit.* **32**, no 9, 1936 (2000).

---

6. A note on some general solutions of the Einstein field equations in a spherically symmetric world
By Paul Kustaanheimo and Bertil Qvist
Originally published: *Societas Scientiarum Fennica. Commentationes Physico-Mathematicae* **XIII** no 16, 1 (1948).
Reprinted in: *Gen. Relativ. Gravit.* **30**, no 4, 663 (1998).
Editorial note: Andrzej Krasiński, *Gen. Relativ. Gravit.* **30**, 659 (1998).
Biographies: Boris Sjöberg, *Gen. Relativ. Gravit.* **30**, 660 (1998) (Qvist) and Juhana Kustaanheimo, *Gen. Relativ. Gravit.* **30**, 661 (1998) (Kustaanheimo).

---

7. The formation of the nebulae
By William B. Bonnor
Originally published: *Zeitschrift für Astrophysik* **39**, 143 (1956).
Reprinted in: *Gen. Relativ. Gravit.* **30**, no 7, 1113 (1998).
Editorial note: Andrzej Krasiński, *Gen. Relativ. Gravit.* **30**, 1111 (1998).
Biography: William B. Bonnor, *Gen. Relativ. Gravit.* **30**, 1112 (1998).

---

8. Isotropic space with discrete gravitational-field sources (On the theory of a nonhomogeneous isotropic Universe)

By Mikhail Fedorovich Shirokov and Iosif Zalmanovich Fisher

Originally published: Izotropnoe prostranstvo s diskretnymi istochnikami gravitacyonnogo polya (K teorii neodnorodnoy izotropnoy vselennoy) *Astronomicheskiy Zhurnal* **39**, 899 (1962) [*Soviet Astronomy A. J.* **6**, 699 (1963)].

Reprinted in: *Gen. Relativ. Gravit.* **30**, no 9, 1411 (1998).

Editorial note: Andrzej Krasiński, *Gen. Relativ. Gravit.* **30**, 1407 (1998).

Biographies: Andrzej Krasiński, *Gen. Relativ. Gravit.* **30**, 1408 (1998) (Shirokov) and Mikhail N. Polozov, *Gen. Relativ. Gravit.* **30**, 1409 (1998) (Fisher).

---

9. Some relativistic cosmological models. A dissertation presented to the faculty of the Graduate School of the University of Texas in partial fulfillment of the requirements for the degree of Doctor of Philosophy

By Ronald Kantowski

This paper was never published before it appeared as a Golden Oldie; it was circulated as a preprint.

First printed in *Gen. Relativ. Gravit.* **30**, no 11, 1665 (1998).

Editorial note: Andrzej Krasiński, *Gen. Relativ. Gravit.* **30**, 1663 (1998).

Biography: Ronald Kantowski, *Gen. Relativ. Gravit.* **30**, 1663 (1998).

---

10. Three papers by Prahalad Chunilal Vaidya, 1943 – 1953

10 (a) The external field of a radiating star in general relativity

Originally published: *Current Science* **12**, 183 (1943).

10 (b) The gravitational field of a radiating star

Originally published: *Proceedings of the Indian Academy of Sciences* **A33**, 264 (1951).

10 (c) "Newtonian" time in general relativity

Originally published: *Nature* **171**, no 4345, 260 (1953).

Reprinted in: *Gen. Relativ. Gravit.* **31**, no 1, 119, 121 and 137 (1999).

Editorial note: Andrzej Krasiński, *Gen. Relativ. Gravit.* **31**, 115 (1999).

Biography: Prahalad Chunilal Vaidya, *Gen. Relativ. Gravit.* **31**, 116 (1999).

---

11. Two papers by Hidekazu Nariai, 1950 – 1951

11 (a) On some static solutions of Einstein's gravitational field equations in a spherically symmetric case

Originally published: *Scientific Reports of the Tôhoku University* **34**, 160 (1950).

11 (b) On a new cosmological solution of Einstein's field equations of gravitation

Originally published: *Scientific Reports of the Tôhoku University* **35**, 46 (1951).

Reprinted in: *Gen. Relativ. Gravit.* **31**, no 6, 951 and 963 (1999).

Editorial note: Andrzej Krasiński, *Gen. Relativ. Gravit.* **31**, 945 (1999).

Biography: Kenji Tomita, *Gen. Relativ. Gravit.* **31**, 948 (1999).

---

12. On a class of solutions of the gravitation equations of relativity

By B. Datt

Originally published: Über eine Klasse von Lösungen der Gravitationsgleichungen der Relativität. *Zeitschrift für Physik* **108**, 314 (1938).

Translated by: Herbert Pfister and John Stewart.

Reprinted in: *Gen. Relativ. Gravit.* **31**, no 10, 1619 (1999).

Editorial note: Andrzej Krasiński, *Gen. Relativ. Gravit.* **31**, 1615 (1999).

Biography: missing **and still wanted**.

---

13. Spherically symmetrical models in general relativity

By Hermann Bondi

Originally published: *Mon. Not. Roy. Astr. Soc.* **107**, 410 (1947).

Reprinted in: *Gen. Relativ. Gravit.* **31**, no 11, 1783 (1999).

Editorial note: Andrzej Krasiński, *Gen. Relativ. Gravit.* **31**, 1777 (1999).

Biography: Hermann Bondi, *Gen. Relativ. Gravit.* **31**, 1780 (1999).

---

14. Two papers by Aleksandr Friedman, 1922 – 1924

14 (a) On the curvature of space

Originally published: Über die Krümmung des Raumes. *Zeitschrift für Physik* **10**, 377–386 (1922).

14 (b) On the possibility of a world with constant negative curvature of space

Originally published: Über die Möglichkeit einer Welt mit konstanter negativer Krümmung des Raumes *Zeitschrift für Physik* **21**, 326–332 (1924).

Both papers translated by: George F. R. Ellis and Henk van Elst.

Reprinted in: *Gen. Relativ. Gravit.* **31**, no 12, 1991 and 2001 (1999).

Editorial note: Andrzej Krasiński and George F. R. Ellis, *Gen. Relativ. Gravit.* **31**, 1985 (1999).

Biography: Andrzej Krasiński, *Gen. Relativ. Gravit.* **31**, 1989 (1999).

See also Addendum in *Gen. Relativ. Gravit.* **32**, 1937 (2000).

---

15. Vacuum quantum fluctuations in curved cpace and the theory of gravitation

By Academician Andriey Dmitrievich Sakharov

Originally published: Vakuumnye kvantovye fluktuacii v iskrivlennom prostranstve i teoria gravitacii. *Doklady Akademii Nauk SSSR* **177**, 70 (1967) [*Soviet Physics – Doklady* 12, 1040-1041 (1968)].

Reprinted in: *Gen. Relativ. Gravit.* **32**, no 2, 365 (2000).

Editorial note: Hans-Jürgen Schmidt, *Gen. Relativ. Gravit.* **32**, 361 (2000).

Biography: Hans-Jürgen Schmidt, *Gen. Relativ. Gravit.* **32**, 362 (2000).

---

16. Relativistic cosmology. I

By Amalkumar Raychaudhuri

Originally published: *Physical Review* **98**, 1123 (1955).

Reprinted in: *Gen. Relativ. Gravit.* **32**, no 4, 749 (2000).

Editorial note: John Earman and Amal Kumar Raychaudhuri, *Gen. Relativ. Gravit.* **32**, 743 (2000).
Biography: A. K. Raychaudhuri, *Gen. Relativ. Gravit.* **32**, 746 (2000).

---

17. Two papers by Kurt Gödel, 1949 – 1952

17 (a) An example of a new type of cosmological solutions of Einstein's field equations of gravitation
Originally published: *Reviews of Modern Physics* **21**, 447 (1949).

17 (b) Rotating universes in general relativity theory
Originally published: in *Proceedings of the International Congress of Mathematicians*. Edited by L. M. Graves et al., Cambridge, Mass. 1952, vol. 1, p. 175.

Reprinted in: *Gen. Relativ. Gravit.* **32**, no 7, 1409 and 1419 (2000).
Editorial note: George F. R. Ellis, *Gen. Relativ. Gravit.* **32**, 1399 (2000).
Biography: Andrzej Krasiński, *Gen. Relativ. Gravit.* **32**, 1407 (2000).

---

18. The Classification of spaces defining gravitational fields
By Aleksei Zinovyevitch Pyetrov
Originally published: Klassifikacya prostranstv opredelyayushchikh polya tyagoteniya. *Uchenye Zapiski Kazanskogo Gosudarstvennogo Universiteta im. V. I. Ulyanovicha-Lenina*, **114**, (8) 55–69 (1954); Jubilee (1804–1954) Collection.
Translated by: Jacek Jezierski, Malcolm A. H. MacCallum and Aleksandr G. Polnarev.
Reprinted in: *Gen. Relativ. Gravit.* **32**, no 8, 1665 (2000).
Editorial note: Malcolm A. H. MacCallum, *Gen. Relativ. Gravit.* **32**, 1661 (2000).
Biography: Alexander Gusev, *Gen. Relativ. Gravit.* **32**, 1663 (2000).

---

19. Two papers by Edward Arthur Milne and William Hunter McCrea, 1934

19 (a) A Newtonian expanding Universe
By Edward Arthur Milne
Originally published: *Quart. J. Math. Oxford* **5**, 64–72 (1934).

19 (b) Newtonian Universes and the curvature of space
By William Hunter McCrea and Edward Arthur Milne
Originally published: *Quart. J. Math. Oxford* **5**, 73–80 (1934).

Reprinted in: *Gen. Relativ. Gravit.* **32**, no 9, 1939 and 1949 (2000).
Editorial note: Andrzej Krasiński, p. 1933 (2000).
Biography: Andrzej Krasiński, *Gen. Relativ. Gravit.* **32**, 1935 (2000) (Milne).
The biography of W. H. McCrea was printed with the GO#5, with a note in *Gen. Relativ. Gravit.* **32**, 1936 (2000).

---

20. Radiation states and the problem of energy in general relativity.
By Louis Bel
Originally published: Les états de Radiation et le Probleme de l'Énergie en Relativité Générale. *Cahiers de Physique* **16**, 59–80 (1962).

Translated by: Malcolm A. H. MacCallum
Reprinted in: *Gen. Relativ. Gravit.* **32**, no 10, 2047 (2000).
Editorial note: José M. M. Senovilla,*Gen. Relativ. Gravit.* **32**, 2043 (2000).
Biography: José M. M. Senovilla,*Gen. Relativ. Gravit.* **32**, 2045 (2000).

---

21. Two papers by Vladimir Afanasyevich Ruban, 1968 – 1969

21 (a) T-models of "sphere" in general relativity theory
Originally published: T-modeli "shara" w obshchey teorii otnositelnosti. *Zhurnal Eksperimentalnoy i Teoreticheskoy Fiziki, Pisma v Redakciyu* **8**, no 11, 669–673 (1968) [*Soviet Physics JETP (Lett.)* **8**, 414–417 (1968)].

21 (b) Spherically symmetric T-models in the general theory of relativity
Originally published: Sfericheski simmetrichnye T-modeli v obshchey teorii otnositelnosti. Zhurnal Eksperimentalnoy i Teoreticheskoy Fiziki **56** 1914–1928 (1969) [*Soviet Physics JETP* **29**, 1027–1034 (1969)].

Reprinted in: *Gen. Relativ. Gravit.* **33**, no 2, 369 and 375 (2001).
Editorial note: Andrzej Krasiński, *Gen. Relativ. Gravit.* **33**, 363 (2001).
Biography: Irina Dymnikova, *Gen. Relativ. Gravit.* **33**, 366 (2001).

---

22. On the three-dimensional spaces which admit a continuous group of motions
By Luigi Bianchi
Originally published: Sugli spazi a tre dimensioni che ammettono un gruppo continuo di movimenti. *Memorie di Matematica e di Fisica della Societa Italiana delle Scienze, Serie Terza* **XI**, 267–352 (1898).
Translated by: Robert Jantzen
Reprinted in: *Gen. Relativ. Gravit.* **33**, no 12, 2171 (2001).
Editorial note: Robert Jantzen, *Gen. Relativ. Gravit.* **33**, 2157 (2001).
Biography: Robert Jantzen, *Gen. Relativ. Gravit.* **33**, 2168 (2001).

---

23. R- and T-regions in a spacetime with a spherically symmetric space
By Igor D. Novikov
Originally published: R- i T-oblasti v prostranstve-vremeni so sphericheski symmetrichnym prostranstvom. *Soobshcheniya GAISh [Communications of the State Shternberg Astronomical Institute]* **132**, 3–42 (1964).
Translated by: S. B. Beisekeev and Malcolm A. H. MacCallum.
Reprinted in: *Gen. Relativ. Gravit.* **33**, no 12, 2259 (2001).
Editorial note: Andrzej Krasiński, *Gen. Relativ. Gravit.* **33**, 2255 (2001).
Biography: Igor D. Novikov, *Gen. Relativ. Gravit.* **33**, 2258 (2001).

---

24. Visual horizons in world models
By Wolfgang Rindler
Originally published: *Mon. Not. Roy. Astr. Soc.* **116**, 662–677 (1956).
Reprinted in: *Gen. Relativ. Gravit.* **34**, no 1, 133 (2002).
Editorial note: Andrzej Krasiński, *Gen. Relativ. Gravit.* **34**, 131 (2002).
Biography: Wolfgang Rindler, *Gen. Relativ. Gravit.* **34**, 132 (2002).

25. Lectures on general relativity
By Andrzej Trautman
This text was never published before it appeared as a Golden Oldie; it was circulated as copies of the typescript.
First printed in: *Gen. Relativ. Gravit.* **34**, no 5, 721 (2002).
Editorial note: Piotr Chruściel, *Gen. Relativ. Gravit.* **34**, 715 (2002).
Biography: Andrzej Trautman, *Gen. Relativ. Gravit.* **34**, 718 (2002).

26. Gravitational collapse: the role of general relativity
By Roger Penrose
Originally published: *Rivista del Nuovo Cimento, Numero Speziale* **I**, 252 (1969).
Reprinted in: *Gen. Relativ. Gravit.* **34**, no 7, 1141 (2002).
Editorial note: Andrzej Królak, *Gen. Relativ. Gravit.* **34**, 1135 (2002).
Biography: Werner Israel, *Gen. Relativ. Gravit.* **34**, 1138 (2002).

27. The field of a single centre in Einstein's theory of gravitation, and the motion of a particle in that field
By Johannes Droste
Originally published: *Koninklijke Nederlandsche Akademie van Wetenschappen Proceedings* **19**, 197 (1917).
Reprinted in: *Gen. Relativ. Gravit.* **34**, no 9, 1545 (2002).
Editorial note: Tony Rothman, *Gen. Relativ. Gravit.* **34**, 1541 (2002).
Biography: Carlo Beenakker, *Gen. Relativ. Gravit.* **34**, 1543 (2002).

28. On the singularities of a Riemannian manifold
By György Szekeres
Originally published: *Publicationes Mathematicae Debrecen* **7**, 285 (1960).
Reprinted in: *Gen. Relativ. Gravit.* **34**, no 11, 2001 (2002).
Editorial note: Peter Szekeres, *Gen. Relativ. Gravit.* **34**, 1995 (2002).
Biography: György Szekeres, *Gen. Relativ. Gravit.* **34**, 1999 (2002).

29. Relativistic hydrodynamics
By John Lighton Synge
Originally published: *Proceedings of the London Mathematical Society* ser. 2 **43**, 376–416 (1937).
Reprinted in: *Gen. Relativ. Gravit.* **34**, no 12, 2177 (2002).
Editorial note: Jürgen Ehlers, *Gen. Relativ. Gravit.* **34**, 2171 (2002).
Biography: Peter Hogan, *Gen. Relativ. Gravit.* **34**, 2174 (2002).

30. The seminar notes of Wolfgang Kundt on the Bianchi classification by the Schücking method
By Wolfgang Kundt, based on a seminar by Engelbert Schücking
This text was never published before it was printed as a Golden Oldie; it had existed only as handwritten notes in German.
First printed in: *Gen. Relativ. Gravit.* **35**, no 3, 491 (2003).

Editorial note: Andrzej Krasiński, Christoph G. Behr, Engelbert Schücking, Frank B. Estabrook, Hugo D. Wahlquist, George F. R. Ellis, Robert Jantzen and Wolfgang Kundt, *Gen. Relativ. Gravit.* **35**, 475 (2003) (listed as "Golden Oldie" with the separate title "The Bianchi classification in the Schücking–Behr approach").

Biography: Engelbert Schücking, *Gen. Relativ. Gravit.* **35**, 487 (2003).

---

31. On the gravitational field of a mass point according to Einstein's theory

By Karl Schwarzschild

Originally published: Über das Gravitationsfeld eines Massenpunktes nach der Einsteinschen Theorie. *Sitzungsberichte der Königlich Preussischen Akademie der Wissenschaften zu Berlin, Phys.-Math. Klasse*, 189–196 (1916).

Translated by: Salvatore Antoci

Reprinted in: *Gen. Relativ. Gravit.* **35**, no 5, 951 (2003).

Editorial note: Salvatore Antoci and Dierck-Ekkehard Liebscher, *Gen. Relativ. Gravit.* **35**, 945 (2003).

Correction to the editorial note: José M. M. Senovilla, *Gen. Relativ. Gravit.* **39**, no 5, 685 (2007).

Biography: missing.

---

32. Empty space-times admitting a three parameter group of motions

By A.H. Taub

Originally published: *Annals of Mathematics* **53**, 472–490 (1951).

Reprinted in: *Gen. Relativ. Gravit.* **36**, no 12, 2699 (2004).

Editorial note: Malcolm A. H. MacCallum, *Gen. Relativ. Gravit.* **36**, 2689 (2004).

Biography: Bahram Mashhoon, *Gen. Relativ. Gravit.* **36**, 2695 (2004).

---

33. On the general spherically symmetric solutions of Einstein's gravitational equations in vacuo

By J.T. Jebsen

Originally published: Über die allgemeinen kugelsymmetrischen Lösungen der Einsteinschen Gravitationsgleichungen im Vakuum. *Arkiv för Matematik, Astronomi och Fysik* **15**, Nr. 18, 1–9 (1921).

Translated by: S. Antoci and D. Liebscher.

Reprinted in: *Gen. Relativ. Gravit.* **37**, no 12, 2253 (2005).

Editorial note: Stanley Deser, *Gen. Relativ. Gravit.* **37**, 2251 (2005).

Correction to the editorial note: Jürgen Ehlers and Andrzej Krasiński, *Gen. Relativ. Gravit.* **38**, no 8, 1329 (2006).

Biography: Nils Voje Johansen and Finn Ravndal, *Gen. Relativ. Gravit.* **38**, no 3, 537 (2006).

---

34. On the definition of distance in general relativity

By Ivor Malcolm Haddon Etherington

Originally published: *Philosophical Magazine*, series 7 **15**, Nr. 18, 761 (1933).

Reprinted in: *Gen. Relativ. Gravit.* **39**, no 7, 1055 (2007).

Editorial note: George F.R. Ellis, *Gen. Relativ. Gravit.* **39**, 1047 (2007).

Biography: Andrzej Krasiński, *Gen. Relativ. Gravit.* **39**, 1053 (2007).

35. Perturbations of a cosmological model and angular variations of the microwave background

By Rainer K. Sachs and Arthur M. Wolfe

Originally published: *Astrophysical Journal* **147**, 73–90 (1967).

Reprinted in: *Gen. Relativ. Gravit.* **39**, no 11, 1944 (2007).

Editorial note: George F.R. Ellis and Jürgen Ehlers, *Gen. Relativ. Gravit.* **39**, 1929 (2007).

Biographies: Rainer Sachs, George Ellis and Andrzej Krasiński, *Gen. Relativ. Gravit.* **39**, 1941 (2007) (Sachs), Arthur M. Wolfe, *Gen. Relativ. Gravit.* **39**, 1942 (2007) (Wolfe).

36. Geometrical theorems on Einstein's cosmological equations

By Edward Kasner

Originally published: *American Journal of Mathematics* **43**, 217–221 (1921).

Reprinted in: *Gen. Relativ. Gravit.* **40**, no 4, 872 (2008).

Editorial note: John Wainwright, *Gen. Relativ. Gravit.* **40**, 865 (2008).

Biography: Andrzej Krasiński, *Gen. Relativ. Gravit.* **40**, 868 (2008).

37. The cosmological constant and the theory of elementary particles

By Yakov Borisovich Zeldovich

Originally published: Kosmologicheskaya postoyannaya i teoria elementarnykh chastits. *Uspekhi Fizicheskikh Nauk* **95**, 209–230 (1968) [*Soviet Physics Uspekhi* **11**, 381–393 (1968)].

Reprinted in: *Gen. Relativ. Gravit.* **40**, no 7, 1562 (2008).

Editorial note: Varun Sahni and Andrzej Krasiński, *Gen. Relativ. Gravit.* **40**, 1557 (2008).

Biography: Andrzej Krasiński, *Gen. Relativ. Gravit.* **40**, 1560 (2008).

38. The dynamics of general relativity

By Richard Arnowitt, Stanley Deser and Charles W. Misner

Originally published: in *Gravitation: an introduction to current research*, Ed. Louis Witten, Wiley, 1962, pp. 227–265, (Chapter 7).

Reprinted in: *Gen. Relativ. Gravit.* **40**, no 9, 1997 (2008).

Editorial note: Jorge Pullin, *Gen. Relativ. Gravit.* **40**, 1989 (2008).

Biographies: Richard Arnowitt, *Gen. Relativ. Gravit.* **40**, 1991 (2008) (Arnowitt); Stanley Deser and Andrzej Krasiński, *Gen. Relativ. Gravit.* **40**, 1992 (2008) (Deser); Charles W. Misner, *Gen. Relativ. Gravit.* **40**, 1993 (2008) (Misner).

39. The redshift of extragalactic nebulae

By Fritz Zwicky

Originally published: Die Rotverschiebung von extragalaktischen Nebeln. *Helv. Phys. Acta* **6**, 110–127 (1933).

Translated by: Anita Ehlers.

Reprinted in: *Gen. Relativ. Gravit.* **41**, no 1, 207 (2009).

Editorial note: Jürgen Ehlers, *Gen. Relativ. Gravit.* **41**, 203 (2009).

Biography: Andrzej Krasiński, *Gen. Relativ. Gravit.* **41**, 204 (2009).

---

40. A theorem on Petrov types

By Joshua N. Goldberg and Rainer K. Sachs

Originally published: *Acta Phys. Polon.* **22** (supplement), 13 (1962).

Reprinted in: *Gen. Relativ. Gravit.* **41**, no 2, 433 (2009).

Editorial note: Andrzej Krasiński and Maciej Przanowski, *Gen. Relativ. Gravit.* **41**, 421 (2009).

Biography: Joshua Goldberg (Goldberg), *Gen. Relativ. Gravit.* **41**, 430 (2009). The biography of Rainer K. Sachs was printed with the GO#35.

---

41. Relativistic cosmology

By George F. R. Ellis

Originally published: in *Proceedings of the International School of Physics "Enrico Fermi", Course 47: General relativity and cosmology*, Ed. Rainer K. Sachs. Academic Press, 1971, pp. 104–182.

Reprinted in: *Gen. Relativ. Gravit.* **41**, no 3, 581 (2009).

Editorial note: William Stoeger, *Gen. Relativ. Gravit.* **41**, 575 (2009).

Biography: George F. R. Ellis, *Gen. Relativ. Gravit.* **41**, 578 (2009).

---

42. Two papers by John Lighton Synge, 1934, and Felix A. E. Pirani, 1956

42 (a) On the deviation of geodesics and null geodesics, particularly in relation to the properties of spaces of constant curvature and indefinite iine-element

By John Lighton Synge

Originally published: *Ann. Math.* **35**, 705–713 (1934).

42 (b) On the physical significance of the Riemann tensor

By Felix A.E. Pirani

Originally published: *Acta Phys. Polon.* **15**, 389–405 (1956).

Reprinted in: *Gen. Relativ. Gravit.* **41**, no 5, 1206 and 1216 (2009).

Editorial note: Andrzej Trautman, *Gen. Relativ. Gravit.* **41**, 1195 (2009).

Biography: Felix Pirani and Andrzej Trautman, *Gen. Relativ. Gravit.* **41**, 1199 (2009) (Pirani). The biography of John Lighton Synge was printed with the GO#29.

---

43 On the general relativity theory

By Hermann Weyl

Originally published: Zur allgemeinen Relativitätstheorie. *Physikalische Zeitschrift* **24**, 230–232 (1923).

Translated by: Anita Ehlers.

Reprinted in: *Gen. Relativ. Gravit.* **41**, no 7, 1661 (2009).

Editorial note: Jürgen Ehlers, *Gen. Relativ. Gravit.* **41**, 1655 (2009).

Biography: Andrzej Krasiński, *Gen. Relativ. Gravit.* **41**, 1657 (2009).

---

44. Exact solutions of the field equations of the general theory of relativity

By Pascual Jordan, Jürgen Ehlers and Wolfgang Kundt

Originally published: Strenge Lösungen der Feldgleichungen der allgemeinen Relativitätstheorie. *Akademie der Wissenschaften und der Literatur, Abhandlungen der Mathematisch-naturwissenschaftliche Klasse* Nr 2, 21–105 (1960).
Translated by: Anita Ehlers, Manfred Trümper and Wolfgang Kundt.
Reprinted in: *Gen. Relativ. Gravit.* **41**, no 9, 2191 (2009).
Editorial note: George Ellis and Jürgen Ehlers, *Gen. Relativ. Gravit.* **41**, 2179 (2009).
Biographies: Andrzej Krasiński, *Gen. Relativ. Gravit.* **41**, 2182 (2009) (Jordan); Biennial Report 2006/2007 of the Albert Einstein Institute, *Gen. Relativ. Gravit.* **41**, 1899, (2009) (Ehlers); Wolfgang Kundt, *Gen. Relativ. Gravit.* **41**, 2184 (2009) (Kundt).

---

45. A new class of vacuum solutions of the Einstein field equations
By Roy Patrick Kerr and Alfred Schild
Originally published: in *Atti del Convegno sulla Relativita Generale: Problemi dell' Energia e Onde Gravitazionali.* G. Barbera Editore, Firenze 1965, pp. 1–12.
Reprinted in: *Gen. Relativ. Gravit.* **41**, no 10, 2485–99 (2009).
Editorial note: Andrzej Krasiński, Enric Verdaguer and Roy Patrick Kerr, *Gen. Relativ. Gravit.* **41**, 2469–84 (2009).
Biographies: Andrzej Krasiński, *Gen. Relativ. Gravit.* **41**, 2482 (2009) (Kerr); and Lawrence Shepley, *Gen. Relativ. Gravit.* **8**, 955 (1977) (Schild).

---

46 (a) Black hole equilibrium states, part I: Analytic and geometric properties of the Kerr solutions
By Brandon Carter
Originally published: in *Black Holes – Les astres occlus.* Edited by C. deWitt and B. deWitt, Gordon and Breach Science Publishers, New York – London – Paris 1973, pp. 61–124.
Reprinted in: *Gen. Relativ. Gravit.* **41**, no 12, 2874–938 (2009).
Editorial note: Niky Kamran and Andrzej Krasiński, *Gen. Relativ. Gravit.* **41**, 2867–71 (2009).
Biography: Brandon Carter, *Gen. Relativ. Gravit.* **41**, 2870–71 (2009).

46 (b) Black hole equilibrium states, part II: General theory of stationary black hole states
By Brandon Carter
Originally published: in *Black Holes – Les astres occlus.* Edited by C. deWitt and B. deWitt, Gordon and Breach Science Publishers, New York – London – Paris 1973, pp. 125–214.
Reprinted in: *Gen. Relativ. Gravit.* **42**, no 3, 653–744 (2010).
Editorial note: Marek Abramowicz, *Gen. Relativ. Gravit.* **42**, 647–651 (2010).

---

47. Two papers by Myron Mathisson, 1931 and 1937

47 (a) The mechanics of a matter particle in general relativity
Originally published: Die Mechanik des Materieteilchens in der allgemeinen Relativitätstheorie. *Zeitschrift für Physik* **67**, 826–44 (1931).

47 (b) New mechanics of material systems

Originally published: Neue Mechanik materieller Systeme. *Acta Phys. Polon.* **6**, 163–200 (1937).

Both papers translated by: Anita Ehlers.
Reprinted in: *Gen. Relativ. Gravit.* **42**, no 4, 989–1010 and 1011–48 (2010).
Editorial note: Andrzej Trautman, *Gen. Relativ. Gravit.* **42**, 985 (2010).
Biography: Andrzej Trautman, *Gen. Relativ. Gravit.* **42**, 986 (2010).

---

48. On the initial value problem of the equations of gravitation

By Karl Stellmacher
Originally published: Zum Anfangswertproblem der Gravitationsgleichungen. *Mathematische Annalen* **115**, 136–52 (1938).
Translated by: Henk van Elst.
Reprinted in: *Gen. Relativ. Gravit.* **42**, no 7, 1769 (2010).
Editorial note: Helmut Friedrich, *Gen. Relativ. Gravit.* **42**, 1765 (2010).
Biography: Hubert Goenner, *Gen. Relativ. Gravit.* **42**, 1766 (2010).

---

49. On Hamilton's canonical equations

By Władysław Ślebodziński
Originally published: Sur les Équations canoniques de Hamilton *Bulletins de la Classe des Sciences, Acad. Royale de Belg.* **5** no 17, 864–70 (1931).
Translated by: Marie-Noëlle Célérier.
Reprinted in: *Gen. Relativ. Gravit.* **42**, no 10, 2529 (2010).
Editorial note: Andrzej Trautman, *Gen. Relativ. Gravit.* **42**, 2525 (2010).
Biography: Witold Roter, *Gen. Relativ. Gravit.* **42**, 2527 (2010).

---

50. Observations in cosmology

By Jerome Kristian and Rainer K. Sachs
Originally published: *Astrophysical Journal* **143**, 379–99 (1966).
Reprinted in: *Gen. Relativ. Gravit.* **43**, no 1, 337 (2011).
Editorial note: George Ellis, *Gen. Relativ. Gravit.* **43**, 331 (2011).
Biography: Andrzej Krasiński, *Gen. Relativ. Gravit.* **43**, 335 (2011) (Kristian).
The biography of Rainer K. Sachs was printed with the GO#35.

---

51. Conformal treatment of infinity

By Roger Penrose
Originally published: in *Relativity, groups and opology*. Edited by B. de Witt and C. de Witt, Gordon and Breach, New York - London 1964, pp. 565 – 584.
Reprinted in: *Gen. Relativ. Gravit.* **43**, no 3, 901 (2011).
Editorial note: Helmut Friedrich, *Gen. Relativ. Gravit.* **43**, 897 (2011).
The biography was printed with the GO#26.

---

52. Two papers by Tullio Levi-Civita, 1917 – 1919

52 (a) The physical reality of some normal spaces of Bianchi

Originally published: Realtá fisica di alcuni spazi normali del Bianchi. *Rendiconti della Reale Accademia dei Lincei* **26**, 519–31 (1917).

52 (b) Einsteinian ds^2 in Newtonian fields. IX: The Analog of the logarithmic potential

Originally published: ds^2 Einsteiniana in campi Newtoniani. IX: l'Analogo del potenziale logaritmico. *Rendiconti della Reale Accademia dei Lincei* **28**, 101–109 (1919).

Both papers translated by: Robert Jantzen.
Reprinted in: *Gen. Relativ. Gravit.* **43**, no 8, 2307–20 and 2321–30 (2011).
Editorial note: Malcolm A. H. MacCallum, *Gen. Relativ. Gravit.* **43**, 2297 (2011).
Biography: Andrzej Krasiński, *Gen. Relativ. Gravit.* **43**, 2301 (2011).

---

53. The beginning of the world from the point of view of quantum theory

By Georges Lemaître
Originally published: *Nature* **127**, 706 (1931).
Reprinted in: *Gen. Relativ. Gravit.* **43**, no 10, 2929 (2011).
Editorial note: Jean-Pierre Luminet, *Gen. Relativ. Gravit.* **43**, 2911 (2011).
The biography was printed with the GO#2.

---

54. Large number coincidences and the anthropic principle in cosmology

By Brandon Carter
Originally published in: *Confrontation of cosmological theories with observational data. IAU Symposium No. 63.* Edited by M. S. Longair. D. Reidel Publishing Company, Dordrecht + Boston 1974, pp. 291–298.
Reprinted in: *Gen. Relativ. Gravit.* **43**, no 11, 3225 (2011).
Editorial note: George F. R. Ellis, *Gen. Relativ. Gravit.* **43**, 3213 (2011).
The biography was printed with the GO#46a.

---

55. Quantum theory of weak gravitational fields

By Matvei P. Bronstein
Originally published: Quantentheorie schwacher Gravitationsfelder. *Phys. Zeitschr. der Sowjetunion* **9**, 140–157 (1936).
Translated by M. A. Kurkov and Stanley Deser.
Reprinted in: *Gen. Relativ. Gravit.* **44**, pp. 267 (2012).
Editorial note: Stanley Deser and Alexei Starobinsky, *Gen. Relativ. Gravit.* **44**, 263 (2012).
Biography by Stanley Deser, *Gen. Relativ. Gravit.* **44**, 265 (2012).

---

56. Three papers by Hermann Weyl and Rudolf Bach, 1917 – 1922

56 (a) 3. On the theory of gravitation

By Hermann Weyl
Originally published: 3. Zur Gravitationstheorie. *Ann. Physik* **54**, 117–145 (1917).

56 (b) 5. Comment on the axially symmetric solutions to Einstein's equations of gravitation

By Hermann Weyl
Originally published: 5. Bemerkung über die axialsymmetrischen Lösungen der Einsteinschen Gravitationsgleichungen. *Ann. Physik* **59**, 185–188 (1919).

56 (c) New solutions of the Einstein field equations. B. Explicit determination of static, axially symmetric fields. With a supplement on the static two-body problem.

By Rudolf Bach and Hermann Weyl

Originally published: Neue Lösungen der Einsteinschen Gravitationsgleichungen. B. Explicite Aufstellung statischer axialsymmetrischer Felder. Mit einem Zusatz ber das statische Zweikörperproblem. *Math. Z.* **13**, 134 – 145 (1922).

All papers translated by Gernot Neugebauer and David Petroff.

Reprinted in: *Gen. Relativ. Gravit.* **44**, 779 - 832 (2012).

Editorial note: Gernot Neugebauer, David Petroff and Bahram Mashhoon, *Gen. Relativ. Gravit.* **44**, 771 - 777 (2012).

Biography by Hubert Goenner, *Gen. Relativ. Gravit.* **44**, 776 (2012) (Bach).

The biography of Hermann Weyl was printed with the GO#43.

---

57. The geometry of free fall and light propagation

By Jürgen Ehlers, Felix A. E. Pirani and Alfred Schild

Originally published in: *General relativity, papers in honour of J. L. Synge.* Edited by L. O'Reifeartaigh. Oxford, Clarendon Press 1972, pp. 63–84.

Reprinted in: *Gen. Relativ. Gravit.* **44**, 1587 – 1609 (2012).

Editorial note: Andrzej Trautman, *Gen. Relativ. Gravit.* **44**, 1581 – 1586 (2012).

The biographies were printed separately earlier, namely:

Jürgen Ehlers: GO#44

Felix A. E. Pirani: GO#42

Alfred Schild: Lawrence Shepley, *Gen. Relativ. Gravit.* **8**, 955 (1977).

---

58. Relativistic cosmology

By Howard Percy Robertson

Originally published: *Rev. Mod. Phys.*, **5**, pp. 62–90 (1933).

Reprinted in: *Gen. Relativ. Gravit.* **44**, 2115 – 2144 (2012).

Editorial note: George Ellis, **44**, 2099 – 2109 (2012).

Biography: Andrzej Krasiński, *Gen. Relativ. Gravit.* **44**, 2109 – 2111 (2012).